T0134612

Advanced Technologies in Modern Robotic Applications

Chenguang Yang · Hongbin Ma
Mengyin Fu

Advanced Technologies in Modern Robotic Applications

Chenguang Yang
Key Lab of Autonomous Systems
and Networked Control,
Ministry of Education
South China University of Technology
Guangzhou
China

and

Centre for Robotics and Neural Systems
Plymouth University
Devon
UK

Hongbin Ma
School of Automation
Beijing Institute of Technology
Beijing
China

and

State Key Lab of Intelligent Control
and Decision of Complex Systems
Beijing Institute of Technology
Beijing
China

Mengyin Fu
School of Automation
Beijing Institute of Technology
Beijing
China

and

State Key Lab of Intelligent Control
and Decision of Complex Systems
Beijing Institute of Technology
Beijing
China

ISBN 978-981-10-9263-3 ISBN 978-981-10-0830-6 (eBook)
DOI 10.1007/978-981-10-0830-6

Jointly published with Science Press, Beijing
ISBN: 978-7-03-047352-3

© Science Press and Springer Science+Business Media Singapore 2016
Softcover reprint of the hardcover 1st edition 2016
This work is subject to copyright. All rights are reserved by the Publishers, whether the whole or part of the material is concerned, specifically the rights of translation, reprinting, reuse of illustrations, recitation, broadcasting, reproduction on microfilms or in any other physical way, and transmission or information storage and retrieval, electronic adaptation, computer software, or by similar or dissimilar methodology now known or hereafter developed.
The use of general descriptive names, registered names, trademarks, service marks, etc. in this publication does not imply, even in the absence of a specific statement, that such names are exempt from the relevant protective laws and regulations and therefore free for general use.
The publishers, the authors and the editors are safe to assume that the advice and information in this book are believed to be true and accurate at the date of publication. Neither the publishers nor the authors or the editors give a warranty, express or implied, with respect to the material contained herein or for any errors or omissions that may have been made.

Printed on acid-free paper

This Springer imprint is published by Springer Nature
The registered company is Springer Science+Business Media Singapore Pte Ltd.

Preface

Today's digital lifestyle is born from the revolutions of personal computer (PC) and Internet. Moving forward, it seems inevitable that robots will find their ways into our daily lives, integrating seamlessly into the fields of manufacturing, construction, household duties, services, medical operations, health care, etc. A new era of Industry 4.0 is coming, and robots will play significant roles in the new waves of technological revolutions. Unfortunately, most of the existing robot technologies are designed for traditional industrial applications for which the robots operate behind safeguarding and for predefined routine tasks, and thus are not able to perform varying tasks in unknown, complex, and dynamic environments. And modern robots are expected to co-habit with our humans and work closely with us in the fields such as manufacturing and health care as well as in other aspects of our daily lives, and are thus inevitably subject to unpredictable and uncertain external disturbances. Therefore, on the one side we focus on advanced control techniques for robots to deal with various uncertainties, and on the other side we pay much attention to visual servoing techniques which enable robot to autonomously operate in a dynamic environment.

In the recent decades, there is a significant trend in the robotic research community to develop advanced technologies that would enable robots to collaborate with humans friendly and naturally. Therefore, human–robot interaction is another important topic to be covered in this book. Much effort has been devoted to human–robot shared control in teleoperation, and advanced techniques involved in human–robot interfaces. Most of the techniques presented in this book have been tested by either simulations or experiments, preferably on human-like robot platforms such as Baxter robot, because humanoid robots are very popular and have been widely accepted by human users due to their appearances. In summary, the objective of this book is to present in a systematic manner the advanced technologies used for various modern robot applications. By bringing fresh ideas, new concepts, novel methods, and tools into robot control, human–robot interaction, robotic teleoperation, and multiple robot collaboration, we are to provide a state-of-the-art and comprehensive treatment of the advanced technologies for a wide range of robotic applications.

This book starts with an introduction to the robot platforms and tools used in this book in Chap. 1, including some popularly used humanoid robot platforms, visual sensors and haptic devices, and the simulation platforms such as MATLAB Robotics Toolbox, Virtual Robot Experiment Platform (V-REP) simulator, and Robot Operating System (ROS). Since robotic kinematics and dynamics are essential for describing the position, orientation, joints motion as well as analyzing and synthesizing the dynamic behavior of a robot, Chap. 2 discusses the robotic kinematics and dynamics modeling procedure of the commercialized Baxter robot and provides a case study of robot modeling. In Chap. 3, we introduce a number of novel intelligent control methods which are useful to deal with the uncertainties associated with the robot manipulators, such as model mismatch, changing of the external environment, and uncertain loads. These control methods include dual adaptive control, optimized model reference control, and discrete-time adaptive control. Machine vision and image processing are very important for advanced robots as they provide comprehensive and abundant information of surrounding environment. To realize the key functions for robots in understanding the surrounding environment, Chap. 4 focuses on vision-based object detection and tracking using pattern recognition and state estimation. And Chap. 5 investigates visual servoing based human–robot interaction which allows users just to perform in the front of the sensor devices without wearing or operating any control devices to achieve their control purposes. Human–robot collaboration is regarded as a key character of next generation robots which allows human and robot to interact physically with guaranteed safety and compliance. In Chap. 6, we investigate a number of robot teleoperation techniques which can be regarded as a straightforward way to achieve human–robot interaction, such as teleoperation using body motion tracking, teleoperation based on adaptive control approach and human–robot interaction using haptic feedback devices. These technologies are further investigated in Chap. 7, with establishment of a human–robot shared control framework in which the automatic obstacle avoidance can be achieved without affecting the intended teleoperation task. Chapter 8 studies several state-of-art techniques for human–robot interfaces, which are key elements for human–robot interactions, such as hand gesture based robot control, Emotiv neuroheadset in the controlling of mobile robot and the EEG signals based robot manipulator control. Robot localization is essential for a wide range of applications, such as navigation, autonomous vehicle, intrusion detection, and so on. Chapter 9 focuses on indoor/outdoor localization of robot, which is crucial for most tasks demanding accuracy. Chapter 10 is dedicated to the theoretical study of multiple robot cooperation in the framework of multi-agents system, including optimal multi-robot formation, hunting activities of a multi-robot system, and multi-robot cooperative lifting control. Finally, some useful technologies developed for other popular robot applications are presented in Chap. 11, including some theoretical and practical results of robot kicking, and reference trajectory adaptation algorithm for motion planning.

This book is featured with a number of attractive and original research studies, including the robot control design in a biomimetic manner, the teleoperation method in a human–robot shared control manner, the novel human–robot interaction interface based on hand gesture recognition and on electromyography (EMG). The publication of this book will systematically bring new knowledge created by the authors to the robotic communities, especially in the topics of biomimetic control, human–robot interaction and teleoperation. The book is primarily intended for researchers and engineers in the robotic and control community. It can also serve as complementary reading for robotics at both graduate and undergraduate levels. The book will help to consolidate knowledge and extend skills of robotic researchers. This book will be extremely useful for early career researchers to be equipped with a comprehensive knowledge of technologies in modern robot applications. This book will also bring fresh new ideas into education, and will benefit students by exposing them to the very forefront of robotics research; preparing them for possible future academic or industrial careers in robotics or robot applications.

The authors would also like to thank the help from our students and collaborators, to name a few, Yiming Jiang, Tao Teng, Xinyu Wang, Peidong Liang, Zhangfeng Ju, Alex Smith, Mei Wu, Xinghong Zhang, Hao Zhou, Dong Wang, Hao Wang, and Sunjie Chen, during the preparation of this book. Appreciation must be made to Prof. Angelo Cangelosi and Dr. Phil. Culverhouse of Plymouth University for their technical support of related research work covered in this book. The authors would also like to thank the generous support from our families. Without their continuous support, this book would not appear in its current form.

The authors would like to acknowledge support from National Natural Science Foundation of China (NSFC) grants 61473120 and 61473038, Guangdong Provincial Natural Science Foundation 2014A030313266 and Beijing Outstanding Talents Programme (2012D009011000003).

Contents

Chapter 1
Introduction of Robot Platforms and Relevant Tools

Abstract This chapter introduces a number of robot platforms and relevant devices used throughout this book, including the humanoid robot platforms such as Baxter robot and iCub robot; visual sensors of Microsoft Kinect, stereo camera Point Grey Bumblebee2 and 3D camera Leap Motion, as well as haptic devices of SensAble Omni and Novint joystick Falcon. Meanwhile, a number of software toolkits useful in robot simulation are also introduced in this chapter, e.g., the MATLAB Robotics Toolbox and the Virtual Robot Experiment Platform (V-REP) simulator. Robot Operating System (ROS) is also briefly introduced by highlighting the ROS characters and ROS level concepts. These devices and toolkits are nowadays becoming more and more popularly used in the study of robotics, as they provide ideal means for the study, design, and test of the robotic technologies.

1.1 Robot Platforms

1.1.1 Baxter® Robot

The increasing pace of introduction of new products, particularly those with many variants and short lifetimes, brings considerable uncertainty to assembly line design. It is therefore very desirable to develop flexible and reconfigurable manufacturing systems. In this context, the humanoid dual-arm Baxter robot with intrinsic safety (e.g., physical compliance created by spring in between driving motors and robot joints) has been developed. The Baxter® humanoid robot made by Rethink Robotics™ offers users an affordable platform with guaranteed safety for both academic and industrial applications. The platform provides the users a good opportunity to carry out research on dual-arm robot manipulation and vision-based control.

To enable compliance on the dual arms, the Baxter® robot is equipped with Series Elastic Actuator (SEA) instead of connecting the motor shaft directly to the joint (usually through a gearbox). The motor in a SEA is coupled to the joint through a spring, so that the torque generated by twist of spring, rather than the torque from the motor directly drives the link. This enables the robot to behave in a human-like elastic manner. Due to the elastic effect of the spring, the SEA will lead to improved shock

© Science Press and Springer Science+Business Media Singapore 2016
C. Yang et al., *Advanced Technologies in Modern Robotic Applications*,
DOI 10.1007/978-981-10-0830-6_1

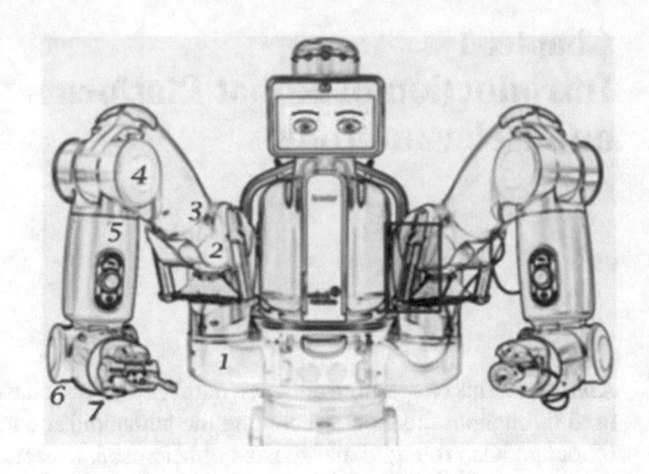

Fig. 1.1 Illustration of the Baxter robot, on which the joints *1, 2, 3* comprise the shoulder, *4 and 5* the elbow, and *6, 7* the wrist. The spring highlighted in *red*, two of which are attached to each arm, generate large forces to improve gravity compensation

tolerance and reduced danger in cases of collision. In addition, the Baxter® robot is able to sense a collision at a very early time instant before it hits badly onto a subject. Therefore, Baxter® robot is a good choice to fulfill safe human–robot interaction.

The Baxter® robot includes a torso based on a movable pedestal and two 7DOF (degree of freedom) arms installed on left/right arm mounts respectively. Each arm has 7 rotational joints and 8 links, as well as an interchangeable gripper (such as electric gripper or vacuum cup) which can be installed at the end of the arm. A head pan with a screen, located on the top of torso, can rotate in the horizontal plane [1].

One of the main features of Baxter is the Series Elastic Actuators (SEAs) which are present in every joint; these are comprised of a spring coupling between the motor and the link, with a built-in hall effect to measure the deflection. This creates a naturally compliant design, but also means that the torque at each joint can be easily estimated by measuring the spring deflection and multiplying by the known stiffness constant. There is also a gravity compensation controller running in a high-frequency loop, which calculates the torques required to counteract gravitational pull on each joint as well as the forces applied to joint 2 from the large external springs highlighted in Fig. 1.1. This is important to note, as the gravity compensation is applied by default, and must be manually disengaged for full torque control application.

1.1.2 iCub Robot

The iCub robot, as shown in Fig. 1.2, is a humanoid robot developed as part of an EU project "robotCub" and subsequently adopted by more than 20 laboratories worldwide. The height of iCub robot is 945 mm (from foot to head) and width of its lower torso is 176 mm (from left to right). The iCub robot has 53 motors that move the head, arms, hands, waist, and legs. The upper body of iCub robot comprises 38 joints: two 7 DOF arms, two 9 DoF hand, and a 6 DoF head. Each leg of the iCub

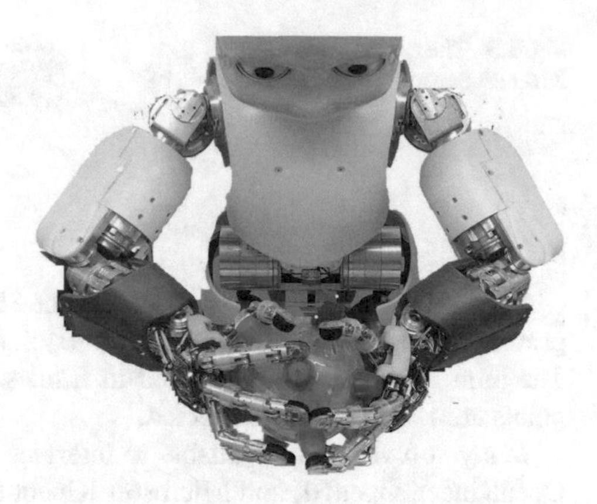

Fig. 1.2 The illustration of a iCub robot holding a ball

consists of 6 joints: three at the hip, one at the knee, and two at the ankle. A 3 DoF waist is installed at the body of iCub robot in order to enhance the workspace of the robot.

The main body of iCub robot is made by aluminum alloy, thus its weight strength ratio is outstanding. And it is appropriately used for highly and medium stressed parts, for example, actuator housing load-bearing parts, and so on. A cable-driven differential mechanism was chosen for iCub robot, in order to provide more stiffness on the hip joint for the implementation of its hip joint, especially for the flexion/extension and abduction/adduction motions of the hip joint.

In [2], the iCub robot is controlled to draw shapes after observing a demonstration by a teacher, by using a series of self-evaluations of its performance. In [3], a technique is proposed for controlling the interaction forces exploiting a proximal six axes force/torque sensor and is tested and validated on the iCub robot. One advantage of iCub robot is that it is an open-source cognitive robotic platform.

1.2 Visual Sensors and Haptic Devices

1.2.1 Microsoft Kinect Sensor

The Kinect sensor, as shown in Fig. 1.3, is a new hands-free and low-cost game controller interface introduced by Microsoft in 2010. It provides RGB video stream with 30 frames/s as well as a monochrome intensity encoded depth map, both in VGA resolution (640×480 pixels). The video offers 8-bit resolution while the depth data is represented in 11 bits in units of millimeters measured from the camera. The depth sensor is novel in the sense that it operates by projecting a known IR pattern onto the environment and measuring the returned IR pattern (using an IR camera). The received pattern is compared to the known projected pattern and the differences are used to extract object depth.

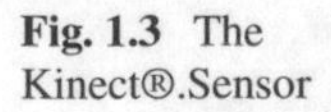

Fig. 1.3 The Kinect®.Sensor

Recently, in the new version of Kinect SDK, a skeletal tracking tool has been provided. This tool aims at collecting the joints as points relative to the device itself. The joint information is collected in frames. For each frame, the positions of 20 points are estimated and collected.

Many softwares are available to interface Kinect with PC, e.g., Libfreenect by OpenKinect, OpenNI, and Microsoft Kinect for windows SDK [4, 5]. OpenKinect is an open community of people interested in making use of the Kinect hardware with PCs and other devices. It releases software Libfreenect which can be used to interface Kinect along with any operating system in PC [4, 6]. OpenNI (open natural interaction) is similar software as Libfreenect [5]. OpenNI can support many other devices other than Kinect. It is open-source software released by Prime Sense, the company which manufactured Kinect for Microsoft. Apart from Kinect the company also manufactured many depth sensors. OpenNI uses middleware called NITE to get the skeleton data of the user [7]. Microsoft Kinect for windows SDK is another similar platform launched by Microsoft for windows systems [8]. SimpleOpenNI that uses OpenNI and NITE [7] is an OpenNI and NITE wrapper for processing. It may not support all the functions that are supported by OpenNI and NITE still, but most of the useful features can be accessed in a simple way [9, 10].

Kinect® is a motion sensing input device for Microsoft for Xbox 360 and Xbox One video game consoles and Windows PCs as shown in Fig. 1.3. It is basically a depth camera. Normal webcams collect light that is reflected by the objects in its field of view and turn them into digital signals, which are turned into an image. Kinect can also measure the distance between the objects and itself. A colored 3D point cloud can be generated from both of the RGB image and the depth image. Thus the Kinect® sensor can be used to detect the environment in front of it.

1.2.2 Point Grey Bumblebee2 Stereo Camera

The Point Grey Bumblebee2 stereo camera is a 2 sensor progressive scan CCD camera with fixed alignment between the sensors. Video is captured at a rate of 20 fps with a resolution of 640×480 to produce dense colored depth maps to assist in tracking and a viable pose estimation of the object. The resolution–speed trade-off has to be managed concisely as an increased frame speed gives a smooth robot trajectory, whereas enhances the processing time. And an increased resolution provides a denser, more accurate point cloud for feature extraction but with increased latency. Figure 1.4 shows the Bumblebee2 camera used in this book.

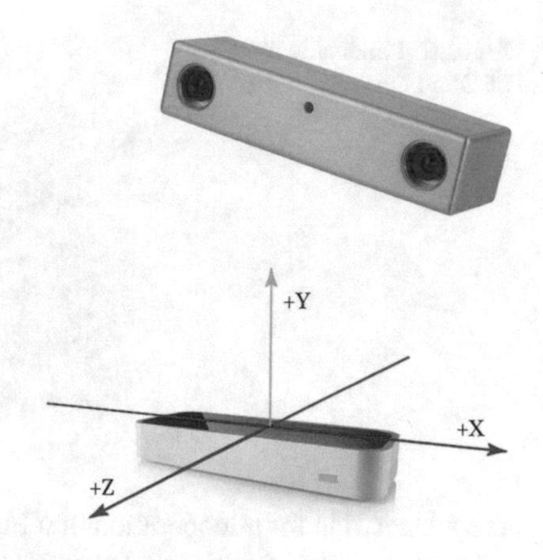

Fig. 1.4 The Point Grey Bumblebee2 camera. Reprinted from Ref. [11], with permission of Springer

Fig. 1.5 Leap Motion Sensor. Reprinted from Ref. [13], with permission of Springer

1.2.3 Leap Motion Sensor

The Leap Motion sensor is equipped with two internal cameras to take photos from different directions to obtain hand action information in a 3D space. The detection range of Leap Motion is between 25 and 600 mm upon the sensor. The shape of the detection space is similar to an inverted cone [12]. Figure 1.5 depicts the coordinate system for the Leap Motion.

Recently, a method of remote interaction was put forward in [14] through detecting the position of palm based on the Leap Motion, and mapping the physical space and information space. As one typical application of the Leap Motion, a MIDI controller using the Leap Motion was reported in the literature [15] and it is also well known that many video games can be played with motion sensors such as Kinect. With the fast development of motion sensing technology, the Leap Motion can serve as an excellent alternative means of Kinect for desktop applications. After extensive testing, one may easily draw a conclusion that the motion sensing technology is more practical and more attractive than traditional way. As to the dynamic hand gesture recognition, in [16], an effective method which can recognize dynamic hand gesture is proposed by analyzing the information of motion trajectory captured by the Leap Motion.

1.2.4 SensAble Omni

The SensAble Omni Haptic Device is made by SensAble® haptic Technologies. This haptic device is of 6 DOFs, in which the first three DOFs contribute to position while the last three form a gimbal contributing to orientation. A stylus equiped with two buttons is also attached to the end effector. Note that the SensAble Omni has been very popularly used in the teleoperation research. The Omni device was utilized on

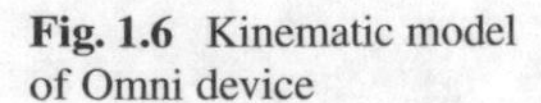

Fig. 1.6 Kinematic model of Omni device

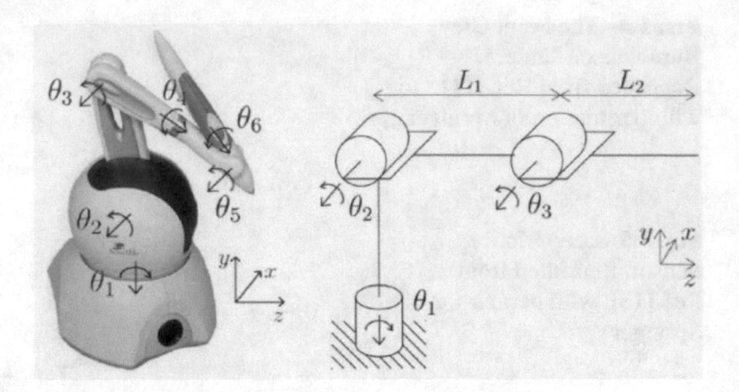

the master side for teleoperation of a Puma 560 robotic arm, and achieved assistance of daily activities for disabled people [17]. In [18], the Omni haptic device is used as a master device to control a virtual fixture in a telerehabilitation system. The Omni could also be used to control a hydraulic machine, e.g., in [19], the Omni device was employed for the boom and stick segments of an excavator used in construction. The kinematics of Omni device including forward kinematics, inverse kinematics, and Jacobian has been well studied in the literature [20, 21]. A Mohammadi *et al* developed a simulation kit to support the Omni using on MATLAB/Simulink platform [22]. We developed a stable and high-performance 32/64bit driver and developed all control strategies on the MATLAB/Simulink platform. The Omni device has a passive 3-DOF stylus and the kinematic model of active 3-DOF is shown in Fig. 1.6.

1.2.5 Novint Falcon Joystick

The haptic device Falcon 3D joystick [23] designed by the Novint Technologies Inc., is able to provide a force feedback up to 1 kg allowing a good illusion of the tactile rendering. Moreover, it has been widely used in many applications such as video games ensures realistic communication capacities, and accurate force feedback giving the illusion of a tactile perception during a simulation.

The 3-DOF haptic device Falcon is shown in Fig. 1.7. This device can be used in research on control and estimation problems for robots involving parallel linkages [24]. In [25], Falcon joystick is introduced into the virtual surgery system as the key device which can realize force feedback in the virtual environment.

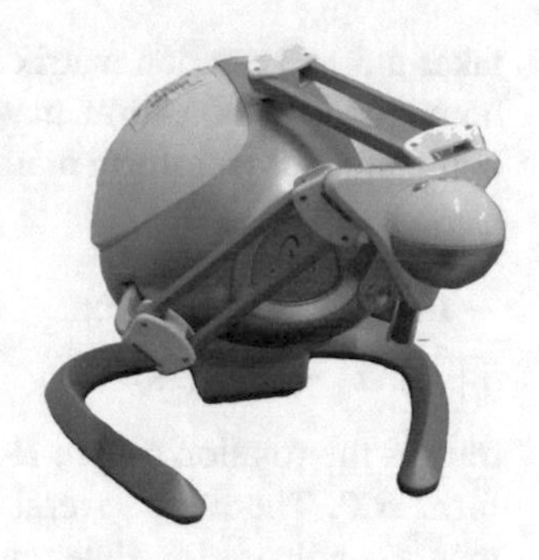

Fig. 1.7 Force feedback device Falcon

1.3 Software Toolkits

1.3.1 MATLAB Robotics Toolbox

Next, we will introduce the MATLAB Robotics Toolbox developed by Peter Corke [26]. It enables us to create a robot model and to manipulate it and to plan its motion trajectory easily. There are plenty of useful built-in functions of forward/inverse kinematics and dynamics for serial-link type manipulators. As we know, the MATLAB/Simulink® software package [27] provides an efficient computation and simulation environment, including a large number of mathematical functions, and a variety of specialized toolboxes. Therefore, MATLAB/Simulink® has been widely used in both academia and industry. For convenience of research, various real-time interface for motion control schemes with MATLAB/Simulink® have been implemented for many robots such as KUKA manipulators [28].

In this subsection, we will briefly introduce some functions which are necessary for simulations of robots. Note that these functions may be extended further or modified in the future, hence the readers are suggested to refer to the documents of the latest Robotics Toolbox available on Peter Corke's web site: http://www.petercorke.com/.

1.3.1.1 Transformation Functions

The Toolbox provides many functions which help to perform matrix operations commonly used in robotics, such as rotations, translations, etc. For example, the functions:

```
R = rotx(theta)
R = roty(theta)
R = rotz(theta)
```

give a 3 × 3 rotation matrix R for a given angle theta around different axes. The function:

```
TR = Rt2tr(R,t)
```

takes a 3×3 rotation matrix R and a 3×1 translation vector t and outputs a 4×4 homogeneous transform matrix. The rotation submatrix can be retrieved from a homogeneous transform matrix using the function:

```
1 R = t2r(T)
```

Likewise, the function:

```
1 [R, t] = tr2rt(TR)
```

returns the rotation matrix R and the translation vector t from a homogenous transform RT. There are several useful methods for representing angles, i.e., rotation matrix, angle vector, Euler angles, quatcrnions, rpy (roll pitch yaw). There are functions within the toolbox to translate in between each type:

```
1 Eul = tr2eul(T)
```

returns a 1×3 vector of Euler angles given a 4×4 homogenous transform matrix or a 3×3 rotation matrix. And

```
1 T = eul2tr(phi, theta, psi)
```

or

```
1 T = eul2tr(eul)
```

returns a homogenous transformation matrix given three Euler angles phi, theta, psi, or a 1×3 vector of Euler angles. And functions “tr2angvec” and “angvec2tr” can be used to convert between rotational matrix and angle-vector format, among which

```
1 [theta, v] = tr2angvec(R)
```

converts a 3×3 rotational matrix or 4×4 homogenous transform matrix to angle-vector format, where theta is the angle and v is a 1×3 vector of axes. Likewise, there is a reciprocal function:

```
1 R = angvec2r(theta, v)
```

or

```
1 T = angvec2tr(theta, v)
```

where the former converts an angle theta and a 1×3 vector to a rotation matrix, and the latter converts an angle theta and 1×3 vector to a homogenous transform matrix. Similarly, there is a function for converting to “rpy” format:

```
1 rpy  = tr2rpy(T)
```

which converts either a 3×3 rotational matrix or a 4×4 homogenous transform matrix to a 1×3 vector consisting of [roll pitch yaw]. The reciprocal function is predictably:

```
1 R = rpy2r(rpy)
```

or

```
1 T = rpy2tr(rpy)
```

The choice of them is dependent on if we need the 3×3 rotation matrix or the 4×4 homogenous transform.

The Toolbox also gives the functionality of plotting coordinate frames on a 3D graph, which allows us to visualize the transforms. The function:

```
trplot(T)
```

plots a 4×4 homogenous transform matrix or 3×3 rotational matrix on a graph. Two frames can be plotted on the same graph using the hold function, like so:

```
trplot(T)
hold on
trplot(T2)
```

There are several options for plotting frames, a full list of which can be accessed by typing:

```
doc trplot
```

The most useful are, for example, to change the color of the frame, type:

```
trplot(T, color, red)
```

To change the thickness, type:

```
trplot(T, thick, 1)
```

where 1 is the required line thickness (default 0.5). Plot options can be stacked, i.e.,

```
trplot(T, color,blue,thick, 0.75, frame,myFrame)
```

which will plot the frame corresponding to the homogenous transform matrix T, with a blue color, line thickness 0.75 and named "myFrame".

Not only can we plot frames, but also the toolbox can animate a frame transformation. The function:

```
tranimate(T, T2)
```

takes two 4×4 homogenous transform matrices or 3×3 rotation matrices T and T2 and creates an animation of pose T moving to pose T2.

1.3.1.2 Robotics Function

There are several key functions to use the toolbox. The first is "Link", which allows us to start creating a robot link using DH parameters. Normally we can type:

```
doc Link
```

To see documentation for a function, but unfortunately for "Link" there is a standard MATLAB function which blocks this. The basic implementation of "Link" is (Fig. 1.8):

```
L1 = Link('d', 0,'a', 1, 'alpha', pi/2);
```

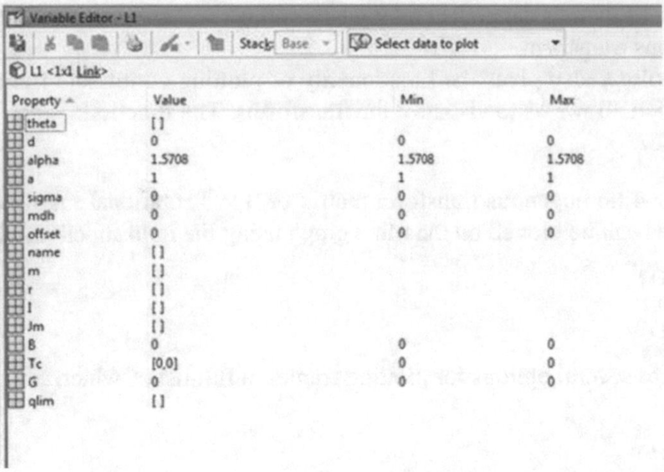

Variable Editor - L1

Stack: Base | Select data to plot

L1 <1x1 Link>

Property	Value	Min	Max
theta	[]		
d	0	0	0
alpha	1.5708	1.5708	1.5708
a	1	1	1
sigma	0	0	0
mdh	0	0	0
offset	0	0	0
name	[]		
m	[]		
r	[]		
I	[]		
Jm	[]		
B	0	0	0
Tc	[0,0]	0	0
G	0	0	0
qlim	[]		

Fig. 1.8 A basic implementation of "Link" (screen snapshot)

where "L1" is the name of the Link class we are creating, and the function "Link" is passed several parameters which describe the feature of link; in the above example we can see that "d" is set to zero, "a" is set to one, "alpha" is set to $\pi/2$. Once we have created the class in MATLAB, we can see the Link class "L1" in the workspace window. If we double click it, it will expand in the main window and we will be able to see all of the different variables that can be set for that link, e.g., offset, name, qlim, etc. We can access these variables in much the same way as in a C++ class, like so:

```
1 L1.qlim
```

which will print out the joint limits for that link; we can set the joint limits using the same method:

```
1 L1.qlim = [-pi/2 pi/2]
```

which will set the minimum limit to $\pi/2$ and the maximum to $\pi/2$. Note that the toolbox is all defined using radians rather than degrees.

Other variables:

```
1 L1.A(q)
```

returns the transform matrix for the link for a given joint angle q.

```
1 L1.sigma
```

can be used to set the joint to be revolute or prismatic (0 for revolute, 1 for prismatic).

```
1 L1.offset
```

sets the offset angle of the joint. This can be used if the joint always starts at an angle other than zero. Any consequent joint commands use this offset as the zero point.

```
1 L1.m
```

can be defined to set the mass of the link.

```
1 L1.r
```

is a 3×1 vector which defines the center of gravity (COG) of the link.

Most of these variables may not be needed explicitly, but they allow us to set the dynamic parameters of the link which is important for system control. Dynamics are outside the scope of this module, so we will not need to worry about these.

To start building a robot we need several links. If we know the DH parameters we can use the "Link" method to create our links. It is possible to create an array of Link classes, from creating the first link following the same procedure above, but use:

```
1 L(1) = Link('d', 0, 'a', 1, 'alpha', pi/2);
```

instead. Our second link can then be defined as L(2), third as L(3) and so on. It does not matter which style we use. To join the Links together, the function:

```
1 SerialLink
```

is required. Its basic implementation is:

```
1 Bot = SerialLink([L1 L2])
```

```
1 Bot = SerialLink(L)
```

depending on which style of Link definition we used earlier. Again, once we have created the SerialLink class (in this case named "Bot"), we can access it in the same way as the Link class, i.e., we can name our robot, which will be displayed in plots, using:

```
1 Bot.name = 'myFirstRobot'
```

Note that for a string, such as the name, we need to pass it variables using the single quote marks to specify that it is a string.

Once we have created our robot, we can plot it. The basic use is:

```
1 Bot.plot([pi/4 pi/4])
```

then we can see the robot in Fig. 1.9. Note that we have to pass the plot function at least a vector of joint angles.

We can rotate around the plot to see it from different angles. Other methods which are part of SerialLink:

```
1 Bot.teach
```

opens a graphic user interface (GUI) and a plot of the robot, which allows us to move the joints using a slider bar as shown in Fig. 1.10:

Angles are displayed in radians. We can press the record button, which will save the slider angles to:

```
1 Bot.qteach
```

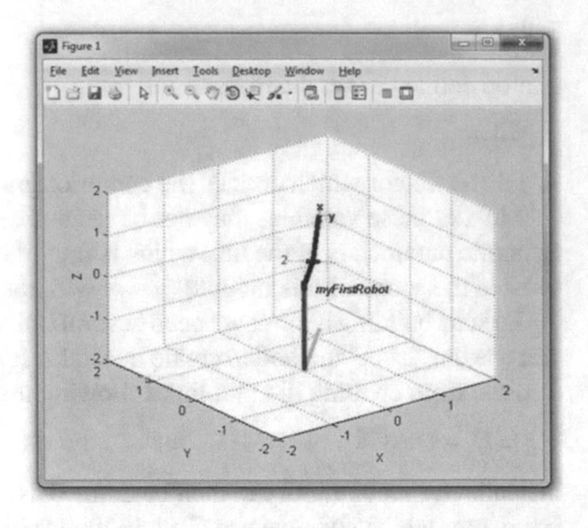

Fig. 1.9 A robot created by the Toolbox (screen snapshot)

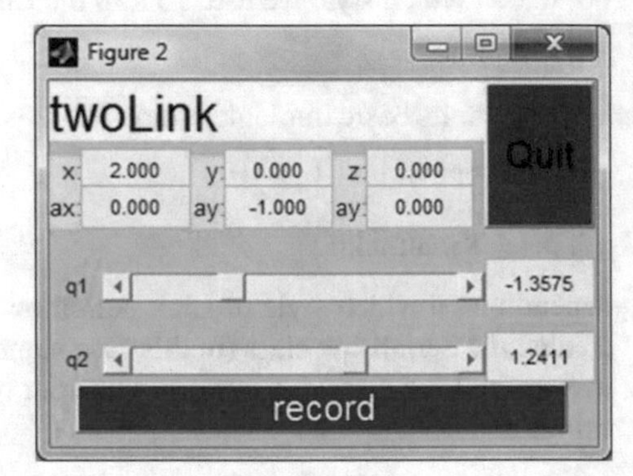

Fig. 1.10 A slider bar in GUI (screen snapshot)

in a matrix form. If we have set "qlim" in the links, then these will be the limits on the slider bars. The forward kinematics of the robot based on the joint angles q can be calculated by

```
1 T = Bot.fkine(q)
```

which returns a 4×4 homogenous transform matrix. And the inverse kinematics can be calculated by

```
1 q = Bot.ikine(T, Q0, M)
```

which returns the joint angles for a desired workspace position. Here T must be a 4×4 homogenous transform, specifying the desired orientation and position in space; Q0 is a vector of estimated joint angles, which can be the robot's rest position; and M is a 1×6 vector which is used as a mask to reduce the computational complexity if the robot does not have 6 joints. For example, if the robot only moves in the X,Y planes, then:

```
1 M= [1 1 0 0 0 0]
```

or if the robot has 5 degrees of freedom, but the rotation around the wrist is not available, then:

```
1 M = [1 1 1 1 1 0]
```

Note that inverse kinematics is not a simple solution, and the solver may not be able to find an answer. We can explore options to try and improve the chances of finding an answer by giving a better Q0 value.

1.3.1.3 Toolbox in Simulink

The toolbox also has many functions which can be used within Simulink to do block diagram designs. Open a new Simulink model and type

```
1 roblocks
```

to get a library of Robot Toolbox blocks. These can be dragged and dropped into a design like as shown in Fig. 1.11. We can double click on boxes to see and change the properties. They can be used with standard Simulink blocks in just the same way. The plot block will animate the robot model given an input of joint angles q. Note that q must be a vector of same length as the number of degrees of freedom (DOF) of the robot.

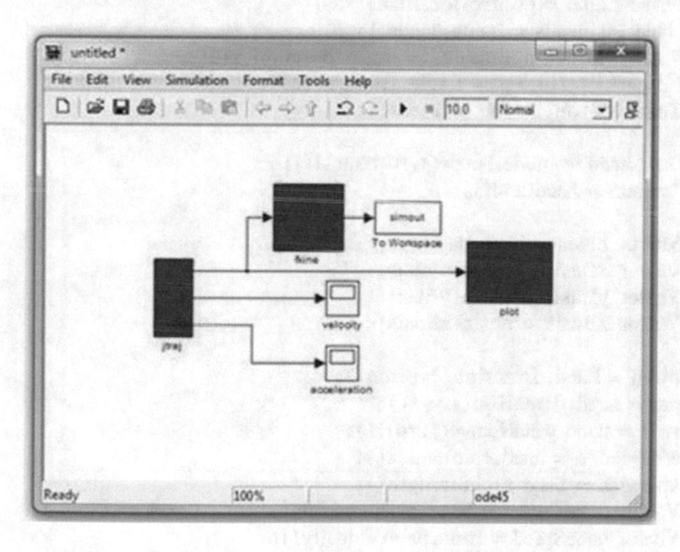

Fig. 1.11 A simulink model of the Toolbox (screen snapshot)

1.3.2 Official SDK of Leap Motion

The software development with the Leap Motion should base on the official SDK and the driver. When it is powered on, the Leap Motion sends hand action information periodically. Every package information is called a frame [29]. The sensor will assign these information with an ID. With the ID, the user may call functions such as Frame::hand(), Frame::finger() to check any object's information. With C++ compiler of Microsoft Visual Studio on Windows or GNU C/C++ Compiler on Linux, we can implement functions of reading and analyzing data from Leap Motion.

One example code within C++ is given in Listing 1.1, which illustrates how to get the basic objects tracked by the Leap Motion system and how to obtain the basic data of the frontmost hand. More details on the Leap Motion API provided by the SDK can be referred to the documentation and the official website of Leap Motion [30]. With powerful functionality provided by Leap Motion SDK, it is possible to develop interesting applications of virtual reality or augmented reality. There are some fantastic video demonstrations on the website of Leap Motion.

Listing 1.1 Leap Motion Example Code

```
1  Leap::Controller controller;
2  // wait until Controller.isConnected() evaluates to true
3  //...
4
5  Leap::Frame frame = controller.frame();
6  Leap::HandList hands = frame.hands();
7  Leap::PointableList pointables = frame.pointables();
8  Leap::FingerList fingers = frame.fingers();
9  Leap::ToolList tools = frame.tools();
10
11 Leap::Hand hand = frame.hands().frontmost();
12 Leap::Arm arm = hand.arm();
13
14 Leap::Matrix basis = hand.basis();
15 Leap::Vector xBasis = basis.xBasis;
16 Leap::Vector yBasis = basis.yBasis;
17 Leap::Vector zBasis = basis.zBasis;
18
19 float pitch = hand.direction().pitch();
20 float yaw = hand.direction().yaw();
21 float roll = hand.palmNormal().roll();
22 float confidence = hand.confidence();
23 float strength = hand.grabStrength();
24 Leap::Vector handCenter = hand.palmPosition();
25 Leap::Vector handSpeed = hand.palmVelocity();
26 std::string handName = hand.isLeft() ? "Left hand" : "Right hand";
```

1.4 V-REP Based Robot Modeling and Simulations

The robot simulation environment can help us to design, control, and test the robot. Therefore, it plays an important role in the field of robotics. In this section and the next section, we will introduce two robot simulation platforms which are popularly used nowadays.

Virtual Robot Experiment Platform (V-REP) is one general purpose robot simulator with integrated development environment. In V-REP, many robot models with built-in controllers are provided in V-REP, so that the robots can be easily controlled without writing many scripts or modifying many parameters. V-REP can be installed and run without powerful graphic card and CPU, and it comes with a large number of robots, sensors, and actuators models. The robot simulator V-REP is based on a distributed control architecture, it means that the object could be controlled individually through a ROS node, a plug-in, an embedded script, a remote API client, or a custom solution. It offers a multitude of functionalities that can be easily integrated and combined through an exhaustive API and script functionality.

Another advantage of V-REP is that controllers can be written in C/C++, Python, Java, Lua, MATLAB, or Urbi. And three physics engines (Bullet Physics, ODE, and Vortex Dynamics) are selected for fast and customizable dynamics calculations, in order to simulate real-world physics and object interactions such as collision response, grasping, etc. Because of all these characteristics, V-REP has been widely used for quick development of new methods, simulation of automation process, as well as fast prototyping and verification, education of robotics, etc. This software is available on Windows, Mac and Linux platforms. Three different versions are available, V-REP pro edu is free for education usage, V-REP pro is for commercial license, and V-REP player is free for scene replay.

The V-REP (main user interface shown in Fig. 1.12) is the result of an effort trying to council all requirements into a versatile and scalable simulation framework.

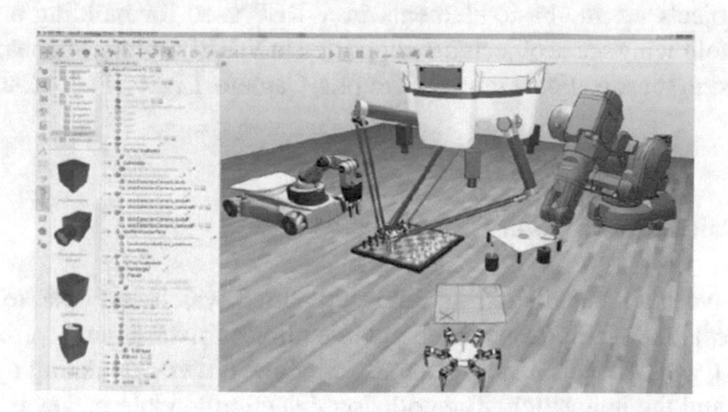

Fig. 1.12 An example V-REP simulation scene showing the diversity of robot types that may be simulated simultaneously. Reprinted from Ref. [31], with permission of Springer

1.4.1 V-REP Simulator

A V-REP simulation model, or simulation scene includes a number of elemental objects or scene objects that are assembled in a tree-like structure.

Scene objects are rarely used on their own, they rather operate on other scene objects, for example, a proximity sensor will detect shapes. In addition, the V-REP allows directly manipulate on one or several scene objects by using several calculation modules. The main calculation modules are as follows:

- Kinematics Module.
- Dynamics Module.
- Collision Detection Module.
- Mesh-mesh Distance Calculation Module.
- Path/motion Planning Module.

Three different physics engines are supported by V-REP's dynamics module currently. The user can quickly switch from one engine to the other in terms of his/her needs. Since the physics simulation is a complex task, the diversity support of physics engines can provide various degrees of precision, speed, or diverse features:

- Bullet physics library.
- Open Dynamics Engine (ODE).
- Vortex Dynamics.

1.4.1.1 Modeling Scene Objects

There are many popular robots models including the mobile and the nonmobile in V-REP. And we can also model our own robot in one V-REP simulation scene which consists of different scene objects.

Scene objects are the basic elements in V-REP used for building a simulation scene. The following scene objects are supported in V-REP: Joints; Shapes; Proximity sensors; Vision sensors; Force sensors; Graphs; Camera; Lights; Paths; Dummies and Mills.

1.4.1.2 Calculation Modules

There are five main calculation modules supported which can operate on one or several objects in V-REP. Kinematics module allows forward and inverse kinematics for any types of mechanism. Dynamics module allows calculating dynamics of rigid body and the interaction. The collision detection module makes it possible to fast check the interference between any shape or collection of shapes. Mesh-mesh distance calculation module allows fast calculated the minimum distance between any shape or collection of shapes. Path planning module processes the holonomic and non-holonomic path planning tasks.

1.4.1.3 Simulation Controllers

It is very convenient to adjust control codes in V-REP. The control code of a simulation or a simulation model is based on three main techniques:

- The control code is executed on another machine.
- The control code is executed on the same machine, but in another process (or another thread) than the simulation loop.
- The control code is executed on the same machine and in the same thread as the simulation.

It is worth to mention that V-REP allows the user to choose among 6 various programming techniques simultaneously.

1.4.2 *Examples of V-REP Simulation*

1.4.2.1 Simulation Model of One Teleoperation System

Figure 1.13 shows a simulated teleoperation system in V-REP. The control signal is skeleton data from OpenNI. The communication between the OpenNI and V-REP has to be real time. V-REP allows the user to choose among various programming techniques simultaneously and even symbiotically. The cooperation system chooses the remote API client method to make V-REP to communicate with OpenNI. The remote API functions are interacting with V-REP via socket communication in a way that reduces lag and network load to a great extent. The remote OpenNI API plugin starts as the "Kinect Server" communicating via a free port set by itself. The temporary remote API server service was chosen to start the "Kinect Server" from within a script at simulation start. After the simulation started, the child script will detect the free serial port set in the "Kinect Server" and receive skeleton data from the "Kinect Server".

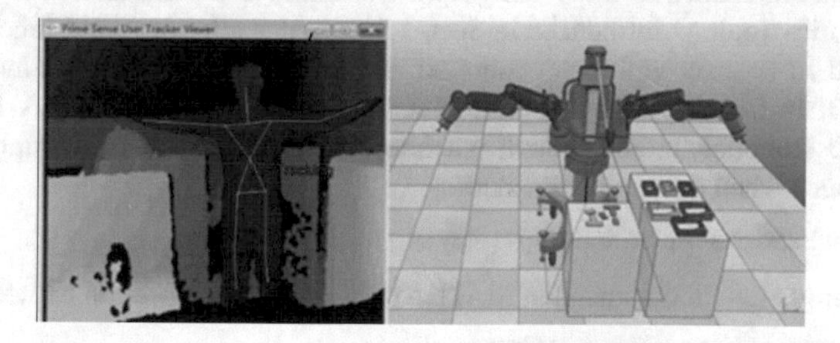

Fig. 1.13 Teleoperation system simulation in V-REP (screen snapshot)

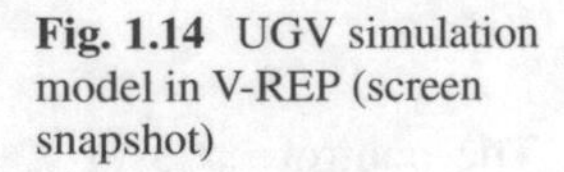

Fig. 1.14 UGV simulation model in V-REP (screen snapshot)

1.4.2.2 Simulation Model of UGV

Figure 1.14 shows one UGV simulation model of an IN^2BOT autonomous vehicle which is modified on Polaris Ranger EV. The vehicle is equipped with a CCD that provides the image of road and traffic signs. A Novatel GPS-aided INS provides the position and attitude of the vehicle. A Velo-dyne 3D LIDAR provides the point cloud of the environment around the vehicle in the range of 60 m. A speedometer provides the measurement of driving speed and mileage. The vehicle is also equipped with throttle, steering, and brake actuators for automatic control.

1.4.2.3 Simulation Model of Baxter Robot

To illustrate the use of V-REP, the inverse kinematics function built in the V-REP is used for simulation. By setting all the joints of the Baxter arm into inverse kinematics, and set the pen as the end effector of inverse kinematics, then send command to V-REP to make the pen moves as the Phantom Omni stylus, joint angles needed for the Baxter to perform the same action as the stylus can be found. Figure 1.15 shows an example of Baxter setup for simulation.

To send commands to the V-REP from Simulink, a few methods can be used. The Robotics Toolbox introduced in Sect. 1.3 provides a lot of useful functions for MATLAB to connect with other robot simulation program, and robot models for MATLAB. In fact, it is very simple to establish connections between V-REP and MATLAB Robotics Toolbox: First, put one command into V-REP script to let it execute once when the simulation start:

```
1    simExtRemoteApiStart(19999)
```

Then, a single line of command in MATLAB can get the connection between them:

```
1 v = VREP(path, directory to V-REP program);
```

Now, ‘v’ is set as a V-REP object which contains the handles about the connection between MATLAB and V-REP. And objects in the V-REP scene can be controlled by only few lines:

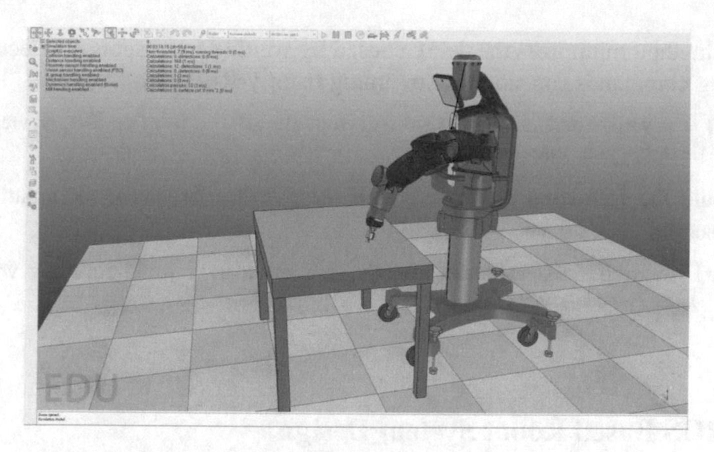

Fig. 1.15 Baxter setup for simulation (screen snapshot)

```
1 joint = v.gethandle({jointname});
2 v.setjointtarget({joint , jointposition});
```

where *joint* is set to be the handle ID of the required *jointname*, and *jointposition* is the joint position of the joint required to move to. These commands are all sent as mode of one shot and wait for response by default, this means the MATLAB will wait for the response of each command returns from the V-REP simulation, it may cause some delay in real-time simulation and could not be avoid if using Robotics Toolbox functions.

To reduce the delay of waiting for response, functions in Robotics Toolbox are not used. The other way to do it is using the original API functions provided by V-REP, so that the mode of sending commands can be selected. In fact, the Robotics Toolbox is also using the original V-REP API functions, it just provides a simple method to do the connection, therefore some of the V-REP API functions such as reading force sensors, are not assessable by only using the V-REP object created by Robotics Toolbox. In fact, there are various available operation modes of sending commands, e.g., one shot, one shot and wait, streaming, streaming and split, buffer.

To prevent causing delay when reading data from V-REP, streaming and buffer mode can be used, and a sample command is shown below:

```
1 [~,~] = v.vrep.simxGetJointPosition(v.vrep.clientID , jointhandle, v.vrep.
      simx_opmode_streaming + alpha);
```

After this command is sent to V-REP, the V-REP server will continuously execute this command and send the joint position to MATLAB. In MATLAB, when the data need to be received, execute the following commands so that the variable in MATLAB will be refreshed.

```
1 [~,~] = v.vrep.simxGetJointPosition(v.vrep.clientID , jointhandle, v.vrep.
      simx_opmode_buffer);
```

This method minimizes the delay caused by waiting for response. Executing the following commands can stop the streaming from V-REP server.

```
1 [~,~] = v.vrep.simxGetJointPosition(v.vrep.clientID , jointhandle, v.vrep.
      simx_opmode_discontinue);
```

Execute the following commands can remove the streaming command in the V-REP server.

```
1 [~,~] = v.vrep.simxGetJointPosition(v.vrep.clientID , jointhandle, v.vrep.
      simx_opmode_remove);
```

1.5 ROS Based Robot System Design

With the rapid development of the field of robotics, the needs of reusability and modularity are more and more intense. But the open-source robot systems available cannot meet these needs. In 2010 Willow Garage Robot Company released an open-source operating system called Robot Operating System or ROS. This operating system for robot has soon launched a boom of learning and using ROS in the field of robotics research [32].

ROS provides a standard operating system which includes hardware abstraction, low-level device control, implementation of commonly used functions, message passing between processes, and package management. Tools and libraries are also available for running the code for construction, writing, and across multiple computers.

ROS is designed to improve the rate of code reuse in robotics research and development. ROS is a distributed processing framework (Nodes). This makes executable files can be individually designed and loosely coupled at runtime. These processes can be encapsulated into data packets (Packages) and stacks (Stacks) in order to share and distribute. ROS also supports a joint system of code library. So that collaboration can also be distributed, from the file system level to the community-level design, making independent decisions on development and implementation possible. All of these above functions can be implemented by basic ROS tools.

Although ROS itself is not a real-time framework, it is possible to integrate ROS with real-time code. Willow Garage PR2 is one typical example of this integration. It uses a system called pr2etherCAT. This system can transport ROS messages in and out of a real-time process. And there are many real-time tool kits available.

Currently, ROS mainly runs on Unix-based platforms. Software for ROS is primarily tested on Ubuntu and Mac OS X systems, however, it has also been supported by Fedora, Gentoo, Arch Linux, and other Linux platforms through the ROS community. In this book, we have developed several robot systems with ROS installed on Arch Linux through the unique feature of Arch Users Repository of Arch Linux by comping the latest sources of ROS.

1.5.1 Main Characteristics of ROS

The runtime architecture of ROS is a processing architecture using ROS communication module to connect peer-to-peer network of processes (potentially distributed across machines) that are loosely coupled. It includes synchronous RPC (remote procedure call) based services communication, based on asynchronous topic data stream communication, and data stored on the server parameters [33].

The main characteristics of ROS can be summarized below:

Distributed communication: A system includes a series of processes that exist in a number of different host and contact during operation by end-topology. Although those software frameworks based on the central server can also realize the benefits of multiprocess and multi-host, but when computers are connected via different networks, central data server problem will arrive.

The point-to-point design and node management mechanisms of ROS can disperse the real-time calculation pressure caused by functions like computer vision and speech recognition and adapt to the challenges encountered.

Language independence: In programming, many programmers will be in favor of a number of programming languages. These preferences are the result of personal experience in each language, debugging effect, grammar, efficiency, and a variety of technical and cultural reasons. To solve these problems, ROS designs a language neutrality framework. It now supports many different languages, such as C ++, Python, Octave and LISP, and also includes a variety of interface in other languages.

The particularity of ROS mainly resides in the message communication layer rather than a deeper level. To-end connectivity and configuration is achieved by XML-RPC mechanism, XML-RPC also contains a description reasonably achievable in most major languages. The designers of ROS want to use a variety of language ROS more natural, more in line with syntax conventions in various languages, rather than based on the interface of C language. However, the existing libraries packaged in some cases support new language conveniently, such as Octave client is packaged through the C ++ library implementation.

In order to support cross-language, ROS uses a simple, language-independent interface definition language to describe the messaging between modules. Interface definition language used to describe the structure of a short text of each message, and message synthesis is also allowed.

Easy code reuse: Most existing robot software projects have included some driver and algorithms which can be reused in other projects. But unfortunately, due to various reasons, the intermediate layer of these codes is confusing and it is very difficult to extract function from its prototype or apply it to other projects.

To cope with these problems, ROS encourages all drivers and algorithms evolved and become libraries independent from ROS. Systems built with ROS have a modular features, each module code can be compiled separately, and CMake tools using in compiling make it very easy to achieve a streamlined concept. ROS encapsulated complex codes in the library, and only creates a few small applications to display the libraries. This allows the transplantation and reused of simple code. As an additional advantage, after code dispersion in the library, unit testing has become very easy, a separate test procedure may test a lot of features of the library.

ROS uses a lot of existing codes from open-source projects, such as the drivers from the Player, motion control and simulation aspects of the code, the visual aspect algorithm from OpenCV, content planning algorithm from OpenRAVE, and so on. In each instance, ROS is used to display the data communication between a variety of configuration options, various softwares, and also on their packaging and minor changes. ROS can continue to be upgraded from the community maintenance, including upgrades ROS from other software libraries, application patch source code.

Elaborated core design: To manage the complex ROS software framework, the authors use a lot of gadgets to compile and run a variety of ROS formation, which become the core of ROS, instead of building a huge development and runtime environment.

These tools serve a lot of tasks, such as organizing the source code structure, access and set configuration parameters, figurative end-to-end connection topology, measure the width of the frequency band, depict the information data, generate document automatically, and so on.

Free and open source: All of the source code of ROS is publicly available. This will certainly facilitate all levels of ROS software debugging, and constantly correct the error. Although non-open-source software like Microsoft Robotics Studio and Webots has many admirable attributes, an open-source platform is also irreplaceable. This is especially true when hardware and software is designing and debugging at the same time.

1.5.2 ROS Level Concepts

According to defenders and distribution, there are two main parts of ROS system code:

(i) Main: the core part, mainly designed and maintained by Willow Garage companies and some developers. It provides the basic tools for a number of distributed computing, as well as a core part of the overall programming of ROS.
(ii) Universe: worldwide code, developed and maintained by ROS community organizations in different countries. One type is the library of code, such as OpenCV, PCL, etc; on the library floor is the other type which is supplied from a functional perspective, such as navigation; the top of the code is the application level code, so that the robot can accomplish a given goal.

From another point of view, ROS has three levels of concepts [34]: the Filesystem level, the Computation Graph level, and the Community level. The rest part of this chapter will introduce these three levels in detail.

1.5.2.1 ROS Filesystem Level

The filesystem level concepts include ROS resources that we can check on our disk. Here are some main filesystem level concepts:

Packages: The software of ROS is organized by many Packages. Packages contain nodes, ROS dependent libraries, data sets, configuration files, third-party software, or any other logical composition. The aim of Package structure is to provide an easy to use structure to facilitate software reuse. Generally speaking, ROS package short but useful.

Stack: Stack is a collection of the package, which provides a fully function, like "navigation stack." Stack is associated with version number, and is also the key to distribute ROS software.

ROS is a distributed processing framework. This makes executable files can be individually designed and loosely coupled at runtime. These processes can be encapsulated into Packages and Stacks in order to share and distribute.

Metapackages: Metapackages are one specialized Packages. This type of Packages only serve to represent a group of related other packages. Metapackages are often used as a backwards compatible place holder for converted rosbuild Stacks.

Package manifests: Manifests (package.xml) offers Package metadata, which including the dependencies between its license information and Package, as well as language features information like compiler flags (compiler optimization parameters).

Stack manifests: Stack manifests (stack.xml) provides Stack metadata, which including the dependencies between its licensing information and Stack.

1.5.2.2 ROS Computation Graph Level

The Computation Graph is the peer-to-peer network of ROS processes that are processing data together [34]. The basic Computation Graph concepts of ROS are nodes, Master, Parameter Server, messages, services, topics, and bags, all of which provide data to the Graph in different ways.

Nodes: Node is a number of processes to perform arithmetic tasks. ROS can use the increasable scale to make the code modular: a system that is typical of the many nodes: a system is one typical example which comprised by many nodes. Here, the node may also be referred to as "software modules." By using the concept of "nodes", the ROS system can run more figurative: when many nodes are running simultaneously, can easily be drawn into an end to end communication chart, in this chart, the process is the nodes in the graph, and the connection relationship is one of the arcs connected end to end.

Parameter server: The Parameter server allows data to be stored by keys in a central location. It is currently part of the Master.

Messages: By transmitting messages between nodes they can communicate to each other. Each message is one data structure. The original standard data types (integer, floating point, boolean, etc.) are supported, and the original array types are also supported. Messages can contain any nested structures and arrays (very similar to the C language structures)

Topics: Message is passed by a publish/subscribe semantics. A node can post messages on a given topic. A node data for a certain thematic concerns and subscribe to specific types. Multiple nodes may simultaneously publish or subscribe to the same theme of the message. Logically, we can think of a topic as a strongly typed message bus. Each bus has its name, and anyone can connect to this bus to send or receive messages as long as they are the right type. Overall, publishers and subscribers are unaware of the existence of each other.

Services: Although the topic-based publish/subscribe model is a very flexible communication mode, but the feature that it broadcasts path planning for joint design is not suitable for simplified synchronous transfer mode. In ROS, we call it a service, with a string and a pair of strictly regulate the message definition: one for the request and one for the response. This is similar to web servers, web servers defined by URIs, but with a full definition of the type of request and response documents. Note that, unlike topic, only one node can be uniquely named to broadcasts a service: Only one service can be called "classification symbol", for example, any URI address given can only own one web server.

On the basis of the above concepts, there is the need for a controller can make all the nodes methodical execution. Such a ROS controller is called the ROS Master.

Master: The ROS Master provides the registration list and search service for other Computing Graph. Nodes are not able to communicate with each other without through the Mater.

Bags: As an important mechanism, Bags are used to store and play back message data of ROS, e.g., sensor data, which may be difficult to acquire and are essential for algorithm development and test. Bags format are used to save and play back message data of ROS.

The ROS master stores the registration information about the topics and services for ROS node. Each node communicates with the Master in order to send to the Master their registration information. These nodes also receive information about other nodes when communicating with the Master. When registration information changes, the Master will return information to these nodes and allow the dynamic creation of connections between nodes and the new node.

The connection between one node and another node is straightforward, and the controller provides only registration information, like a DNS server. When nodes subscribe to a topic, a connection will be established to enable the publication of the topic node on the basis of the consent of the Connection Agreement.

1.5.2.3 ROS Community Level

ROS community-level concept is a form of codes which are released on ROS network. By using these joint libraries, collaboration can be distributed too. This design from the file system level to the community level allows independent development and implementation possible. Because of this distributed architecture, ROS gains rapid development with the exponentially increasing number of package deployment.

References

1. Baxter Product Datasheet. http://rr-web.s3.amazonaws.com/assets/Baxter_datasheet_5.13.pdf
2. Mohan, V., Morasso, P., Zenzeri, J., Metta, G., Chakravarthy, V.S., Sandini, G.: Teaching a humanoid robot to draw shapes. Auton. Robot. **31**(1), 21–53 (2011)
3. Fumagalli, M., Randazzo, M., Nori, F., Natale, L., Metta, G., Sandini, G.: Exploiting proximal f/t measurements for the icub active compliance. In: Proceedings of the 2010 IEEE/RSJ International Conference on Intelligent Robots and Systems (IROS), pp. 1870–1876. IEEE (2010)
4. Fabian, J., Young, T., Peyton Jones, J.C., Clayton, G.M.: Integrating the microsoft kinect with simulink: real-time object tracking example. IEEE/ASME Trans. Mech. **19**(1), 249–257 (2014)
5. Colvin, C.E., Babcock, J.H., Forrest, J.H., Stuart, C.M., Tonnemacher, M.J., Wang, W.-S.: Multiple user motion capture and systems engineering. In: Proceedings of the 2011 IEEE Systems and Information Engineering Design Symposium (SIEDS), pp. 137–140. IEEE (2011)
6. http://www.openkinect.org/wiki/Main_Page
7. Chye, C., Nakajima, T.: Game based approach to learn martial arts for beginners. In: Proceedings of the 2012 IEEE 18th International Conference on Embedded and Real-Time Computing Systems and Applications (RTCSA), pp. 482–485. IEEE (2012)
8. Soltani, F., Eskandari, F., Golestan, S.: Developing a gesture-based game for deaf/mute people using microsoft kinect. In: Proceedings of the 2012 Sixth International Conference on Complex, Intelligent and Software Intensive Systems (CISIS), pp. 491–495. IEEE (2012)
9. Borenstein, G.: Making Things See: 3D Vision with Kinect, Processing, Arduino, and MakerBot, vol. 440. O'Reilly, Sebastopol (2012)
10. Cruz, L., Lucio, D., Velho, L.: Kinect and rgbd images: challenges and applications. In: Proceedings of the 2012 25th SIBGRAPI Conference on Graphics, Patterns and Images Tutorials (SIBGRAPI-T), pp. 36–49. IEEE (2012)
11. Yang, C., Amarjyoti, S., Wang, X., Li, Z., Ma, H., Su, C.-Y.: Visual servoing control of baxter robot arms with obstacle avoidance using kinematic redundancy. In: Proceedings of the Intelligent Robotics and Applications, pp. 568–580. Springer (2015)
12. The principle of leap motion. http://www.3dfocus.com.cn/news/show-440.html
13. Chen, S., Ma, H., Yang, C., Fu, M.: Hand gesture based robot control system using leap motion. In: Proceedings of the Intelligent Robotics and Applications, pp. 581–591. Springer (2015)
14. Xu, C.B., Zhou, M.Q., Shen, J.C., Luo, Y.L., Wu, Z.K.: A interaction technique based on leap motion. J. Electron. Inf. Technol. **37**(2), 353–359 (2015)
15. Pan, S.Y.: Design and feature discussion of MIDI. controller based on leap motion. Sci. Technol. China's Mass Media **10**, 128–129 (2014)
16. Wang, Q.Q., Xu, Y.R., Bai, X., Xu, D., Chen, Y.L., Wu, X.Y.: Dynamic gesture recognition using 3D trajectory. In: Proceedings of 2014 4th IEEE International Conference on Information Science and Technology, Shenzhen, Guanzhou, China, pp. 598–601, April 2014
17. Veras, E., Khokar, K., Alqasemi, R., Dubey, R.: Scaled telerobotic control of a manipulator in real time with laser assistance for adl tasks. J. Frankl. Inst. **349**(7), 2268–2280 (2012)
18. Chi, P., Zhang, D.: Virtual fixture guidance for robot assisted teleoperation. In: Bulletin of advanced technology, Vol 5, No. 7, Jul 2011
19. Hayn, H., Schwarzmann, D.: Control concept for a hydraulic mobile machine using a haptic operating device. In: Proceedings of the 2009 Second International Conferences on Advances in Computer-Human Interactions, ACHI'09, pp. 348–353. IEEE (2009)
20. Sansanayuth, T., Nilkhamhang, I., Tungpimolrat, K.: Teleoperation with inverse dynamics control for phantom omni haptic device. In: 2012 Proceedings of the SICE Annual Conference (SICE), pp. 2121–2126. IEEE (2012)

21. Silva, A.J., Ramirez, O.A.D., Vega, V.P., Oliver, J.P.O.: Phantom omni haptic device: Kinematic and manipulability. In: Proceedings of the 2009 Electronics, Robotics and Automotive Mechanics Conference, CERMA'09, pp. 193–198. IEEE (2009)
22. Mohammadi, A., Tavakoli, M., Jazayeri, A.: Phansim: a simulink toolkit for the sensable phantom haptic devices. In: Proceedings of the 23rd Canadian Congress of Applied Mechanics, pp. 787–790. Vancouver, BC, Canada (2011)
23. N.W.S.: Company specialized in 3d haptic devices. http://www.novint.com/index.php (2012)
24. Martin, S., Hillier, N.: Characterisation of the novint falcon haptic device for application as a robot manipulator. In: Proceedings of the Australasian Conference on Robotics and Automation (ACRA), pp. 291–292. Citeseer (2009)
25. Distante, C., Anglani, A., Taurisano, F.: Target reaching by using visual information and q-learning controllers. Auton. Robot. **9**(1), 41–50 (2000)
26. Corke, P., et al.: A computer tool for simulation and analysis: the robotics toolbox for MATLAB. In: Proceedings of the National Conference of the Australian Robot Association, pp. 319–330 (1995)
27. MATLAB and Simulink for technical computing. http://www.mathworks.com/
28. Chinello, F., Scheggi, S., Morbidi, F., Prattichizzo, D.: KCT: a MATLAB toolbox for motion control of kuka robot manipulators. In: Proceedings of the 2010 IEEE International Conference on Robotics and Automation (ICRA), pp. 4603–4608. IEEE (2010)
29. Elons, A.S., Ahmed, M., Shedid, H., Tolba, M.F.: Arabic sign language recognition using leap motion sensor. In: Proceedings of the 2014 9th International Conference on Computer Engineering and Systems (ICCES), Cairos, pp. 368–373, December 2014
30. Leap motion—mac and pc motion controller for games, design, virtual reality and more
31. Ma, H., Wang, H., Fu, M., Yang, C.: One new human-robot cooperation method based on kinect sensor and visual-servoing. In: Proceedings of the Intelligent Robotics and Applications, pp. 523–534. Springer (2015)
32. O'Kane, J.M.: A gentle introduction to ros (2014)
33. Martinez, A., Fernández, E.: Learning ROS for Robotics Programming. Packt Publishing Ltd, Birmingham (2013)
34. ROS official. http://www.ros.org

Chapter 2
Robot Kinematics and Dynamics Modeling

Abstract The robotic kinematics is essential for describing an end-effector's position, orientation as well as motion of all the joints, while dynamics modeling is crucial for analyzing and synthesizing the dynamic behavior of robot. In this chapter, the kinematics and dynamics modeling procedures of the Baxter robot are investigated thoroughly. The robotic kinematics is briefly reviewed by highlighting its basic role in analyzing the motion of robot. By extracting the parameters from an URDF file, the kinematics model of the Baxter robot is built. Two experiments are performed to verify that the kinematics model matches the real robot. Next, the dynamics of robot is briefly introduced by highlighting its role in establishing the relation between the joint actuator torques and the resulting motion. The method for derivation of the Lagrange–Euler dynamics of the Baxter manipulator is presented, followed by experimental verification using data collected from the physical robot. The results show that the derived dynamics model is a good match to the real dynamics, with small errors in three different end-effector trajectories.

2.1 Kinematics Modeling of the Baxter® Robot

2.1.1 Introduction of Kinematics

The robotic kinematics studies the motion of a robot mechanism regardless of forces and torque that cause it. It allows to compute the position and orientation of robot manipulator's end-effector relative to the base of the manipulator as a function of the joint variables. Robotic kinematics is fundamental for designing and controlling a robot system. In order to deal with the complex geometry of a robot manipulator, the properly chosen coordinate frames are fixed to various parts of the mechanism and then we can formulate the relationships between these frames. The manipulator kinematics mainly studies how the locations of these frames change as the robot joints move.

Kinematics focuses on position, velocity, acceleration, and an accurate kinematics model must be established in order to investigate the motion of a robot manipulator. Denavit–Hartenberg (DH) notations are widely used to describe the kinematic model

© Science Press and Springer Science+Business Media Singapore 2016

C. Yang et al., *Advanced Technologies in Modern Robotic Applications*,
DOI 10.1007/978-981-10-0830-6_2

of a robot. The DH formation for describing serial-link robot mechanism geometry has been established as a principal method for a roboticist [1]. Standard DH notations are used to create a kinematics model of robot. The fundamentals of serial-link robot kinematics and the DH notations are well explained in [2, 3].

In DH convention notation system, each link can be represented by two parameters, namely the link length a_i and the link twist angle α_i. The link twist angle α_i indicates the axis twist angle of two adjacent joints i and $i-1$. Joints are also described by two parameters, namely the link offset d_i, which indicates the distance from a link to next link along the axis of joint i, and the joint revolute angle θ_i, which is the rotation of one link with respect to the next about the joint axis [4]. Usually, three of these four parameters are fixed while one variable is called joint variable. For a revolute joint, the joint variable is parameter θ_i, while for a prismatic joint, it will be d_i. The DH convention is illustrated in Fig. 2.1.

The four parameters of each link can be specified using DH notation method. With these parameters, link homogeneous transform matrix which transforms link coordinate frame $i-1$ to frame i is given as Eq. (2.1).

$$
\begin{aligned}
{}^{i-1}A_i(\theta_i, d_i, a_i, \alpha_i) &= R_z(\theta_i)T_z(d_i)T_x(a_i)R_x(\alpha_i) \\
&= \begin{bmatrix} c\theta_i & -s\theta_i c\alpha_i & s\theta_i s\alpha_i & a_i c\theta_i \\ c\theta_i & c\theta_i c\alpha_i & -c\theta_i s\alpha_i & a_i s\theta_i \\ 0 & s\alpha_i & c\alpha_i & d_i \\ 0 & 0 & 0 & 1 \end{bmatrix}
\end{aligned} \tag{2.1}
$$

where $\sin(\theta)$ is abbreviated as $s\theta$ and $\cos(\theta)$ as $c\theta$, R_j denotes rotation about axis j, and T_j denotes translation along axis j. For an n-link robot arm, the overall arm transform namely forward kinematics, in terms of the individual link transforms can

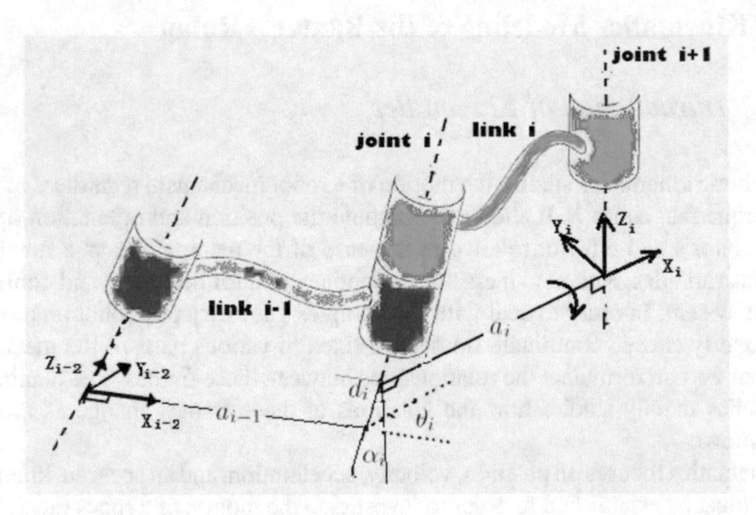

Fig. 2.1 Definition of standard DH link parameters

be expressed in Eq. (2.2). The Cartesian position of the end-effector can be calculated from Eq. (2.3).

$$^{0}A_n = {}^{0}A_1 {}^{1}A_2 \ldots {}^{n-1}A_n \tag{2.2}$$

$$^{n}X_0 = {}^{0}A_n X_n \tag{2.3}$$

where $X = [x, y, z, 1]^T$ is an augmented vector of Cartesian coordinate.

Using the DH description of a robot and the Robotics Toolbox as introduced in Sect. 1.1, we can easily simulate the motion of a robot and calculate the kinematics, e.g., joints configuration for a certain pose and the Jacobian matrix [4].

2.1.2 Kinematics Modeling Procedure

Recalling Sect. 1.1, the Baxter robot consists of a torso based on a movable pedestal and two 7 DOF arms installed on left/right arm mounts, respectively. Next, we will build the kinematics model, the Baxter robot. The modeling procedure of kinematics of the Baxter® robot includes the following steps [5]. First, we perform analysis of mechanical structure of the Baxter® robot, and the main elements in URDF file and a 3D visual model will be described. Second, the method to obtain the DH parameters from a link of the left arm is presented. Finally, we will tabulate the DH parameters and create the kinematic model of dual arms with corresponding DH parameters. This completes the kinematic modeling of Baxter® robot.

2.1.2.1 Structural Analysis

Baxter® robot has two arms and a rotational head on its torso. The arm is shown in Fig. 2.2. The arm of Baxter robot comprises a set of bodies, called links, in a chain and connected by revolute joints. There are seven rotational joints on the arm, namely s0, s1, e0, e1, w0, w1, w2, respectively. Each joint has 1DOF. The *"arm mount"* assembly, is fixed on the *"torso"* assembly. Joint s0 is connected to the *"arm mount"* assembly. Between joint s0 and s1, there is *"upper shoulder"* assembly. *"lower shoulder"* assembly is located between joint s1 and joint e0, and so on. This structural connection is illustrated in Fig. 2.2.

2.1.2.2 URDF Description

URDF is a file in XML format which describes a robot, detailing its parts, joints, dimensions, and so on. It is a fundamental robot model file in ROS, which is a flexible framework for developing robot software. As introduced in Sect. 1.5, ROS is a collection of tools, libraries, and conventions that aims to simplify the task of creating

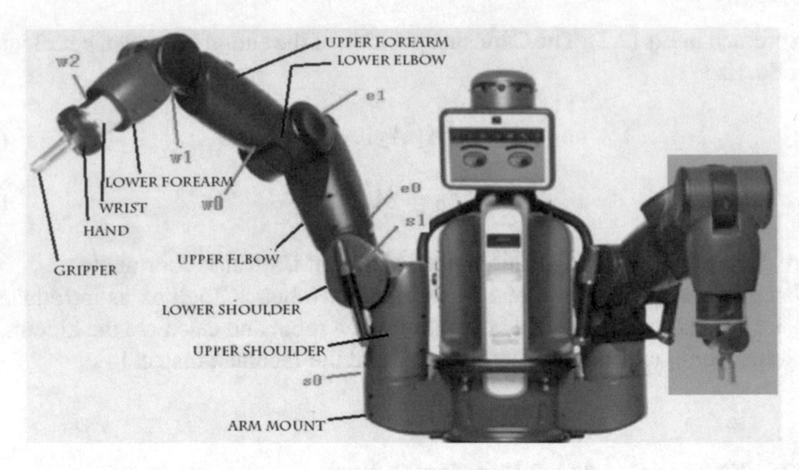

Fig. 2.2 The arms of Baxter robot, on which there are seven joints

complex and robust robot behavior across a wide variety of robotic platforms [6, 7]. The way ROS uses a 3D model of a robot or its parts to simulate them or to simply help the developers in their daily work is by the means of the URDF files. In ROS, a 3D robot such as the Baxter is always associated with a URDF file which describes the kinematic information of the robot.

The Baxter® robot URDF file consists of a series of frames and transform description between frames. A package about Baxter® robot mechanics construction can be found in [8]. This package includes all files that we need to analyze key factors of Baxter® robot, such as URDF file and other necessary files.

There are two major elements in an URDF file which describe the geometry of a robot: "link" and "joint." These two elements are not the same as joints and links normally used in DH notation system. In fact, a "link" in URDF represents a frame. The "link" code section means a frame predefined in the robot: it describes a rigid body with *inertia*, *visual* features, and so on. Below is an example of a link element named *left_upper_shoulder*:

```
1  <link name="left_upper_shoulder">
2   <visual>
3    <origin rpy = "0,0,0" xyz = "0,0,0"/></visual>
4    <inertial>
5    <origin rpy="0,0,0"
6     xyz="0.01783,0.00086,0.19127"/>
7    <mass value="5.70044"/>
8    <inertia
9     ixx="0.04709102262"...izz="0.03595988478"/>
10   </inertial>
11 </link>
```

The *visual* property of the link specifies the shape of the body (such as box, cylinder) for visualization purposes. *Origin* section in *visual* indicates the reference frame of the visual element with respect to the reference frame of the link.

A "joint" in URDF represents a transform between two frames. It does not necessarily mean a motional (translational or rotational) joint. In fact, most of these "joints" are fixed. For example, the "joint" named *torso_s0* and *left_torso_arm_mount* are both fixed, as described above. Below is an example of a "joint" element.

```
1 <joint name="left_s0" type="revolute">
2   <origin rpy="0,0,0" xyz="0.055695,0,0.011038"/>
3   <axis xyz="0,0,1"/>
4   <parent link="left_arm_mount"/>
5   <child link="left_upper_shoulder"/>
6   <limit lower="−1.7" upper="1.7" />
7 </joint>
```

The *type* property specifies the type of "joint": revolute, prismatic, or fixed, whereas the former two types are motional joint. The *origin* property is the transform from the parent link to the child link. The "joint" is located at the origin of the child link, "above. *rpy*" represents the rotation around the axis of the parent frame: first roll around x, then pitch around y and finally yaw around "z. *xyz*" represents the position offset from the "origin.*lower*" and "origin.*upper*" describe the minimum and maximum joint angles in radians. In Fig. 2.3, the first link has name *"base"* which is represented by a rectangle, also representing a coordinate frame named *"base"*. The ellipse represents a coordinate transformation between two frames, for example, through *torso_t0*, frame *base* can be translated to frame *torso*. In order to get the transform between frames, we will use the visual properties of link, and we can get a transform path of left arm, which is depicted in Fig. 2.3 in a bold line.

2.1.2.3 Kinematic Model

Now, we are ready to create a forward kinematics model of Baxter® robot. We will take left arm as an example, and create the left arm simulation model using DH notation.

From the frame of *base*, the "joint" *torso_s0* and *left_torso_arm_mount* are fixed, the first motional joint we can find is *left_s0*, so we consider the *left_s0* as the first joint and *left_upper_shoulder* as the first link, *left_s1* and *left_lower_shoulder* as the second joint and link, respectively. Let us continue in this manner until the seventh joint *left_w2*, the last rotational joint, and link *left_wrist*, are calculated.

The first four frame transformations can be found in URDF file (refer to Fig. 2.3). Here is the description of *left_s0*, *left_s1*, *left_e0*, and *left_e1* in URDF:

```
1 <joint name="left_s0" type="revolute">
2   <origin rpy="0␣0␣0" xyz="0.056␣0␣0.011"/>
3 <joint name="left_s1" type="revolute">
4   <origin rpy="−1.57␣0␣0" xyz="0.069␣0␣0.27"/>
5 <joint name="left_e0" type="revolute">
6   <origin rpy="1.57␣0␣1.57" xyz="0.102␣0␣0"/>
7 <joint name="left_e1" type="revolute">
8   <origin rpy="−1.57␣−1.57␣0"xyz="0.069␣0␣0.262"/>
```

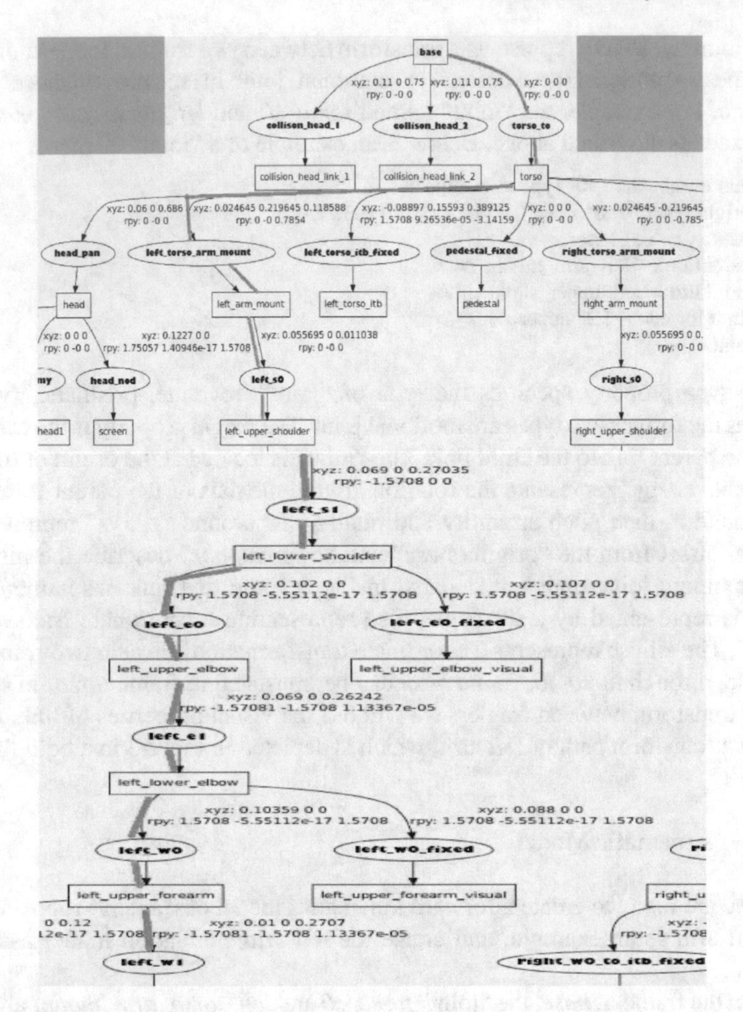

Fig. 2.3 Parts of frames and transform relationship between frames in Baxter® robot, generated from ROS instruction *$rosrun urdfdom urdf_to_graphiz baxter.urdf* [7]. Rectangle("link") represents a frame and ellipse("joint") represents a translation/rotation between frames

These frame transforms are shown in Fig. 2.4.

From *left_s0* to *left_s1*, the axis of joint rotates $-\frac{\pi}{2}$ along X_0 axis (all revolution abides by right-hand law), so link twist $\alpha_1 = -\frac{\pi}{2}$ .The translation of *left_s1* in URDF file is $x = 0.069$ m and $z = 0.27$ m, so $d_1 = 0.27$ m and $a_1 = 0.069$ m, let $\theta_1 = 0$ just because this is a revolute joint.

From *left_s1* to *left_e0*, the axis of joint rotates $\frac{\pi}{2}$ along axis Z_1 first and then rotates $\frac{\pi}{2}$ along axis X_1. Therefore, we should let $\theta_2 = \frac{\pi}{2}$ and let link twist $\alpha_2 = \frac{\pi}{2}$. Although the translation of *left_e0* is $x = 0.102$ m, we have to assign $d_2 = 0$ m and

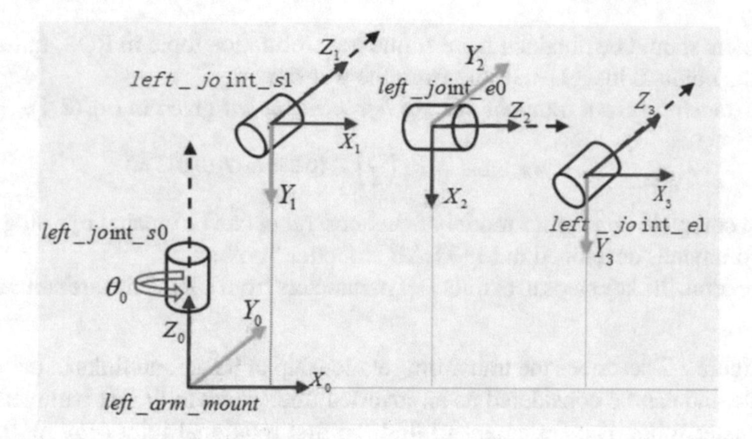

Fig. 2.4 First three joint frame transformational relationship

$a_2 = 0\,\text{m}$ because the $x = 0.102\,\text{m}$ is not the offset along the axis of joint *left_s1*. It is the key to make rotation occur only at the end of a link in this step.

From *left_e0* to *left_e1*, the axis of joint rotates $-\frac{\pi}{2}$ along X_2 axis, so link twist $\alpha_3 = -\frac{\pi}{2}$. The translation of *left_e0* is $x = 0.069\,\text{m}$ and $z = 0.262\,\text{m}$, so $a_3 = 0.069\,\text{m}$. The offset along the revolution axis should be $d_3 = 0.262\,\text{m} + 0.102\,\text{m} = 0.364\,\text{m}$, including the $x = 0.102\,\text{m}$ of joint *left_s1*. Repeating these steps, DH parameter table can be developed as shown in Table 2.1.

From DH notation table, we can get the transform matrix from the frame of *left_arm_mount* to the frame of *left_gripper* (end point). The model of the right arm of Baxter® robot can be created by using the same procedures above. DH notation table of right arm is the same as the left arm except $d_7 = 0.275\,\text{m}$, because in our configuration the gripper on right arm is an electric gripper, and a vacuum cup is installed on the left arm. According to different types of gripper mounted on each arm, the value of d_i would be different. In fact, each Baxter® robot has a unique URDF parameter configuration. Before creating the simulation model, the actual

Table 2.1 DH notation table of the left arm

Link i	θ_i (deg)	d_i (m)	a_i (m)	α_i (rad)
1	q_1	0.27	0.069	$-\frac{\pi}{2}$
2	$q_2 + \frac{\pi}{2}$	0	0	$\frac{\pi}{2}$
3	q_3	$0.102 + 0.262$	0.069	$-\frac{\pi}{2}$
4	q_4	0	0	$\frac{\pi}{2}$
5	q_5	$0.104 + 0.271$	0.01	$-\frac{\pi}{2}$
6	q_6	0	0	$\frac{\pi}{2}$
7	q_7	0.28	0	0

parameters should be obtained from frame transformation topic in ROS, rather than using the nominal ideal URDF file from the website.

The transform from frame of *base* to *left_arm_mount* given in Eq. (2.4).

$$^{base}T_{left\_arm\_mount} = R_z\left(\frac{\pi}{4}\right) T_x(0.056\,\mathrm{m}) T_z(0.011\,\mathrm{m}) \tag{2.4}$$

Then, a complete kinematics model of Baxter® robot can be created by using MATLAB commands developed in MATLAB Robotics Toolbox.

In general, the key steps to extract DH parameters from URDF file are summarized below

(i) Figure 2.3 describes the transform relationship of joints and links in the URDF file and can be considered as an inverted tree. To identify DH parameters for Baxter robot, it is necessary to find the first rotational joint ($rpy \neq 0$) from the root of this tree. Mark it as joint 1, and recognize four elements θ_1, d_1, a_1, α_1 and fill them in the DH table. Denote the next link as link 1, and assign properties includes *mass*, *inertia*, and *origin* to the model. Then continue the above procedure for joint 2, joint 3, and so on.

(ii) When the frame rotates not only about X axis (*r* in *rpy*), but also Z axis(*y* in *rpy*), an additional rotation ($\frac{\pi}{2}/-\frac{\pi}{2}$) needs to be added to the joint revolution angle θ_i.

(iii) The location of a joint seen on the model built in MATLAB Robotics Toolbox may be different from its location on the physical robot. For example, Baxter robot's joint e0 is located toward the middle of the upper arm link as shown in Fig. 2.5a, while on the model developed by the MATLAB Robotics Toolbox, it is located in such a manner that it shares origin with the previous joint s1, as shown in Fig. 2.5b. This is because that the origin of the coordinate frame associated with joint e0 is same as the origin of coordinate frame associated with joint s1, according to DH convention. Consequently, in the MATLAB Robotics Toolbox model the link offset should be modified.

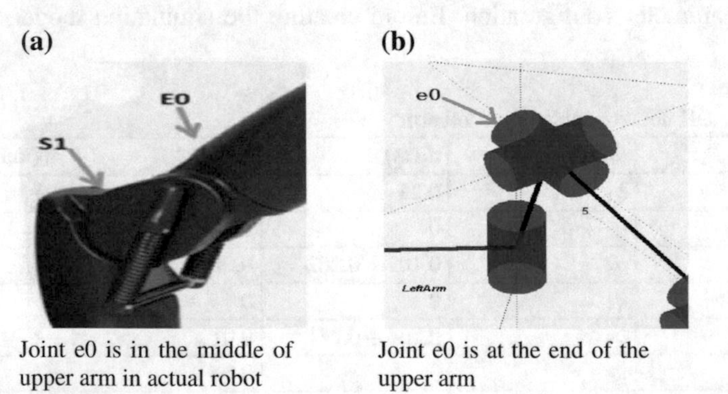

Joint e0 is in the middle of upper arm in actual robot

Joint e0 is at the end of the upper arm

Fig. 2.5 Joint rotates can only be represented at the end of link in model

2.1.3 Experimental Tests on Kinematics Modeling

Two experimental tests have been carried out to demonstrate the effectiveness of the kinematics model of the Baxter® robot. The first experiment shows that robot postures generated by the kinematic model are same as the real robot. The second experiment demonstrates the accuracy of the kinematic model.

2.1.3.1 Experiment Test 1

Given a Baxter robot manipulator, if all the seven joint angels $(\theta_1, \theta_2, \ldots \theta_7)$ are known, then we can calculate the Cartesian position and orientation of the robot. To test the accuracy of the kinematic model, we create the Baxter® robot arm that are put to several postures, retrieving the joint angular value $(\theta_1, \theta_2, \ldots \theta_7)$ and the position and orientation in Cartesian space (x_r, y_r, z_r). Then, input joint angles to our mathematical model and calculate Cartesian position (x_m, y_m, z_m) using forward kinematics. The postures of the robot and generated by model are shown in Fig. 2.6. It can be seen that when same joint angles are given, the model has the same posture as the robot.

2.1.3.2 Experiment Test 2

The pose that Baxter is shipping in or in power-off state is referred to as "shipping pose" [9] (Fig. 2.6d). Baxter's arms should be un-tucked (Fig. 2.6b) before subsequent movements. During the period when robot's arms move from tucked pose to un-tucked pose, and when it moves back to tucked pose, we retrieved a series of joint angles and Cartesian positions (x_r, y_r, z_r) and orientations $(\eta_{Rr}, \eta_{Pr}, \eta_{Yr})$, and calculate our Cartesian position (x_m, y_m, z_m) and orientations $(\eta_{Rm}, \eta_{Pm}, \eta_{Ym})$ using Eqs. (2.1)–(2.4) based on the model built. The deviation between them can be calculated from Eq. (2.5).

$$\begin{aligned} \Delta X &= (\Delta x, \Delta y, \Delta z) = (x_m, y_m, z_m) - (x_r, y_r, z_r) \\ \Delta\eta &= (\Delta\eta_R, \Delta\eta_P, \Delta\eta_Y) = (\eta_{Rm}, \eta_{Pm}, \eta_{Ym}) - (\eta_{Rr}, \eta_{Pr}, \eta_{Yr}) \end{aligned} \tag{2.5}$$

Figure 2.7 shows the position and orientation of left end point during the period of arm is tucking and untucking. Position and orientation of the end point of model (red curve) follows the trajectory of the actual robot (light gray curve) very closely.

The maximum error (unit of *x/y/z* is mm and unit of roll/pitch/yaw is rad) of position between our model and the real robot is sufficiently small and the maximum of errors (absolute values) for right arm is listed as below

$$[e_x, e_y, e_z, e_R, e_P, e_Y]_{max} = [4.2, 3.3, 6.8, 5.6, 7.6, 7.3]$$

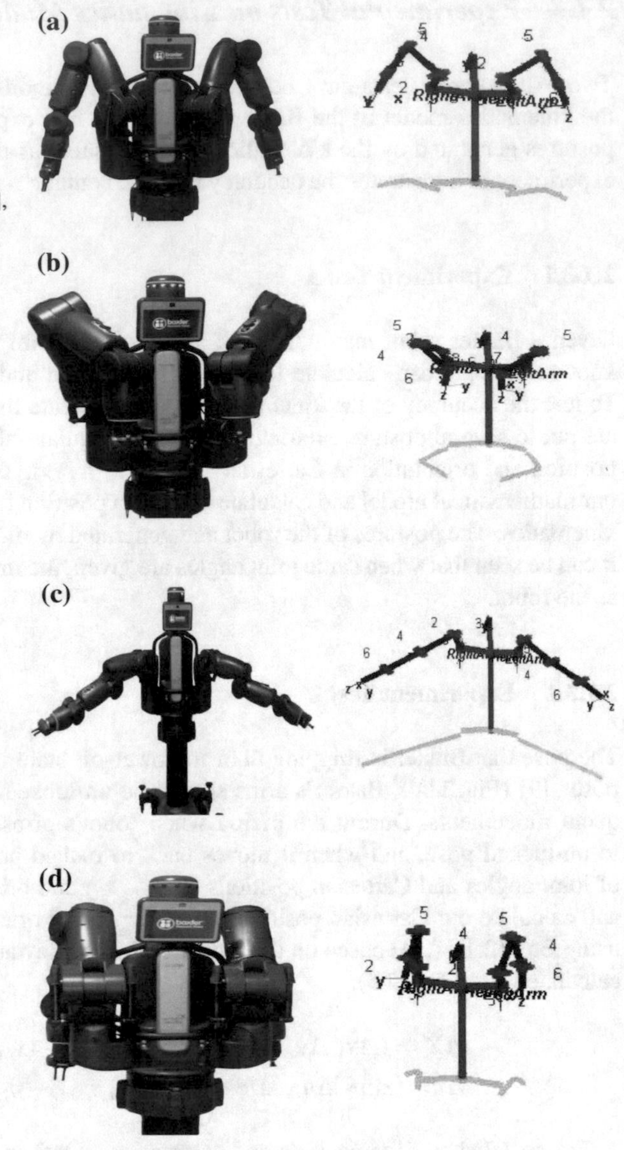

Fig. 2.6 Comparison of the posture of robot generated from model when same joint angles are given. The '*err*' is $(e_x, e_y, e_z, e_R, e_P, e_Y)$. They are calculated from Eq. (2.5), **a** [ready, $err = (0.6, 0.4, -2.4, 2.8, 0.6, 1.3)$], **b** [untacked, $err = (1.2, -0.8, -3.1, -1.0, 1.6, 1.8)$], **c** [extended, $err = (0.6, 0.7, -3.9, -2.8, 0.2, -1.8)$], **d** [tacked, $err = (0.5, 0.4, -1.6, -0.9, 2.3, 0.2)$]

for left arm is

$$[e_x, e_y, e_z, e_R, e_P, e_Y]_{max} = [4.4, 3.8, 6.1, 5.1, 7.2, 7.6]$$

The average error calculated from the data for right arm is

$$[e_x, e_y, e_z, e_R, e_P, e_Y]_{avg} = [0.7, -0.6, -3.8, 2.9, -3.2, -1.2]$$

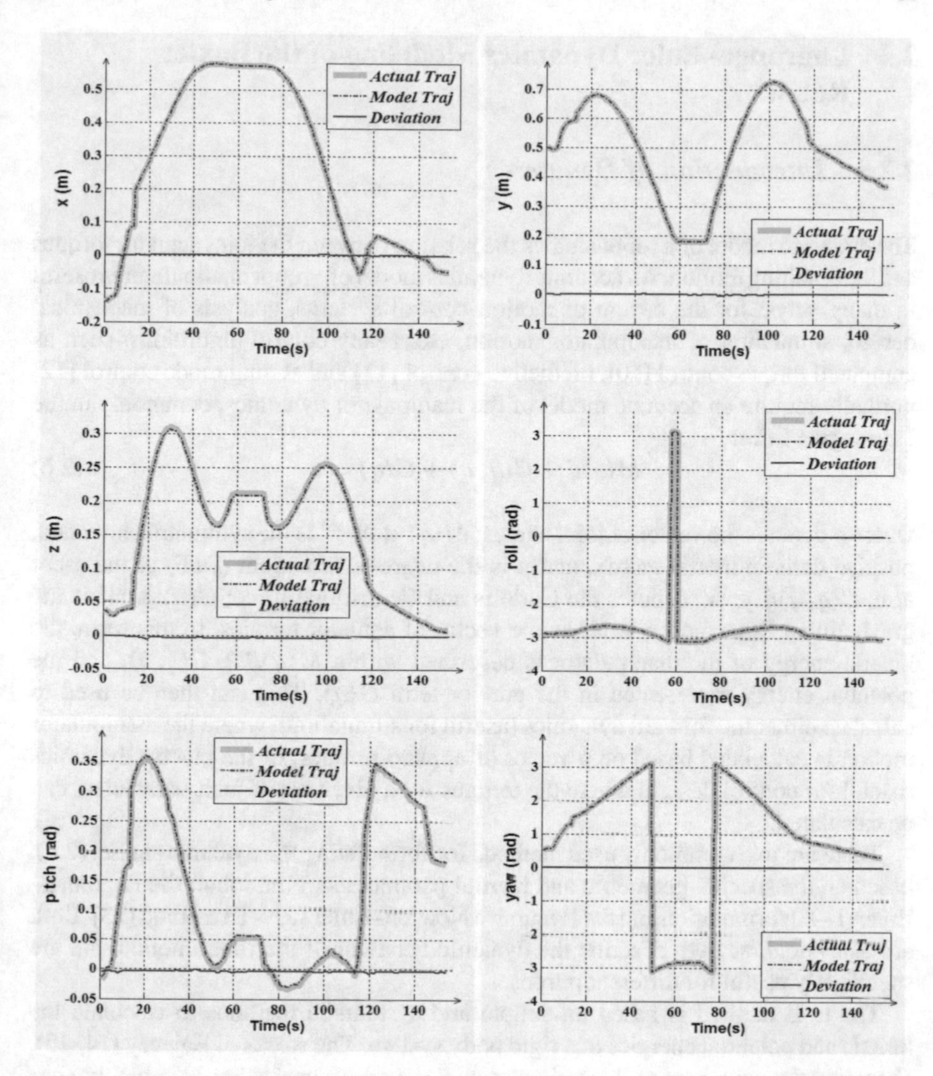

Fig. 2.7 Comparison of the trajectories of robot and those generated from model during the robot's left arm is tucked/un-tucked

for left arm is

$$[e_x, e_y, e_z, e_R, e_P, e_Y]_{avg} = [1.1, 0.6, -2.8, -1.6, -3.5, 1.1]$$

Thus, the accuracy of the kinematic model is satisfactory for simulation purpose.

2.2 Lagrange–Euler Dynamics Modeling of the Baxter Robot

2.2.1 Introduction of Dynamics

The dynamic model of a robot studies the relation between the joint actuator torques and the resulting motion. An accurate dynamics model of a robot manipulator is useful in many ways: for the design of motion control systems, analysis of mechanical design, simulation of manipulator motion, etc. Many control algorithms, such as computed torque control [10], predictive control [11] and sliding mode control [12] normally require an accurate model of the manipulator dynamics, commonly in the Lagrangian form:

$$M(q)\ddot{q} + C(q, \dot{q}) + G(q) = \tau \tag{2.6}$$

where q denotes the vector of joint angles; $M(q) \in \Re^{n \times n}$ is the symmetric, bounded, positive definite inertia matrix, and n is the degree of freedom (DoF) of the robot arm; $C(q, \dot{q})\dot{q} \in \Re^n$ denotes the Coriolis and Centrifugal force; $G(q) \in \Re^n$ is the gravitational force, and $\tau \in \Re^n$ is the vector of actuator torques. In this form, the kinetic energy of the manipulator is described within $M(q)\ddot{q} + C(q, \dot{q})$, and the potential energy represented in the gravity term $G(q)$. This can then be used to calculate either the forward dynamics (useful for simulation), where the manipulator motion is calculated based on a vector of applied torques, or the inverse dynamics (useful for control design) where the torques for a given set of joint parameters can be calculated.

There are two commonly used methods for formulating the dynamics in Eq. (2.6), based on the specific geometric and inertial parameters of the robot: the Lagrange–Euler (L–E) formulation and the Recursive Newton–Euler (RN–E) method [13]. Both are equivalent, as both describe the dynamic behavior of the robot motion, but are specifically useful for different purposes.

The L–E method is based on simple and systematic methods to calculate the kinetic and potential energies of a rigid body system. The works of Bajeczy [14, 15], show that the equations of dynamic motion for a robot manipulator are highly non-linear, consisting of inertial and gravity terms, and are dependent on the link physical parameters and configuration (i.e., position, angular velocity and acceleration). This provides the closed form of the robot dynamics, and is therefore applicable to the analytical computation of robot dynamics [16], and therefore can be used to design joint-space (or task-space, using transformation via the Jacobian) control strategies.

The L–E formulation may also be used for forward and inverse dynamic calculation, but this requires the calculation of a large number of coefficients in $M(q)$ and $C(q, \dot{q})$ from Eq. (2.6), which may take a long time. This makes this method somewhat unsuitable for online dynamic calculations, especially as other methods such as RN–E (described in the next section), or Lee's Generalized d'Alembert Equations (GAE) [17] produce more simple, albeit messy, derivations which are much faster [13]. A recursive L–E method has also been described [18] which greatly reduces the computational cost of the L–E formulation and brings it into line with RN–E methods.

The N-E formulation is based on a balance of all the forces acting on the generic link of the manipulator; this forms a set of equations with a recursive solution [19], and was developed. A forward recursion propagates link velocities and accelerations, then a backward recursion propagates the forces and torques along the manipulator chain. This is developed as a more efficient method than L–E, and is based on the principle of the manipulator being a serial chain; when a force is applied to one link, it may also produce motion in connected links. Due to this effect, there may be considerable duplication of calculation [20], which can be avoided if expressed in a recursive form. This reduction in computational load greatly reduces calculation time, allowing the forward and inverse dynamics calculations to be performed in real time; therefore, it can enable real-time torque control methods of robot manipulators.

2.2.2 *Dynamics Modeling Procedure*

To establish the dynamics model of a robot, as stated above, two common methods have been developed: one is the Lagrange formulation, which gives a closed form of the dynamic equations for ease of theoretical analysis; and the other is the Newton–Euler method, which uses a recursive form with the easy-to-compute merit. In this section, a formulation of the Lagrange–Euler (L–E) equations representing the dynamics of the Baxter manipulator is presented for analysis the dynamics parameters in controller design.

Next, we will show the dynamics modeling of the Baxter robot which is introduced in Sect. 1.1. To accomplish the dynamics modeling of the Baxter Robot, some parameters are specified at first. The common inertial and geometrical parameters of the Baxter robot are shown in Table 2.2.

Table 2.2 Nomenclature

n	Degrees of freedom (DoF) of the manipulator
$q, \dot{q}, \ddot{q} \in \Re^{n\times 1}$	Vector of joint position, angular velocity and acceleration, respectively
a, d, α, θ	Variables denoting the Denavit–Hartenberg parameters
$I_i \in \Re^{3\times 3}$	Inertia tensor of link i
m	Link mass
$\bar{r}_i \in \Re^{4\times 1}$	Center of mass of link i
${}^iT_j \in \Re^{4\times 4}$	Homogeneous transform from link i to j

2.2.2.1 Parameters Specification

The D–H parameters and link masses of the Baxter manipulator are given in Table 2.3 and are derived from the Universal Robot Descriptor File (URDF) [21]. These parameters describe the configuration of the links, and form the basis of the Lagrange–Euler formulation. The homogeneous link transform matrices are formed from the D–H parameters as such below:

$$^{i-1}T_i = \begin{bmatrix} \cos\theta_i & -\cos\alpha_i \sin\theta_i & \sin\alpha_i \sin\theta_i & a_i \cos\theta_i \\ \sin\theta_i & \cos\alpha_i \cos\theta_i & -\sin\alpha_i \cos\theta_i & a_i \sin\theta_i \\ 0 & \sin\alpha_i & \cos\alpha_i & d_i \\ 0 & 0 & 0 & 1 \end{bmatrix} \tag{2.7}$$

where $^0T_i = ^0T_1\, ^1T_2 \ldots ^{i-1}T_i$.

The center of mass (CoM) for each link is given in Table 2.4, which forms the homogeneous column vector $\bar{r}_i = [\bar{x}_i\ \bar{y}_i\ \bar{z}_i\ 1]^T$. The inertia tensors of each joint are given in Table 2.5, represented by the inertias working in each axis I_{xx}, I_{yy}, I_{zz} and cross-talk inertia between axes I_{xy}, I_{yz}, I_{xz}. Here, it is represented as a row vector, but is also commonly found in $I^{3\times3}$ symmetric matrix form. These information were obtained from https://github.com/RethinkRobotics

Table 2.3 D–H parameters of the Baxter robot

Link	θ	d (m)	a (m)	α (rad)	m (kg)
1	θ_1	0.2703	0.069	$-\pi/2$	5.70044
2	θ_2	0	0	$\pi/2$	3.22698
3	θ_3	0.3644	0.069	$-\pi/2$	4.31272
4	θ_4	0	0	$\pi/2$	2.07206
5	θ_5	0.3743	0.01	$-\pi/2$	2.24665
6	θ_6	0	0	$\pi/2$	1.60979
7	θ_7	0.2295	0	0	0.54218

Table 2.4 Center of mass (all units in m)

Link	$\bar{x}$	$\bar{y}$	$\bar{z}$
1	−0.05117	0.07908	0.00086
2	0.00269	−0.00529	0.06845
3	−0.07176	0.08149	0.00132
4	0.00159	−0.01117	0.02618
5	−0.01168	0.13111	0.0046
6	0.00697	0.006	0.06048
7	0.005137	0.0009572	−0.06682

Table 2.5 Link inertia tensors (all units kg × m^2)

Link	I_{xx}	I_{yy}	I_{zz}
1	0.0470910226	0.035959884	0.0376697645
2	0.027885975	0.020787492	0.0117520941
3	0.0266173355	0.012480083	0.0284435520
4	0.0131822787	0.009268520	0.0071158268
5	0.0166774282	0.003746311	0.0167545726
6	0.0070053791	0.005527552	0.0038760715
7	0.0008162135	0.0008735012	0.0005494148
Link	I_{xy}	I_{yz}	I_{xz}
1	−0.0061487003	−0.0007808689	0.0001278755
2	−0.0001882199	0.0020767576	−0.00030096397
3	−0.0039218988	−0.001083893	0.0002927063
4	−0.0001966341	0.000745949	0.0003603617
5	−0.0001865762	0.0006473235	0.0001840370
6	0.0001534806	−0.0002111503	−0.0004438478
7	0.000128440	0.0001057726	0.00018969891

2.2.2.2 Lagrange–Euler Formulation

The Lagrange–Euler equations of motion for a conservative system [13] are given by

$$L = K - P, \qquad \tau = \frac{\mathrm{d}}{\mathrm{d}t}\left(\frac{\partial L}{\partial \dot{q}}\right) - \frac{\partial L}{\partial q} \tag{2.8}$$

where K and P are the total kinetic and potential energies of the system, respectively, $q \in \Re^n$ is the generalized robot coordinates equivalent to θ in Table 2.3, and τ is the generalized torque at the robot joints [13]. The kinematic and potential energies are given by:

$$\begin{aligned} K &= \frac{1}{2}\sum_{i=1}^{n}\sum_{j=1}^{i}\sum_{k=1}^{i}\left[\mathrm{Tr}\left(U_{ij}\ J_i\ U_{ik}^T\right)\dot{q}_j\dot{q}_k\right] \\ P &= \sum_{i=1}^{n} -m_i\ \mathbf{g}\left({}^0T_i\ \bar{r}_i\right) \end{aligned} \tag{2.9}$$

which, when substituted into Eq. (2.8), gives the expression:

$$\tau_i = \frac{\mathrm{d}}{\mathrm{d}t}\left(\frac{\partial L}{\partial \dot{q}}\right) - \frac{\partial L}{\partial q}$$

$$= \sum_{j=i}^{n} \sum_{k=1}^{j} \mathrm{Tr}(U_{jk} J_j U_{ji}^T) \ddot{q}_k - \sum_{j=i}^{n} m_j \mathbf{g} U_{ji} \bar{r}_j$$
$$+ \sum_{j=i}^{n} \sum_{k=1}^{j} \sum_{m=1}^{j} \mathrm{Tr}(U_{jkm} J_j U_{ji}^T) \dot{q}_k \dot{q}_m. \tag{2.10}$$

This can be expressed more simply in the form given in Eq. (2.6), as a sum of the inertia, Coriolis/centrifugal and gravity terms. The elements of the symmetric matrix $M(q)$ are given by

$$M_{i,k} = \sum_{j=\max(i,k)}^{n} \mathrm{Tr}(U_{jk}\ J_j\ U_{ji}^T) \quad i, k = 1, 2, \ldots n, \tag{2.11}$$

the Coriolis/centrifugal force vector $C(q, \dot{q})$

$$C_i = \sum_{k=1}^{n} \sum_{m=1}^{n} h_{ikm} \dot{q}_k \dot{q}_m$$
$$h_{ikm} = \sum_{j=\max(i,k,m)}^{n} \mathrm{Tr}(U_{jkm}\ J_j\ U_{ji}^T) \tag{2.12}$$

and the gravity vector $G(q)$

$$G_i = \sum_{j=i}^{n} (-m_j \mathbf{g} U_{ji} \bar{r}_j) \tag{2.13}$$

where $\mathbf{g} = [0,\ 0,\ -9.81,\ 0]$ is the gravity row vector. The matrix U_{ij} is the rate of change of points on link i relative to the base as the joint position q_j changes

$$U_{ij} \equiv \frac{\partial T_i^0}{\partial q_j} = \begin{cases} {}^0T_{j-1}\ Q_j\ {}^{j-1}T_i & j \le i \\ 0 & j > i \end{cases} \tag{2.14}$$

which allows derivation of the interaction effects between joints, U_{ijk}

$$U_{ijk} \equiv \frac{\partial U_{ij}}{\partial q_k}$$
$$= \begin{cases} {}^0T_{j-1}\ Q_j\ {}^{j-1}T_{k-1}\ Q_k\ {}^{k-1}T_i & i \ge k \ge j \\ {}^0T_{k-1}\ Q_k\ {}^{k-1}T_{j-1}\ Q_j\ {}^{j-1}T_i & i \ge j \ge k \\ 0 & i < j \text{ or } i < k \end{cases} \tag{2.15}$$

where, for Baxter, as all joints are revolute,

$$Q_j = \begin{bmatrix} 0 & -1 & 0 & 0 \\ 1 & 0 & 0 & 0 \\ 0 & 0 & 0 & 0 \\ 0 & 0 & 0 & 0 \end{bmatrix}. \tag{2.16}$$

The J_i matrices are independent of link position or motion, and therefore only needs to be calculated *once* from the inertia tensors, link masses and link CoMs:

$$J_i = \begin{bmatrix} \frac{-I_{xxi}+I_{yyi}+I_{zzi}}{2} & I_{xyi} & I_{xzi} & m_i\bar{x}_i \\ I_{xyi} & \frac{I_{xxi}-I_{yyi}+I_{zzi}}{2} & I_{yzi} & m_i\bar{y}_i \\ I_{xzi} & I_{yzi} & \frac{I_{xxi}+I_{yyi}-I_{zzi}}{2} & m_i\bar{z}_i \\ m_i\bar{x}_i & m_i\bar{y}_i & m_i\bar{z}_i & m_i \end{bmatrix} \tag{2.17}$$

This concludes the calculations required to form the L–E dynamics of the Baxter arm.

2.2.3 Experimental Studies

To collect data from the Baxter robot, a PID position controller is employed in a dual loop configuration, as shown in Fig. 2.8. A desired velocity $\dot{q}_d$ is generated from the outer loop

$$\begin{aligned} e = q_r - q, \quad \dot{e} = \frac{\mathrm{d}}{dt}e \\ \dot{q}_d = K_p e + K_d \dot{e} \end{aligned} \tag{2.18}$$

which is then used to generate torque in the inner loop

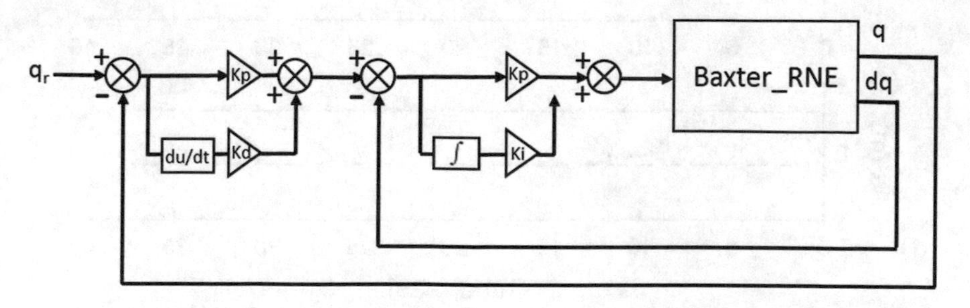

Fig. 2.8 Block diagram of torque control system

$$\dot{\epsilon} = \dot{q}_d - \dot{q}_r, \quad \epsilon = \int \dot{\epsilon}\, \mathrm{d}t$$

$$\tau_r = K_p \dot{\epsilon} + K_i \epsilon. \tag{2.19}$$

The trajectories of q_r were created in two ways: generated using sine and cosine patterns or using a touchpad input, both in Cartesian space. Inverse kinematics are performed using the inverse Jacobian method, that is

$$\dot{q}_r = J^{\dagger}(q)\dot{x}_r \tag{2.20}$$

where $\dot{x}_r$ is the reference Cartesian velocity and $J^{\dagger}$ is the pseudo-inverse of the Jacobian matrix. The selected test trajectories in Fig. 2.9, show the actual Cartesian test trajectories x which are calculated from $x = F(q)$, where $F(q)$ is the forward kinematics of the robot. For the experiment, the right-hand manipulator of the Baxter was driven through these three trajectories and data collected at 50 Hz, including joint positions and velocities q, $\dot{q}$, Cartesian position x and the estimated torques applied to the motors τ. This is calculated on board of the Baxter robot from the deflection of the internal springs (and large external springs at joint 2), summed with the automatic gravity compensation. The joint accelerations $\ddot{q}$ were estimated through first order numerical differentiation of $\dot{q}$, which accounts for the noise in the calculated results.

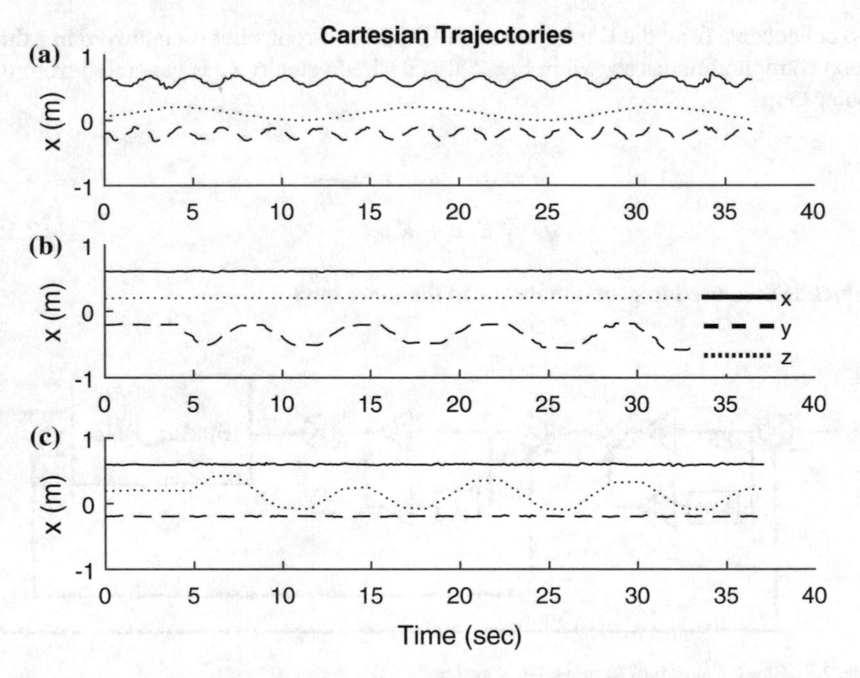

Fig. 2.9 Cartesian trajectories selected for experimentation. In **a** all three dimensions follow a sinusoidal pattern, with a lower frequency in the z-axis. For the trajectories in **b** and **c** the arm was moved only in the y and z-axes, respectively

To find the explicit form of Baxter manipulator dynamics, MATLAB's symbolic toolbox was utilized. In their raw state, the symbolic representations of the elements of $D(q)$, $C(q, \dot{q})$ and $G(q)$ have many coefficients (over half a million), they cannot be printed here.

To confirm the accuracy of the process, a numerical form was also created. Joint positions and velocities were recorded from the Baxter moving in three different trajectories, and the joint accelerations estimated through numerical differentiation, i.e., $\ddot{q}_i = \frac{d\dot{q}_i}{dt}$, where $dt = 0.02$ is the sampling period of the trajectory recorder. The results from the L–E form are compared against torques recorded from the Baxter, and torque trajectories generated using the RN–E method from Peter Corke's Robotics Toolbox [3]. It is possible to generate the analytical representation of Eq. (2.6) using this RN–E method, but only if n is small due to heavy memory consumption. Due to the way Baxter is controlled, the recorded torques are a sum of the actuator torques measured via internal spring deflection and two torque vectors acting on joint 2 (hysteresis and crosstalk) to compensate for the large external springs, mentioned previously. All results that are collected from the right-hand manipulator, with the end-effector aligned with the z-axis. No external forces were applied to the arm during testing.

In Fig. 2.10, the arm was moving in all three planes. It is noticeable that the torques generated from L–E and RN–E are much noisier; this is due to the numerical differentiation of the joint accelerations $\ddot{q}$. This could be reduced by passing it through a low pass filter. However, we can see that the shape of the trajectories is very similar.

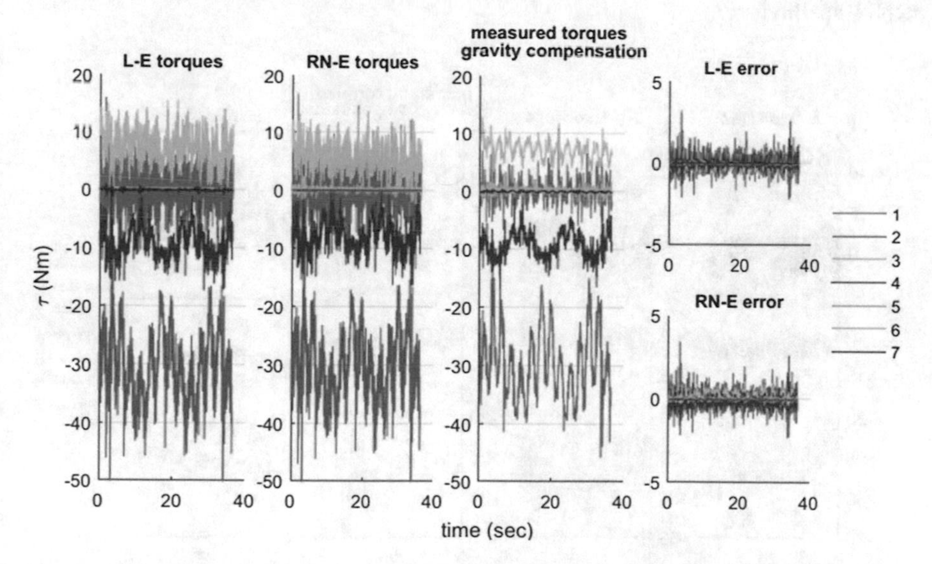

Fig. 2.10 Comparison of torque generated through L–E and RN–E methods with torques recorded from the Baxter robot during the trajectory from Fig. 2.9a. The trajectory for this sample was moving the end-effector in a circular trajectory in the x, y planes and in a cosine pattern in the z-axis, where $x = 0.6 + 0.1\sin(t)$, $y = -0.2 + 0.1\cos(t)$ and $z = 0.1 + 0.1\cos(0.2t)$. The errors (*far right*) are the modeled torques subtracted from the recorded torques

The first joint and distal joints 5–7 only require very small torque input as they are mostly unaffected by gravity. Examining the L–E error plot in Fig. 2.10, the noise dominates the largest errors but it is centered around zero for all joints, confirming that there are no bias errors. The RN–E error plot shows a similar range of error, but it is noticeable that the error for joint 3 has some positive bias.

In Fig. 2.11, we have similar results, with an even smaller error result. In this case, the arm was moved in a way to generate higher accelerations by quickly switching the target position in the y-axis only. This movement is mostly achieved using joint 2 at the shoulder, noticeable in the plots. The low error result in this case confirms a good match for the kinetic part of the dynamic model. Again, looking at the RN–E there is an obvious positive bias in the torques calculated for joint 3.

A slower trajectory was applied to the robot for the results in Fig. 2.12, moving primarily in the z-axis, which is evident by the large changes occurring in joint 4. Noise is reduced due to minimal acceleration in the trajectory. The error results again show no bias errors, and within a good tolerance of around ± 1.5 Nm which mostly can be accounted for by noise from the acceleration trajectory derivation.

A good comparison of the methods is shown in Table 2.6, where the average (and a sum total) error of both methods for each joint are shown side by side. These are calculated from each trajectory set, i.e., set 1 in Table 2.6 corresponds to the first trajectory results in Fig. 2.10. Looking through the table, it can be seen that the average error for each joint is comparable between the methods, apart from in joints 3–4 which have significantly larger errors in every set. This gives a good indication that the L–E method is not only accurate, but also slightly more accurate than the RN–E method.

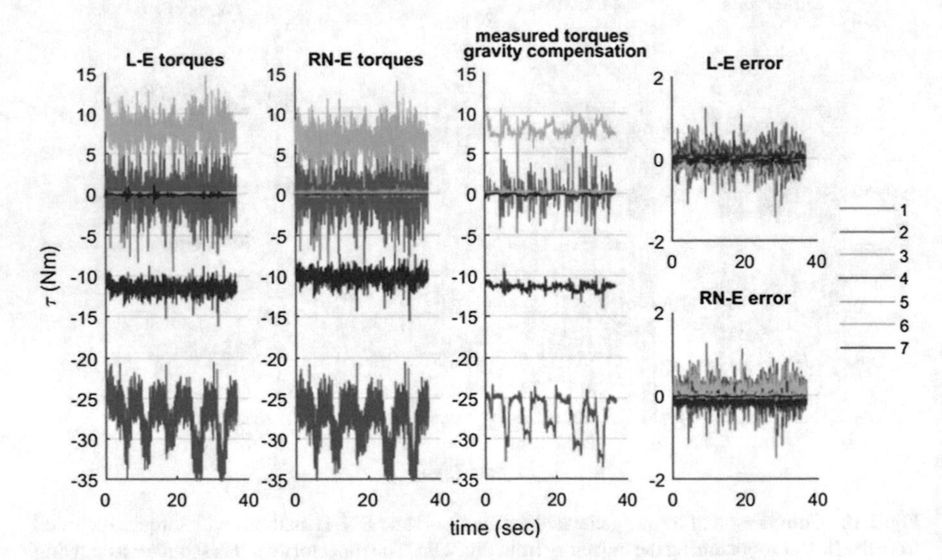

Fig. 2.11 Second comparison of torque trajectories from Fig. 2.9b; for this sample the end-effector is fixed in the x, z plane and is switched quickly between two positions in the y-axis

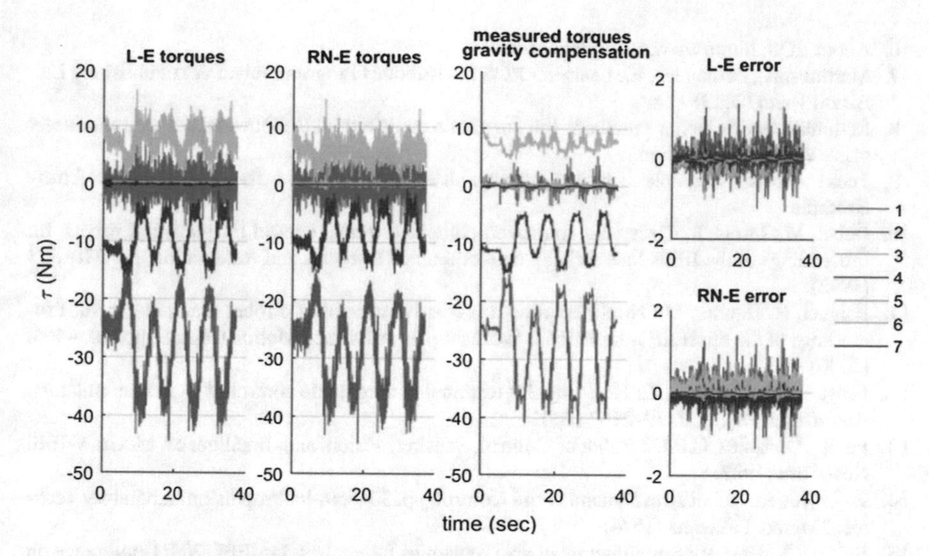

Fig. 2.12 Final comparison of torque trajectories from Fig. 2.9c input. The end-effector was moved up and down in the z-axis, and held constant in the x, y plane

Table 2.6 Averages of calculation errors for L–E and RN–E methods

Joint	Set					
	1		2		3	
	L–E	RN–E	L–E	RN–E	L–E	RN–E
$\lvert\bar{e}_1\rvert$	0.0105	0.0105	0.0148	0.0149	0.0061	0.0058
$\lvert\bar{e}_2\rvert$	0.1002	0.0665	0.0464	0.0931	0.0703	0.0872
$\lvert\bar{e}_3\rvert$	0.0475	0.1382	0.0083	0.1355	0.0367	0.1358
$\lvert\bar{e}_4\rvert$	0.0151	0.1231	0.0079	0.1210	0.0064	0.1148
$\lvert\bar{e}_5\rvert$	0.0068	0.0120	0.0099	0.0145	0.0079	0.0140
$\lvert\bar{e}_6\rvert$	0.0006	0.0103	0.0047	0.0136	0.0003	0.0113
$\lvert\bar{e}_7\rvert$	0.0012	0.0036	0.0003	0.0055	0.0013	0.0024
$\sum_{i=1}^{n} \bar{e}_i$	0.0710	0.0869	0.0527	0.1161	0.0366	0.1089

References

1. Denavit, J.: A kinematic notation for lower-pair mechanisms based on matrices. ASME J. Appl. Mech. **22**, 215–221 (1955)
2. Nethery, J.F., Spong, M.W.: Robotica: a mathematica package for robot analysis. Robot. Autom. Mag. IEEE **1**(1), 13–20 (1994)
3. Corke, P.: A robotics toolbox for matlab. Robot. Autom. Mag. IEEE **3**(1), 24–32 (1996)
4. Corke, P.: A simple and systematic approach to assigning denavit-hartenberg parameters. IEEE Trans. Robot. **23**(3), 590–594 (2007)
5. Ju, Z., Yang, C., Ma, H.: Kinematics modeling and experimental verification of baxter robot, Chinese Control Conference (CCC). **33**, 8518–8523 (2014)

6. About ROS: http://www.ros.org/about-ros/
7. Martinez, A., Fernández, E.: Learning ROS for Robotics Programming. Packt Publishing Ltd, Birmingham (2013)
8. RethinkRobotics baxter common: https://github.com/RethinkRobotics/baxter_common/tree/master/baxter_description
9. Tuck Arms Example: https://github.com/RethinkRobotics/sdk-docs/wiki/Tuck-Arms-Example
10. Uebel, M., Minis, I., Cleary, K.: Improved computed torque control for industrial robots, In: Proceedings of the IEEE International on Conference Robotics and Automation, pp. 528–533 (1992)
11. Poignet, P., Gautier, M.: Nonlinear model predictive control of a robot manipulator. In: Proceedings of the 6th IEEE International Workshop on Advanced Motion Control, pp. 401–406. (2000)
12. Feng, Y., Yu, X., Man, Z.: Non-singular terminal sliding mode control of rigid manipulators. Automatica **38**(12), 2159–2167 (2002)
13. Fu, K., Gonzalez, C.R.C.: Robotics Control, Sensing, Vision, and, Intelligence. McGraw-Hill, New York (1987)
14. Bejczy, A.K.: Robot Arm Dynamics and Control, pp. 33–669. Jet Propulsion Laboratory Technical Memo, Pasadena (1974)
15. Bejczy, A., Paul, R.: Simplified robot arm dynamics for control. In: IEEE 20th Conference on Decision and Control including the Symposium on Adaptive Processes (1981)
16. Megahed, S.M.: Principles of Robot Modelling and Simulation. Wiley, Hoboken (1993)
17. Lee, C., Lee, B., Nigam, R.: Development of the generalized d'alembert equations of motion for mechanical manipulators. In: 22nd IEEE Conference on Decision and Control, pp. 1205–1210 (1983)
18. Hollerbach, J.M.: A recursive lagrangian formulation of maniputator dynamics and a comparative study of dynamics formulation complexity. IEEE Trans. Syst. Man Cybern. **10**(11), 730–736 (1980)
19. Siciliano, B., Sciavicco, L., Villani, L., Oriolo, G.: Robotics: Modelling, Planning and Control. Springer, Heidelberg (2009)
20. Mckerrow, P.: Introduction to Robotics. Addison-Wesley Longman Publishing, Boston (1991)
21. Ju, Z., Yang, C., Ma, H.: Kinematic modeling and experimental verification of Baxter robot. In: Proceedings of the 33rd Chinese Control Conference Nanjing, China, pp. 8518–8523. 28–30 July 2014

Chapter 3
Intelligent Control of Robot Manipulator

Abstract This chapter presents a series of intelligent control schemes for robot manipulator control. A controller using fuzzy logic to infer the learning parameters is developed to improve the performance of the dual-adaptive controller. A new task space/joint space hybrid control scheme for bimanual robot with impedance and force control is also introduced. The task space controller adapts end-point impedance, to compensate for interactive dynamics, and the joint space controller adapts the impedance to improve robustness against external disturbances. An adaptive model reference control is designed for robots to track desired trajectories and for the closed-loop dynamics to follow a reference model, which is derived using the LQR optimization technique to minimize both the motion tracking error and the transient acceleration for a smooth trajectory. A new discrete-time adaptive controller for robot manipulator with uncertainties from the unknown or varying payload is introduced, based on the idea of one-step guess. The history information is used to estimate the unknown fixed or time-varying payload on the end effector.

3.1 Dual-Adaptive Control of Bimanual Robot

Robot control is a practical but challenging topic demanded by both academical and industrial communities [1–3]. Many methods such as proportion–integral–derivative (PID) control, sliding mode control [1, 2], fuzzy control [4, 5], neural network (NN) control [6], adaptive control [7–9], etc. have been applied in robot control. A critical problem in robot control lies in the uncertainties associated with the robot manipulators, such as the model mismatch, changing of the external environment, and uncertain loads. Therefore, the development of adaptive and intelligent learning control methods has gained more attention. The intelligent control methods including adaptive control, fuzzy control, and NN control have been widely adopted as a viable alternative to traditional control methods.

In [10], a model reference adaptive control design [11] is proposed for impedance control of robot manipulators, which is motivated by the model reference adaptive control design for motion control as in [12]. By studying how human beings interact with unstable dynamics, [13] has shown that humans are able to learn to stabilize

© Science Press and Springer Science+Business Media Singapore 2016

C. Yang et al., *Advanced Technologies in Modern Robotic Applications*,
DOI 10.1007/978-981-10-0830-6_3

control by adapting the mechanical impedance in size and geometry independently on the force. The fact that the central nervous system is able to minimize instability, error and effort, was further shown in [14] where the electromyography (EMG) signals of muscles showed that the co-contraction of muscle was first increased, after which hand-path error is decreased and later metabolic cost. A model for motor learning was described in [15], where the learning function is characterized by a V-shaped adaptation in each muscle independently. The controller has been implemented in [16], where it was shown that in its application the robot system, the controller produces behaviors very similar to that observed in humans.

The subject of bimanual control is fairly young in the literature. Early research involving bimanual control was aimed at rehabilitation devices [17–20]. The work in [21] describes a leader–follower type configuration, where the follower robot(s) maintain coordination with the leader through a compliant position controller, similar to the one described in [22], as well as passive compliance from spring-damper manipulators. They show that good cooperation can be obtained through this method, with tasks involving multiple robots perform well.

The control method presented in [23] forms the basis of the dual-adaptive control of bimanual robot. And this method has been extended to the task with both perturbation at the endpoint and on the limbs in a previous work [24]. The Baxter bimanual robot has been shown in Sect. 1.1 and its kinematic and dynamic model are also well developed in Chap. 2. In this section, we investigate the dual-adaptive control for Baxter robot by using a fuzzy controller translating the control engineer experience in tuning these meta-parameters [25]. The control gains are initially set to some low values to ensure stable learning, which are also used as a baseline for fuzzy inference of gain values.

3.1.1 Preliminaries

3.1.1.1 Fuzzy Control

The process for fuzzy inference of an output Y in Fig. 3.1 requires several steps. Fuzzification maps a real scalar value (for example, temperature) into fuzzy space; this is achieved using membership functions. X is a space of points, with elements $x \in X$ [26]. A fuzzy set A in X is described by a membership function $\mu_A(x)$ which associates a grade of membership $\mu_A(x_i)$ in the interval [0, 1] to each point x in A. Several different choices exist for the membership function, such as Gaussian, trapezoidal, etc. But in the scope of this chapter triangular functions are used for the sake of simplicity and low sensitivity as described in [27]. Membership functions are set so that the completeness ϵ of all fuzzy sets is 0.4 to minimize uncertainty and ensure that one rule is always dominant. Several definitions are required. A union, which corresponds to the connective OR, of two sets A and B is a fuzzy set C

$$C = A \cup B; \quad \mu_C(x) = \max\big[(\mu_A(x), \mu_B(x)\big], \; x \in X \tag{3.1}$$

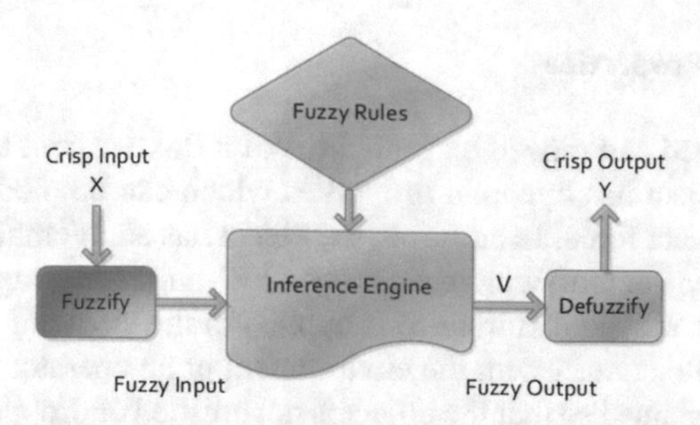

Fig. 3.1 Diagram illustration of fuzzy system. The input X is *fuzzified* into sets. The output sets V are generated by the inference engine based on the fuzzy rules. Y is the "real world" output which is retrieved through the *defuzzification* of V

An intersection, which corresponds to connective AND, can similarly be described

$$C = A \cap B; \quad \mu_C(x) = \min\left[(\mu_A(x), \mu_B(x)\right], \; x \in X \tag{3.2}$$

The Cartesian product can be used to describe a relation between two or more fuzzy sets; let A be a set in universe X and B a set in universe Y [28]. The Cartesian product of A and B will result in a relation R

$$A \times B = R \subset X \times Y \tag{3.3}$$

where the fuzzy relation R has a membership function

$$\mu_R(x, y) = \mu_{A\times B}(x, y) = \min\left[\mu_A(x), \mu_A(y)\right] \tag{3.4}$$

This is used in the Mamdani Min Implication to relate an input set to an output set, i.e.,

$$\text{IF } x \text{ is } A \quad \text{THEN } y \text{ is } B \tag{3.5}$$

The rule set is used to implicate the output, which is then aggregated for all rules. Defuzzification is then performed again, for which several methods exist but we chose the centroid method due to its accuracy [29]. The defuzzified value y^* is given by

$$y^* = \frac{\int \mu_B(y)\, y \, \mathrm{d}x}{\int \mu_B(y)\, \mathrm{d}x} \tag{3.6}$$

which computes the center of mass of the aggregated output membership function and relates the μ value back to a crisp output.

3.1.1.2 Object Properties

An object is grasped and moved by both arms of a Baxter robot to follow a desired trajectory. The object has dynamic properties which can be modeled by a spring–mass–damper contact force. In addition, the object has an internal disturbance (as if it were a running power tool) which generates a low-amplitude, high-frequency force on the arms, and a second disturbance is applied to the elbow of the arms, as if the robot had come into contact with the environment or an operator. Minimal tracking error must be maintained so that the object is not crushed or dropped, and the control effort must be minimized so that the system remains compliant.

The object is modeled as a mass–spring–damper system, as shown in Fig. 3.2. For simplicity but without loss of generality, the shape of the object is assumed to be a sphere. The interaction forces F_R and F_L resulting from the object dynamics are described with reference to the Cartesian position and velocity of the central mass, and left and right end effectors $X, \dot{X}$ respectively.

$$\begin{aligned} F_R &= -k\big(l-(X_m-X_R)\big)-d(\dot{X}_R-\dot{X}_m) \\ F_L &= k\big(l-(X_L-X_m)\big)-d(\dot{X}_m-\dot{X}_L) \end{aligned} \tag{3.7}$$

That is, the forces acting at each of the left and right hand, are the sum of forces arising from interaction of the mass of the object and the hand upon the spring-damper system distributed equally across the surface of the object. From Newton's Second Law of motion, we have

$$-m\ddot{X}_m = F_L + F_R \tag{3.8}$$

which together with Eq. (3.7) yields

$$-m\ddot{X}_m = 2d\dot{X}_m + 2kX_m - d(\dot{X}_L+\dot{X}_R) - k(X_L+X_R) \tag{3.9}$$

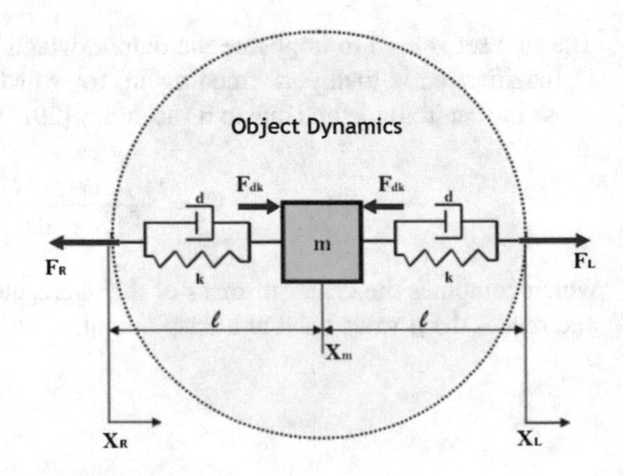

Fig. 3.2 Object dynamics modeled as a mass–spring–damper system, where F_R and F_L are the interaction forces acting on the left and right arms as a result of the Cartesian positions and velocities of the left and right end effectors

For convenience, the object dynamics are formed into a state space representation with X_m and $\dot{X}_m$ as the states, and X_L, X_R, $\dot{X}_L$ and $\dot{X}_R$ as the inputs:

$$\begin{bmatrix} \dot{X}_m \\ \ddot{X}_m \end{bmatrix} = \begin{bmatrix} 0 & 1 \\ \frac{-2k}{m} & \frac{-2d}{m} \end{bmatrix} \begin{bmatrix} X_m \\ \dot{X}_m \end{bmatrix} + \begin{bmatrix} 0 & 0 & 0 & 0 \\ \frac{k}{m} & \frac{k}{m} & \frac{d}{m} & \frac{d}{m} \end{bmatrix} \begin{bmatrix} X_L \\ X_R \\ \dot{X}_L \\ \dot{X}_R \end{bmatrix} \tag{3.10}$$

The formulation of the outputs F_L and F_R is given by:

$$\begin{bmatrix} F_L \\ F_R \end{bmatrix} = \begin{bmatrix} k & d \\ k & d \end{bmatrix} \begin{bmatrix} X_m \\ \dot{X}_m \end{bmatrix} + \begin{bmatrix} -k & 0 & -d & 0 \\ 0 & -k & 0 & -d \end{bmatrix} \begin{bmatrix} X_L \\ X_R \\ \dot{X}_L \\ \dot{X}_R \end{bmatrix} + \begin{bmatrix} kl \\ -kl \end{bmatrix} \tag{3.11}$$

The dynamics of the object shown in Eqs. 3.29 and 3.30 can be changed by specifying the mass, stiffness and damping m, k, d respectively.

3.1.1.3 Robot Dynamics

In Sect. 2.2, the dynamics of the Baxter robot is introduced and well established. Here the dynamics of the manipulators are given in a compact form as:

$$\begin{aligned} M_L(q_L)\ddot{q}_L + C_L(q_L, \dot{q}_L)\dot{q}_L + G_L(q_L) &= \tau_{u_L} + J^T(q_L)F_L + \tau_{dist_L} \\ M_R(q_R)\ddot{q}_R + C_R(q_R, \dot{q}_R)\dot{q}_R + G_R(q_R) &= \tau_{u_R} + J^T(q_R)F_R + \tau_{dist_R} \end{aligned} \tag{3.12}$$

where the subscripts "L" and "R" stand for left and right robot arms, respectively. The notation $\tau_u \in \mathbf{R}^n$ is the vector of control input torque, and $\tau_{dist} \in \mathbf{R}^n$ is the disturbance torque caused by friction, disturbance or load etc. The control torques τ_u are generated by the designed controllers, in order to achieve desired performance in terms of motion tracking and disturbance rejection. Forces F_L, F_R are as described in Eq. 3.11, which couple the dynamics between the arms through the object.

3.1.1.4 Trajectory Generation

A smooth trajectory [30] is defined in Cartesian space as

$$\begin{aligned} y^*(t) &= y^*(0) + \big(y^*(T) - y^*(0)\big)G \\ G &= 10\left(\frac{2t}{T}\right)^3 - 15\left(\frac{2t}{T}\right)^4 + 6\left(\frac{2t}{T}\right)^5 \end{aligned} \tag{3.13}$$

where $y^*(0)$ is the start point of the curve, $y^*(T)$ is the final point, and T is the period of the trajectory. The desired Cartesian trajectory of the left (leading) arm is represented by a vector $X_L^* \in \mathbf{R}^{6\times 1}$:

$$X_L^* = \left[x_L^*\ y_L^*\ z_L^*\ \vartheta_L^*\ \phi_L^*\ \psi_L^*\right]^T \tag{3.14}$$

The initial position of the end effector of the left arm, $X_L(0) = fkine\big(q_L(0)\big)$, with $fkine(\cdot)$ denoting the forward dynamics, defines the $x_L, z_L, \vartheta_L, \phi_L$ and ψ_L components of (3.33) so that the arm moves in the y-plane along the trajectory defined from Eq. 3.13:

$$X_L^*(t) = \left[x_L(0)\ y_L^*(t)\ z_L(0)\ \vartheta_L(0)\ \phi_L(0)\ \psi_L(0)\right]^T \tag{3.15}$$

The Cartesian velocity, required for inverse kinematics in Eq. (3.17), is the first differential of $X_L^*(t)$ defined in Eq. (3.15):

$$\dot{X}_L^*(t) = \frac{\mathrm{d}}{\mathrm{d}t} X_L^*(t) \tag{3.16}$$

Joint space angular velocity $\dot{q}_L^*$ is defined through the inverse Jacobian and the first differential of the Cartesian trajectory:

$$\dot{q}_L^* = J^{-1}(q_L)\dot{X}_L^* \tag{3.17}$$

Joint angles and accelerations are the integration and differentiation of joint velocity, respectively:

$$q_L^* = \int \dot{q}_L^*\,\mathrm{d}t, \quad \ddot{q}_L^* = \frac{\mathrm{d}}{\mathrm{d}t}\dot{q}_L^* \tag{3.18}$$

The desired trajectory of the right (following) hand, X_R^* is defined as an offset in the y-plane from the left hand:

$$X_R^* = \left[x_L(t)\ \big(y_L(t) + p^*\big)\ z_L(t)\ \vartheta_L(t)\ \phi_L(t)\ \psi_L(t)\right]^T \tag{3.19}$$

where p^* is the desired distance between the end effectors of left and right manipulators. Following the same calculation as above, we obtain q_L^*, $\dot{q}_L^*$ and $\ddot{q}_L^*$.

3.1.2 Adaptive Control

3.1.2.1 Biomimetic Impedance and Force Control

Input torques τ_{u_L} and τ_{u_R} are the online-adapted inputs to the left and right manipulators defined as

$$\begin{aligned}\tau_{u_L} &= \tau_{r_L} + \tau_{x_L} + \tau_{j_L}\\ \tau_{u_R} &= \tau_{r_R} + \tau_{x_R} + \tau_{j_R}\end{aligned} \quad (3.20)$$

For brevity, in the following equations the subscripts $(\cdot)_L$ and $(\cdot)_R$ are excluded unless required, although it must be noted that each manipulator is controlled separately; the only interaction between the two is through trajectory generation. In Eq. (3.20), the reference torque $\tau_r \in \Re^n$ and adaptive feedback terms $\tau_x \in \Re^n$ and $\tau_j \in \Re^n$, where n is the number of degrees of freedom (DoF), are defined as

$$\begin{aligned}\tau_r &= M(q)\ddot{q}^* + C(q,\dot{q})\dot{q}^* + G(q) - L(t)\varepsilon(t)\\ \tau_x &= J^T(q)\big(-F_x(t) - K_x(t)e_x(t) - D_x(t)\dot{e}_x(t)\big)\\ \tau_j &= \Gamma\big(-\tau(t) - K_j(t)e_j(t) - D_j(t)\dot{e}_j(t)\big)\end{aligned} \quad (3.21)$$

where $M(q) \in \Re^{n\times n}$, $C(q,\dot{q}) \in \Re^n$ and $G(q) \in \Re^n$ form the inertia, coriolis/centrifugal and gravitational forces respectively. The $L(t) \in \Re^{n\times n}$ term corresponds to the desired stability margin, and Γ is a reduction matrix of the form $\Gamma \in \Re^{(n\times n)} = \text{diag}\,[1, 1, \ldots, 0]$ where the number of ones in the diagonal relate to the number of shoulder joints. Torques τ_x and τ_j are adaptive task space torques and joint space torques, respectively; the adaptive rules are defined as

$$\begin{aligned}\delta\tau(t) &= Q_\tau(|\varepsilon_j|, |\tau_u|)\ \varepsilon_j(t) - \gamma_j\tau(t)\\ \delta K_j(t) &= Q_{Kj}(|\varepsilon_j|, |\tau_u|)\ \varepsilon_j(t)e_j^T(t) - \gamma_j K_j(t)\\ \delta D_j(t) &= Q_{Dj}(|\varepsilon_j|, |\tau_u|)\ \varepsilon_j(t)\dot{e}_j^T(t) - \gamma_j D_j(t)\end{aligned} \quad (3.22)$$

similarly, in the task space

$$\begin{aligned}\delta F_x(t) &= Q_{Fx}(|\varepsilon_x|, |F_u|)\ \varepsilon_x(t) - \gamma_x F_x(t)\\ \delta K_x(t) &= Q_{Kx}(|\varepsilon_x|, |F_u|)\ \varepsilon_x(t)e_x^T(t) - \gamma_x K_x(t)\\ \delta D_x(t) &= Q_{Dx}(|\varepsilon_x|, |F_u|)\ \varepsilon_x(t)\dot{e}_x^T(t) - \gamma_x D_x(t)\end{aligned} \quad (3.23)$$

where the position, velocity, and tracking errors are defined in both task and joint spaces:

$$
\begin{aligned}
e_j(t) &= q(t) - q^*(t), \quad e_x(t) = X(t) - X^*(t), \\
\dot{e}_j(t) &= \dot{q}(t) - \dot{q}^*(t), \quad \dot{e}_x(t) = \dot{X}(t) - \dot{X}^*(t), \\
\varepsilon_j(t) &= \dot{e}_j(t) + \kappa e_j(t), \quad \varepsilon_x(t) = \dot{e}_x(t) + \kappa e_x(t)
\end{aligned}
\tag{3.24}
$$

and γ_j, γ_x are the forgetting factors

$$
\begin{aligned}
\gamma_j &= \alpha_j(\bar{\varepsilon}_j, \bar{\tau}_j)\ \exp\left(-\frac{\varepsilon_j(t)^2}{0.1\alpha_j(\bar{\varepsilon}_j, \bar{\tau}_j)^2}\right) \\
\gamma_x &= \alpha_x(\bar{\varepsilon}_x, \bar{F}_u)\ \exp\left(-\frac{\varepsilon_x(t)^2}{0.1\alpha_x(\bar{\varepsilon}_x, \bar{F}_u)^2}\right)
\end{aligned}
\tag{3.25}
$$

The gains Q_τ, Q_K, Q_D and α are derived from a fuzzy relation of $|\tau_u(t)|$, which is used as an indicator of control effort, and $|\varepsilon_j(t)|$ to gauge tracking performance; similarly in task space Q_{Fx}, Q_{Kx}, Q_{Dx} are derived from fuzzy relation of the task space control effort and tracking error F_u and ε_x where $F_u = (J^T)^{-1}(q)\ \tau_u$. The gain α is adapted to change the magnitude and shape of γ so that if tracking performance is satisfactory, the forgetting factor is high and effort is minimized. Figure 3.3 demonstrates how the value of α changes the shape of γ.

3.1.2.2 Fuzzy Inference of Control Gains

Before fuzzification can be performed on ε_x and τ_u, they must first be normalized in such a manner that, so that the same inference engine is made generic and not sensitive to input magnitude. Average tracking errors $\hat{\varepsilon}_j, \hat{\varepsilon}_x \in \Re^6$, input torque $\hat{\tau}_u \in \Re^n$ and input force $\hat{F}_u \in \Re^6$ are calculated for each DoF.

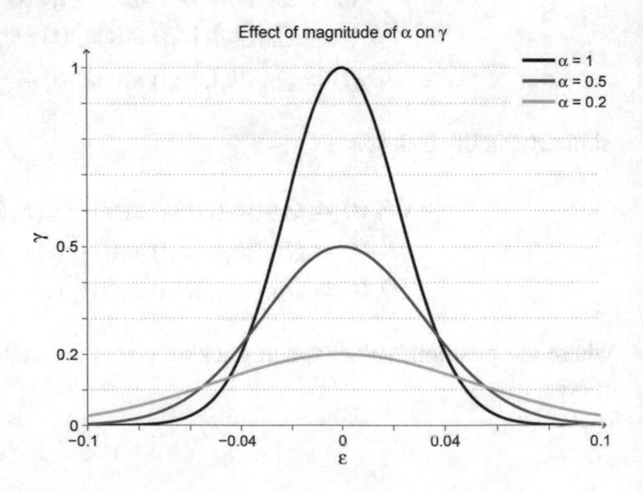

Fig. 3.3 Example of how the magnitude of α affects the forgetting factor γ. Higher values of α have a high narrow shape, so that when tracking performance is good the control effort is reduced maximally. When tracking performance is poor, the forgetting factor is minimized

$$\hat{\varepsilon}_{x_i} = \frac{\sum |\varepsilon_{x_i}(t)|}{t_f/\delta t} \quad \hat{\varepsilon}_{j_i} = \frac{\sum |\varepsilon_{j_i}(t)|}{t_f/\delta t}$$
$$\hat{\tau}_{u_i} = \frac{\sum |\tau_{u_i}(t)|}{t_f/\delta t} \quad \hat{F}_{u_i} = \frac{\sum |F_{u_i}(t)|}{t_f/\delta t} \tag{3.26}$$

This initial data is obtained by collecting data from the controller using fixed values for the gains Q_T, Q_K, Q_D and α. These values can be set to some low values, to ensure stability with no tuning needed to improve tracking error or control effort, as would normally be the case. To compute input to the fuzzy systems $\bar{\varepsilon}_x$, $\bar{\varepsilon}_j$, $\bar{\tau}_u$, $\bar{F}_u$, these baselines are compared against feedback from the system, to allow some recognition of performance:

$$\bar{\varepsilon}_{j_i}(t) = \frac{|\varepsilon_{j_i}(t)|}{\hat{\varepsilon}_{j_i}}\sigma, \quad \bar{\varepsilon}_{x_i}(t) = \frac{|\varepsilon_{x_i}(t)|}{\hat{\varepsilon}_{x_i}}, \sigma$$
$$\bar{\tau}_{u_i}(t) = \frac{|\tau_{u_i}(t)|}{\hat{\tau}_{u_i}}\sigma, \quad \bar{F}_{u_i}(t) = \frac{|F_{u_i}(t)|}{\hat{F}_{u_i}}\sigma \tag{3.27}$$

As a result of this, for all inputs to fuzzy systems, the value of σ indicates that the controller is performing the same as the fixed-value controller. The values less than σ indicate an improvement, and values greater than σ indicate that performance is worse. Here, we set $\sigma = 0.5$, which effectively normalizes the result between 0 and 1 (where a result of 1 would indicate that current feedback is twice as high as the baseline). There is no upper limit to the variables generated in Eq. (3.27), so any input above 1 returns a maximum membership in the "high" classification. This allows a generic set of input membership functions to be applied to all systems.

The rules for fuzzy inference of the control gains are set using expert knowledge, although in general: IF control effort is too high THEN gain is set low; IF tracking error is poor THEN gain is set high. The fuzzy truth table (Table 3.1) for the Q gains demonstrates this. The truth table for the forgetting factor gain is slightly different and can be found in Table 3.2. How changes in control effort and tracking error affect the $Q_{(\cdot)}$ gains are shown in Fig. 3.4a. It can be seen that in general, gain increases when tracking error is high and control effort is low, and minimal gain occurs when tracking error is low and control effort high. The surface of fuzzy inference of α is shown in Fig. 3.4b.

Table 3.1 Fuzzy truth table for inference of control gains $Q_T, Q_{Fx}, Q_K, Q_{Kx}, Q_D$ and Q_{Dx}

		$\bar{\tau}_u, \bar{F}_u$		
		Low	Same	High
$\bar{\varepsilon}_x, \bar{\varepsilon}_j$	Low	L	L	L
	Same	L	M	M
	High	H	H	M

Table 3.2 Fuzzy truth table for inference of α_j and α_x

		$\bar{\tau}_u, \bar{F}_u$		
		Low	Same	High
$\bar{\varepsilon}_x, \bar{\varepsilon}_j$	Low	M	H	H
	Same	M	M	H
	High	L	L	M

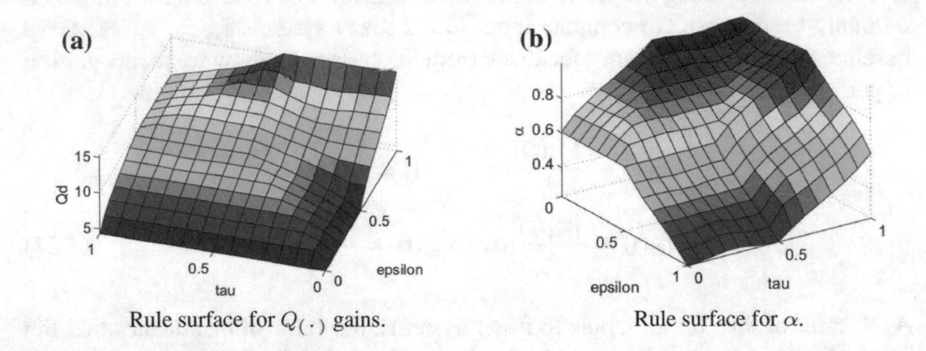

Rule surface for $Q_{(\cdot)}$ gains. Rule surface for α.

Fig. 3.4 Surface plots showing rule surfaces for **a** adaptation gains, and **b** value of α, based on inputs $\bar{\varepsilon}$ and $\bar{\tau}_u$ described in Eq. (3.27)

3.1.3 Simulation Studies

The simulations are carried out using a full kinematic and dynamic model of the Baxter bimanual robot, as well as the MATLAB Robotics Toolbox introduced in Sect. 1.3. The task is to move an object of uncertain dynamics through a trajectory without crushing or releasing the bimanual grasp. The object is modeled as described in Sect. 3.1.1.2 with contact forces $F_{L,R}$, as well as high-frequency internal disturbance forces F_{int_L} and F_{int_R}. In addition to this, a low-frequency, high-amplitude external disturbance F_{ext} is applied to both arms, which simulates interaction with the environment. The internal and external forces F_{int} and F_{ext} are introduced in four phases: In *Phase I* only F_{ext} is applied; *Phase II* both F_{ext} and F_{int}; *Phase III* only F_{int}, and *Phase IV* no disturbance is applied.

The proposed controller is compared against the controller with fixed values for $Q_{(\cdot)}$ gains and $\alpha_{(\cdot)}$ to confirm the benefit of the fuzzy inference of the aforementioned values. To test this, a performance index q is computed as:

$$q = \int_{t_s}^{t_f} F_u(t)^T Q F_u(t) + \varepsilon_x^T(t) R\, \varepsilon_x(t)\, \mathrm{d}t \tag{3.28}$$

where $Q, R \in \Re^{6\times 6}$ are positive diagonal scaling matrices, and t_s and t_f can be set to obtain q for each phase of the simulation. A lower performance index is better, as the aim is to minimize both tracking error and control effort.

Simulations are run on both the fixed-value controller and the fuzzy inferred controller with the same disturbance being applied in both cases. For F_{int} amplitude is set to 10 N, frequency 50 Hz, and for F_{ext} amplitude and frequency are set to 50 N and 0.5 Hz, respectively. The object dynamic properties are set to $k = 10, d = 5, m = 1$. The period of the trajectory, T, is set to $2.4s$, so that in each phase the arms will travel through the trajectory twice. Performance is measured over each of these phases by setting t_s, t_f in Eq. (3.28) accordingly. In addition, the integrals of 2-norm of input torque and tracking error, $\int ||\tau_u||\ \mathrm{d}t$ and $\int ||\varepsilon_x||\ \mathrm{d}t$ are presented to evaluate each individually.

Results for tracking error performance are shown in Fig. 3.5, from which we see that it is clear that fuzzy inference of control gains significantly improves tracking for all phases of the simulation, and on both arms; in particular, in *Phase* II (when disturbance forces are greatest) it can be seen that the right arm has a much small error when gains are fuzzy inferred. Performance of control torque Fig. 3.6 shows that slight improvement can be seen in all phases except *Phase* II, where a slight increase in control effort is observed for the fuzzy inferred controller. It can also be noted that input torque for the left arm is markedly less than the right; this is expected due to the following (right) arm having to constantly adjust not only to the disturbance forces but also the uncertain position of the leading (left) arm. Comparing Figs. 3.5 and 3.6, it can be seen that the rule base is weighted towards improving the tracking error rather than reducing control effort; an important consideration when the task is to move a dynamic object through a trajectory, and system compliance may be of less concern. Figure 3.7 shows performance indexes in each phase for the left arm of the robot, comparing the controller with fuzzy inferred gains to the fixed gain controller.

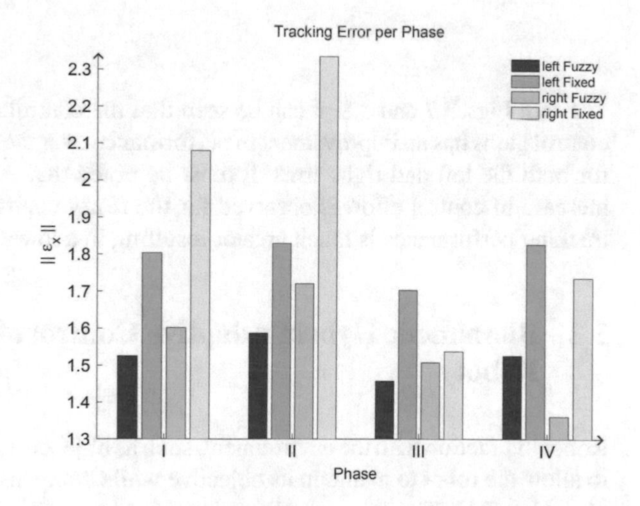

Fig. 3.5 Integral of tracking error $\int ||\varepsilon_x||\ \mathrm{d}.t$ over each phase, for *left* (*blue*, *green*) and *right* (*cyan*, *yellow*) arms, comparing controller with fuzzy inference of control gains to controller with fixed gains

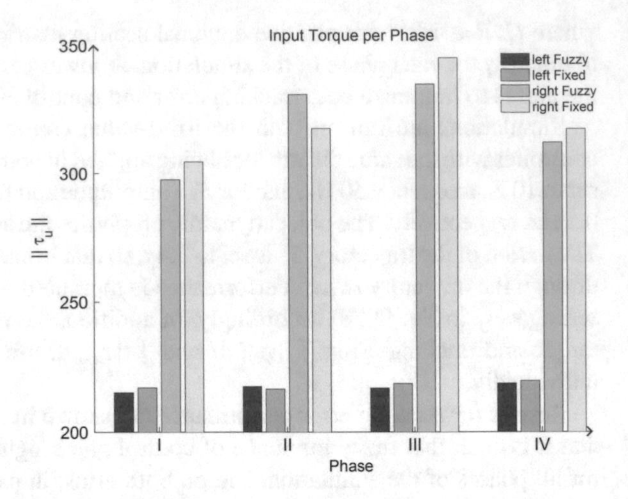

Fig. 3.6 Comparing the integral of control effort $\int \|\tau_u\|\, \mathrm{d}.t$ over each phase

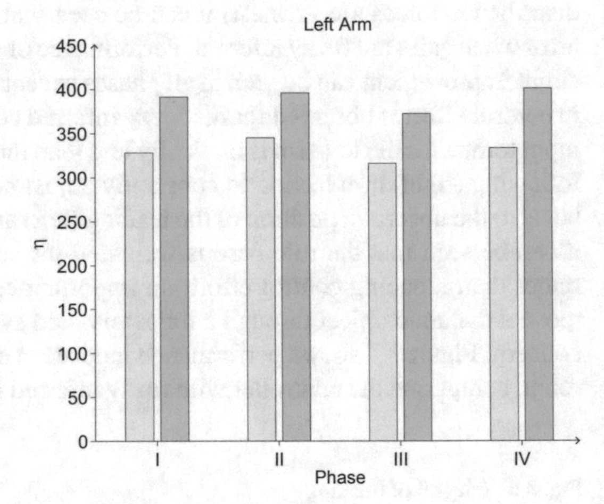

Fig. 3.7 Comparison of performance index for the left arm. Indices for the controller using the fuzzy inferred adaptive gains are shown in *yellow*, fixed gain controller in *cyan*

From Figs. 3.7 and 3.8, it can be seen that the controller with fuzzy inference of control gains has an improvement in performance over the controller with fixed gains for both the left and right arms. It must be noted that even though in *Phase* II an increase in control effort is observed for the fuzzy controller. The improvement in tracking performance is much greater resulting in a lower performance index score.

3.2 Biomimetic Hybrid Adaptive Control of Bimanual Robot

Robot interaction with the environment, such as objects or humans, requires feedback to allow the robot to maintain its objective while compensating for disturbance from obstacles [31]. The future proliferation of robots within our natural environment

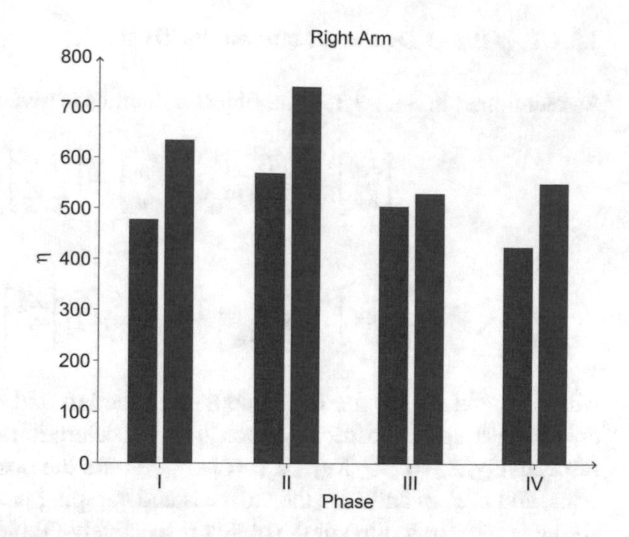

Fig. 3.8 Comparison of performance index for the right arm. Indices for the controller using the fuzzy inferred adaptive gains are shown in *red*, fixed gain controller in *blue*

requires manipulator design to follow that of a human, i.e., it is easier to change the robot, than to change our entire environment. Hence, a robot with two arms gives the ability to perform a much wider range of tasks than if it only had the use of one arm, such as lifting heavy objects or carrying delicate items. Recently, research into impedance control has been focused on adaptation, i.e., online adjustment of stiffness and damping of the manipulator [23]. In [23], a complete theoretical framework of human-like adaptation of force and impedance has been established. In this section, we design a joint/task space hybrid adaptive controller in which impedance and force are adapted online using a novel biomimetic algorithm and minimizes both tracking error and control effort, as observed in humans [32].

3.2.1 *Preliminaries and Problem Formulation*

A scenario is posed where two robot manipulators are placed next to each other, so that a large portion of their workspaces are shared. The task is for the two robots to work cooperatively, to move a object with unknown internal dynamics, e.g., vibration, through a trajectory. The object dynamics is modeled as a mass–spring–damper system, as shown in Fig. 3.2. The object exerts a high-frequency, low-amplitude disturbance force to the end effectors, as the vibrating tool does. In addition, a second disturbance force is applied proximally; that is, an external collision force. This second force is modeled as a low-frequency, high-amplitude disturbance acting upon the first link of the robot. The controller is required to closely follow the trajectory, i.e., minimize tracking error, as well as minimize "metabolic" cost. This is achieved through the use of a biomimetic controller.

3.2.1.1 Object Dynamics and Robot Dynamics

As mentioned in Sect. 3.1.1, the object dynamics is given as

$$\begin{bmatrix} \dot{X}_m \\ \ddot{X}_m \end{bmatrix} = \begin{bmatrix} 0 & 1 \\ \frac{-2k}{m} & \frac{-2d}{m} \end{bmatrix} \begin{bmatrix} X_m \\ \dot{X}_m \end{bmatrix} + \begin{bmatrix} 0 & 0 \\ \frac{k}{m} & \frac{d}{m} \end{bmatrix} \begin{bmatrix} X_A \\ \dot{X}_A \end{bmatrix} \quad (3.29)$$

$$\begin{bmatrix} F_L \\ F_R \end{bmatrix} = \begin{bmatrix} k & d \\ k & d \end{bmatrix} \begin{bmatrix} X_m \\ \dot{X}_m \end{bmatrix} + \begin{bmatrix} -k\mathbf{I} & -d\mathbf{I} \end{bmatrix} \begin{bmatrix} X_A \\ \dot{X}_A \end{bmatrix} + \begin{bmatrix} kl \\ -kl \end{bmatrix} \quad (3.30)$$

where F_L and F_R are the forces applied to the left and right arms, X_m, $\dot{X}_m$ and $\ddot{X}_m$ are the task space position, velocity, and acceleration of the mass of the object, respectively, $X_A = [X_L\ X_R]^T$, $\dot{X}_A = [\dot{X}_L\ \dot{X}_R]^T$ are the positions and velocities of the arms, and k, d, m and l are the stiffness and damping coefficients, mass and natural spring length (or radius) of the object respectively (Table 3.3).

The robot dynamics of the left and right arms used here are the same in Sect. 3.1.1 which are described as:

$$\begin{aligned} M_L(q_L)\ddot{q}_L + C_L(q_L, \dot{q}_L)\dot{q}_L + G_L(q_L) &= \tau_{u_L} + f_L \\ M_R(q_R)\ddot{q}_R + C_R(q_R, \dot{q}_R)\dot{q}_R + G_R(q_R) &= \tau_{u_R} + f_R \end{aligned} \quad (3.31)$$

where $q_{L,R}$ denote the joint position of left and right robot arm, respectively. $M_{L,R}(q) \in \mathbf{R}^{n \times n}$ is the symmetric-bounded positive definite inertia matrix, and n

Table 3.3 Nomenclature

Symbol	Description
q_L, q_R	Joint angle: left and right arm
$\dot{q}_L, \dot{q}_R$	Joint velocity: left and right arm
X_L, X_R	Cartesian position: left and right arm
$\dot{X}_L, \dot{X}_R$	Cartesian velocity: left and right arm
l	Natural ball spring length
m, k, d	Object mass, stiffness, and damping, respectively
F_L, F_R	Forces at left and right hand, respectively
F_{int}, F_{ext}	Internal and external forces, respectively
e, e_x	Joint and task space position error, respectively
ε	Tracking error
$\Vert \cdot \Vert$	Euclidian vector norm
$0_{m \times m}$	$m \times m$-dimensional zero matrix
n	Number of manipulator joints

is the degree of freedom (DoF) of the robot arm; $C_{L,R}(q,\dot{q})\dot{q} \in \mathbf{R}^n$ denotes the Coriolis and Centrifugal force; $G_{L,R}(q) \in \mathbf{R}^n$ is the gravitational force; $\tau_{u_{L,R}} \in \mathbf{R}^n$ is the vector of control input torque; and $f_{L,R} \in \mathbf{R}^n$ is the external force caused by friction, disturbance or load etc. The control torque $\tau_{u_{L,R}}$ is generated by the controller to be designed, in order to achieve desired performance in terms of motion tracking and disturbance rejection.

A smooth trajectory which defined in Sect. 3.1.1 is described as

$$\begin{aligned} f(t) &= f(0) + \big(f(T) - f(0)\big)g \\ g &= 10\left(\frac{2t}{T}\right)^3 - 15\left(\frac{2t}{T}\right)^4 + 6\left(\frac{2t}{T}\right)^5 \end{aligned} \tag{3.32}$$

where $f(0)$ and $f(T)$ are the start and end points, and T is the time period of the trajectory. This is used to define the desired task space trajectory of the leading left arm $X_L^* \in \Re^6$

$$X_L^* = \big[x_L(0)\; f(t)\; z_L(0)\; \vartheta_L(0)\; \phi_L(0)\; \psi_L(0)\big]^T \tag{3.33}$$

such that the y-component of the trajectory moves through a minimum-jerk trajectory while all other components maintain their initial position. The trajectory of the right arm X_R^* is defined to maintain a constant offset from the left arm in all degrees of freedom in the task space, that is

$$X_R^* = \Big[x_L(t)\; \big(y_L(t) + p^*\big)\; z_L(t)\; \vartheta_L(t)\; \phi_L(t)\; \psi_L(t)\Big]^T \tag{3.34}$$

where p^* is the desired distance between the end effectors of left and right manipulators. Inverse kinematics are performed using the inverse Jacobian method. For the leading arm, it is trivial to define the first derivative of X_L^* so that $\dot{X}_L^*$ can be converted to joint angles q_L^* as so:

$$q_L^* = \int \dot{q}_L^* \,\mathrm{d}t = \int J^{-1}(q_L)\dot{X}_L^* \,\mathrm{d}t \tag{3.35}$$

Similarly, q_R^* and $\dot{q}_R^*$ can be calculated using the inverse Jacobian method in Eq. (3.35).

3.2.2 Adaptive Bimanual Control with Impedance and Force

3.2.2.1 Adaptive Law

The adaptive law has been designed to minimize motion tracking error, whilst also minimising control effort. The adaptation of stiffness and damping follows the work in [15], which shows that human motor control is based on simultaneous optimisation

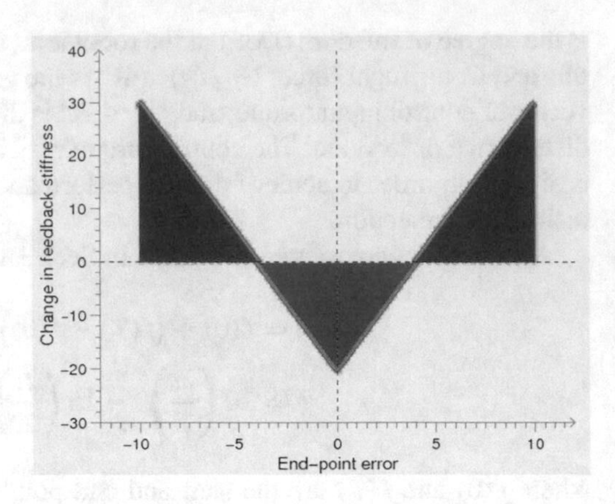

Fig. 3.9 Change in endpoint stiffness, i.e., δK follows a "V-shaped" adaptation. The *lighter shaded areas* indicate where the error is above an error threshold, so stiffness increases; the *darker shaded area* indicates where the error is acceptable, so stiffness is reduced, which in turn reduces applied effort

of stability, accuracy and efficiency; the "V-shaped" learning function can be seen in Fig. 3.9. Here this adaptation is continuous so that error and effort are continuously minimized. The controller developed in [23] has been adapted in this section to be applied to dual-arm manipulation, as an extension of the previous single-robot arm application. It should be noted that the controllers defined below will be applied to both left and right manipulators, and thus the subscripts "L" and "R" are omitted for convenience.

3.2.2.2 Feedforward Compensation and Minimum Feedback

Given the dynamics of a robot manipulator in Eq. 3.31, we employ the following controller as a basic torque input

$$\tau_r(t) = M(q)\ddot{q}^* + C(q, \dot{q})\dot{q}^* + G(q) - L(t)\varepsilon(t) \tag{3.36}$$

where $L(t)\varepsilon(t)$ corresponds to the desired *stability margin*, and the first three terms are feedforward compensation. As widely used in sliding mode control, we define tracking error $\varepsilon(t)$ as below

$$\varepsilon(t) = \dot{e}(t) + \kappa e(t)$$

where $e(t)$ and $\dot{e}(t)$ are joint errors to be defined in Eq. 3.38. On top of the above control input $\tau_r(t)$, we develop two adaptive controllers in joint space and task space, respectively.

3.2.2.3 Joint Space Adaptive Control

We apply joint space controller from [23] as below:

$$\tau_j(t) = -\tau(t) - K(t)e(t) - D(t)\dot{e}(t) \tag{3.37}$$

with

$$e(t) = q(t) - q^*(t), \ \dot{e}(t) = \dot{q}(t) - \dot{q}^*(t) \tag{3.38}$$

where $-\tau(t)$ is the learned *feedforward* torque, $-K(t)e(t)$ and $-D(t)\dot{e}(t)$ are *feedback* torques due to stiffness and damping, respectively. We apply the adaptive law developed in [23] (but here in the time domain, rather than iteration, making the controller applicable to nonperiodic tasks) as follows:

$$\begin{aligned}
\delta\tau(t) &= \tau(t) - \tau(t-\delta t) = Q_\tau\varepsilon(t) - \gamma\tau(t) \\
\delta K(t) &= K(t) - K(t-\delta t) = Q_K\varepsilon(t)e^T(t) - \gamma K(t) \\
\delta D(t) &= D(t) - D(t-\delta t) = Q_D\varepsilon(t)\dot{e}^T(t) - \gamma D(t)
\end{aligned} \tag{3.39}$$

where γ is defined as the forgetting factor

$$\gamma = \frac{a}{1 + b\|\varepsilon(t)\|^2} \tag{3.40}$$

given that $K(0) = 0_{[n,n]}$ and $D(0) = 0_{[n,n]}$, and Q_K, Q_D are diagonal positive-definite constant gain matrices.

3.2.2.4 Task Space Adaptive Control

From Eqs. 3.15 and 3.34, we can see that the movement of the object is described by a Cartesian trajectory, so it is more useful to describe the control in terms of Cartesian rather than joint errors. This makes control of impedance in Cartesian space much more straightforward. First, we redefine our error terms to be in Cartesian, rather than joint space

$$\begin{aligned}
e_x(t) &= X(t) - X^*(t) \\
\dot{e}_x(t) &= \dot{X}(t) - \dot{X}^*(t) \\
\varepsilon_x(t) &= \dot{e}_x(t) + \kappa e_x(t)
\end{aligned} \tag{3.41}$$

This leads to a change in the *feedforward* and *feedback* terms described in Eq. 3.39 with γ_x defined similarly as in Eq. 3.40

$$\begin{aligned}
\delta F_x &= Q_F\ \varepsilon_x - \gamma_x F_x \\
\delta K_x &= Q_{K_x}\ \varepsilon_x e_x - \gamma_x K_x \\
\delta D_x &= Q_{D_x}\ \varepsilon_x \dot{e}_x - \gamma_x D_x
\end{aligned} \tag{3.42}$$

These terms can be described as a sum of forces F_{xkd} where

$$F_{xkd} = F_x + K_x e_x + D_x \dot{e}_x \tag{3.43}$$

which can be transformed into joint space using the equation below:

$$\tau_x = J^T(q)\, F_{xkd} \tag{3.44}$$

where $J(q)$ is the Jacobian satisfying $\dot{x} \equiv J(q)\dot{q}$.

Combination of the basic controller Eq. (3.36), the joint space controller Eq. (3.37) and the task space controller Eq. (3.44) gives us

$$\tau_u(t) = \tau_j(t) - \tau_x(t) + \tau_r(t) \tag{3.45}$$

3.2.3 Adaptive Control with Internal Interaction

In this part, we focus on the case in which only an internal interaction force is applied, that is, the two arms are subjected to forces at the end effector, arising from the dynamic properties of the object modeled in Eqs. (3.29) and (3.30), as well as a high-frequency, low-amplitude perturbation F_{int}. This perturbation force is defined as a sine wave acting in the plane of movement, described in Eq. (3.15)

$$\begin{aligned} F_{int} &= [0 \;\; y_{int} \;\; 0 \;\; 0 \;\; 0 \;\; 0]^T \\ y_{int} &= A_{int} \sin(2\pi f_{int} t) \end{aligned} \tag{3.46}$$

where $0 < A_{int} \le 10$ is the amplitude and $100 < f_{int} \le 1000$ is the frequency of oscillation in Hertz. The forces applied to each arm F_{int_L}, F_{int_R} are only the negative and positive components, respectively

$$\begin{aligned} F_{int_L} &= \begin{cases} y_{int} & \text{if } y_{int} < 0 \\ 0 & \text{if } y_{int} \ge 0 \end{cases} \\ F_{int_R} &= \begin{cases} 0 & \text{if } y_{int} \le 0 \\ y_{int} & \text{if } y_{int} > 0 \end{cases} \end{aligned} \tag{3.47}$$

Effectively, this simulates the object bouncing between the hands, only *pushing against* the hands, not *pulling*, as shown in Fig. 3.13. To compensate for the effect of internal interaction, we only consider task space adaptive control together with basic control, i.e., $\tau_u(t) = -\tau_x(t) + \tau_r(t)$.

Simulations are carried out using the MATLAB Robotics Toolbox (see Sect. 1.3) with twin Puma 560 models, which can be seen in Fig. 3.10. In this section, we focus on the manipulation control along the task trajectory, and thus the effect of gravity acting on the object can be ignored (as if the object is moving on top of a table),

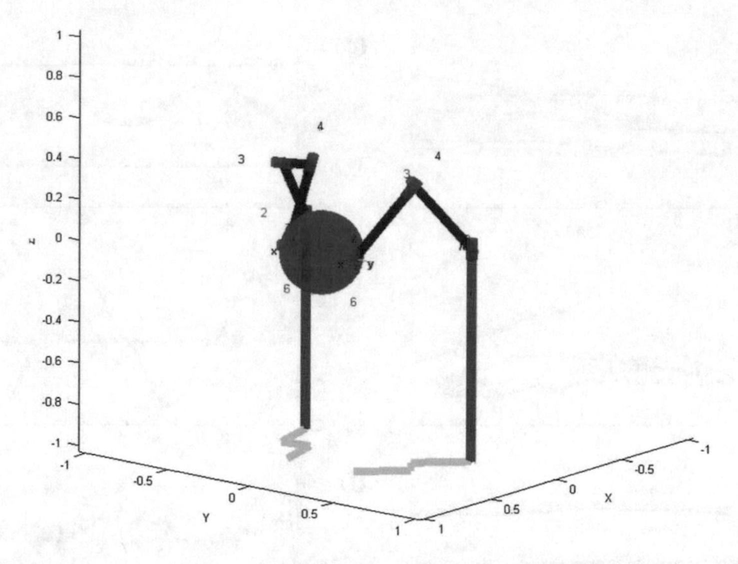

Fig. 3.10 Modelling of twin *Puma 560* manipulators holding the object (in *blue*)

but the object can be assumed to be dropped if the distance between the end effectors of the manipulators becomes larger than the diameter of the object. The controller parameters are selected as

$$\begin{aligned} Q_F &= diag([10\ 10\ 10\ 1\ 1\ 1]) \\ Q_{K_x} &= diag([5\ 20\ 1\ 10\ 0.1\ 0.1]) \times 10^3 \\ Q_{D_x} &= diag([1\ 1\ 1\ 1\ 1\ 5]) \\ a &= 0.2, \quad b = 3, \quad \kappa = 20 \end{aligned} \tag{3.48}$$

and

$$L = diag([5\ 10\ 20\ 1\ 1\ 1]) \tag{3.49}$$

The parameters for the trajectory defined in Eq. (3.13), $y^*(0) = y_L(0)$, $y^*(T) = y_L(0) - 0.2\,\text{m}$, $T = 4.8\,\text{s}$. The radius of the object is set at $r = 0.25\,\text{m}$, with the initial positions of the arms $X_L(0)$ and $X_R(0)$ set so the Euclidian distance between the hands, that is, $h = \|X_L(0) - X_R(0)\|$ is slightly smaller than the diameter of the object. This value also sets the desired distance between the hands throughout the trajectory, p^*, as shown in Eq. (3.34). The value of h can therefore be used to determine if the object is in the grasp; if $h \geq 2r$ then the object is dropped. In addition, if the controller becomes unstable, the simulation is stopped. In Fig. 3.11 we compare a controller with no adaptation to our adaptive method. As can be seen from the time scales, without adaptive feedback the controller is unable to complete the task, and tracking errors increase very rapidly. In contrast, the proposed adaptive controller is able to complete the task with very small tracking error from the desired

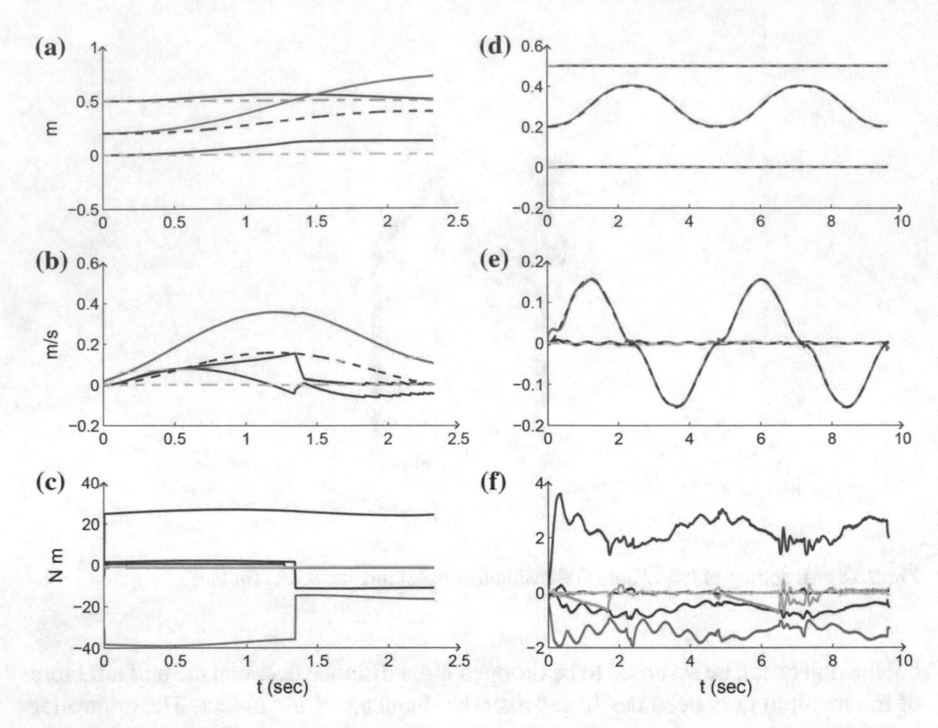

Fig. 3.11 $A_{int} = 10\,\mathrm{N}, f_{int} = 50\,\mathrm{Hz}$. *Solid lines* in (**a**, **b**, **d**, **e**) are actual values while *dashed lines* are desired. Graphs **a**–**c** show the Cartesian position, velocity, and input torque to the arm without adaptation; **d**–**f** show the corresponding values with adaptation

trajectory. Figure 3.12 shows the adapted parts of the controller. It can be noted that the green trace in Fig. 3.12 is the y-axis, so a larger increase in gain is expected due to the majority of disturbance forces being encountered in this plane. Also, it can be seen that these values will tend towards zero when there is little error, minimising effort and maintaining stability.

3.2.4 Adaptive Control with Both Internal and External Interaction

Further to the previous simulation study, a second disturbance force F_{ext} is applied, as discussed in Sect. 3.2.1 and can be seen in Fig. 3.13, where

$$
\begin{aligned}
F_{ext} &= [x_{ext} \quad y_{ext} \quad z_{ext} \quad 0 \quad 0 \quad 0]^T \\
x_{ext} &= y_{ext} = z_{ext} = A_{ext} \sin(2\pi f_{ext} t)
\end{aligned}
\tag{3.50}
$$

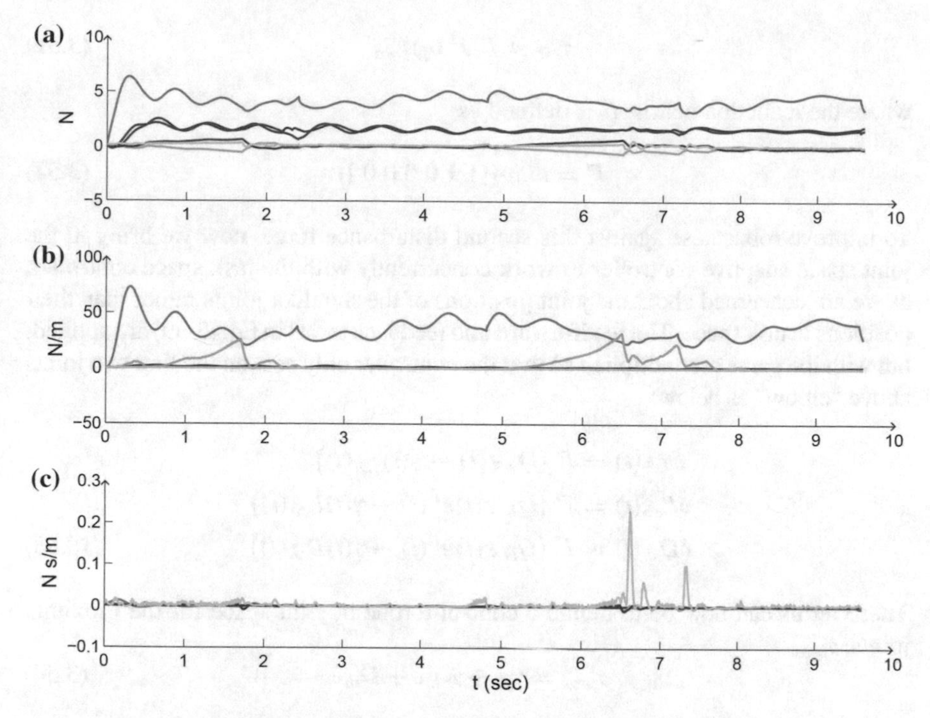

Fig. 3.12 Adapted feedforward force F_x (**a**) and the diagonal elements of the stiffness matrix K_x (**b**), and damping matrix D_x (**c**) for the left arm; the x, y, z components in *blue*, *green*, *red*, respectively, *roll*, *pitch*, *yaw* components in *cyan*, *purple* and *orange*, respectively

where $10 < A_{ext} \leq 100$ is the amplitude of perturbation and $0.1 < f_{ext} \leq 10$ is the frequency. To simulate this force being applied at the "elbow", the following equation is used to calculate the torque caused by the external force F_{ext}

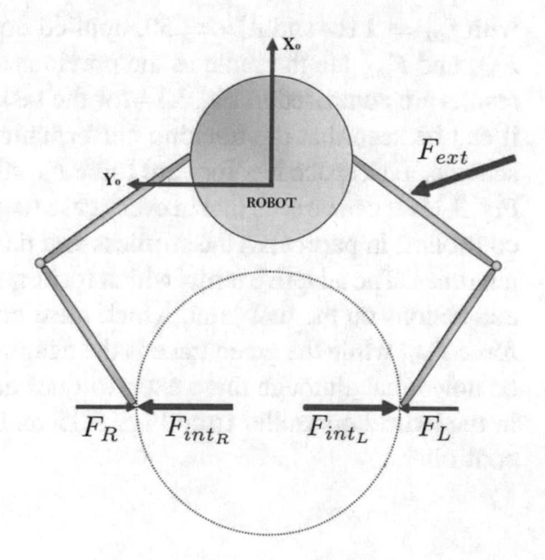

Fig. 3.13 Diagram of interaction forces acting on the manipulator pair. F_L and F_R are the *passive* forces caused by dynamic interaction with the object. F_{int_L} and F_{int_R} are the high-frequency, low-amplitude forces caused by *active* vibration of the object. F_{ext} is the external force modelling contact with the environment

$$\tau_{ext} = \Gamma\ J^T(q)F_{ext} \tag{3.51}$$

where the reduction matrix Γ is defined as

$$\Gamma = diag\big([1\ 1\ 0\ 0\ 0\ 0\ ]\big) \tag{3.52}$$

To improve robustness against this second disturbance force, now we bring in the joint space adaptive controller to work concurrently with the task space controller, as we are concerned about the joint positions of the shoulder joints rather than their positions in task space. The feedforward and feedback terms in Eq. (3.39) are applied, but with the gains premultiplied so that the controller only acts on the first two joints above "elbow" as below:

$$\begin{aligned}\delta\tau_{sh}(t) &= \Gamma\ \big(Q_\tau\ \varepsilon(t) - \gamma(t)\tau_{sh}(t)\big)\\ \delta K_{sh}(t) &= \Gamma\ \big(Q_K\ \varepsilon(t)e^T(t) - \gamma(t)K_{sh}(t)\big)\\ \delta D_{sh}(t) &= \Gamma\ \big(Q_D\ \varepsilon(t)\dot{e}^T(t) - \gamma(t)D_{sh}(t)\big)\end{aligned} \tag{3.53}$$

These terms can now be to define a control torque in joint space for the proximal joints, τ_{prox}

$$\tau_{prox} = \tau_{sh} + K_{sh}e + D_{sh}\dot{e} \tag{3.54}$$

The overall controller now becomes

$$\tau_u(t) = \tau_j(t) - \tau_{prox}(t) + \tau_r(t) \tag{3.55}$$

Gain and trajectory parameters are the same as in Sect. 3.2.3 with the addition of the joint space control gains, which are defined as

$$Q_\tau = Q_F,\ Q_K = Q_{K_x},\ Q_D = Q_{D_x} \tag{3.56}$$

with $f_{ext} = 1$ Hz and $A_{ext} = 50$, applied equally to both arms. The internal forces F_{int_L} and F_{int_R} are the same as the previous simulation in Sect. 3.2.3. Tracking error results are compared in Fig. 3.14 for the task space controller and hybrid controller; it can be seen that the tracking performance is improved using the hybrid control scheme. Task space feedforward force F_x, stiffness K_x and damping D_x are shown in Fig. 3.15. It can be seen that in every case these values are much reduced in the hybrid controller; in particular, the stiffness and damping, which are reduced by more than ten times. The adaptive terms which form τ_j are shown in Fig. 3.16. Blue traces show adaptations on the first joint, which must compensate the most against the external force F_{ext}, while the green trace is the adaptation on the second shoulder joint. It can be noted that although these extra torques are applied to the arms, the sum of effort in the hybrid controller from Figs. 3.15 and 3.16 is less than that of the task space controller.

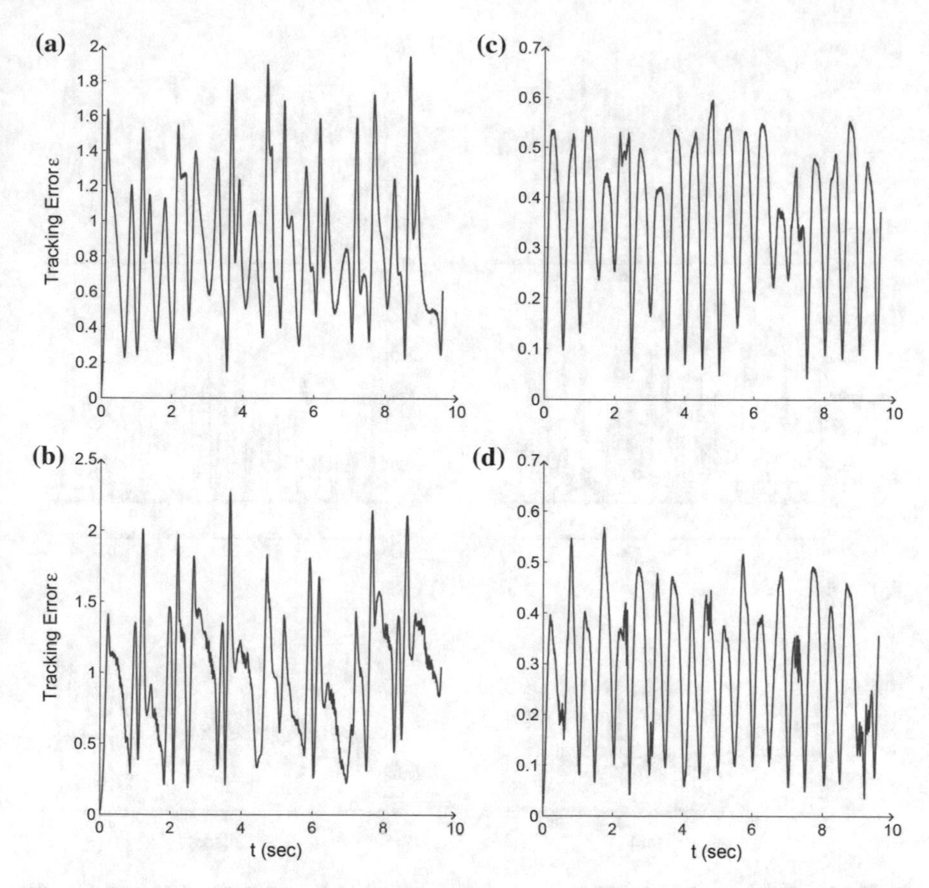

Fig. 3.14 **a, b** show the 2-norm of the tracking error, $||\varepsilon_x||$, from the purely task space controller, for the left and right end effectors, respectively. Graphs **c, d** show the respective tracking error norm when simulated with the hybrid controller

3.2.4.1 Discussion

In this section, two simulation studies have been carried out to test the effectiveness of the biomimetic, bimanual adaptive controller, in a task to move a vibrating object through a trajectory while under the influence of

- dynamics of the object and internal disturbance forces caused by vibration;
- dynamics of the object, internal disturbance forces and external perturbation forces.

The simulated experiments show that

- The proposed controller can perform under conditions which are not possible without adaptation.
- Control effort and motion tracking error are minimized, creating inherent stability.

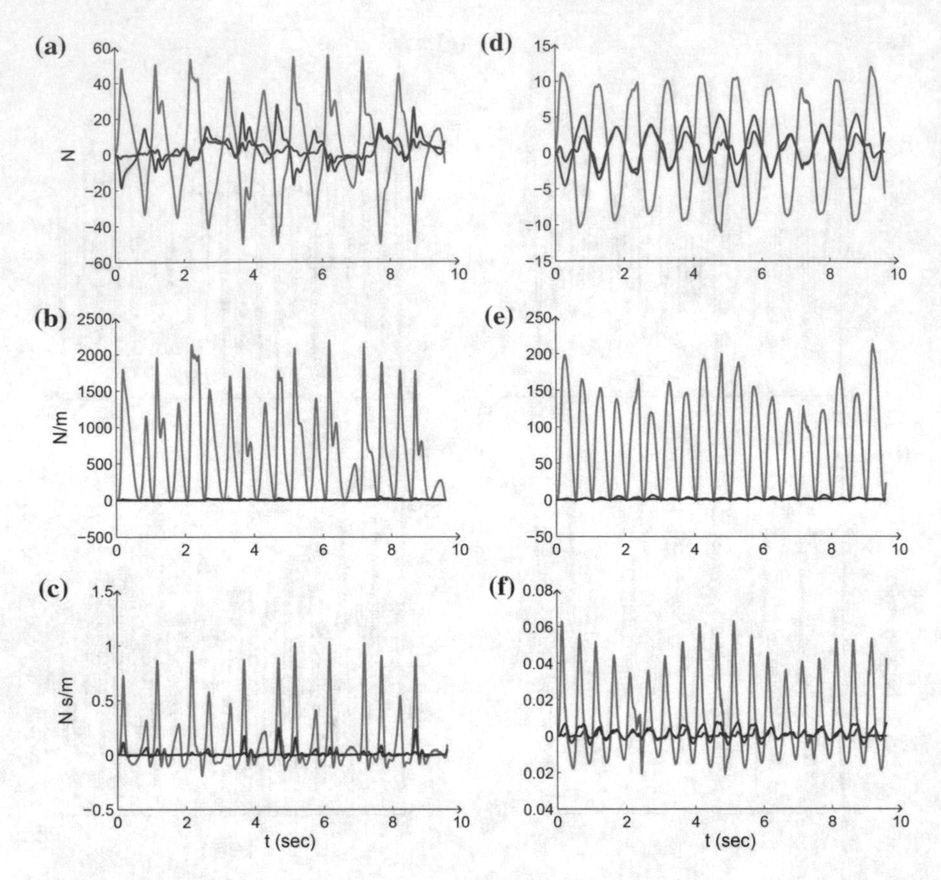

Fig. 3.15 Graphs **a–c** show F_x and diagonal components of the K_x and D_x matrices for the purely task space controller, **d–f** the respective adaptations for the hybrid controller, both taken from the left arm. The *blue*, *green* and *red lines* denote the *x*, *y*, *z* planes, respectively

- The hybrid task space/joint space controller is robust when an external collision force is applied.
- Performance is improved when using the hybrid controller, in terms of control effort and tracking error.

3.3 Optimized Motion Control of Robot Arms with Finite Time Tracking

In Sects. 3.1 and 3.2, the dual-adaptive control and hybrid motion/force control have been well developed. But in these results, the control schemes are designed without the optimization of the motion tracking in terms of a given performance index.

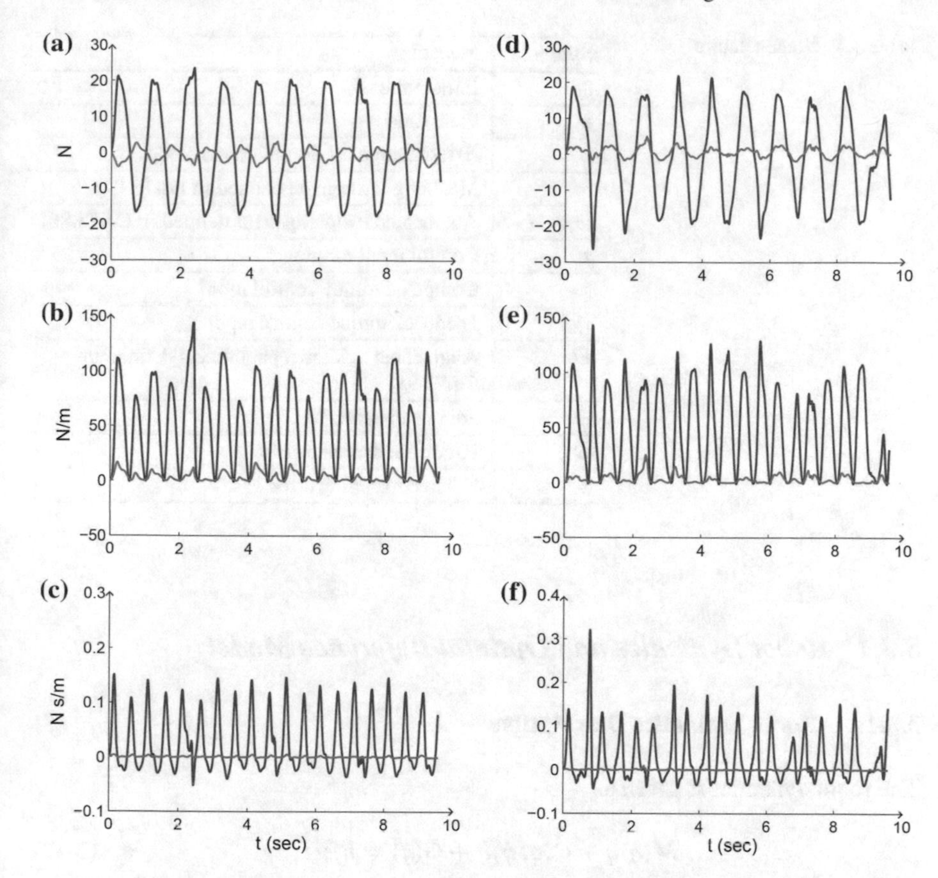

Fig. 3.16 Adaptive terms forming τ_j: **a** shows the progression of nonzero components of τ_{sh}, **b** the adaptation of nonzero diagonal components of K_{sh} and **c** the nonzero diagonal components of damping D_{sh}. for the left arm. Graphs **d–f** show the same values for the right arm. The *blue line* is related to the first joint, and the *green line* the second joint

This section investigates motion control of robot arms to make the closed-loop dynamics match a reference model in finite time, and the reference model is obtained using LQR optimization such that the motion tracking error together with transient acceleration will be minimized [33]. The property of minimized acceleration not only reduces energy cost but also makes the motion smoother. Based on the obtained reference model, an adaptive controller, which is capable of coping with both parametric uncertainties and functional uncertainties, is shown to guarantee the closed-loop stability and asymptotic convergent filtered matching error (Table 3.4).

Table 3.4 Nomenclature

Symbol	Meaning
t_0	Initial time
t_f	Final time
η	External input of the reference model
w	Matching error given defined in Eq. (3.84)
$\bar{w}$	Augmented matching error defined in Eq. (3.86)
τ	Control input torque
τ_{ct}	Computed torque control input
τ_{fb}	Feedback torque control input
Θ	Augmented unknown parameters defined in Eq. (3.98)
$\hat{\Theta}$	Parameter estimate
$\tilde{\Theta}$	Parameter estimate error
z	Filtered matching error

3.3.1 Robot Dynamics and Optimal Reference Model

3.3.1.1 Robot Dynamics Description

The robot dynamics is given as

$$M(q)\ddot{q} + C(q, \dot{q})\dot{q} + G(q) = u + f \tag{3.57}$$

where $u \in \mathbf{R}^n$ is the vector of control input, and $f \in \mathbf{R}^n$ is the external force caused by friction, disturbance, or load etc. It is well-known that the model Eq. (3.57) has the following properties [34], which will be used later in the control design in Sect. 3.3.2:

Property 3.1 *The matrix $M(q)$ is symmetric and positive definite*

Property 3.2 *The matrix $2C(q, \dot{q}) - \dot{M}(q)$ is a skew-symmetric matrix if $C(q, \dot{q})$ is in the Christoffel form, i.e., $q^T(2C(q, \dot{q}) - \dot{M}(q))q = 0,\ \forall q \in \mathbf{R}^n$.*

Property 3.3 $\|M(q)\| \leq \theta_M$, $\|C(q, \dot{q})\| \leq \theta_C\|\dot{q}\|$ *and* $\|G(q)\| \leq \theta_G$, *where* $\theta_M,\ \theta_C,\ \theta_G$ *are positive scalars*

For the sake of theoretical analysis, we adopt the following reasonable assumption, which is valid in most practical cases:

Assumption 3.1 The external force satisfies $\|f\| \leq \theta_{f1}\|q\|\|\dot{q}\| + \theta_{f2}$, i.e., the external force is dependant on robot position and velocity, as well as a bias.

3.3.1.2 Finite Time Linear Quadratic Regulator

For a linear system with completely sterilizable pair (A, B) [35]

$$\dot{x} = Ax + Bu, \quad x(t_0) = x_0, \quad x \in \mathcal{R}^n, \quad u \in \mathcal{R}^n \tag{3.58}$$

Consider the following performance index aiming to minimize the difference between actual output y and desired output y_d

$$J = \int_{t_0}^{t_f} ((y - x_d)^T Q'(y - y_d) + u^T Ru)dt \tag{3.59}$$

where

$$R = R^T > 0, \quad Q = Q^T \leq 0 \tag{3.60}$$

It is easy to show that performance index J can be rewritten as

$$J = \int_{t_0}^{t_f} ((x - x_d)^T Q(x - x_d) + u^T Ru)dt \tag{3.61}$$

where

$$Q = C^T Q'C, x_d = C^T (CC^T)^{-1} \tag{3.62}$$

The optimal control input $u^*(t)$, $t > 0$, that minimizes the above performance index Eq. (3.61), is given by

$$u^* = -R^{-1}B^T(Px + s) \tag{3.63}$$

where P is the solution of the following differential Riccati Equation

$$-\dot{P} = PA + A^T P - PBR^{-1}B^T P + Q \tag{3.64}$$

with terminal condition

$$P(t_f) = 0_{n\times n}; \tag{3.65}$$

and s is the solution of

$$-\dot{s} = (A - BR^{-1}B^T P)^T s - Qx_d \tag{3.66}$$

with terminal condition

$$s(t_f) = 0_{n\times 1} \tag{3.67}$$

and the minimized performance index is

$$\begin{aligned} J^* &= \min J \\ &= x^T(t_0)P(t_0)x(t_0) + 2x^T(t_0)s(t_0) + c(t_0) \end{aligned} \tag{3.68}$$

where $c(t)$ is decided by the following differential equation

$$\dot{c} = s'BR^{-1}B^Ts - x_d^TQx_d \tag{3.69}$$

with terminal condition

$$c(t_f) = 0 \tag{3.70}$$

3.3.1.3 Optimal Model Parameters

In this section, we aim to construct a reference model, which is to be optimized by minimizing both motion acceleration and reference tracking error. Thereafter, the designed controller will shape the closed-loop vessel dynamics to match the constructed reference model such that vessel dynamic response will follow the optimized motion. Denote the desired position and velocity as q_d and $\dot{q}_d$, respectively. Let us consider a second-order reference model as follows:

$$M_d\ddot{q} + C_d\dot{q} + K_dq = -F_\eta(q_d, \dot{q}_d) \tag{3.71}$$

where F_η can be regarded as an artificial force that which drives q to follow q_d. Next, we will chose optimal mass matrix M_d, damping matrix C_d and stiffness matrix K_d by minimizing a properly defined performance index.

To guarantee smooth motion, let us consider to reduce unnecessary angular accelerations, while at the same time to ensure motion tracking performance. The parameters in the matrix M_d, C_d and K_d should be suitably chosen in order to make sure motion tracking and to shape transient acceleration profile as well. Due to the limitation of actuator capacity, too much effort should be avoided during the transient. Consequently, huge acceleration should also be avoided. For this purpose, let us consider the following performance index:

$$I_P = \int_{t_0}^{t_f} \left(\bar{e}^TQ\bar{e} + \ddot{q}^TM_d\ddot{q}\right)dt. \tag{3.72}$$

which minimizes both the motion tracking error

$$\bar{e} = [e^T, \dot{e}^T]^T \in \mathcal{R}^{2n}, e = q - q_d \in \mathcal{R}^n \tag{3.73}$$

and the motion accelerations $\ddot{q}$. The desired inertia matrix M_d is to be set by the designer by his/her experience, and typically it can be chosen as a diagonal matrix. For, e.g., $M_d = \text{diag}(m_1, m_2, \ldots, m_n)$, with $m_1, m_2, \ldots,$ and m_n the desired apparent inertial value. The Q matrix is a weighting matrix that penalize on the tracking errors, both of position and of velocity. When the desired position trajectory is just a set point, this Q matrix would tend to reduce extra speed. Typically, the Q matrix can be chosen also as a diagonal matrix as $Q = \text{diag}(q_1, q_2, q_3, q_{v1}, q_{v2}, q_{v3})$, with q_1, q_2, q_3 as the weighting factors again position errors and q_{v1}, q_{v2}, q_{v3} are the weighting factors for velocity errors.

Following the procedure, we developed in [36, 37], now let us minimize the performance index I_P by suitably choosing C_d, K_d and F_η. In order to apply the LQR optimization technique detailed in Sect. 3.3.1.2, we rewrite the reference model (3.71) as

$$\dot{\bar{q}} = A\bar{q} + Bu \tag{3.74}$$

with

$$\bar{q} = [q,\ \dot{q}] \in \mathcal{R}^6, \quad \bar{q}_d = [q_d,\ \dot{q}_d] \in \mathcal{R}^{2n} \tag{3.75}$$

$$A = \begin{bmatrix} 0_{3\times3} & I_{3\times3} \\ 0_{3\times3} & 0_{3\times3} \end{bmatrix}$$

$$B = \left[0_{3\times3}, I_{3\times3}\right]^T$$

$$u = -M_d^{-1}[K_d,\ C_d]\bar{q} - M_d^{-1}F_\eta(q_d, \dot{q}_d) \tag{3.76}$$

Noting that in the above equation u is defined as $u = \ddot{q}$, we can then rewrite the performance index Eq. (3.72) as

$$I_P = \int_{t_0}^{t_f} (\bar{e}^T Q e + u^T M_d u)dt \tag{3.77}$$

Let us regard u as the control input to system Eq. (3.74), then the minimization of Eq. (3.77) subject to dynamics constraint Eq. (3.74) becomes the standard LQR control design problem. According to the LQR optimal control technique reviewed in Sect. 3.3.1.2, the solution of u that minimizes Eq. 3.77 is

$$u = -M_d^{-1}B^T P\bar{q} - M_d^{-1}B^T s \tag{3.78}$$

where P is the solution of the following differential equation

$$-\dot{P} = PA + A^T P - PBM_d^{-1}B^T P + Q,$$
$$P(t_f) = 0_{6\times6} \tag{3.79}$$

and s is the solution of the following differential equation

$$-\dot{s} = (A - BM_d^{-1}B^T P)^T g - Q\bar{q}_d,$$
$$s(t_f) = 0_{6\times 1} \tag{3.80}$$

Comparing Eqs. (3.76) and (3.78), we can see that the matrices K_d and C_d can be calculated in the following manner:

$$[K_d, C_d] = B^T P, \quad F_\eta = B^T s \tag{3.81}$$

From the above procedure, we see that to obtain a reference model with a minimized performance index defined in Eq. (3.72), we first need to specify the value of Q matrix and M_d matrix; then we can calculate K_d and C_d according to Eqs. (3.79)–(3.81).

3.3.2 Adaptive Model Reference Control Design

3.3.2.1 Model Matching Errors

By using $e = q - q_d$, the reference model Eq. (3.71) can be rewritten as

$$M_d\ddot{e} + C_d\dot{e} + K_d e = -\sigma \tag{3.82}$$

where

$$\sigma = F_\eta + M_d\ddot{q}_d + C_d\dot{q}_d + K_d q_d \tag{3.83}$$

Next, we intend to seek a proper control input f in Eq. (3.57), such that the controlled dynamics given in Eq. (3.57) match the desired reference model dynamics given in Eq. (3.82). In order to measure the difference between the controlled dynamics of Eq. (3.57) and the reference model dynamics given in Eq. (3.82), we introduce the following matching error defined as [36]

$$w = M_d\ddot{e} + C_d\dot{e} + K_d e + q \in \mathcal{R}^3 \tag{3.84}$$

such that if the following condition holds

$$w(t) = 0_{n\times 1}, \ t > t_f \tag{3.85}$$

then the dynamics given in Eq. (3.57) will exactly match the desired reference model dynamics given in Eq. (3.82).

For the convenience of the following analysis, we define an augmented matching error as

$$\bar{w} = K_q w = \ddot{e} + C_m \dot{e} + K_m e + K_q \sigma \tag{3.86}$$

where $C_m = M_d^{-1} C_d$, $K_m = M_d^{-1} K_d$ and $K_\sigma = M_d^{-1}$.

Remark 3.4 The virtual mass matrix M_d should always be chosen as positive definite such that it is invertible and $\bar{w}$ in Eq. (3.86) is well defined.

Consider that acceleration measurement is usually not available, so does the signal $\bar{\omega}$. Therefore, let us design a filtered matching error z by using Laplace operator "s" as below:

$$\begin{aligned} z = & \operatorname{diag}\left(1 - \frac{\gamma_1}{s+\gamma_1}, 1 - \frac{\gamma_2}{s+\gamma_2}, \ldots, 1 - \frac{\gamma_n}{s+\gamma_n}\right) \dot{e} \\ & + \operatorname{diag}\left(\frac{1}{s+\gamma_1}, \frac{1}{s+\gamma_2}, \ldots, \frac{1}{s+\gamma_n}\right) \varepsilon \end{aligned} \tag{3.87}$$

where ε is defined as below

$$\varepsilon = C_m \dot{e} + K_m e + K_\sigma \sigma \tag{3.88}$$

For convenience of following computation, we rearrange z:

$$z = \dot{e} - \varepsilon_h + \varepsilon_l \tag{3.89}$$

where ϵ_h and ϵ_l are defined as below

$$\begin{aligned} \varepsilon_h &= \operatorname{diag}\left(\frac{\gamma_1 s}{s+\gamma_1}, \frac{\gamma_2 s}{s+\gamma_2}, \ldots, \frac{\gamma_n s}{s+\gamma_n}\right) e \\ \varepsilon_l &= \operatorname{diag}\left(\frac{1}{s+\gamma_1}, \frac{1}{s+\gamma_2}, \ldots, \frac{1}{s+\gamma_n}\right) \varepsilon \end{aligned} \tag{3.90}$$

i.e., high-pass and low-pass filtered, respectively, we see that the augmented matching error $\bar{w}$ can be written as

$$\bar{w} = \dot{z} + \Gamma z \tag{3.91}$$

where

$$\Gamma = \operatorname{diag}(\gamma_1, \gamma_2, \ldots, \gamma_n) \tag{3.92}$$

It implies that z could be obtained by passing $\bar{w}$ through a filter. From Eqs. (3.91) and (3.86), we see that $z = 0_{n\times1}$ and subsequently $\dot{z} = 0_{n\times1}$ will lead to $w = 0_{n\times1}$, i.e., matching error minimized. Then, the closed-loop dynamics (3.57) would exactly match the reference model (3.71). In the next section, we shall design an adaptive controller to guarantee that there exists a finite time $t_0 < t_z \ll t_f$ such that $z(t) = 0_{n\times1}$, $\forall t > t_z$.

3.3.2.2 Adaptive Controller Design

This section is devoted to discuss the details of the adaptive control design. The control input of the manipulator described in Eq. (3.57) is proposed as below:

$$\tau = J^T f$$
$$\tau = f_{ct} + f_{fb} \tag{3.93}$$

where τ_{ct} and τ_{fb} are the computed torque control input and the feedback torque control input, respectively. The feedback torque control input is given by

$$f_{fb} = -K_z z - f_s \tag{3.94}$$

where K_z is a diagonal positive definite matrix, and

$$f_s = K_s \frac{z}{\|z\|} \tag{3.95}$$

with K_s being another diagonal positive definite matrix. The computed torque control input is designed as

$$f_{ct} = -Y(\ddot{q}_r, \dot{q}_r, \dot{q}, q, z)\hat{\Theta} \tag{3.96}$$

where the trajectories $\dot{q}_r$ and $\ddot{q}_r$ are defined as follows:

$$\dot{q}_r = \dot{q}_d + e_h - \varepsilon_l$$
$$\ddot{q}_r = \ddot{q}_d + \dot{e}_h - \dot{\varepsilon}_l \tag{3.97}$$

and $\hat{\Theta}$, referring to Property 3.3 is the estimate of

$$\Theta = [k_M, k_C, k_D, k_G + K_L]^T \tag{3.98}$$

and

$$Y(\ddot{q}_r, \dot{q}_r, \dot{q}, q, z)$$
$$= \frac{z}{\|z\|}\left[\|\ddot{q}_r\|, \|\dot{q}\|\|\dot{q}_r\|, \|\dot{q}\|^2\|q\|, 1\right] \in \mathcal{R}^{3\times 4} \tag{3.99}$$

In the following, for convenience, we use Y instead of $Y(\ddot{q}_r, \dot{q}_r, \dot{q}, q, z)$ where it does not result in any confusion.

To obtain $\hat{\Theta}$, we design the following parameter estimation update law:

$$\dot{\hat{\Theta}} = \Gamma_\Theta^{-1} Y^T z \tag{3.100}$$

where Γ_Θ is a diagonal positive definite matrix.

3.3.2.3 Closed-Loop System Performance Analysis

Lemma 3.5 *If a function $V(t) \leq 0$ with initial value $V(0) > 0$ satisfies the following condition*

$$\dot{V} \leq -\kappa V^p, \ \ 0 < p < 1.$$

Then, $V(t) \equiv 0$, $\forall t \geq t_c$, for a certain t_c that satisfies

$$t_c \leq \frac{V^{1-p}(0)}{\kappa(1-p)}$$

Consider a scalar functional U_1 defined as

$$U_1(t) = \frac{1}{2} z^T M(q) z \tag{3.101}$$

If the following inequality holds

$$\dot{U}_1 \leq -\kappa \|z\|^{\delta} \tag{3.102}$$

where $\kappa_0 > 0, 0 < \delta < 2$ and $\|\cdot\|$ is matrix 2-norm. Denote $\lambda_{\min}(M)$ and $\lambda_{\max}(M)$ as the maximum and minimum eigenvalues of matrix M, respectively. Then, we see that there must exist a time instant $t_c \leq \frac{2(\lambda_{\max}(M))^{\frac{\delta}{2}} U_1^{1-\delta/2}(0)}{\kappa(2-\delta)}$ such that $U_1(t) \equiv 0, \forall t > t_c$.

Theorem 3.6 *Consider the closed-loop control system with manipulator dynamics given in Eq. (3.57) and the controller Eq. (3.93), under Assumption 3.1. If the parameters in controller are properly chosen, i.e., K_s is large enough, then the following results hold: (i) all the signals in the closed-loop are uniformly bounded, and (ii) the filtered matching error z and the original matching error w will be made zeros, i.e., $z = w = 0_{3\times 1}$ in a finite t_z, $t_0 < t_z \ll t_f$.*

Proof To prove the theorem above, let us consider the following Lyapunov-like composite energy function:

$$V_1(t) = U_1(t) + U_2(t) \tag{3.103}$$

where

$$U_2(t) = \frac{1}{2} \tilde{\Theta}^T(t) \Gamma_{\Theta}^T \tilde{\Theta}(t) \tag{3.104}$$

and

$$\tilde{\Theta} = \hat{\Theta} - \Theta \tag{3.105}$$

Note that according to the definition in Eq. (3.97), obviously we have $\dot{q} - \dot{q}_r = z$, and $\ddot{q} - \ddot{q}_r = \dot{z}$. Substituting the controller Eq. (3.93) into vessel dynamics, we have

$$M\dot{z} + Cz = -K_z z - K_s \frac{z}{\|z\|} + \mathcal{C} \tag{3.106}$$

where

$$\mathcal{C} = -Y(\ddot{q}_r, \dot{q}_r, \dot{q}, q, z)\hat{\Theta} - M\ddot{q}_r - C\dot{q}_r - G - \Delta \tag{3.107}$$

which according to Property 3.3 and the definition of Y in Eq. (3.99) satisfies

$$\begin{aligned} z^T \mathcal{C} &\leq \|z\|(\|M\|\|\ddot{q}_r\| + \|C\|\|\dot{q}_r\| + \|G + \Delta\|) - z^T Y\hat{\Theta} \\ &\leq \|z\|([\|\ddot{q}_r\|, \|\dot{q}\|\|\dot{q}_r\|, \|q\|\|\dot{q}\|, 1]\Theta - z^T Y\hat{\Theta} \\ &= \|z\|[\|\ddot{q}_r\|, \|\dot{q}\|\|\dot{q}_r\|, \|q\|\|\dot{q}\|, 1](\Theta - \hat{\Theta}) \\ &= -z^T Y\tilde{\Theta} \end{aligned} \tag{3.108}$$

Then, by considering the closed-loop dynamics given in Eq. (3.106) and Property 3.2, we obtain

$$\begin{aligned} \dot{U}_1 &= z^T M\dot{z} + \frac{1}{2} z^T \dot{M} z \\ &= z^T M\dot{z} + z^T Cz \\ &= z^T(-M\ddot{q}_r - C\dot{q}_r + G + \Delta + f) \\ &\leq z^T\left(\mathcal{C} - K_z z - K_s \frac{z}{\|z\|}\right) \\ &= z^T\left(-Y\tilde{\Theta} - K_z z - K_s \frac{z}{\|z\|}\right) \end{aligned} \tag{3.109}$$

From Eq. (3.100), we have

$$\dot{U}_2 = \dot{\tilde{\Theta}}^T \Gamma_{\Theta}^T \tilde{\Theta} = z^T Y\tilde{\Theta} \tag{3.110}$$

According to Eqs. (3.103), (3.109), and (3.110), we obtain the boundedness of $V_1(t)$ according to the following derivation:

$$\dot{V}_1 = \dot{U}_1 + \dot{U}_2 \leq -K_z\|z\|^2 - z^T K_s \frac{z}{\|z\|} \leq 0 \tag{3.111}$$

This implies that both U_1 and U_2 are bounded and consequently z and $\tilde{\Theta}$ are bounded, and this leads to the boundedness of all the other closed-loop signals. Consider that K_s is large enough such that

$$\|K\| - \|Y\tilde{\Theta}\| \geq \kappa_0 > 0 \tag{3.112}$$

Then, Eq. (3.109) would lead to $\dot{U}_1 \leq -\kappa_0\|z\|$. Compared with Eq. (3.102), we immediately obtain that $\|z\| \equiv 0$ for $t > t_c$ and so does w. This completes the proof. ■

3.4 Discrete-Time Adaptive Control of Manipulator with Uncertain Payload

The robotic manipulator systems are complicated and coupling nonlinear systems, and it is inevitable that there exist modeling error and disturbances in the systems. Therefore, designing a controller, which is robust to the uncertainty of the modeling error and disturbances in the systems/environment and posses satisfactory tracking performance, is of great importance. For most industry robots and family service robots, the influence of the payload on the end effector can be neglected when the value is small. However, in some situations, the payload may greatly influence the dynamic of the robot and hence the motion of the robot. In fact, the payload on the end effectors of the robot manipulator is not always known in advance or unchanging, even it may change from one value to another or is time-varying as time goes by. For example, the manipulator in the factory may move some goods of different mass, the payload of the manipulator is not known and may change from one to another.

For another example, in the welding factory, the welding robot will face that the mass of the welding rod slowly change in the working period. In these cases, it postulates that the controller of the system can identify the unknown payload and adjust it to the change of payload.

In these existing work, the discrete-time adaptive control methods for the manipulators with uncertainties from the unknown or time-varying load were seldom studied. And in the modern factories, the computer aided control is mainly used and the essential of the computer control is the discrete-time digital control. Thus the research on the discrete-time adaptive control of the manipulator is of great significance. On the other hand, the payload of the end effector is not always known or may vary in different work periods, so that estimation of the uncertainties is very important in the robot control.

In this section, we consider unknown fixed or varying payloads in the end effector [38]. The payload uncertainty is addressed through a new discrete-time adaptive controller based on the so-called "one-step guess" [39, 40] estimation of the payload, and investigate the effects of the new control scheme for a two-link robot manipulator.

3.4.1 Problem Formulation

Generally speaking, the robot dynamics given in Eq. (3.113) (see [33]):

$$M(q)\ddot{q} + C(q, \dot{q})\dot{q} + G(q) = u + f \tag{3.113}$$

where $M(q)$, $C(q, \dot{q})$ $G(q)$ were defined in Sect. 3.3. In this section, we suppose that f is the external payload of the robot manipulator and ignore the influence the other external disturbance factors.

In practice, the external payload f is not always known in advance, what we can use are some history observed values of the joint angles. So we can discretize the dynamic equation and calculate the payload at past time. Assuming that the payload vary small in the next sampling time, consulting the one-step-guess idea of the article [39], we can use the neighboring values and using as the estimated value at the present time to design the controller so that the end effector can track a desired reference trajectory. So the discrete-time adaptive controller mainly include the following three parts:

(i) The discretization of the robot dynamic equation;
(ii) The estimation of the unknown fixed or time-varying payload according to the history information of joint angles;
(iii) Designing the current time control signal based on the estimated payload to track the desired reference trajectory at next time.

3.4.2 *Discrete-Time Adaptive Control*

In this section, the dynamic equation of robot manipulator is reconstructed as in the continuous-time state space form and discretized state equation. Then, we calculate the estimated value of the unknown payload according to the previous joint angles information. At last, we design the controller based on the estimated payload and the desired joint angles curve.

3.4.2.1 Discretization of the Robot Dynamic Equation

Now we set $\bar{q} = [q \ \ \dot{q}]^T$, $q \in R^n$ and assuming that the matrix $M(q)$ is invertible, then we can rewrite Eq. (3.113) in a new state space form

$$\dot{\bar{q}} = A(q, \dot{q})\bar{q} + B(q)u + F(q)f - Q(q)G(q) \tag{3.114}$$

with

$$A(q, \dot{q}) = \begin{bmatrix} 0_{n\times n} & I_{n\times n} \\ 0_{n\times n} & -M^{-1}(q)C(q, \dot{q}) \end{bmatrix}$$

$$B(q) = F(q) = Q(q) = \begin{bmatrix} 0_{n\times n} \\ M^{-1}(q) \end{bmatrix} \tag{3.115}$$

Then, we discretize the state equation using the theory of discretization. Let the sampling time interval be T, and the sampled joint angles at time $t_k = kT$ be υ_k, i.e. $\upsilon_k = q(t_k)$. Correspondingly, we define $\dot{\upsilon}_k = \dot{q}(t_k)$. Then, we can obtain that

$$\bar{\upsilon}_{k+1} = L_k \bar{\upsilon}_k + H_k u_k + R_k f_k - S_k G(\upsilon_k) \tag{3.116}$$

where $\bar{\upsilon}_k = [\upsilon_k \quad \dot{\upsilon}_k]^T$, L_k, H_k, R_k and S_k are counterpart matrices in the discretized state equation which correspond to the continuous-time matrices $A(q, \dot{q}), B(q), F(q)$, and $Q(q)$. Their values can be calculated as Eq. (3.117) using the theory of the discretization for the state space equation:

$$\begin{aligned} L_k &= e^{A(q(t_k),\dot{q}(t_k))T} \\ H_k &= \int_{(k-1)T}^{kT} e^{A(q,\dot{q})t} B(q) dt \\ R_k &= \int_{(k-1)T}^{kT} e^{A(q,\dot{q})t} F(q) dt \\ S_k &= \int_{(k-1)T}^{kT} e^{A(q,\dot{q})t} Q(q) dt \end{aligned} \tag{3.117}$$

Theoretically speaking, we can calculate the formula $e^{A(q,\dot{q})T}$ using the MATLAB function `expm` as soon as the matrix $A(q, \dot{q})$ and the sampling time interval T are known; besides, the matrices L_k, H_k, R_k, and S_k can be obtained via the MATLAB function `ode45` via Runge–Kutta method.

However, as to the controller designer, in fact we can only get the sampled values $\upsilon_k = q(t_k)$ and $\dot{\upsilon}_k = \dot{q}(t_k)$ at the sampling time instant t_k. So the estimated values of L_k, H_k, R_k, and S_k are as the following:

$$\begin{aligned} \hat{L}_k &= e^{A_k T} \\ \hat{H}_k &= \int_{(k-1)T}^{kT} e^{A_k t} B(\upsilon_k) dt \\ \hat{R}_k &= \int_{(k-1)T}^{kT} e^{A_k t} F(\upsilon_k) dt \\ \hat{S}_k &= \int_{(k-1)T}^{kT} e^{A_k t} Q(\upsilon_k) dt \end{aligned} \tag{3.118}$$

where

$$A_k = A(\upsilon_k, \dot{\upsilon}_k) \tag{3.119}$$

Let us defined the estimated errors as below:

$$\tilde{L}_k = L_k - \hat{L}_k, \quad \tilde{H}_k = H_k - \hat{H}_k, \quad R_k = R_k - \hat{R}_k, \quad \tilde{S}_k = S_k - \hat{S}_k \tag{3.120}$$

which should be small enough if the sampling time interval is small enough. In our later simulation studies, the sampling frequency is taken as 100 Hz, i.e., $T = 0.01$ s, so the matrix A_k is determined at each sampling time and the estimate of L_k can be calculated.

For the matrices H_k, R_k and S_k, we can use the Runge–Kutta method with the MATLAB function "ode45" to calculate their estimated value at sampling time $t_k = kT$.

3.4.2.2 Estimation of the Payload

At time $t_{k-1} = (k-1)T$, the discretized difference equation in ideal case is as follows:

$$\bar{v}_k = L_{k-1}\bar{v}_{k-1} + H_{k-1}u_{k-1} + R_{k-1}f_{k-1} - S_{k-1}G(v_{k-1}) \tag{3.121}$$

Then we can obtain that

$$R_{k-1}f_{k-1} = \bar{v}_k - L_{k-1}\bar{v}_{k-1} - H_{k-1}u_{k-1} + S_{k-1}G(v_{k-1}) \tag{3.122}$$

Here we set the right-hand side of Eq. (3.122) as

$$P(\bar{v}_k, \bar{v}_{k-1}) = \bar{v}_k - L_{k-1}\bar{v}_{k-1} - H_{k-1}u_{k-1} + S_{k-1}G(v_{k-1}) \tag{3.123}$$

To solve Eq. (3.122), noting that R_{k-1} is generally not a square matrix and hence not invertible, we may use the least square method or the regularized least square method to estimate f_{k-1} at time instant $t_{k-1} = (k-1)T$. That is to say, the estimated value of the payload may be obtained by

$$\bar{f}_{k-1} = (R_{k-1}^T R_{k-1} + Q_T^T Q_T)^{-1} R_{k-1}^T P(\bar{v}_k, \bar{v}_{k-1}) \tag{3.124}$$

where Q_T is one matrix for fine-tuning the estimation such that $Q_T\bar{f}_{k-1} = 0$. Noting that R_{k-1} and $P(\bar{v}_k, \bar{v}_{k-1})$ are in fact not available, we may only use their estimated values. Thus, the estimated value of the payload at time instant $t_{k-1} = (k-1)T$ can be given by

$$\hat{f}_{k-1} = (\hat{R}_{k-1}^T \hat{R}_{k-1} + Q_T^T Q_T)^{-1} \hat{R}_{k-1}^T \hat{P}(\bar{v}_k, \bar{v}_{k-1}) \tag{3.125}$$

Here $\hat{R}_{k-1}^T$, $\hat{P}(\bar{v}_k, \bar{v}_{k-1})$ are the estimated values of R_{k-1}^T, $P(\bar{v}_k, \bar{v}_{k-1})$ at the sampling time instant t_{k-1}, where $\hat{P}(\bar{v}_k, \bar{v}_{k-1})$ can be calculated by Eq. (3.126):

$$\hat{P}(\bar{v}_k, \bar{v}_{k-1}) = \bar{v}_k - \hat{L}_{k-1}\bar{v}_{k-1} - \hat{H}_{k-1}u_{k-1} + \hat{S}_{k-1}G(v_{k-1}) \tag{3.126}$$

wherein $\hat{L}_{k-1}$, $\hat{H}_{k-1}$ and $\hat{S}_{k-1}$ are calculated from Eq. (3.118).

At the next sampling time $t_k = kT$, the unknown load f_k is not available, hence we may only use the previously calculated $\hat{f}_{k-1}$ to serve as the estimate of f_k for the purpose of designing the control signal u_k at time instant t_k. This idea of one-step guess [39] is natural and simple since the change of payload on the end effector in each sampling period would be very small at most time instants when the sampling interval T is small enough. That is to say, the posterior estimate of the payload at time $t_k = kT$ can be given by

$$\check{f}_k = \hat{f}_{k-1} = (\hat{R}_{k-1}^T \hat{R}_{k-1} + Q_T^T Q_T)^{-1} \hat{R}_{k-1}^T \hat{P}(\bar{v}_k, \bar{v}_{k-1}) \tag{3.127}$$

3.4.2.3 Adaptive Controller Design

According to the discrete-time dynamics, given in Eq. (3.116), of the robot manipulator, we can design the controller at time $t_k = kT$ to track the desired trajectory $\bar{v}_{k+1}^*$ at time $t_{k+1} = (k+1)T$. The ideal controller should make control input u_k at time instant t_k lead to $\bar{v}_{k+1} = \bar{v}_{k+1}^*$. Here, the desired joint angle vector $\bar{v}_{k+1}^*$ denotes the desired reference trajectory at time t_{k+1}, while the joint angle vector v_{kT} denotes the actual angles of the robot manipulator at the sampling time t_k.

From Eq. (3.116), we have

$$H_k u_k = \bar{v}_{k+1} - L_k \bar{v}_k - R_k f_k + S_k G(v_k) \tag{3.128}$$

Since the unknown payload f_k and exact values of matrices H_k, L_k, R_k, S_k are not available at timc instant t_k, in order to approximate the right hand side of the above equation, denoted by

$$V(\bar{v}_k, \bar{v}_{k+1}) = \bar{v}_{k+1} - L_k \bar{v}_k - R_k f_k + S_k G(v_k) \tag{3.129}$$

we may only use

$$\hat{V}(\bar{v}_k, \bar{v}_{k+1}^*) = \bar{v}_{k+1}^* - \hat{L}_k \bar{v}_k - \hat{R}_k \check{f}_k + \hat{S}_k G(v_k) \tag{3.130}$$

to serve as the estimated value of $V(\bar{v}_k, \bar{v}_{k+1})$ at time instant $t_k = kT$.

Thus, in order to track the desired trajectory at time instant t_{k+1}, we may design the discrete-time adaptive controller by the regularized least squares method

$$u_k = (\hat{H}_k^T \hat{H}_k + Q_U^T Q_U)^{-1} \hat{H}_k^T \hat{V}(\bar{v}_k, \bar{v}_{k+1}^*) \tag{3.131}$$

where Q_U is a matrix for fine-tuning the components of vector u_k. In this way, it is possible to improve the control performance, which may be seen in the later simulations.

3.4.3 Simulation Studies

Consider a two-link planar manipulator as the control object, the payload of the end effector of the robot manipulator is fixed or slowly degrade, and the exact value of the payload at present is unknown, therefore there are some uncertainties in the robot control system.

For a two-link planar robot manipulator, the dynamics can be given by the following equation [41]

$$\begin{bmatrix} M_{11} & M_{12} \\ M_{21} & M_{22} \end{bmatrix} \begin{bmatrix} \ddot{q}_1 \\ \ddot{q}_2 \end{bmatrix} + \begin{bmatrix} C_{11} & C_{12} \\ C_{21} & C_{22} \end{bmatrix} \begin{bmatrix} \dot{q}_1 \\ \dot{q}_2 \end{bmatrix} + \begin{bmatrix} g_1 \\ g_2 \end{bmatrix} = \begin{bmatrix} u_1 \\ u_2 \end{bmatrix} + \begin{bmatrix} 0 \\ f \end{bmatrix} \tag{3.132}$$

where

$$\begin{aligned} M_{11} &= m_1 l_{c1}^2 + I_1 + m_2(l_1^2 + l_{c2}^2 + 2l_1 l_{c2} \cos q_2) + I_2 \\ M_{12} &= M_{21} = m_2 l_1 l_{c2} \cos q_2 + m_2 l_{c2}^2 + I_2 \\ M_{22} &= m_2 l_{c2}^2 + I_2 \\ C_{11} &= -m_2 l_1 l_{c2} \sin q_2 \dot{q}_2 \\ C_{12} &= -m_2 l_1 l_{c2} \sin q_2 (\dot{q}_1 + \dot{q}_2) \\ C_{21} &= -m_2 l_1 l_{c2} \sin q_2 \dot{q}_1 \\ C_{22} &= 0 \\ G_1 &= m_1 l_{c1} g \cos q_1 + m_2 g [l_{c2} \cos(q_1 + q_2) + l_1 \cos q_1)] \\ G_2 &= m_2 l_{c2} g \cos(q_1 + q_2) \\ I_1 &= \frac{1}{3} m_1 l_{c1}^2 \\ I_2 &= \frac{1}{3} m_2 l_{c2}^2 \end{aligned} \tag{3.133}$$

In our study, the two-link robot manipulator with an unknown payload on the end effector will track a desired sine curve. First, we determine the desired sine curve of the end effector in the joint space, then we can estimate the unknown payload according to the known history joint angles information based on Eq. (3.125); assuming that the payload at the next time is not varying when the sampling time interval is small enough, we can design the controller of the robot manipulator in order to track the desired curve.

In our simulations, we consider the following two situations:

(i) The payload of the end effector is unknown and fixed (time-invariant) during a period, for example, a constant payload $f = 150$ is applied to the end effector from step 30 to step 70;

(ii) The payload of the end effector is unknown and time-varying as time goes by, for example, a linearly reducing payload starting from $f = 100$ at step 1 to $f = 90$ at step 100.

As to the choice of matrix Q_T, we take

$$Q_T = \begin{bmatrix} 10 & 0 \\ 0 & 0 \end{bmatrix} \tag{3.134}$$

which can make the first component of the estimated payload vector close to 0 (noting that the first component of the true payload is exactly 0). Further, the matrix Q_U is chosen as or close to a 2×2 zero matrix so that the magnitude of control signals is not too large. In the situation of unknown fixed payload, Q_U is simply chosen as a 2×2 zero matrix; while in the situation of unknown time-varying payload, Q_U is taken as

$$Q_U = \begin{bmatrix} 0.00015 & 0 \\ 0 & 0 \end{bmatrix} \tag{3.135}$$

which can yield a satisfactory control signal.

Figures 3.17, 3.18, 3.19, 3.20, and 3.21 illustrate the results of the tracking performance for two-link robot manipulator with an unknown fixed payload on end effector. In the experiment, the desired trajectory of the two links are the sine curves with different amplitudes and same frequencies. Figure 3.17 depicts the position tracking trajectory, from which we can see that the controller has a good tracking performance for the sine curve and there is jumping shock on the payload, the tracking error of link 1 varying between [0.0554 0.1130], and the tracking error of link 2

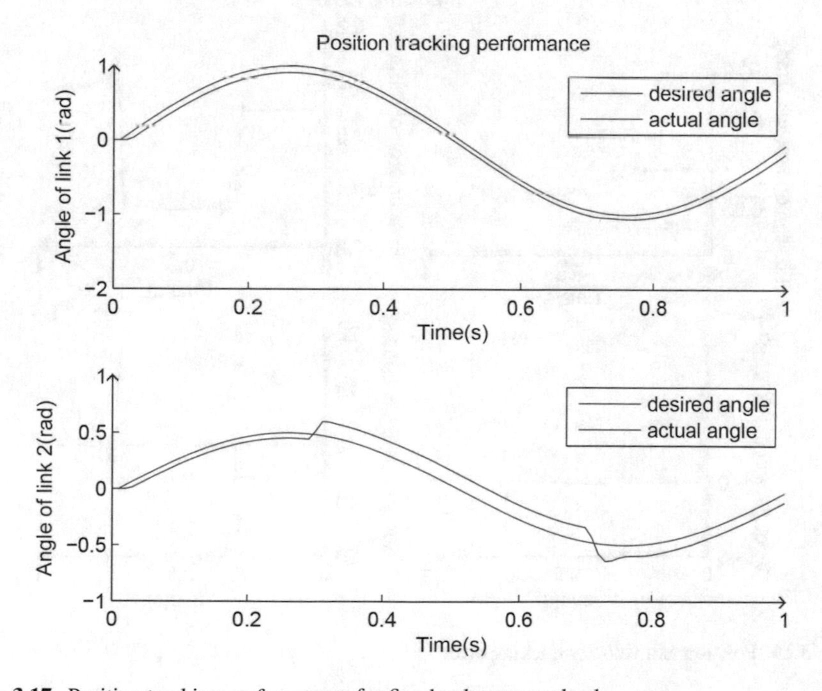

Fig. 3.17 Position tracking performances for fixed unknown payload

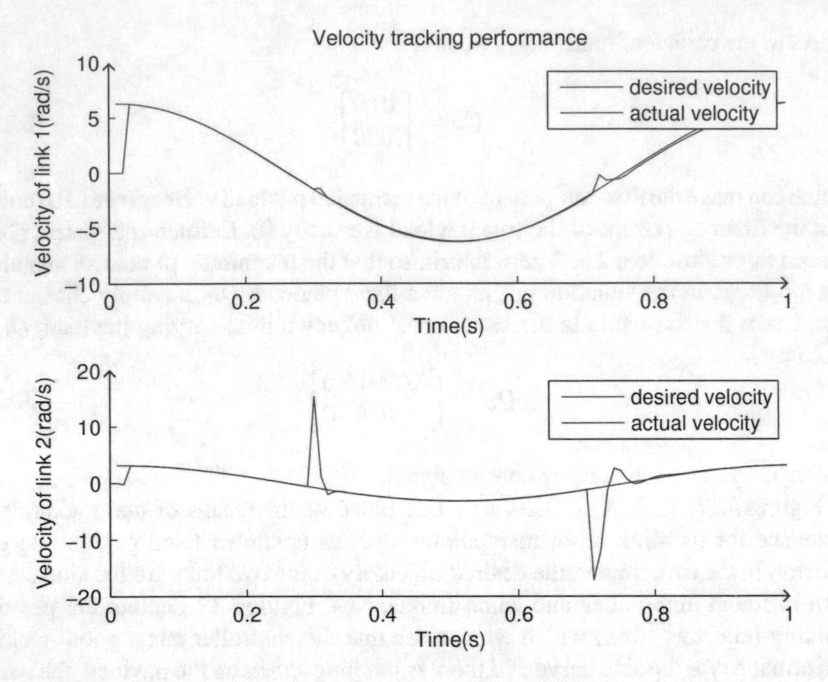

Fig. 3.18 Velocity tracking performances for fixed unknown payload

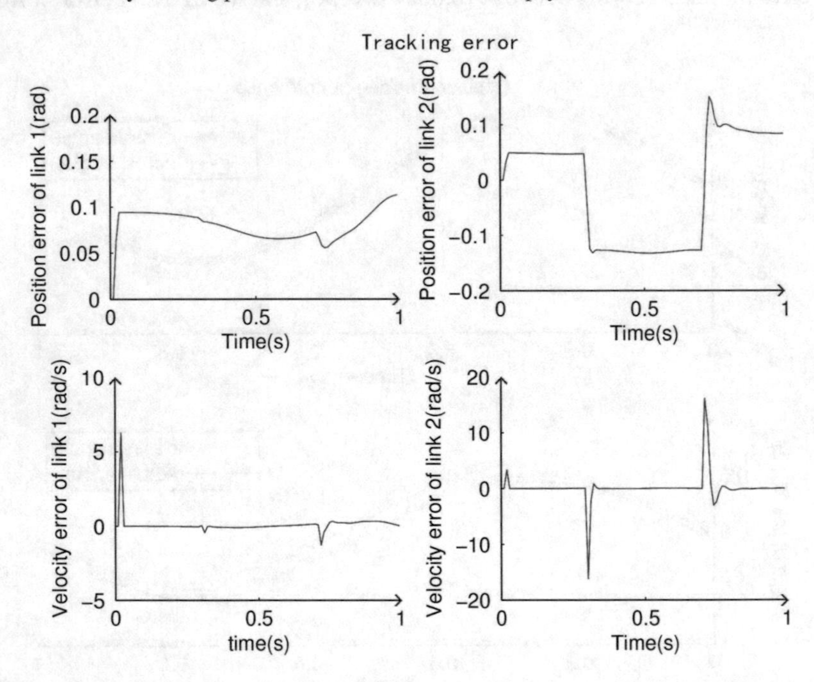

Fig. 3.19 Position and velocity tracking error

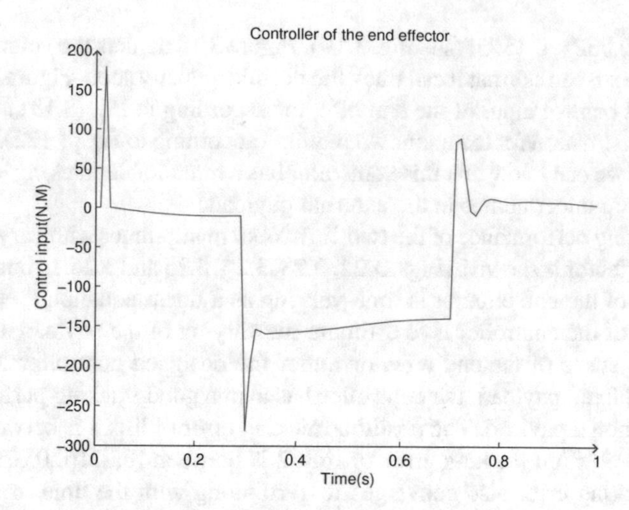

Fig. 3.20 Designed controller for fixed unknown payload

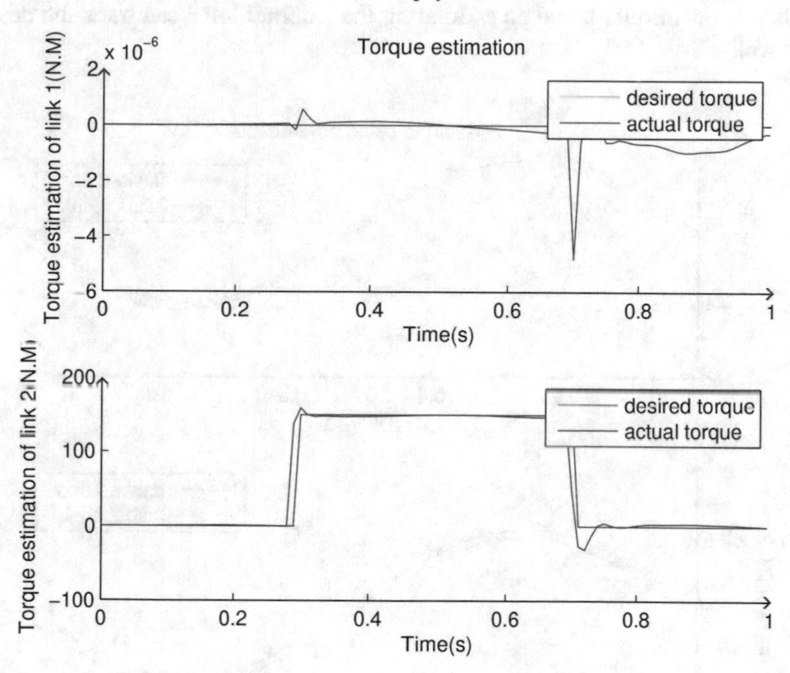

Fig. 3.21 Torque estimation for fixed unknown payload

is between [0.0324, 0.1505] (see Fig. 3.19). Figure 3.18 depicts the velocity tracking curve, where we can see that it can track the desired velocity soon. Figure 3.20 depicts the designed control input of the end effector according to Eq. (3.131). Figure 3.21 depicts the estimation for the unknown payload according to Eq. (3.122). From these four figures, we can know that this controller has a remarkable tracking performance while there are uncertainties in the external payload.

The tracking performance of the two-link robot manipulator with varying payload on the end effector is shown in Figs. 3.22, 3.23, 3.24, 3.25 and 3.26. In our simulation, the payload of the end effector is time-varying as a linear equation $f = 100 - 10t$, the purpose of the controller is to estimate the varying of the payload and track the desired sine curve of the end effector under the designed controller. Just like the situations of fixed payload, the controller has also a good tracking performance for varying unknown payload. The position tracking error of link 1 is between [0.0383, 0.1585], the position tracking error of link 2 is between [0.0216, 0.0707], and the velocity tracking error also converges to zero along with the time (see Fig. 3.24). Figure 3.26 shows that the algorithm has an satisfactory estimation performance. So the adaptive controller based on estimating the external force can track the desired curve well.

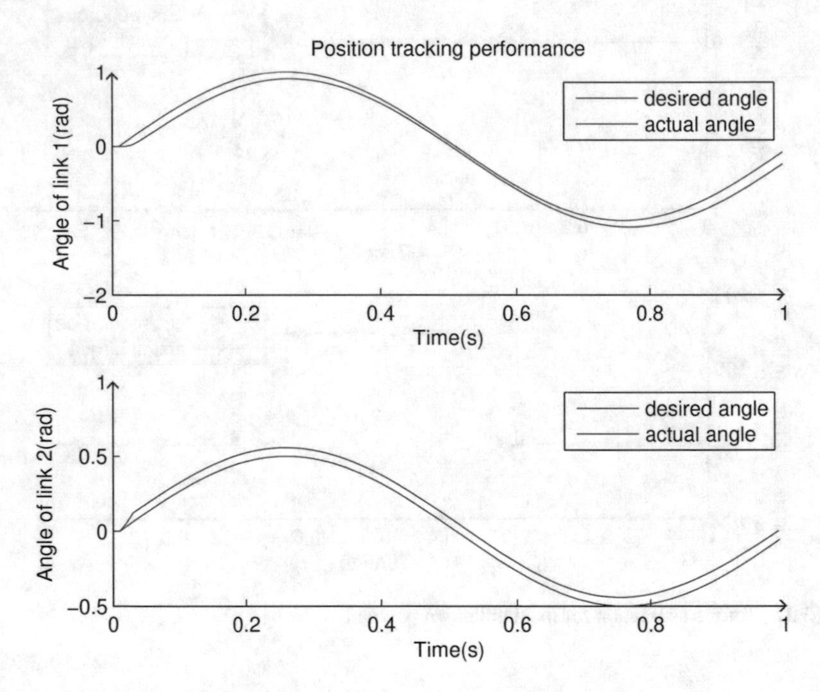

Fig. 3.22 Position tracking performances for varying unknown payload

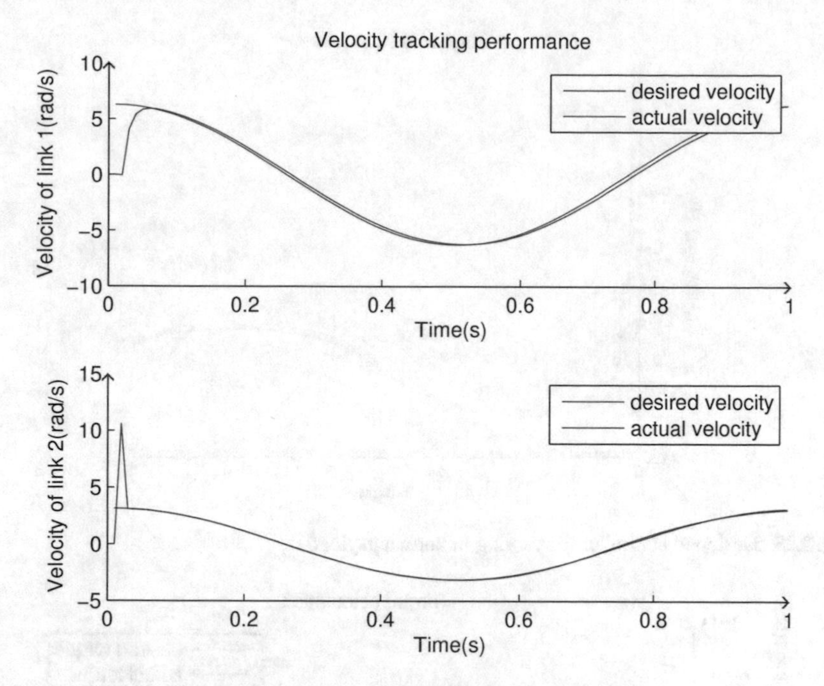

Fig. 3.23 Velocity tracking performances for varying unknown payload

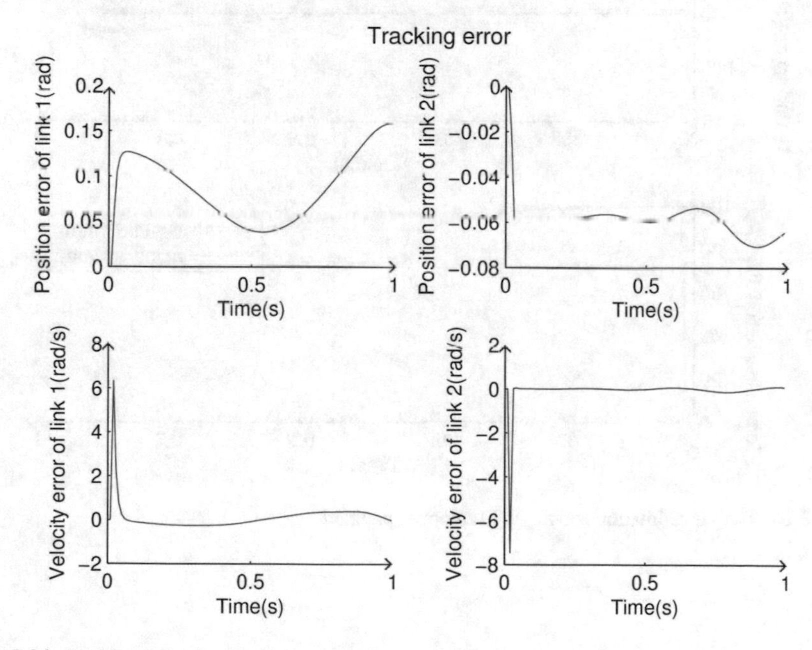

Fig. 3.24 Position/velocity tracking errors

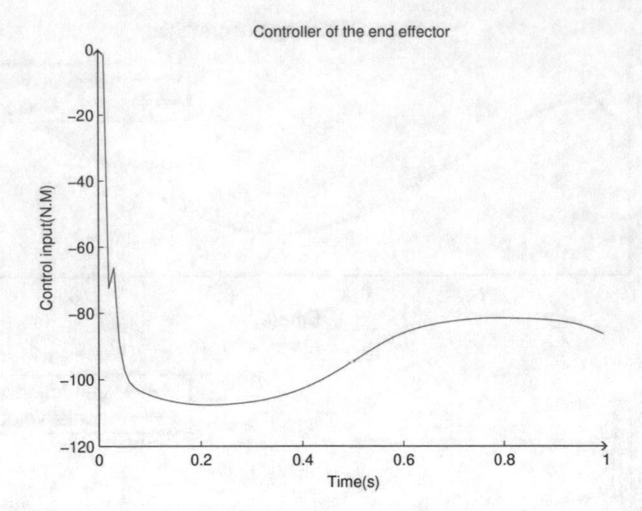

Fig. 3.25 Designed controller for varying unknown payload

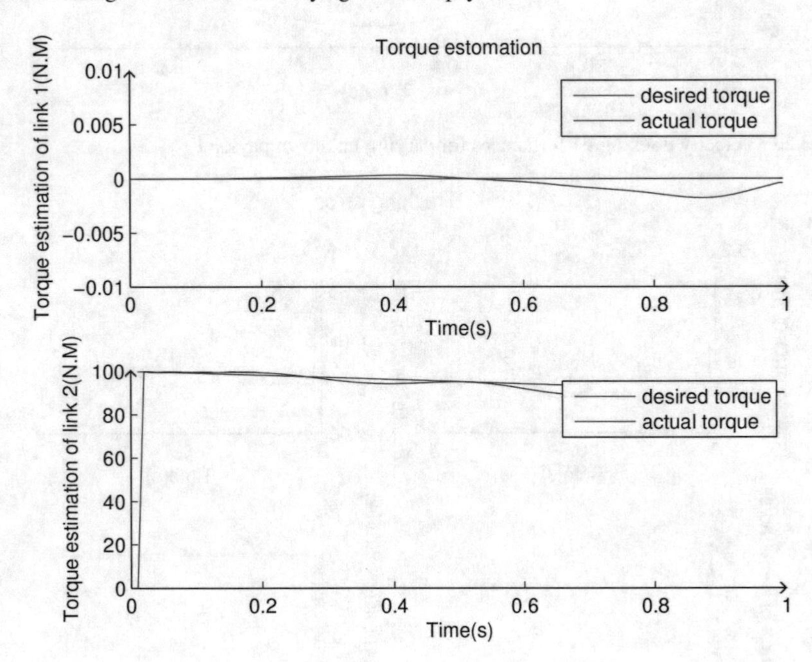

Fig. 3.26 Torque estimation for varying unknown payload

References

1. Slotine, J.J., Sastry, S.S.: Tracking control of non-linear systems using sliding surfaces with application to robot manipulators. Am. Control Conf. **1983**, 132–135 (1983)
2. Parra-Vega, V., Arimoto, S., Liu, Y.H., Hirzinger, G., Akella, P.: Dynamic sliding pid control for tracking of robot manipulators: theory and experiments. IEEE Trans. Robot. Autom. **19**(6), 967–976 (2004)
3. Chen, C., Liu, Z., Zhang, Y., Chen, C.L.P., Xie, S.: Saturated nussbaum function based approach for robotic systems with unknown actuator dynamics. IEEE Trans. Cybern. **PP**(99), 1–1 (2015)
4. Takeuchi, T., Nagai, Y., Enomoto, N.: Fuzzy control of a mobile robot for obstacle avoidance. Inf. Sci. **45**(88), 231–248 (1988)
5. Liu, Z., Chen, C., Zhang, Y.: Decentralized robust fuzzy adaptive control of humanoid robot manipulation with unknown actuator backlash. IEEE Trans. Fuzzy Syst. **23**(3), 605–616 (2015)
6. Lewis, F.L., Yesildirak, A., Jagannathan, S.: Neural Network Control of Robot Manipulators and Nonlinear Systems. Taylor & Francis Inc, Lindon (1998)
7. Slotine, J.-J.E., Li, W.: On the adaptive control of robot manipulators. Int. J. Robot. Res. **6**(3), 49–59 (1987)
8. Chen, C., Liu, Z., Zhang, Y., Chen, C.P.: Adaptive control of robotic systems with unknown actuator nonlinearities and control directions. Nonlinear Dyn. **81**(3), 1289–1300 (2015)
9. Chen, C., Liu, Z., Zhang, Y., Chen, C.P., Xie, S.: Adaptive control of mimo mechanical systems with unknown actuator nonlinearities based on the nussbaum gain approach. IEEE/CAA J. Autom. Sin. **3**(1), 26–34 (2016)
10. Soitrov, Z.M., Botev, R.G.: A model reference approach to adaptive impedance control of robot manipulators. In: IEEE/RSJ International Conference on Intelligent Robots and Systems, IROS Proceedings of the 1993, pp. 727–733 (1993)
11. Landau, Y.D.: Adaptive Control. Marcel Dekker, New York (1979)
12. Slotine, J.-J.E., Li, W.: On the adaptive control of robotic manipulators. Int. J. Robot. Res. **6**(3), 49–59 (1987)
13. Burdet, E., Osu, R., Franklin, D., Milner, T., Kawato, M.: The central nervous system stabilizes unstable dynamics by learning optimal impedance. Nature **414**(6862), 446–449 (2001)
14. Franklin, D.W., Osu, R., Burdet, E., Kawato, M., Milner, T.E., et al.: Adaptation to stable and unstable dynamics achieved by combined impedance control and inverse dynamics model. J. Neurophysiol. **90**(5), 3270–3282 (2003)
15. Franklin, D.W., Burdet, E., Tee, K.P., Osu, R., Chew, C.M., Milner, T.E., Kawato, M.: CNS learns stable, accurate, and efficient movements using a simple algorithm. J. Neurosci. **28**(44), 11165–11173 (2008)
16. Ganesh, G., Albu-Schaffer, A., Haruno, M., Kawato, M., Burdet, E.: Biomimetic motor behavior for simultaneous adaptation of force, impedance and trajectory in interaction tasks. 2010 IEEE International Conference on Robotics and Automation (ICRA), pp. 2705–2711. IEEE (2010)
17. Burgar, C.G., Lum, P.S., Shor, P.C., Hf, M.V.D.L.: Development of robots for rehabilitation therapy: the palo alto va/standford experience. J. Rehabil. Res. Develop. **37**(6), 663–673 (1999)
18. Lum, P.S., Burgar, C.G., Shor, P.C., Matra, M., Machiel, V.D.L.: Robot-assisted movement training compared with conventional therapy techniques for the rehabilitation of upper-limb motor function after stroke. Arch. Phys. Med. Rehabil. **83**(7), 952–959 (2002)
19. Volpe, B.T., Krebs, H.I., Hogan, N., Edelstein, L., Diels, C., Aisen, M.: A novel approach to stroke rehabilitation: robot-aided sensorimotor stimulation. Neurology **54**(10), 1938–1944 (2000)
20. Riener, R., Nef, T., Colombo, G.: Robot-aided neurorehabilitation of the upper extremities. Med. Biol. Eng. Comput. **43**(1), 2–10 (2005)
21. Kumar, V.: Control of cooperating mobile manipulators. IEEE Trans. Robot. Autom. **18**(1), 94–103 (2002)
22. Hogan, N.: Impedance control: An approach to manipulation: Part iiimplementation. J. Dyn. Syst. Meas. Control **107**(1), 8–16 (1985)

23. Yang, C., Ganesh, G., Haddadin, S., Parusel, S., Albu-Schäeffer, A., Burdet, E.: Human-like adaptation of force and impedance in stable and unstable interactions. IEEE Trans. Robot. **27**(5), 918–930 (2011)
24. Smith, A., Yang, C., Ma, H., Culverhouse, P., Cangelosi, A., Burdet, E.: Biomimetic join/task space hybrid adaptive control for bimanual robotic manipulation. In: IEEE International Conference on Control and Automation, 2014. ICCA '14. IEEE (2014)
25. Smith, A., Yang, C., Ma, H., Culverhouse, P., Cangelosi, A., Burdet, E.: Dual adaptive control of bimanual manipulation with online fuzzy parameter tuning. In: 2014 IEEE International Symposium on Intelligent Control (ISIC), pp. 560–565. IEEE (2014)
26. Zadeh, L.A.: Fuzzy sets. Inf. Control **8**(3), 338–353 (1965)
27. Bouchon-Meunier, B., Dotoli, M., Maione, B., et al.: On the choice of membership functions in a mamdani-type fuzzy controller. In: Proceedings of the First Online Workshop on Soft Computing, Nagoya, Japan. Citeseer (1996)
28. Ross, T.J.: Fuzzy Logic with Engineering Applications. Wiley, West Sussex (2009)
29. Takagi, T., Sugeno, M.: Fuzzy identification of systems and its applications to modeling and control. IEEE Trans. Syst. Man Cybern. **15**(1), 116–132 (1985)
30. Pattacini, U., Nori, F., Natale, L., Metta, G., Sandini, G.: An experimental evaluation of a novel minimum-jerk cartesian controller for humanoid robots. In: 2010 IEEE/RSJ International Conference on Intelligent Robots and Systems. Taipei, Taiwan (2010)
31. Hogan, N.: Impedance control: an approach to manipulation. American Control Conference, 1984, pp. 304–313. IEEE (1984)
32. Smith, A., Yang, C., Ma, H., Culverhouse, P., Cangelosi, A., Burdet, E.: Biomimetic joint/task space hybrid adaptive control for bimanual robotic manipulation. In: 11th IEEE International Conference on Control & Automation (ICCA), pp. 1013–1018. IEEE (2014)
33. Yang, C., Ma, H., Fu, M., Smith, A.M.: Optimized model reference adaptive motion control of robot arms with finite time tracking. In: 2012 31st Chinese Control Conference (CCC), pp. 4356–4360. IEEE (2012)
34. Tayebi, A.: Adaptive iterative learning control for robot manipulators. Automatica **40**(7), 1195–1203 (2004)
35. Anderson, B., Moore, J.B.: Optimal Control: Linear Quadratic Methods. Prentice-Hall Inc, Upper Saddle River (1990)
36. Yang, C., Li, Z., Li, J.: Trajectory planning and optimized adaptive control for a class of wheeled inverted pendulum vehicle models. IEEE Trans. Cybern. **43**(1), 24–36 (2013)
37. Yang, C., Li, Z., Cui, R., Xu, B.: Neural network-based motion control of underactuated wheeled inverted pendulum models. IEEE Trans. Neural Netw. Learn. Syst. **25**(11), 2004–2016 (2014)
38. Li, J., Ma, H., Yang, C., Fu, M.: Discrete-time adaptive control of robot manipulator with payload uncertainties. In: 2015 IEEE International Conference on Cyber Technology in Automation, Control, and Intelligent Systems (CYBER), pp. 1971–1976. IEEE (2015)
39. Rong, L., Ma, H., Wang, M.: Nontrivial closed-loop analysis for an extremely simple one-step-guess adaptive controller. In: 2011 Chinese Control and Decision Conference (CCDC), pp. 1385–1390 (2011)
40. Rong, L., Ma, H., Yang, C., Wang, M.: Decentralized adaptive tracking with one-step-guess estimator. In: 2011 9th IEEE International Conference on Control and Automation (ICCA), pp. 1320–1325 (2011)
41. Huang, A.-C., Chien, M.-C.: Adaptive Control of Robot Manipulators: A Unified Regressor-Free Approach. World Scientific, Singapore (2010)

Chapter 4
Object Detection and Tracking

Abstract Visual sensors provide comprehensive and abundant information of surrounding environment. In this chapter, we will first give a brief introduction of visual recognition with basic concepts and algorithms, followed by the introduction of the useful software toolkit JavaScript Object Notation (JSON) framework. Then we will review a series of vision-based object recognition and tracking techniques. The detection and classification of pedestrians in infrared thermal images is investigated using deep learning method. And the algorithm for tracking single moving objects based on JSON visual recognition framework is also introduced. As the extension of single moving objects tracking, we will also show the visual tracking to multiple moving objects with the aid of the particle swarm optimization (PSO) method.

4.1 Introduction of Machine Vision Recognition

From our human's perspective, recognition is about searching and comparison. Our brain has a huge knowledge base, storing tens of thousands of objects. Sometimes images themselves are stored in the brain, in other cases only abstract features are stored. Those images and corresponding features are called references. Then, our eyes are opened, new images are transmitted to the brain. Those test images are compared with references, and conclusions come out.

The visual recognition in machine is in some extent similar to humans' recognition. In order to recognize an object in a given image, machine needs to know what the object looks like. It learns the knowledge by loading some references before hand. Then, given a test image, matches between the references and the test samples are constructed. The machine then decides if there exists the wanted object in the test image, and locates it if possible, by analyzing the result of matches. Generally speaking, recognition is divided into two parts: detecting and locating. Detecting checks whether there exists a specific kind of object in the image. Then, locating marks regions of the detecting objects in the image.

There are three key concepts that are helpful to understand the recognition problem: feature space, similarity metric, and search space and strategy.

© Science Press and Springer Science+Business Media Singapore 2016

C. Yang et al., *Advanced Technologies in Modern Robotic Applications*,
DOI 10.1007/978-981-10-0830-6_4

Feature space: Feature space determines the information used in matching. It may be the image itself, but specific features are used more often: edges, contours, surfaces, corners, etc. Intrinsic structures which imply the invariant properties of an image are preferred features.

Similarity metric: The second key concept is related to the selection of a similarity metric. Since the metric measures the similarity among feature vectors, it is closely related to the selection of features. Typical similarity metrics include cross-correlation, sum of absolute difference, Fourier phase correlation, sum of distance of nearest points, and many others. The choice of similarity metric is also thought to be an important task to determine the transformation between two images.

Search space and strategy: The third key concept is search space and strategy. If the test image can be obtained by some transformations and distortions from a reference, a search space could be set up to match two images, if those transformations and distortions are included in the space. A search strategy is a guidance to move in the search space. In most cases, search space appears to be very large, so a proper search strategy plays an important role to control the computational complexity.

Under those key concepts, several methods are developed:

Cross-correlation: Cross-correlation is a simple statistical method to detect and locate objects. The reference is an image fragment which contains the object, called template.

For a template T and an image I (T is smaller than I), the two-dimensional normalized cross-correlation measuring the similarity under certain translation (u, v) is

$$C(u, v) = \frac{\sum_x \sum_y T(x, y) I(x - u, y - v)}{[\sum_x \sum_y I^2(x - u, y - v)]^{\frac{1}{2}}}. \tag{4.1}$$

If the template matches the image approximately at a translation of (i, j), then the cross-correlation will have its peak at $C(i, j)$ [1].

Fourier Transform: Another method of object detection is Fourier transform. Its main advantages over cross-correlation method rely on robustness of frequency domain against frequency-dependent noise. Also, using Fast Fourier Transform (FFT), the algorithm can be efficiently implemented [2]. However, Fourier method is based on invariant properties of transformations of the image, so it is only applicable for some well-defined transformations, such as translation and rotation. If there are some more general transformations and distortions, it cannot recognize the object effectively.

Compared to cross-correlation and Fourier transform, point mapping is a universal method. It can be used to detect objects even if the transformation between the reference and the test image is arbitrary and unknown [3]. This method consists of three stages. First, compute features in both images. Second, feature points in the reference image are matched with feature points in the test image based on least square regression or similar techniques. Finally, the result of matching is used to determine objects in the test image. It is the primary method used in this project.

Besides, there are some other recognition techniques, and many of them are related to machine learning. Among the well known methods, machine vision recognition based on features such as histogram of oriented gradients (HOG) are widely used. Besides, in the last decade, new approach of deep learning emerged, which outperforms many existing methods for huge-size database samples. Convolutional neural network (CNN) has been regarded as a seminal break- through towards deeper understanding to machine intelligence or artificial intelligence.

4.1.1 Tools for Machine Vision

Fast progress of machine vision hardly happens without excellent software packages and tools. For researchers, their first choice may be MATLAB computer vision system toolbox. This toolbox offers extremely fast coding and prototyping, with most machine vision functions and algorithms built in.

For application developers, OpenCV (Open Source Computer Vision) is a free yet complete solution. Since its first release in 1999, more than 500 optimized algorithms for image and video analysis have been included. Most functions in OpenCV are written in C++, and many of them are of C, Java, and Python interfaces, so they are portable and extremely fast. Providing common interfaces, OpenCV accelerates the development for both desktop and embedded system. Using CUDA and OpenCL for parallel computing, the latest version of OpenCV reinforces its support on GPU.

Except for MATLAB and OpenCV, there are other computer vision libraries. SimpleCV and Pillow are two popular libraries in Python. They have the ability to deal with image processing tasks. VLFeat is a library that focuses on local features detection and matching. Besides, ccv is a modern vision library containing some state-of-the-art algorithms.

This project utilizes functions in OpenCV. The latest stable version of OpenCV, OpenCV 2.4 contains dozens of modules. The commonly used modules include (i) core functionality; (ii) image processing; (iii) GUI and media I/O; and (iv) 2D features. The core module defines basic data structures and operations on images. The image processing module serves as a preprocessing step. For example, a denoising operation may be helpful if we want to remove noise from the captured images. Also, luminance balancing is a powerful technique to improve the sense of images. To visualize the result of matching, we would like to view both the reference and the test image simultaneously, so a GUI interface is necessary in this case. Corresponding functions are in GUI and media I/O module.

In order to increase speed and to reduce computational complexity, we do not calculate the similarity of images in pixels. In contrast, we extract features from images, and then calculate the similarity over them. Finding keypoints, computing features, and matching keypoints of two images, those functions are packed in 2D features module.

In object recognition, features are abstraction of image pixels. They not only reduce the complexity of computation, but also keep key information of original image. Some of features have two properties: locality and invariance. A local feature can be computed using only a small fragment of an image, while a global feature is extracted from the entire image. Compared to global features, local features are more representative when the image is suffering from aliasing and overlap. An invariant feature means that some transformations that are applied to an image have no effect on the computational result of the feature. This property is powerful to deal with affine transformation and geometric distortion of images. If a feature is both local and invariant, it is called a local invariant feature, which is a key component of point mapping method in object recognition.

4.1.2 Blob/Edge Detection

In machine vision, a blob is defined as a regular area (usually circular) whose gray scale or color is significantly different from its surrounding areas. Blob detection is known as a special case of region detection. A key step in blob detection is to determine the boundary of blobs. In this problem, a basic idea is to examine the response of the convolution of the image signal and the derivative of Gaussian function.

In one-dimensional case, if we calculate the convolution of signal f and the first derivative of Gaussian function $\frac{d}{dx}g$, then at the jumping edge of f there is a large response. The position of maximal response corresponds to the position of the jumping edge.

Another method to obtain the jumping edge utilizes the property of zero-crossing of second derivative. For the convolution of signal f and $\frac{d^2}{dx^2}g$, there is a null point in the position of the edge. In the case of two successive jumps (one-dimensional blob), the size of blob is the size of normalized Gaussian function with the biggest response. In two-dimensional image, Laplace of Gaussian (LoG) is used in blob detection [4].

A powerful blob detection algorithm is Determinant of Hessian (DoH). For an image I and a point $\mathbf{x}(x, y)$, at a given scale σ, the definition of the Hessian matrix $\mathbf{H}(\mathbf{x}, \sigma)$ is given below

$$\mathbf{H}(\mathbf{x}, \sigma) = \begin{bmatrix} L_{xx}(\mathbf{x}, \sigma) & L_{xy}(\mathbf{x}, \sigma) \\ L_{xy}(\mathbf{x}, \sigma) & L_{yy}(\mathbf{x}, \sigma) \end{bmatrix} \tag{4.2}$$

where $L_{xx}(\mathbf{x}, \sigma)$ is the convolution of the second derivative $\frac{\partial^2}{\partial x^2}g(\sigma)$ and the image at the point $\mathbf{x}$. The definition of $L_{xy}(\mathbf{x}, \sigma)$ and $L_{yy}(\mathbf{x}, \sigma)$ are similar [5]. The determinant of Hessian is therefore

$$det(\mathbf{H}) = \sigma^4(L_{xx}L_{yy} - L_{xy}^2). \tag{4.3}$$

The step of detection is similar to LoG. Using different scale σ to generate a series of kernels of $\frac{\partial^2 g}{\partial x^2}$, $\frac{\partial^2 g}{\partial y^2}$, and $\frac{\partial^2 g}{\partial x \partial y}$, then searching the maximal responses in spatial and scale space to locate blobs.

Like the Gaussian function in blob detection, there are two kinds of edge detection methods: first-order differential operator and second-order differential operator [6]. First-order differential operator utilizes the step change at the edge, and detects the local maximum of the gradient at the edge.

4.1.3 Feature Point Detection, Description, and Matching

Since blob detection, or more generally, feature point detection, serves as a pre-processing step of machine vision recognition, it is required to be computed quickly. Scale-Invariant Feature Transform (SIFT) and Speeded Up Robust Features (SURF) are two such solutions.

SIFT Feature Detection: The key idea of SIFT algorithm relies on the concept of Difference of Gaussian (DoG). The detection of feature points in a specific scale can be done by the difference of two successive scales.

In fact, DoG is an approximation of LoG. The advantage of DoG over LoG is obvious in the aspect of computational complexity. The computing of LoG requires second derivative of Gaussian kernel in two directions. In contrast, DoG uses the Gaussian kernel directly. Also, since LoG algorithm is more robust than DoH, as an approximation of LoG, DoG is robust against noise as well.

DoG is computed by generating a Gaussian image pyramid. This pyramid has O octaves, and each octave has S levels. Next octave is obtained by downsampling the previous one. DoG is the difference of successive levels in each octave. All the DoG construct a new pyramid, which has O octaves but only $S-1$ levels.

SURF Feature Detection: SIFT is an approximation of LoG, while SURF is an approximation of DoH. SURF simplifies the second derivative of Gaussian kernel, so that convolution can be done with only addition and subtraction. With the help of parallel computing, SURF algorithm is generally faster than SIFT.

Computation of SURF relies on integral image. At each point (i, j), the value of integral image $ii(i, j)$ is equal to the sum of gray values from upper left point to point (i, j).

$$ii(i,j) = \sum_{i' \leq i, j' \leq j} p(i', j') \tag{4.4}$$

where $p(i', j')$ is the gray value of point (i', j') in original image.

One important property of integral image is that the sum of values in a window can be calculated from the integral image of its four vertices. If (i_1, j_1), (i_2, j_2), (i_3, j_3), and (i_4, j_4) are the upper left, upper right, lower left, and lower right vertices of the window, then

$$\sum_{w} = ii(i_1, j_1) - ii(i_2, j_2) - ii(i_3, j_3) + ii(i_4, j_4). \tag{4.5}$$

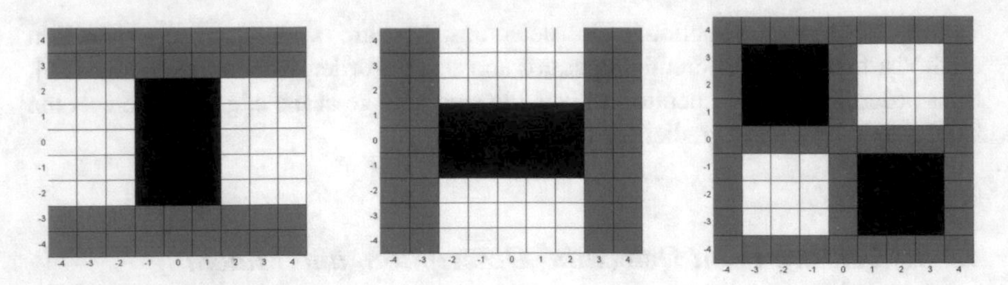

Fig. 4.1 Simplified Gaussian kernels

In order to convert the Gaussian convolution to box filter operation, we need to simplify the second derivatives to some very simple kernels. After simplification, three new kernels are generated as shown in Fig. 4.1, where a white block represents 1, a gray block represents 0, and a black block represents -1. Those kernels are much simpler than the second derivatives and can be computed via integral image very fast.

Given different scale σ, the simplified kernels have different sizes. In practice, kernels of size 9×9, 15×15, 21×21, and 27×27 are usually used. Similar to SIFT algorithm, after searching the local maximum in each scale, the algorithm finds out the wanted feature points.

4.1.3.1 Feature Point Description

After locating feature points in an image, the next step is to describe the local pattern of each feature point. A descriptor is a kind of description about the local structure of an image. Up to now, a plenty of descriptors have been proposed, based on gradient distribution, space frequency, filters, moment invariance, etc. The most famous descriptor is SIFT descriptor. It is in the category of gradient distribution and has outstanding performance. In addition to SIFT, SURF descriptor is also introduced in this section.

SIFT Descriptor: In order to satisfy rotation invariance, the first consideration of SIFT descriptor is to find a main direction. For a feature point (x, y) and its scale σ, the Gaussian image of this scale is

$$L(x, y) = G(x, y, \sigma) * I(x, y). \tag{4.6}$$

In this Gaussian image, select each feature point as the center, set $3 \times 1.5\ \sigma$ as the radius, and compute the length and angle of the gradient of image fragment. The equations are

$$m(x, y) = \sqrt{(L(x+1, y) - L(x-1, y))^2 + (L(x, y+1) - L(x, y-1))^2} \quad (4.7)$$

$$\theta(x, y) = \arctan\left(\frac{L(x, y+1) - L(x, y-1)}{L(x+1, y) - L(x-1, y)}\right). \quad (4.8)$$

Then, the directions of gradient are separated into 36 columns, each column takes up 10°. The peak value in this histogram represents the main direction of the feature point.

Hoping that the gradients near the feature point have larger weights, each point in the histogram needs to be weighted. A proper weighting function is circular Gaussian function, whose standard deviation is usually selected as 1.5 σ.

When the main direction of the feature point is obtained, it can be fully defined as a tuple (x, y, σ, θ). Where x and y determine the position of the feature point, σ is the scale of the point and θ is the main direction.

SURF Descriptor: Similar to SIFT algorithm, the first step is to find the main direction of the feature point. But rather than using a Gaussian function, a Haar wavelet operator is used in SURF. Set the feature point as the center, and compute the response of Haar wavelets in a 6 σ circle area. Then, each point in the response is weighted. After normalized, a fan that centered at the feature point and has an angle of $\pi/3$ is constructed and rotated. For each small rotation step (e.g., $\pi/6$), the sum of the Haar response dx and dy is calculated, so we obtain the following vector (m_w, θ_w):

$$m_w = \sum_w dx + \sum_w dy \quad (4.9)$$

$$\theta_w = \arctan\left(\frac{\sum_w dx}{\sum_w dy}\right). \quad (4.10)$$

The main direction is defined as the maximum sum of response

$$\theta = \theta_w \text{ corresponds to } \max(m_w). \quad (4.11)$$

After that, the generation of SURF descriptor is described below. First, according to the main direction of the Haar wavelet response, select an image fragment of the size 20 σ × 20 σ. This part is separated into 4 × 4 regions, and in each region a 2 σ × 2 σ Haar template is used to calculate the response. As a result, for each region 25 samples are obtained. After weighted by a Gaussian function, four statistical variables are computed in each region, which are $\sum dx$, $\sum |dx|$, $\sum dy$, and $\sum |dy|$. So a SURF descriptor vector has a total length of 64 dimensions.

4.1.3.2 Feature Point Matching

As discussed in previous sections, feature points are representation of local stable areas of an image. If working properly, two images of the same scene can be matched.

If the distance between two feature vectors is small, it is likely that two points are the same position of the scene; otherwise, if the distance is large, there are more likely to be two different positions. Feature point matching techniques use that idea and reduce the image matching problem to feature point matching problem.

The fundamental idea of matching is searching for similar vectors in high-dimensional space via a distance function. Generally, there are two types of searching, range search, and K-neighbor search. Range search looks for all the vectors with distance less than a threshold to the reference vector. K-neighbor search returns the nearest k vectors given by the reference.

Brute-Force Matching: Brute-force search is the easiest searching method. If A is the point set of all feature points from image 1, and B is the point set from image 2, then for each point in A, the distance for every point in B is calculated, and the point with minimal distance is considered as the matched point.

K-dimension Tree Matching: Instead of direct scanning, it is also possible to construct a data structure to accelerate the matching process. It is potentially faster than brute-force search since real data has some properties like clustering, which may benefit from index. However, we should also notice that building an index structure costs a lot, so the final running time may not be faster than brute-force search.

K-dimension tree, abbreviated as Kd-tree, is a kind of data structure to separate a point in a k-dimension space [7]. Kd-tree is a binary tree, it has these elements in each node of the tree:

- data: the data vector
- range: a vector that marks the space range of the node
- split: an integer that represents the index of direction axis which is perpendicular to the split surface
- left: a pointer to a Kd-tree which consists of all the nodes in the left surface
- right: a pointer to a Kd-tree which consists of all the nodes in the right surface
- parent: a pointer to the parent node

In this structure, *range* is the occupied space of a node, and *data* is a vector of original feature description. There is a surface that crosses the data vector and split the space into two parts. If $split = i$, then every node inside the *range* with its i dimension less than $data[i]$ is separated to the left subspace; likewise, every node inside the same range but its i dimension is greater than $data[i]$ is separated to the right subspace.

Matching Refinement: In the previous computation, we have obtained the feature point set $\{p_i\}, i = 1, 2, \ldots, m$, from image 1, set $\{q_j\}, j = 1, 2, \ldots, n$ from image 2, and the nearest matching between p_i and q_j: $\langle p_i, q_j \rangle$. Although it is the nearest matching, it does not ensure that all matched points are in the same position of the scene. In fact, the correctness of matching should be further evaluated. Matching refinement is the process to deal with the problem.

Briefly speaking, there are two kinds of refinement methods: ratio-based refinement and consistency refinement.

The algorithm searching for the nearest point in Kd-tree is mentioned above. It is easy to expand the algorithm to search K-nearest points. Let us consider the case where $K = 2$. The algorithm returns a list of two points, in which D_1 is the distance to the nearest point, and D_2 is the distance to the second nearest point. From experiment, the following fact can be discovered: if the ratio $\frac{D_1}{D_2}$ is less than some value, then the probability of correct matching is large; otherwise the probability of wrong matching is large. So we can choose a threshold and remove all the matches whose ratios are bigger than the threshold. Commonly the threshold is chosen as 0.49.

Different from ratio-based refinement, consistency refinement relies on the fact that it is possible to determine the transformation of two images first, and then check if the correctness of matches.

To determine the transformation between two images in a 2D scenario, at least four matches are required. Assume that images $I(x, y, 1)$ and $I'(x', y', 1)$ satisfy the perspective relationship

$$\mathbf{I}' \sim \mathbf{HX} = \begin{bmatrix} h_0 & h_1 & h_2 \\ h_3 & h_4 & h_5 \\ h_6 & h_7 & h_8 \end{bmatrix} \tag{4.12}$$

Then the coordinates of x' and y' can be expressed by [8].

$$x' = \frac{h_0 x + h_1 y + h_2}{h_6 x + h_7 y + h_8} \tag{4.13}$$

$$y' = \frac{h_3 x + h_4 y + h_5}{h_6 x + h_7 y + h_8}. \tag{4.14}$$

4.2 JavaScript Object Notation (JSON)-Based Vision Recognition Framework

This section reports a vision recognition framework—JSON vision recognition framework. At first, let us introduce some properties of the JSON.

JSON is a standardized data interchange format [9]. Its standard is based on JavaScript but the implementation is language independent. When JSON is integrated into vision development, it is advantageous to store and retrieve image labels, and accelerate tuning procedures. The JSON data format is built on two structures: (a) a collection of key–value pairs and (b) an ordered list of values. The combination of two results in a hierarchical structure of the data. Generating or parsing JSON data is supported by a lot of packages on dozens of languages. Here is an illustration of processing JSON data by a C++ library RapidJSON [10]. It has a group of small yet powerful API functions. The basic functions are simple and intuitive.

First, include some header files to the project.

```
1 #include <rapidjson/document.h>
2 #include <rapidjson/writer.h>
3 #include <rapidjson/prettywriter.h>
4 #include <rapidjson/stringbuffer.h>
```

Second, declare a JSON document object you want to use. For example,

```
rapidjson::Document doc
```

If a user has read the file via IO stream, and stored it into a "std::string variable str", construct the JSON object that needs only one operation:

```
doc.Parse(str.c_str());
```

where "c_str()" method returns the C-style character array.

In rapidJSON package, Document class is the top structure of the JSON file, and it should always be a single object. A more general class is called "Value", which can refer to any object in the JSON structure, including top structure. By given a path of the object we want to retrieve, it is possible to get its value recursively.

To read and write values in JSON, let us see a function to retrieve the reference to an arbitrary object in the JSON structure.

```
const rapidjson::Value&
JsonHandler::getReference(const rapidjson::Value& doc, vector<string> position)
{
    assert(position.size() >= 1);
    int checkVal; // convert string to number if necessary
    if ((checkVal = checkDigit(position[0])) == -1) {
        // position[0] is a string
        const rapidjson::Value& tmp = doc[position[0].c_str()];
        if (position.size() == 1) {
            return tmp;
        }
        else {
            vector<string>::iterator it = position.begin();
            it++;
            vector<string> tailPosition(it, position.end());
            return getReference(tmp, tailPosition);
        }
    }
    else {
        // position[0] is a number
        const rapidjson::Value& tmp = doc[checkVal];
        if (position.size() == 1) {
            return tmp;
        }
        else {
            vector<string>::iterator it = position.begin();
            it++;
            vector<string> tailPosition(it, position.end());
            return getReference(tmp, tailPosition);
        }
    }
}
```

In this function, "*position*" is a vector of string which locates the object that we want to retrieve. Since both key–value pairs and ordered list may exist in the structure, we need to judge either to get the value by key or search the element by subscript. Each time we step one layer down the structure, then call the function with new "*root*" node and the tail of path. Finally, the specific reference of the object will be returned.

Using the same pattern, it is not hard to retrieve any information from JSON file if the structure and query path are known. Some useful helper functions can be developed, such as

```
bool getBoolVal(const rapidjson::Value& doc, vector<string>
    position);
void setBoolVal(rapidjson::Value& doc, vector<string>
    position, bool newVal);
int getIntVal(const rapidjson::Value& doc, vector<string>
    position);
void setIntVal(rapidjson::Value& doc, vector<string> position
    , int newVal);
double getDoubleVal(const rapidjson::Value& doc, vector<
    string> position);
void setDoubleVal(rapidjson::Value& doc, vector<string>
    position, double newVal);
string getStrVal(const rapidjson::Value& doc, vector<string>
    position);
void setStrVal(rapidjson::Value& doc, vector<string> position
    , string newVal);
size_t getArraySize(const rapidjson::Value& doc, vector<
    string> position);
```

Using those helper functions, developers are able to deal with the testing data with ease.

4.2.1 JSON in Image Labels

Now let us see how JSON can be helpful for collecting image label information. Suppose a developer has a set of images to test his object detection program, and there are some objects that he wants to annotate. One may use a labeling tool and come out with a JSON format file below.

```
{
  "train": [
      {
          "filename": "1.jpg",
          "folder": "images",
          "size": {
              "nrows": 256,
              "ncols": 256
          },
          "objects": [
              {
                  "name": "car",
                  "id": 0,
                  "region": {
                      "xmin": 27,
```

```
16                      "xmax": 166,
17                      "ymin": 155,
18                      "ymax": 221
19                  }
20              }
21          ]
22      },
23      /* More lines here */
24  ],
25  "test": [
26  /* More lines here */
27  ]
28 }
```

There is a single-root object in this data, which is denoted by two curly braces. The root object has two key–value pairs, called "*train*" and "*test*" separately. The value of both pairs is an ordered list, which contains zero or more image description objects. Each image description object contains four key–value pairs, including "*filename*," "*folder*," "*size*," and "*objects*," and some of them have even sublayer structures. By constructing a JSON file like this, the whole image dataset is well labeled.

So how could JSON be helpful in vision development? Here is the first point. Many open image datasets, such as SUN database [11] for scene recognition, Barcelona dataset [12] for semantic labeling, only provide labels in proprietary formats, i.e., MATLAB matrix data. If they distribute data in JSON format, and if developers choose to store their image labels in JSON format, it would be beneficial in two aspects.

First, developers have more freedom to generate and collect data. They could collect the data from different sources, or use different program for labeling, and this will not be a problem. In MATLAB, Python, and many other environments, there are tools to convert data into JSON format, so all the data collected will have a universal structure.

Second, developers have more freedom to use the data in different applications and different platforms. JSON data is language independent, it can be used by a PC program written in C#, as well as an android app written in Java. There is no need to consider compatibility problem.

More and more researchers have realized the importance of JSON data format. Some dataset, such as OpenSurfaces [13] for material recognition, and Princeton Tracking Benchmark [14] for visual tracking, are available by JSON format.

```
1  "detector": "SURF",
2     "extractor": "SURF",
3     "matcher": "BFMatcher",
4     "drawing": true,
5     "siftConfig": {
6         "nfeatures": 0,
7         "nOctaveLayers": 3,
8         "contrastThreshold": 0.04,
9         "edgeThreshold": 10.0,
10        "sigma": 1.6
11    },
```

```
12      "surfConfig": {
13          "hessianThreshold": 400.0,
14          "nOctaves": 4,
15          "nOctaveLayers": 2,
16          "extended": true,
17          "upright": false
18      },
19      /* More lines here */
20      "BFMatcherConfig": {
21          "normType": "NORM_L2",
22          "crossCheck": false
23      },
24      /* More lines here */
```

The JSON data indicates the program to use SURF algorithm for keypoint detection and feature extraction, then match the keypoints of two images by brute-force matching. All the parameters of each algorithm are also stored in JSON structure.

The design mentioned above enables automatic tuning. Figure 4.2 shows how it can be achieved.

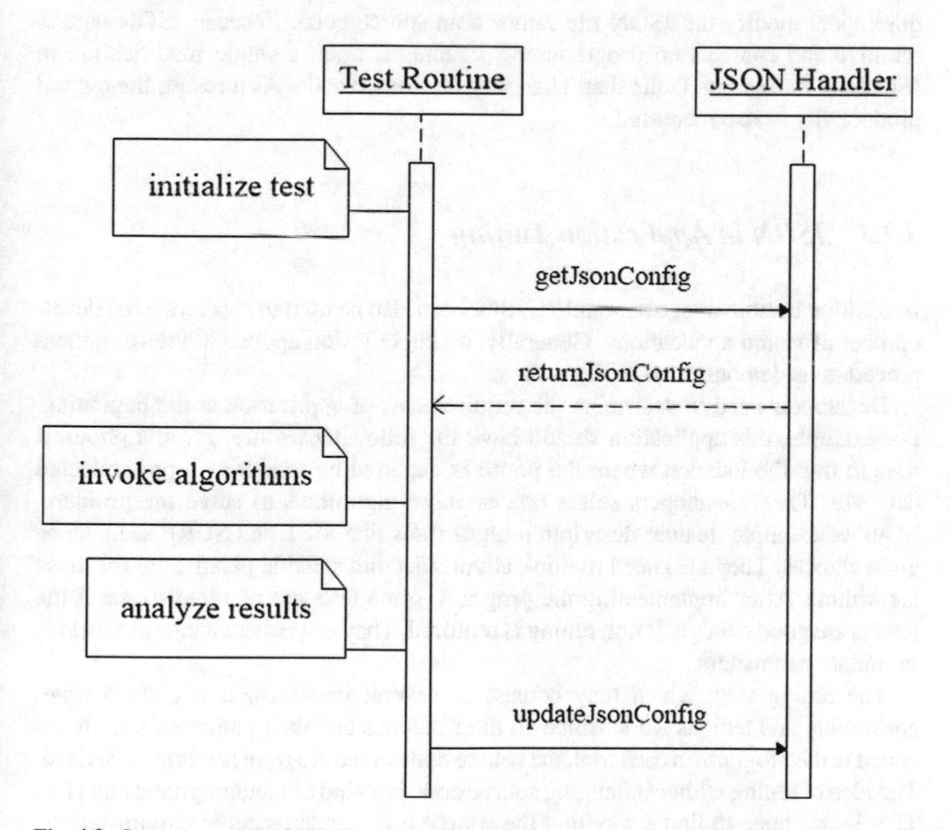

Fig. 4.2 Sequence graph of JSON in application

Figure 4.2 demonstrates one iteration of a testing loop. In each iteration, two classes of instances are active: a test routine and a JSON handler. When the test routine is ready, it asks for the JSON configuration from the JSON handler. Based on returned data, the test routine invokes selected algorithm with given parameters. Then it compares the test results with the desired output. Based on analysis and some adaptive rules, the test routine updates the JSON configuration, and prepares for the next round's test.

By applying this method, there is no need to adjust the source code each time in tuning step. All the modifications are done in the JSON file. As a result, every iteration has its behavior determined by JSON configuration, and tuning is achieved without even stopping the testing program. Under this circumstances, developers do not need to search the reference for algorithms and parameters, since all of them are written in a JSON template file, with both names and default values. Also there is no need to try a different algorithm or function manually, because it is possible to try a bunch of candidates at the same time, and report the best configuration.

Sometimes, developers know which parameter is the key of performance, so it is better for humans to advice machines in the tuning process. In this scenario, developers modify the JSON file rather than source code. Because JSON data is intuitive and contains no programming structures, often a single modification in JSON file makes less faults than changing the source code. As a result, the overall productivity is also increased.

4.2.2 JSON in Application Tuning

In addition to store image annotation, JSON can also be used to accelerate the development of vision applications. Generally, a simple vision application development procedure is demonstrated in Fig. 4.3:

Developers need to determine the requirements of application at the beginning. For example, this application should have the following feature: given a photo, it tries to find the location where the photo is captured by searching a pre-collected database. Then, developers select one or more algorithms to solve the problem. In above example, feature description algorithms like SIFT and SURF seem to be good choices. They also need to think about selecting suitable parameters for those algorithms. After implementing the program, some tests are required to see if the results are good enough. If not, tuning is required. They can either change algorithms or mutate parameters.

The tuning step, which may consist of several iterations, is usually a time-consuming and tedious work. Since all the functions and their parameters are hard-coded in the program, in each trial, the source code of the program needs to be revised. The idea of tuning without changing source code is a kind of metaprogramming [15]. That is, we hope to find a way that the source code, under specific circumstances, can adapt itself. As a result, a developer does not need to repeat changing source code until a satisfied solution is found. To solve the tuning problem, we can apply

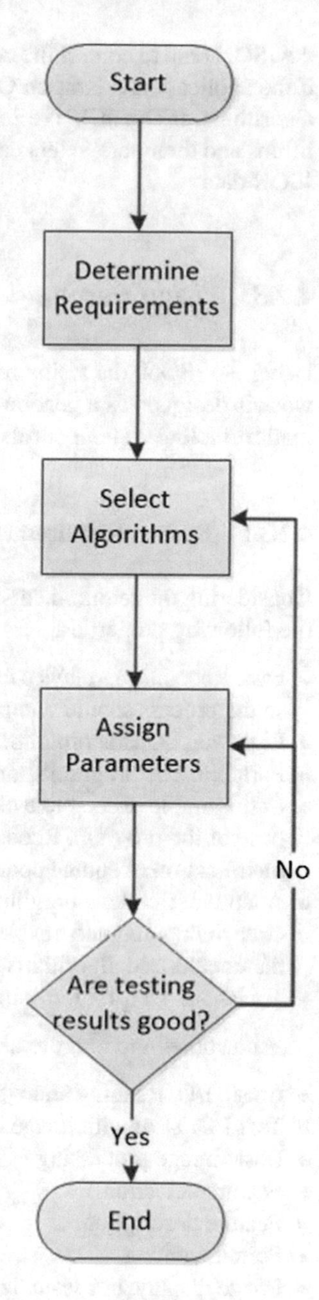

Fig. 4.3 Vision application development procedure

the JSON data to control the behavior of the vision application. Life would be easier if the application is built on OpenCV. One advantage to use OpenCV is that all the algorithms in OpenCV are inherited from cv::Algorithm class, so naturally algorithms and their parameters can be determined in running time, under the control of JSON data.

4.2.3 Vision Recognition Framework

Using the JSON, the vision recognition framework could be designed. This framework is designed as a general vision recognition library, suitable for not only automatic refueling systems but also face detection, scene locating, and other applications.

4.2.3.1 System Requirement

Considering the requirements, the vision recognition framework is required to have the following properties:

- Fast. Recognition plays a crucial role in closed-loop motion control of the robot, so the process should complete in seconds.
- Lightweight. This program runs on embedded system, which is resource limited.
- Portable. The program should be able to run on different devices.
- Configurable. Recognize objects in refueling process should not be the only purpose of the program. Recognize other targets, if required, can be done with little modification of source code.
- Adaptive. Update a program on an embedded system is harder than PC. To accelerate debugging and enable the ability to adjust parameters in testing procedure, parameters and algorithms can be determined at runtime.
- Toolchain. Helpful tools are also required.

And some functions are used in the program design:

- Image I/O. Reading and writing images of various formats.
- Text I/O. Deal with image labels and configuration.
- Basic image processing.
- Feature detection.
- Feature description.
- Point matching.
- Image display. For visualization purpose.
- Label generator. To accelerate image annotation.

4.2.3.2 Annotation Tool

We use OpenCV to build our framework, but some components are still missing. For those images served as references, the program needs to know the area of each target, and compares feature points within that area when identifying an object in test images. And for test images, it is also necessary to locate each target manually, so we can compare the results from our program to the "groundtruth." Make it short, all the images are required to be annotated in some way.

First, the size of image, including the width (number of columns) and height (number of rows). In a dataset the size of image may be consistent so there is no need to compute one by one.

Second, information of objects. Since there may exist multiple objects in a single image, this record should be stored as a list. Each element in the list consists of several sub-records.

- name: the class of the object, such as "station," "machine," "keyboard," etc.
- id: an identity number for that object. It should be unique in this image.
- region: a rectangle which indicates the area of the object. Four values, i.e., "xmin", "ymin", "xmax", and "ymax" are enough.

Despite the annotation for each image, some global-level information is required as well. We would like to provide some hints for the program to find and open images, so folder and file names should be provided. Also notice that the whole image set is separated into two groups, one for reference and the other for testing. So each image's annotation is placed in one of the three directories: "*train*", "*validate*", and " *test*", where "*validate*" is a special testing set for debugging and tuning purpose only.

The operation is simply to add an image, choose a label for the target, and then drag on the image to select a region. If the result is not desired, we can simply delete it or adjust it using arrow keys in the keyboard. When the job is complete for all images, save the results in a text file. Using a script file, a JSON annotation file is saved with well-defined structure. A detailed explanation of JSON is presented in the next section.

In order to parse and modify JSON data in the program, the library RapidJSON is used. RapidJSON is a JSON parser and generator written in C++. It is small and fast. All its functions are packed in header files, so it can be integrated into other programs easily.

In order to retrieve a value in JSON file, the file should be loaded and converted into an internal document structure. Then, via the APIs provided by RapidJSON, any legal value can be retrieved recursively. The process is illustrated in Algorithm 1.

4.2.3.3 Structure of Framework

Based on JSON and OpenCV, the structure of the vision recognition framework can be designed. This framework has a hierarchical structure of four layers. From bottom

Algorithm 1 parse_json_val

Require: A pointer to JSON document *doc*, a vector of string *path*
Ensure: An value in JSON
1: $sub\_doc \leftarrow doc[path.head]$ ▷ go next layer
2: **if** *length*(*path*) is 1 **then**
3: return $sub\_doc.value$
4: **else**
5: return *parse_json_val*(*sub_doc*, *position.tail*)
6: **end if**

to top, they are dependence layer, abstract layer, business logic layer, and application layer.

The dependence layer of the framework consists of OpenCV, RapidJSON, and some other utilities. They provide basic support and functionality. To interact with JSON document, helper functions are written to convert integer, float, and boolean variables to their string representation. Next, to access image information, several small classes such as "ImageSize", "ImageRegion", and "ImageObject" are defined. Finally, some functions are added to filter the matching results.

The second layer is the abstract layer. As we have discussed in the previous section, this framework is highly configurable by adjusting JSON parameters. The abstract layer provides a complete set of interfaces to access and modify data in a JSON document. Functions in this layer include

- File I/O;
- JSON document constructor;
- getter and setter of boolean variable;
- getter and setter of integer variable;
- getter and setter of float variable;
- getter and setter of string variable;
- getter of array size; and
- getter of reference of arbitrary node.

Next is business logic layer. Functions in this layer closely cooperate with two JSON configuration files: a program configuration file and an image annotation file. Information in the first file controls the behavior of the program, and the second one provides metadata and labels. The design of interface and JSON structure must be consistent, otherwise, the framework may not work.

On the top is application layer. It connects to the recognition task directly, and its source code may be changed for different recognition algorithms. Some simple function calls, such as "*loadInfo()*" and "*runTest()*" are provided for automatic testing of the program. Now the JSON-based framework is built and is available in vision recognition.

4.3 Deep Learning-Based Object Recognition

Based on the visual recognition algorithms and software introduced in Sect. 4.1, this section explores how to detect and to classify pedestrian in infrared thermal images. The infrared thermal images can be applied in robots perception systems, as well as the advanced driver assistance system (ADAS) [16] and the surveillance systems. This section demonstrates the use of deep learning in object recognition, which can be also extended to other applications [17].

4.3.1 Logistic Regression-Based Classification

We address the problem of building a pedestrian detector as classification problem. Therefore, given an input as a patch image, the model predicts the probability whether the patch contains a human being or not. We start from logistic regression, which is a simple learning algorithm to make such decision. Then, we introduce an algorithm for features learning based on convolutional neural networks.

As shown in Fig. 4.4, the classifier's input, the *i*th example, represents the image patch, while the output represents the corresponding label.

We use the LSI Far Infrared Pedestrian Dataset (LSIFIR) [18]. In the classification training set each example is a 64×32 image patch. The first step is to unroll it into a vector representation $x \in R^{2048}$, then we add the intercept term (*bias*), so that we have $x \in R^{2049}$. Then we preprocess the data by applying feature normalization, to obtain data with 0 mean and 1 as standard deviation.

$$h_\theta(x) = \frac{1}{1 + \exp(-\theta^\top x)} \tag{4.15}$$

In logistic regression, we use the hypothesis defined in Eq. (4.15) to try to predict the probability that a given example belongs to the 1 class (presence of pedestrian) versus the probability that it belongs to the 0 class (background).

For a set of training examples with binary labels $\{(x^{(i)}, y^{(i)}) : i = 1, \ldots, m\}$ the following cost function measures how well a given hypothesis $h_\theta(x)$ fits our training data:

$$J(\theta) = -\sum_i \left(y^{(i)} \log(h_\theta(x^{(i)})) + (1 - y^{(i)}) \log(1 - h_\theta(x^{(i)}))\right) \tag{4.16}$$

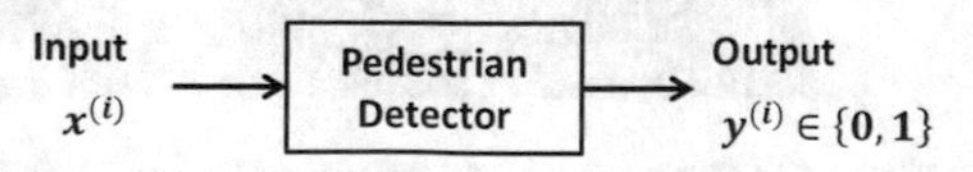

Fig. 4.4 Pedestrian detector as a classifier. Reprinted from Ref. [17], with permission of Springer

The parameters of the model ($\theta \in R^{2049}$) can be obtained by minimizing the cost function $J(\theta)$, using different tools of optimization, as well as we provide the cost function and the gradient defined in Eq. (4.17).

$$\nabla_\theta J(\theta) = \sum_i x^{(i)}(h_\theta(x^{(i)}) - y^{(i)}) \tag{4.17}$$

After the training phase, we use the learned θ to predict an unseen input x by computing the hypothesis $h_\theta(x)$. The class label is predicted as 1 if $h_\theta(x) \geq 0.5$, and as 0 otherwise.

4.3.1.1 Training and Testing the Logistic Regression Model

The pedestrian infrared dataset is divided into detection dataset and classification dataset, which contains 81 592 images. See Table 4.1 for detail.

In Fig. 4.5, we show some positive examples (pedestrian) and negative examples (background) from the classification training dataset.

To evaluate the learning algorithm, we use F score defined in Eq. (4.18), because an evaluation based on accuracy is not significant as well as the number of positive and negative examples we have is very different.

$$F = \frac{2 \cdot prec \cdot rec}{prec + rec} \tag{4.18}$$

Table 4.1 LSIFIR classification dataset

Data type	Positive examples	Negative examples
Training set	10 208	43 390
Test set	5 944	22 050

Fig. 4.5 Positive (*top*) and negative (*bottom*) training examples. Reprinted from Ref. [17], with permission of Springer

We compute *prec* (precision) and *rec* (recall) as follows:

$$prec = \frac{tp}{tp + fp} \tag{4.19}$$

$$\text{rec} = \frac{tp}{tp + fn} \tag{4.20}$$

where

- *tp* is the number of true positives: the ground truth label says it is a pedestrian and our algorithm correctly classified it as a pedestrian.
- *fp* is the number of false positives: the ground truth label says it is not a pedestrian, but our algorithm incorrectly classified it as a pedestrian.
- *fn* is the number of false negatives: the ground truth label says it is a pedestrian, but our algorithm incorrectly classified it as nonpedestrian.

After training the model, we compute different evaluation parameters as shown in Table 4.2. We conclude that the model predicts well for both training and test sets.

We have trained different models based on the maximum number of iterations, then the evaluation parameters are computed for training and test sets. The results are shown in Fig. 4.6.

4.3.2 Convolutional Neural Network (CNN)-Based Classification

CNN is very similar to ordinary neural networks with a specialized connectivity structure. Convolutional neural networks were inspired by the visual system structure. In addition to this, CNN-based pattern recognition systems are among the best performing systems [19]. Unlike standard multilayer neural networks, CNN is easier to train [19], LeCun et al. [20] designed and trained a convolutional network using the error gradient, obtaining state-of-the-art performance on several pattern recognition tasks. Recently, very deep convolutional neural networks are designed and trained successfully and won many computer vision challenges [21–23].

In logistic regression model, we use the raw pixels of the image patch as a feature vector to make decision (pedestrian or no). There exist many kind of features, extracted from images, which can be used as feature vector for a variety of learning algorithms. Features descriptors like SIFT (Scale-Invariant Feature Transform) or

Table 4.2 Logistic regression results

Data type	Accuracy	tp	fp	fn	prec	rec	*F* score
Training set	98.04	9540	380	668	0.9616	0.9345	0.9479
Test set	96.94	5236	146	708	0.9728	0.8808	0.9245

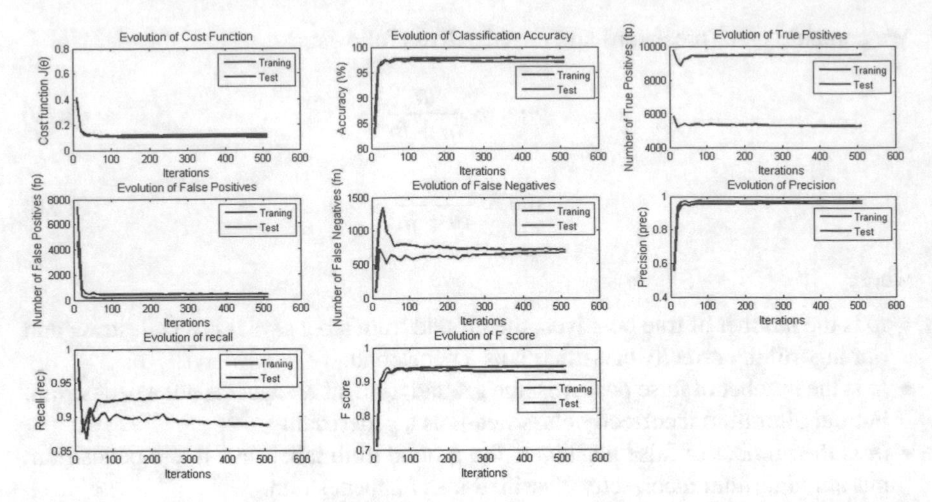

Fig. 4.6 Evaluation parameters for logistic regression-based classification. Reprinted from Ref. [17], with permission of Springer

HOG (Histogram of Oriented Gradients) are widely used in many different computer vision tasks.

Unlike traditional object recognition systems which use hand-designed features, CNN model learns new features, hence hopefully it may be suitable for infrared images. The reasons why we adopt a CNN-based learning features approach are twofold: First, we can learn more higher representations as well as we go deep through the network; Second, infrared images are different from visible images, thus we need appropriate features. Now we ignore how to design such features, and instead, we try to learn them.

The input image will be transformed from raw pixels to new feature maps as long as it passes through different layers [24]. Figure 4.7 shows the neurons activations in each layer of our trained CNN model, starting from the input image patch, going through different feature maps (we plot only the first six feature maps of each layer), ending by the class score given by the output unit.

4.3.2.1 CNN Layers

To build a CNN architecture, there are mainly three different types: convolutional layer (CONV), subsampling layer (SUBS), and fully connected layer (FC). Then we stack these layers to form a full CNN architecture.

- INPUT [$64 \times 32 \times 1$] holds the input image, in this case it is one-channel 64×32 image.

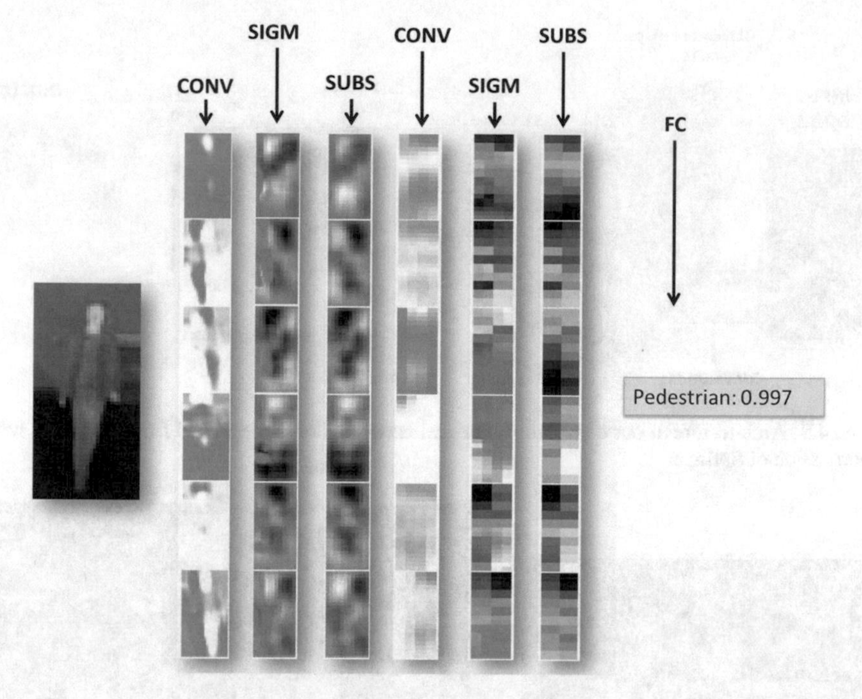

Fig. 4.7 The activations of an example CNN architecture. Reprinted from Ref. [17], with permission of Springer

- CONV layer convolves the input image with the learned convolutional filters. In our case we have six kernels (9×9), so the resulting feature maps will have the dimension of [$56 \times 24 \times 6$].
- SIGM layer will apply nonlinear function to the previous layer elements such as sigmoid function. Therefore, this layer has the same size as the CONV layer.
- SUBS layer will perform a downsampling operation to the input feature maps. The resulting size is [$28 \times 12 \times 6$].
- FC layer will compute the class score, like regular neural networks. In this layer, each neuron is connected to all the neurons of previous layer.

4.3.2.2 Architecture of Our CNN Model

We adopt a similar architecture to the one developed by LeCun for document recognition [20]. In our model shown in Fig. 4.8, we use a [6c 2s 12c 2s] architecture. In the first convolution layer, we use six filters with 9×9 kernel size. While in the second convolution layer, we use 12 filters with same kernel size. For both subsampling layers, we downsample the input one time. So that one feature map of size 20×4 became a 10×2 feature map. The following paragraphs describe in detail our architecture.

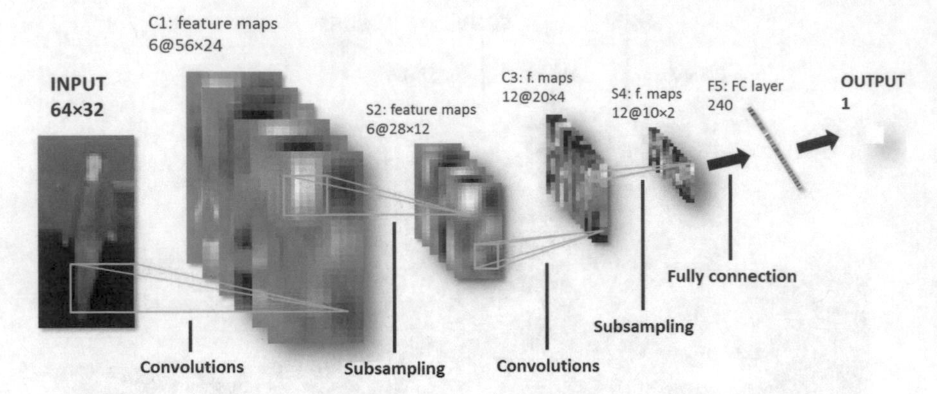

Fig. 4.8 Architecture of our convolutional neural network model. Reprinted from Ref. [17], with permission of Springer

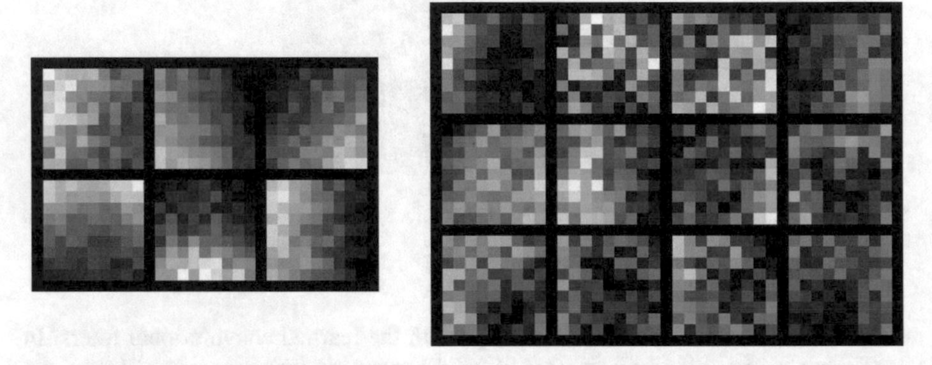

Fig. 4.9 Visualization of learned kernels example, first convolution filters (*left*), second convolution filters (*right*). Reprinted from Ref. [17], with permission of Springer

Layer C1 is a convolutional layer with six feature maps. Each unit in each feature map is connected to a 9×9 neighborhood in the input. The size of the feature maps is 56×24 which prevents connection from the input from falling off the boundary. An example of learned filters associated to this layer is illustrated in Fig. 4.9, the left picture.

Layer S2 is a subsampling layer with six feature maps. Each unit in each feature map is connected to a 2×2 neighborhood in the corresponding feature map C1. The size of feature map is 28×12, and it has half the number of width and height as feature maps in C1 (downsampling operation).

Layer C3 is a convolutional layer with 12 feature maps. Each unit in each feature map is connected to several 9×9 neighborhoods. The size of the feature map is 20×4. Figure 4.9, the right picture shows an example of learned filters in this layer.

Layer S4 is subsampling layer with 12 feature maps. Each unit in each feature map is connected to 2×2 neighborhood in the corresponding feature map C3. The size of the feature map is 10×2.

Layer F5 is a fully connected layer with 240 units (in C4 there is 12 10×2 feature maps). The output represents the class score or the probability that an input image patch is considered as pedestrian.

4.3.2.3 Training and Testing the CNN Model

The research in [20] explained how to train a convolutional networks by back-propagating the classification error. In our work, we use the same classification dataset (LSIFIR) which is used to train the logistic regression model, described in Sect. 4.3.1.1. The results for the training and the test set are described in Table 4.3.

The results show that the model fits well our training set and also generalizes well in the test set, which proves that the model does not overfit the training data. Moreover, the CNN-based classification performs better than logistic regression model for both training and test set.

The evaluation parameters for the CNN-based classification are computed for both training and test sets, after training different models based on the number of iterations. The results are shown in Fig. 4.10.

4.3.3 Detection

Given an infrared thermal image with any size contains or not one or several pedestrians, the goal is to detect all the persons presented in the image and bound boxes (Fig. 4.11).

To get advantage of the pedestrian detector developed previously, we use a technique of sliding window. We fix the widow size to 64×32, the same size as the patch used for classification. Then we classify the patch using CNN or logistic regression. The output of a classifier is stored in an image at the same location of the patch, an example is shown in Fig. 4.12.

In the detection image the white regions correspond to the pedestrian. Therefore, the whiter the region the larger the probability that it contains a pedestrian is big,

Table 4.3 CNN based classification results

Data type	Accuracy	tp	fp	fn	prec	rec	*F* score
Training set	99.60	10047	51	161	0.9949	0.9842	0.9895
Test set	99.13	5730	29	214	0.9949	0.9639	0.9792

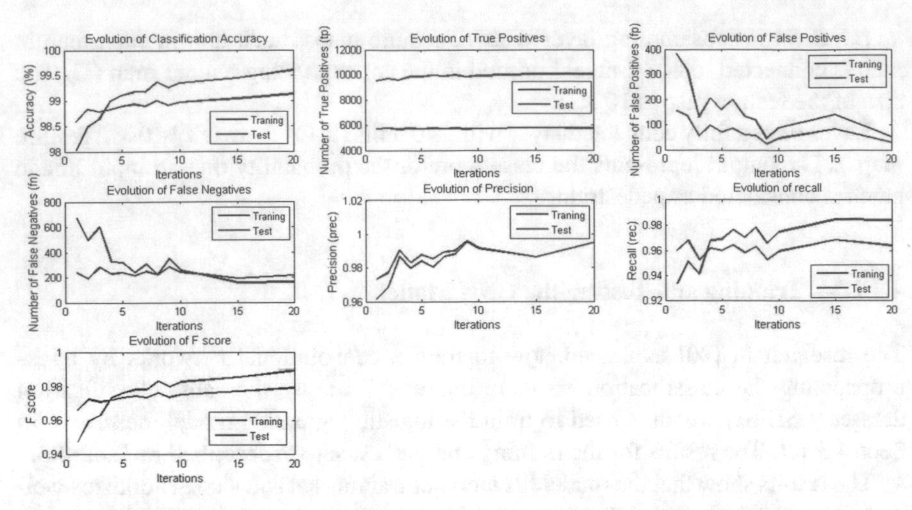

Fig. 4.10 Evaluation parameters for CNN-based classification. Reprinted from Ref. [17], with permission of Springer

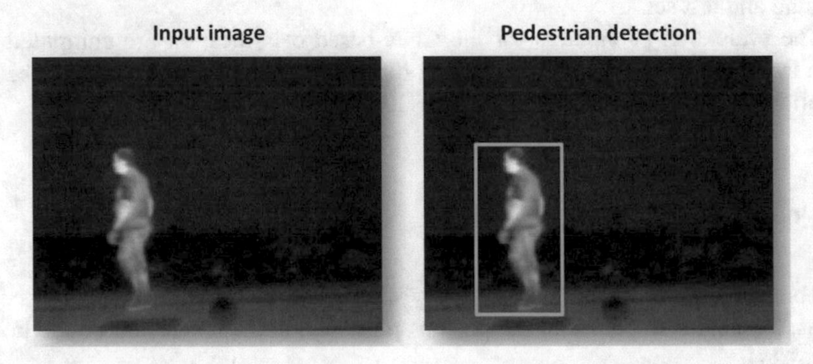

Fig. 4.11 Example of pedestrian detection, (*left*) the input image, (*right*) pedestrian detection ground truth

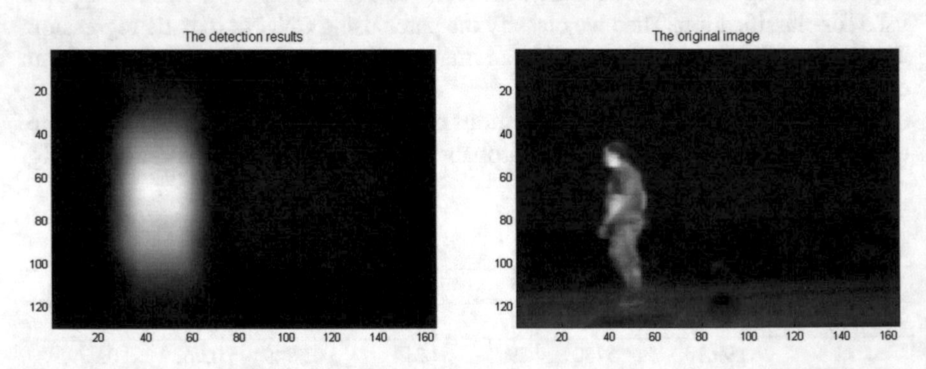

Fig. 4.12 An example of detection using sliding window. Reprinted from Ref. [17], with permission of Springer

and more the region is black more the pedestrian detection probability is low (the background probability is high).

To localize the pedestrian in the image, we estimate the contour of the white region in the detection image and we approximate it to a bound box as shown in Fig. 4.13. Then the coordinates of this box will be projected in the original image to get the location of region of interest.

In general, for the images where the region of interest is bigger than the patch size, we first perform a downsampling to the input image to get different images at different scales. Then, for each resulting image we repeat the procedure (sliding window and classification), so that we can detect pedestrian at different scales.

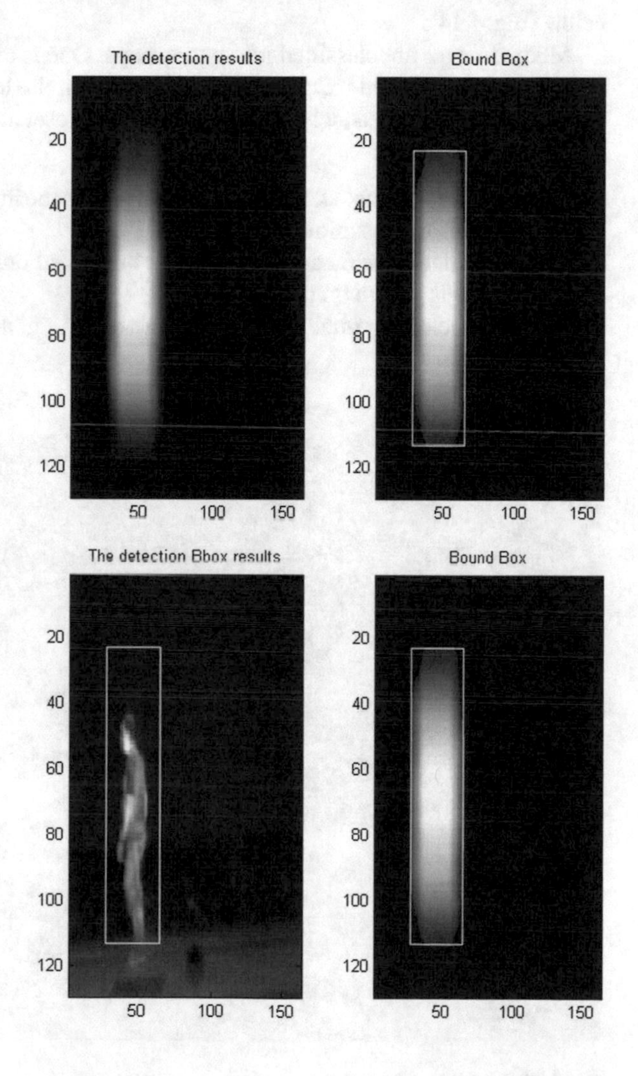

Fig. 4.13 An example of detected pedestrian localization. Reprinted from Ref. [17], with permission of Springer

4.4 Tracking a Single Moving Object

4.4.1 Data Collection

In Sect. 4.3, we have investigated the object detection problems using the deep learning algorithms. This section will show an approach for tracking single moving objects based on JSON visual recognition framework described in Sect. 4.2.

To test the object recognition performance, a total of 179 images collected from five gas stations are used. These images are captured with a portable camera of 1920×1080 resolution. The locations of those gas stations are shown in the map below (Fig. 4.14).

All the images are classified into two groups. One is reference group, containing eight manually selected images, all the others are in the test group.

The following targets, whose numbers in the dataset are counted in Table 4.4, are labeled in each image:

- Gas station, labeled as "station." In order to reduce the interference of other objects, only the roof of the station is marked.
- Oiling machine, labeled as "machine." It is marked only when the full outline of the machine is shown on the image.
- Gas gun, labeled as "gun." To simplify the matching, a group of guns are marked in one region.

Fig. 4.14 Locations of gas station

Table 4.4 Number of objects in dataset

Gas station	Oiling machine	Gas gun	Card slot	Number keyboard
23	66	59	42	44

- Card slot, labeled as "slot." Only one slot is marked in each image.
- Number keyboard, labeled as "keyboard." Only one keyboard is marked in each image.

4.4.2 Recognition Algorithm

The recognition Algorithms 2 and 3 are designed for the task.

Algorithm 2 recog_test

Require: A set of reference images *q_mats*, a set of test images *t_mats*, regions of targets for reference images *q_rois*, regions of targets for test images *t_rois*.

```
1: for each image in q_mats do                                      ▷ q_mats[i]
2:    f_refs[i] ← feature points
3:    d_refs[i] ← corresponding feature vectors
4: end for
5: start timer
6: for each image in t_mats do                                      ▷ t_mats[i]
7:    f_tests[i] ← feature points
8:    d_tests[i] ← corresponding feature vectors
9:    for j in range(len(d_refs)) do
10:        test_regions ←
11:        recog_each_pair(f_refs[j], d_refs[j], q_rois[j], f_tests[i], d_tests[i])
12:        compare the results between test_region and t_rois
13:    end for
14: end for
15: stop timer
```

From the algorithm, it is clear to see that six parameters affect the result of judgment.

(i) *nmatches*. It represents the maximum number of reserved matching pairs for analysis. In practice, there are usually hundreds of feature points in an image region. It is not a good idea to take all of them into account, since there are too many redundancies. So in this step, only a small portion of best matches are reserved for later analysis. In the experiment, *nmatches* is set to 100. If there are less than 100 matches in a region, *nmatches* is set to that value.

(ii) *min_matches*. Although we do not need too many matches, we do need a few to come out with the result. If the number of matches between the region in

Algorithm 3 recog_each_pair

Require: Points and description of reference image *f_ref*, *d_ref*, points and description of test image *f_test*, *d_test*, regions of targets in reference image *query_roi*.

Ensure: A vector of image region *roi* for each target in test image.

```
1: set variable nmatches                                 ▷ max number of matching for analysis
2: set constant min_matches                              ▷ min number of matching for analysis
3: set constant max_var                                  ▷ max variance allowed for matches
4: set constant match_ratio                              ▷ min matching ratio after refinement
5: set constant dense_ratio                              ▷ percent of points ignored in margin
6: set constant scale_const                              ▷ max scale change between reference and test
7: ntargets ← len(query_roi)                             ▷ number of targets in reference image
8: (xminq, xmaxq, yminq, ymaxq) ← unpack(query_roi)
9: idx = 0                                               ▷ loop variable
10: while idx < ntargets do
11:     restore nmatches to default value
12:     f_query ← feature points in query_rio[idx]
13:     d_query ← corresponding vectors to f_query
14:     if d_query is empty then                         ▷ judge 1: any feature point in region?
15:         idx ← idx + 1                                ▷ skip this target
16:         continue
17:     end if
18:     (matches, scores) ← match(d_query, d_test)       ▷ match two images
19:     nmatches ← min(nmatches, len(matches))
20:     if nmatches < min_matches then                   ▷ judge 2: enough matches?
21:         idx ← idx + 1
22:         continue
23:     end if
24:     reserved ← sort(scores)[0 : nmatches]            ▷ find best nmatches pairs
25:     matches ← filter(matches, reserved)
26:     (xq, yq) ← coordinates of points in f_ref
27:     (xt, yt) ← coordinates of points in f_test
28:     k ← abs((yt − yq)/(xt − xq))                     ▷ compute the slopes of matching
29:     median_k ← median(k)                             ▷ compute the median of slopes
30:     var_k ← var(k)                                   ▷ compute the variance of slopes
31:     if var_k > max_var then                          ▷ judge 3: large variant?
32:         idx ← idx + 1
33:         continue
34:     end if
35:     up_k ← median_k + 2 ∗ var_k
36:     down_k ← median_k − 2 ∗ var_k
37:     (xqr, xtr, yqr, ytr) points in range(down_k, up_k)
38:     if len(xqr) < match_ratio ∗ nmatches then        ▷ judge 4: few left?
39:         idx ← idx + 1
40:         continue
41:     end if
```

```
42:     xt_sort ← sort(xtr)
43:     yt_sort ← sort(ytr)
44:     xmint ← x coordinate of which(xt_sort, quantile = dense_ratio)
45:     xmaxt ← x coordinate of which(xt_sort, quantile = 1 − dense_ratio)
46:     ymint ← y coordinate of which(xt_sort, quantile = dense_ratio)
47:     ymaxt ← y coordinate of which(xt_sort, quantile = 1 − dense_ratio)
48:     test_roi ← (xmaxt − xmint, ymaxt − ymint)
49:     if test_roi is larger than scale_const times of query_roi[idx] or smaller than 1/scale_const
    times of that then                                        ▷ judge 5: scale change too large?
50:         idx ← idx + 1
51:         continue
52:     else
53:         roi.append([idx, xmint, xmaxt, ymint, ymaxt])
54:     end if
55:     idx ← idx + 1
56: end while
57: return roi
```

reference image and the test image is less than *min_matches*, it is confident to regard them as dissimilar. min *_matches* is 10 in the experiment.

(iii) *max_var*. Using parallel criterion of matching, an important judgment is to determine if the slopes of matched lines are too scattered, which usually means that the alignment of two are not very well. If the variance of slopes is small, it is generally believed that there is a proper matching between two images. *max_var* is set to 0.01 in the experiment.

(iv) *match_ratio*. When all the nonparallel matching lines have been deleted, the deletion should not affect too many of them. The more the matches left, the higher the probability of correctly detecting an object. *match_ratio* is 0.2 in the experiment.

(v) *dense_ratio*. This parameter is set to deal with the wild points while locating. The concept of "quantile" is used to exclude possible wild points without examining them. For example, if all the points are sorted from left to right, then we can define the points in 10 % → 90 % region as the dense area, and points outside both ends are excluded. In the experiment *filter_ratio* is set to 0.05, which corresponds to a 5 % → 95 % region.

(vi) *scale_const*. The size of reference object and the detected one should be comparable, otherwise it might be a fake target. *scale_const* is set to 3 in this experiment to approve a maximum scale change of 3 times. It means that the target is at most three times of the reference and at least one-third times of that.

To evaluate the performance of the system, various indices are evaluated.

First, detecting objects is a classification problem. For each test image and each object, the program makes two predictions: positive or negative. At the same time, the image holds the ground truth: true or false. The prediction and the ground truth form four quadrants: true positive, true negative, false positive, and false negative. Based on the four quadrants, three important indices are defined [25].

$$accuracy = \frac{true\ positive + false\ negative}{all\ samples} \tag{4.21}$$

$$precision = \frac{true\ positive}{true\ positive + false\ positive} \tag{4.22}$$

$$recall = \frac{true\ positive}{true\ positive + false\ negative}. \tag{4.23}$$

Here, *accuracy* evaluates the percentage of correct predictions of the algorithm. *precision* is a description of percentage of relevant results versus irrelevant ones. Lastly, *recall* shows the total percentage of relevant results returned by the algorithm. In the dataset, for each object, the proportion of images that include the object is small, so *recall* is not that meaningful and thus deleted.

Second, when estimating the correctness of locating, a simple indicator called *coverage* is proposed. If the area of labeled region is A_r, the area of located region is A_p, and the intersection between two is A_i, then coverage is defined as

$$coverage = \frac{A_i^2}{A_r \times A_p}. \tag{4.24}$$

The bigger the coverage is, the better the performance will be achieved.

Third, since our program runs on embedded system, the time of computation is a critical indicator.

$$time\ elapse = \frac{total\ running\ time}{number\ of\ images}. \tag{4.25}$$

This indicator is only comparable under certain hardware.

4.4.3 Analysis of Results

The test group contains 171 images, each of them contains one or more objects.

Using the SIFT algorithm for feature detection and description, and the recognition algorithm described in Sect. 4.4.2, the results in Table 4.5 are achieved.

Figures 4.15, 4.16 and 4.17 are some visualized images of the results.

Now conclusions can be made by summarizing the experimental results. First, point mapping method has strong invariance over different conditions, so a small portion of typical reference images is able to recognize objects in a large test dataset.

Second, larger objects are easier to be recognized. This is a property of both point mapping method and our recognition algorithm. Big objects such as gas station and oiling machine occupy large areas of images, so the number of feature points in a region are usually large. As a result, the number of matches between the region in reference and the test image are considerable. Notice that *min_matches* sets a lower limit for the number of matches, it is possible that small objects like card slot and keyboard are easier to be ignored.

Table 4.5 Test results

Class	TP	TN	FP	FN	accuracy (%)	precision (%)	coverage(%)
Gas station	21	2	0	148	98.8	100	64.22
Oiling machine	61	5	6	99	93.6	91.0	51.41
Gas gun	39	20	3	109	86.5	92.8	57.77
Card slot	8	34	0	129	80.1	100	37.29
Number keyboard	32	12	2	125	91.8	94.1	43.34

Fig. 4.15 Matching of oiling machine

Fig. 4.16 Matching of gas guns

Fig. 4.17 Matching of number keyboard

Third, all the objects have high precision index. This is an advantage of point mapping method. For two images, due to a variety of reasons, a good matching of the same object may not be found; however, if a good matching can be found, the similarity of two regions is guaranteed, so it is confident that they are the same object. This property is useful in practice since the result of recognition rarely confuses the system. It means that if an object is not found, the system will change its position and try again; if an object is found, the system will hardly act wrongly.

Fourth, the area of recognition is usually smaller than the desired one. It is a direct result from *dense_ratio*. If all the matched points are examined and brought back to the image, the result of locating should be much better.

Fifth, efficiency is not bad. The time to recognize can be separated into two parts. The first part is loading and computing the feature points and vectors for the test image. On average, it takes about 2.163 s. The second part is running the algorithm for each pair of query image and test image. Each run takes about 0.101 s an average of 0.101 s. Considering there are eight reference images, the second part runs on 0.808 s. The total running time for each judgment is 2.971 s, or approximately 3 s.

The experiment is conducted on a PC with an Intel Core i7-3517U processor and 4 GB memory. A typical embedded system does not have such computation resource. However, the size of all images are 1920×1080 pixels. If the images are a quarter of the original and the speed of processor is also a quarter of this one, the time consumed should be the same. So with small images, it is possible to complete the recognition in just several seconds.

4.5 Tracking Multiple Moving Objects

In Sect. 4.4, based on the recognition algorithms, tracking of single moving object is performed. This section will extend the tracking to multiple moving objects using the particle swarm optimization (PSO) method [26].

4.5.1 PSO Algorithms

The *PSO* [27], developed by Kennedy and Eberhart in 1995, originates from the simulation of the behavior of birds and fish. Because of easy implementation and quick convergence into the acceptable solution, *PSO* has received increasing attention in the community of evolutionary computation and computer science, and has been particularly utilized in nonlinear function optimization, constrained optimization, feedforward and recurrent neural networks [28–30], multi-objective optimization [31], multi-model optimization [32–34], dynamic objective optimization [35, 36], PID control [37], electricity optimization and control [38–40], electromagnetic [41, 42], classification and feature selection [43, 44], etc. Recently, *PSO* has been successfully applied to address the problems of locating and tracking the targets in the dynamic environment.

To illustrate the standard *PSO* method in a better manner, we first introduce the following notations. In an n-dimensional solution space Ω, let the position and the velocity of the ith particle be $X_i = (x_{i1}, x_{i2}, \ldots, x_{in})$ and $V_i = (v_{i1}, v_{i2}, \ldots, v_{in})$, respectively. Let $P_i = (p_{i1}, p_{i2}, \ldots, p_{in})$ and $G = (g_1, g_2, \ldots, g_n)$ be the personal best previous position of the ith particle and the best previous position of all particles, respectively. Cognitive factor c_1 affects the impact by the previous best position of each particle, and social factor c_2 influences the impact by the best position of all particles. In addition, the uniformly distributed random numbers including r_1 and r_2 are in the range from 0 to 1.

The canonical *PSO* algorithm [45] can be briefly summarized as

(i) Initialize the parameters including the particle's position, the particle's velocity, inertia weight, acceleration coefficients, etc.

(ii) Calculate every particle's fitness $F(X_i(t))$ and modify the best previous position $P_i(t+1)$ of each particle by

$$P_i(t+1) = \begin{cases} X_i(t) & \text{if } F(X_i(t)) < F(P_i(t)) \\ P_i(t) & \text{if } F(X_i(t)) \geq F(P_i(t)) \end{cases} \tag{4.26}$$

where t is denoted as the number of generation.

(iii) Update the currently best position of all particles by

$$G(t+1) = \arg\min\{F(P_1(t)), F(P_2(t)), \ldots, F(P_n(t))\}. \tag{4.27}$$

(iv) As is well known, the velocity and the position of each particle are generally modified by inertia weight method or constriction factor method.

First, without loss of generality, the velocity and the position in the inertia weight method are described by

$$V_{ij}(t+1) = \omega(t)V_{ij}(t) + c_1r_1(P_i(t) - X_{ij}(t)) + c_2r_2(G(t) - X_{ij}(t)) \tag{4.28}$$

$$X_{ij}(t+1) = X_{ij}(t) + V_{ij}(t+1) \tag{4.29}$$

where the inertia weight ω denotes the momentum coefficient of the previous velocity. Large ω in the early search stage promotes the powerful exploration ability, while small ω in the latter search stage promotes the powerful exploitation ability. Moreover, the first part in Eq. (4.28) represents the momentum of the previous velocity, while the second part denotes the personal thinking of each particle, and the third part denotes the cooperation and the competition among all particles.

Second, the velocity and the position in the constriction factor method are described by

$$V_{ij}(t+1) = \chi(V(t)_{ij} + c_1 r_1 (P_i(t) - X_{ij}(t)) + c_2 r_2 (G(t) - X_{ij}(t))) \quad (4.30)$$

$$X_{ij}(t+1) = X_{ij}(t) + V_{ij}(t+1) \quad (4.31)$$

where constriction factor χ is $\frac{2}{|2-\varphi-\sqrt{\varphi^2-4\varphi}|}$ with $\varphi = c_1 + c_2$. Typically, it is in the low-dimensional search space that the good setting parameters are set to $\chi = 0.729$, $\varphi = 4.1$, etc.

(v) Go to (ii) until the stopping criterion is satisfied.

The trade-off between the global search ability and the local search ability plays an important role in searching for thc acceptable or global optimum in particle swarm optimization method.

Concerning the practical optimization problem, it is very difficult to design the normal balance between the global search ability and the local search ability. If the local search ability is very powerful in the whole process, premature convergence, which is a challenging problem in particle swarm optimization method, may appear in the search process. The phenomenon can be summarized as follows. In the early search process, all particles get trapped into the local optimum and do not escape from the local optimum in the later generations. Additionally, the hypersurface of optimization function greatly affects premature convergence. If the adjacent region of global optimum in the optimization surface is large, the probability of appearing premature convergence may be very low, so deflecting technology and stretching technology, which enlarge the adjacent region of the global optimum, are prevalently utilized to improve the efficiency. Therefore, the trade-off between exploration ability and exploitation ability associating to the hypersurface of objective function plays a great role in avoiding premature convergence. Particularly, the proposed objective function in this section can reduce the number of local optimum and smooth the optimization surface, probably overcoming premature convergence in the swarm optimization.

Furthermore, convergence analysis of *PSO* algorithm is crucial for parameter selection and the convergence speed of the particles. In order to analyze the convergence of PSO method by the state space method, we assume that $\phi_1(t) = c_1 r_1$ and $\phi_2(t) = c_2 r_2$, and then $\phi(t) = \phi_1(t) + \phi_2(t)$. Let $P_d(t)$ and $y(t)$ be $\frac{\phi_1(t)p_1(t)+\phi_2(t)p_2(t)}{\phi(t)}$ and $P_d(t) - X(t)$, respectively. Notice that Eqs. (4.28) and (4.29) can be rewritten as

$$v(t+1) = \omega(t)v(t) + \phi(t)y(t) \quad (4.32)$$

$$y(t+1) = -\omega(t)v(t) + (1 - \phi(t))y(t). \quad (4.33)$$

For the sake of simplicity, Eqs. (4.32) and (4.33) can be expressed as

$$Y(t+1) = M(t)Y(t) \quad (4.34)$$

where

$$Y(t+1) = \begin{bmatrix} v(t+1) \\ y(t+1) \end{bmatrix}, \ M(t) = \begin{bmatrix} w(t) & \phi(t) \\ -w(t) & 1-\phi(t) \end{bmatrix}, \ Y(t) = \begin{bmatrix} v(t) \\ y(t) \end{bmatrix}. \tag{4.35}$$

Essentially, the mathematical model of *PSO* method is a second-order linear random dynamic system and the stability of *PSO* method depends on the products of the state matrices $M(t)(1 \leq t \leq \infty)$. However, it is hard to give the sufficient and necessary convergence conditions because of the complicated computing products of $M(t)(1 \leq t \leq \infty)$ including the random variable. Poli [46] considers the random factor in particle swarm optimization method to conclude the convergence condition from the expectation and the variation point. Generally speaking, the researchers generally simplify the *PSO*'s model to overcome this challenging problem by considering the random value as a constant value and setting the attractor of dynamic system to the constant attractor [47–49]. Therefore, the simplified model of *PSO* method is

$$Y(t+1) = M\,Y(t). \tag{4.36}$$

Under the assumptions of the constant matrix M and the constant attractor P_d, the eigenvalues of matrix M are calculated by

$$\lambda_{1,2} = \frac{1+\omega-\phi \pm \sqrt{(\phi-\omega-1)^2-4\omega}}{2}. \tag{4.37}$$

If $\Delta = (\phi-\omega-1)^2 - 4\omega \geq 0$ and $|\lambda_{1,2}| < 1$, the eigenvalues $\lambda_{1,2}$ of matrix M are real numbers, resulting in the convergence of all particles. In other words, inertia weight and acceleration coefficients should be yielded by

$$\begin{cases} \omega < \phi + 1 - 2\sqrt{\phi} \\ \omega > \frac{\phi}{2} - 1 \\ \omega < 1 + \phi. \end{cases} \tag{4.38}$$

If $\Delta = (\phi-\omega-1)^2 - 4\omega < 0$ and $|\lambda_{1,2} < |1$, the eigenvalues $\lambda_{1,2}$ are the conjugate complex numbers, leading to oscillation convergence, that is to say, inertia weight and acceleration coefficients should be

$$\phi + 1 - 2\sqrt{\phi} < \omega < 1. \tag{4.39}$$

Note that the particles finally converge into the adjacent region of the acceptable or global solution, the conditions are

$$\begin{cases} \omega < 1 \\ \omega > \frac{\phi}{2} - 1 \\ \phi > 0. \end{cases} \tag{4.40}$$

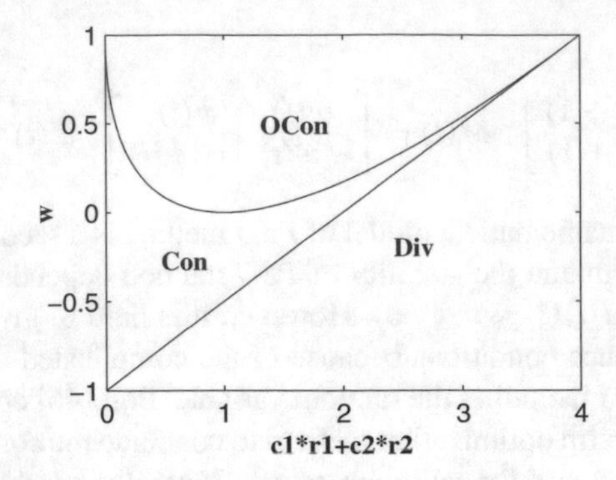

Fig. 4.18 The convergence region of the simplified *PSO*.method. Con region in the figure is the convergence region, that is to say, when the parameters including inertia weight and acceleration coefficients are selected in this region, the simplified *PSO* method can converge into one attractor. OCon region is the oscillation convergence region, and Div region is the divergence region in simplified *PSO* method

According to Eq. (4.40), the convergence region, which denotes the stability condition between inertia weight and acceleration coefficient, is plotted in Fig. 4.18. In terms of convergence conditions of the simplified *PSO* method, several crucial remarks which benefit parameter selection and understand the essence of the simplified *PSO* method, can be concluded as follows.

4.5.2 Objective Function of the Irregular Shape Target

Before locating and tracking the target, target identification is the crucial step in the following experiments. In order to locate irregular targets in the specific environment, objective function, which extracts some valuable pixels of the target's window, only consists of some pixels in the graph. Generally speaking, the general objective function F_{ij} between the subgraph at point (i, j) and benchmark graph is

$$F_{ij}(Y, Y_d) = \sum_{i=1}^{RL} \sum_{j=1}^{RW} \sum_{k=1}^{3} |Y(i, j, k) - Y_d(k)| \tag{4.41}$$

where $Y(i, j, 1)$, $Y(i, j, 2)$, and $Y(i, j, 3)$ denote the pixel's RGB values of the point (i, j), and $Y_d(1)$, $Y_d(2)$, and $Y_d(3)$ denote the pixel's RGB values of the target, respectively. Obviously, $F_{ij}(Y, Y_d)$ denotes the similarity between the target's window and benchmark window.

According to Eq. (4.41), it is hard to discern the target in the noisy environment because the invalid information plays an impact on objective fitness of the target. To cope with this problem, the novel objective function is proposed to select the pixels which are similar to the RGB values of the target.

First, to calculate the similarity of the pixel (p, q), the single pixel's similarity $f_{pq}(1 \leq p \leq RW, 1 \leq q \leq RL)$ is defined as

$$f_{pq} = \sum_{k=1}^{3} |Y(p, q, k) - Y_d(k)|. \tag{4.42}$$

Second, for the sake of extracting the valid pixels, objective fitness of each pixel in the window is sorted by the ascending order, that is,

$$f_1' < f_2' < f_3' < \cdots < f_{RW \times RL}'. \tag{4.43}$$

Then, the modified objective function $F_{i,j}'$ can be expressed as

$$F_{i,j}' = \sum_{k'=1}^{\text{Number}_1} f_{k'}' \tag{4.44}$$

that is to say, this objective function only selects from 1 pixel to Number_1 pixels in target's window, where

$$\text{Number}_1 = \text{coefficient}_1 \times RW \times RL. \tag{4.45}$$

Note that coefficient_1 primarily deletes the unrelated information of the targets and the noisy signal since both of them have a great impact on the target's objective fitness. In summary, information extraction in the target's window can deal with the identification of the eccentric target in the noisy environment.

To demonstrate the effectiveness of the improved objective function, the error surface between both objective functions is compared. First, the optimization surface of the general objective function is plotted in Fig. 4.19, where the size of the window is set to 20×20.

Moreover, the surface of the modified objective function is plotted in Fig. 4.20. Finally, the error surface between the general objective surface and the improved objective surface is illustrated in Fig. 4.21.

4.5.3 *Locating Multiple Targets by Adaptive* PSO *Method*

For the sake of the cooperation in one subgroup and the competition among the subgroups, the interaction matrix is determined by the distances between their positions and the center points of other subgroups, and each subgroup can adjust the particle's

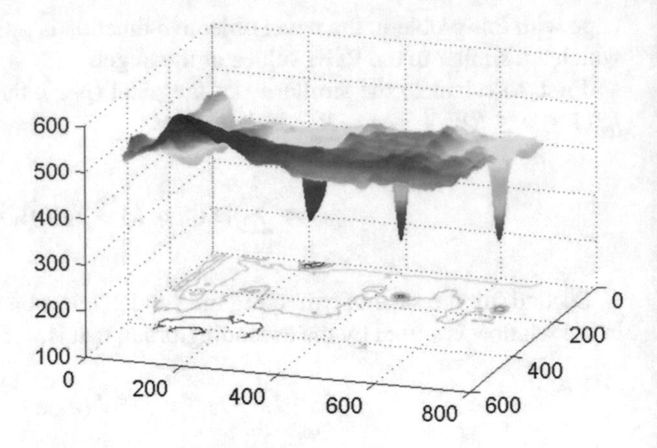

Fig. 4.19 The general objective fitness's surface of the first frame from the camera. According to the pixel's RGB of the ant, the optimization surface, which is shown according to the general *PSO* method, does not own many local optima

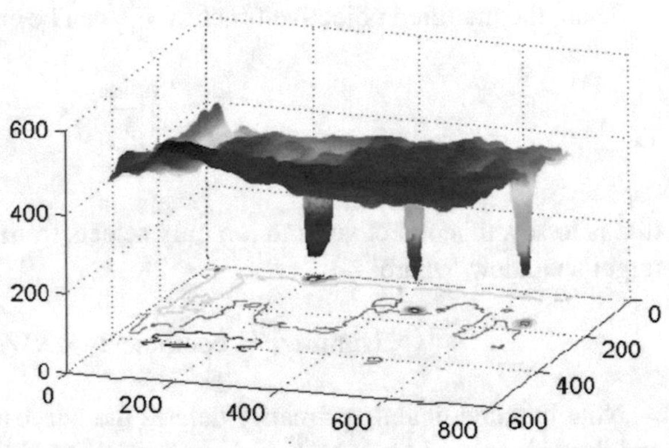

Fig. 4.20 The improved optimization surface calculation by Eq. (4.44). The surface of the improved objective function is better than that of the general objective function since it can enlarge the adjacent region of the targets, leading to high efficiency of locating the targets

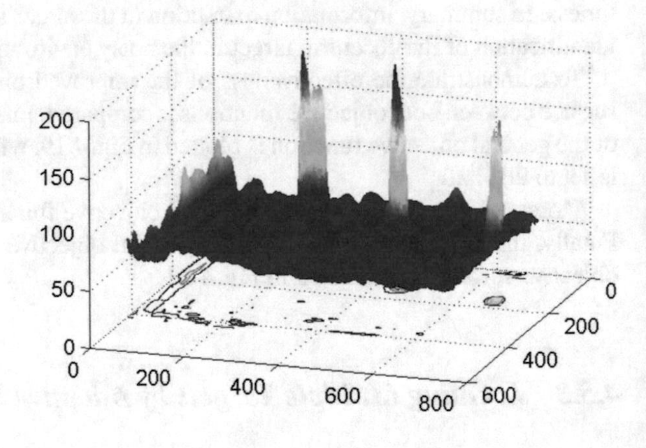

Fig. 4.21 The difference between both fitness landscapes employing by Eqs. (4.41) and (4.44). Compared with Figs. 4.19 and 4.20, the latter method has larger adjacent region than the former one. Similar to deflection operator and stretching operator [35], this method is helpful for locating the targets in the noisy environment

number to improve the efficiency of locating multiple targets from the theoretical and experimental perspectives.

In the case of searching for multiple targets in parallel, the whole swarm is divided into a number of subgroups and each subgroup is in charge of finding one target. To employ smallest number of subgropus in this method, the particles in different subgroups can exclude each other to find all targets. Therefore, the framework among the subgroups from the particles' topology perspective can be expressed as Fig. 4.22.

In summary, the task of locating multiply targets can be divided into one subtask of converging into one target and another subtask of excluding among the subgroups. First, by employing the standard *PSO* method, the particles in one subgroup have the ability to converge into one target. Second, to exclude the particles in the different subgroups, the particles should be initialized by the random value if the particles get into the threshold region of another subgroup.

To converge into the target by one subgroup and repel each other among the subgroups, the interaction matrix A denoting the cooperative or competitive relationship among the subgroups is defined as

$$A = \begin{bmatrix} a_{11} & a_{12} & a_{13} & \cdots & a_{1n} \\ a_{21} & a_{22} & a_{23} & \cdots & a_{2n} \\ a_{31} & a_{32} & a_{33} & \cdots & a_{3n} \\ \cdot & \cdot & \cdot & \cdots & \cdot \\ a_{m1} & a_{m2} & a_{m3} & \cdots & a_{mn} \end{bmatrix} \tag{4.46}$$

where m represents the subgroup number and n represents the whole swarm population number, respectively.

Then the element a_{ij}, which denotes the interaction between the ith particle and the jth central position of each subgroup, is defined as

$$a_{ij} = \begin{cases} 0 & X_i \in G_j \\ 1 & ||(X_i, G_{cj})||_2 < BenchmarkDistance \ and \ X_i \notin G_j \end{cases} \tag{4.47}$$

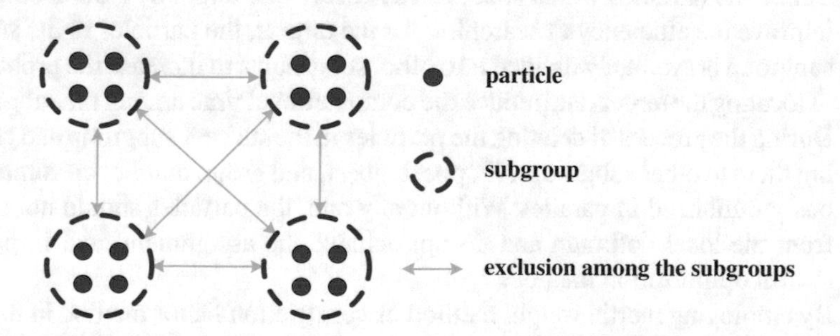

Fig. 4.22 The adaptive PSO's framework among the subgroups. The *red point* only denotes one particle in the subgroup and the *blue circle* denotes one subgroup. By applying *PSO*.method in the subgroup, the particles can search for one target in the first graph. Moreover, the *green arrow* denotes the exclusion interaction among the subgroups

where G_j denotes the set of the jth subgroup and G_{cj} the central position of the jth subgroup. The Euclid-norm distance is denoted by $||.||_2$ and BenchmarkDistance denotes the exclusion threshold distance of each subgroup. If BenchmarkDistance is too large, the efficiency of locating the targets should be low since the particles, having the high probability, are initialized by the random value; if BenchmarkDistance is too small, the goal of excluding among the subgroups will not be achieved and the particles in the different subgroups may converge into one target. Therefore, BenchmarkDistance is vital for achieving the high efficiency of searching for multiple targets.

In terms of the above-mentioned fundamental technologies, we give the detailed description on the adaptive particle swarm optimization method. For one thing, the elements in the particle are composed of the center point of the graph, the window size, objective fitness, and subgroup number.

(i) Initialize the whole population number and subgroup number, the pixel's RGB of benchmark target, the threshold central distance of each subgroup, the size of benchmark target's graph, the range of inertia weight, the ranges of the particle's velocity and position, acceleration coefficients and the maximum iteration time, etc.

(ii) If the particles in the different subgroups converge into one target, the particles in one subgroup should be deleted to achieve no overlapping area between all subgroups.

(iii) According to DistanceLength between the particle's position and the best point of each subgroup, the interaction matrix A is calculated by Eq. (4.47), by denoting the exclusion among the subgroups.

(iv) Calculate the fitness of each particle according to the formula Eq. (4.44). If the fitness of the new position X_i is better than that of $pbest_i$ and the interaction matrix a_{ij} is equal to 0, then $pbest_i$ will be replaced by the new position X_i.

(v) Select the best position $gbest_{1,2...GN}$ of all subgroups in terms of their personal position pbest of each subgroup, where GN is the subgroup number in the whole swarm.

(vi) Delete the particles which successfully locate one target by *PSO* method. To improve the efficiency of searching for the targets, the particles in the success subgroup is averagely divided into other subgroups, to increase the probability of locating the targets and reduce the computational time among the subgroups. During the process of deleting the particles in the success subgroup and assigning them to other subgroups, X_i, pbest, gbest, and group number are simultaneously initialized in parallel. Without any one, the particles should not escape from the local optimum and do not achieve the assignment goal in particle swarm optimization method.

(vii) By employing inertia weight method or constriction factor method in the subgroup, the velocity V and the position X of the particles are updated by Eqs. (4.28) and (4.29), respectively. In general, the inertia weight is linearly decreased in the whole search process.

(viii) Testify the velocity and the position of the particles beyond the setting range $[V_{\min}, V_{\max}]$ and $[X_{\min}, X_{\max}]$. If the position or the velocity of the particles are beyond the range, they should be initialized by the random value.
Notice that some researchers set the maximum or minimum value of the particle's velocity and position when the particles go beyond the setting range. Sometimes, the particles under this condition should converge into the edge of the setting region, leading to the low efficiency. Therefore, the method of initializing random value is selected in adaptive particle swarm optimization method.

(ix) If the exit condition is satisfied, break; otherwise, go to (iii).

4.5.4 *Tracking Multiple Targets by Swarm Optimization*

In order to better understand the adaptive swarm optimization of tracking multiple targets, it is necessary to introduce the mathematical model of tracking one target and the method of tracking the nearby targets. The former can solve the problem of tracking the random behavior objective, while the latter mainly copes with the problem of tracking the conflicting targets.

4.5.4.1 Tracking Mathematical Model

As to the canonical particle swarm optimization, it is difficult to track multiple targets because the particles only converge into the single global optimum corresponding to losing the diversity during the tracking process. Inspired by particle swarm optimization, the mathematical model of tracking the target has the ability of predicting the target's movement and covering the random behavior target. More specifically, the particle's new position is mainly determined by the last best position of its subgroup and the covered radius of the subgroup, hence, the mathematical model can be calculated as

$$X_i(t+1) = C_j(t) + R_0(t) \cdot (2\,\mathrm{rand}() - 1) + \mathrm{coefficient}_2 \cdot (C_j(t) - C_j(t-1)) \tag{4.48}$$

where $C_j(t)$ is the best position of the jth subgroup and $X_i(t)$ is the ith particle's position classified in the jth subgroup. The function rand() can generate the uniform distribution random number between 0 and 1. $R_0(t)$ denotes the covered radius of the subgroup and coefficient$_3$ denotes the inertia weight of the target's previous movement, respectively.

In order to not overlap the region between the subgroups, $R_0(t)$ is less than its smallest distances among other subgroups, and it is defined as

$$R_0(t) = \min_{i,j=1,\ldots,n} 0.5 \cdot \mathrm{Dist}(C_i(t), C_j(t)) \tag{4.49}$$

where $\mathrm{Dist}(C_i(t), C_j(t))$ function is the distance between point $C_i(t)$ and point $C_j(t)$, while n denotes the subgroup number.

Moreover, $R_0(t)$ should be selected between the minimum covered radius $R_{\min}$ and the maximum covered radius $R_{\max}$. The minimum radius $R_{\min}$ is not set to very small because the target may easily escape from the covered region in one subgroup, and $R_{\max}$ plays an important role in scouting the target's behavior. In addition, under the constant particle's number of each subgroup, large $R_{\max}$ may result that the particles do not track the target because the constant particles cannot cover the region of one subgroup. Additionally, coefficient_2 mainly influences the next position of the particle by the previous experience.

According to the success rates in the experiments of tracking the targets, it is a challenging problem to track the conflicting targets since the particles in two subgroups merely converge into one target. To overcome this problem, we propose a novel method for tracking two conflicting targets.

First, the particles in one subgroup should be in charge of tracking the target closing to the center point of two subgroups and the mathematical model is

$$X_i(t+1) = C_j(t) + R_1 \cdot (2\,\mathrm{rand}() - 1) \tag{4.50}$$

where R_1 represents the covered radius of closing subgroup.

Second, the particles in another subgroup search for the target which is far from the center point of two subgroups and prevent the target from escaping the covered region. And this mathematical model can be written as

$$X_i(t+1) = C_j(t) + R_2 \cdot (2\,\mathrm{rand}() - 1) \tag{4.51}$$

where R_2 represents the covered radius of farther subgroup.

Generally speaking, R_2 is much larger than R_1 since the overlapping area between two conflicting subgroups is generally very small. According to the overlapped covered areas of both subgroups, the overlapping ratio ρ between the inner covered region and the outer covered region is calculated by

$$\rho = \frac{R_1^2}{R_2^2} \tag{4.52}$$

where small ρ results in small overlapping area between two subgroups, therefore, one subgroup can search for the closing target and another subgroup can scout the farther target from the center point.

From a graphical perspective, the overlapping region of two groups is influenced by the ratio ρ. The following graphs chiefly concentrate on the comparison on the overlapping region.

According to Fig. 4.23, the overlapping region is mainly influenced by ρ calculated by R_1 and R_2. If the ratio ρ is too large, it is difficult to identify both targets in the chaos region because of the large overlapping region. If the ratio ρ is too small,

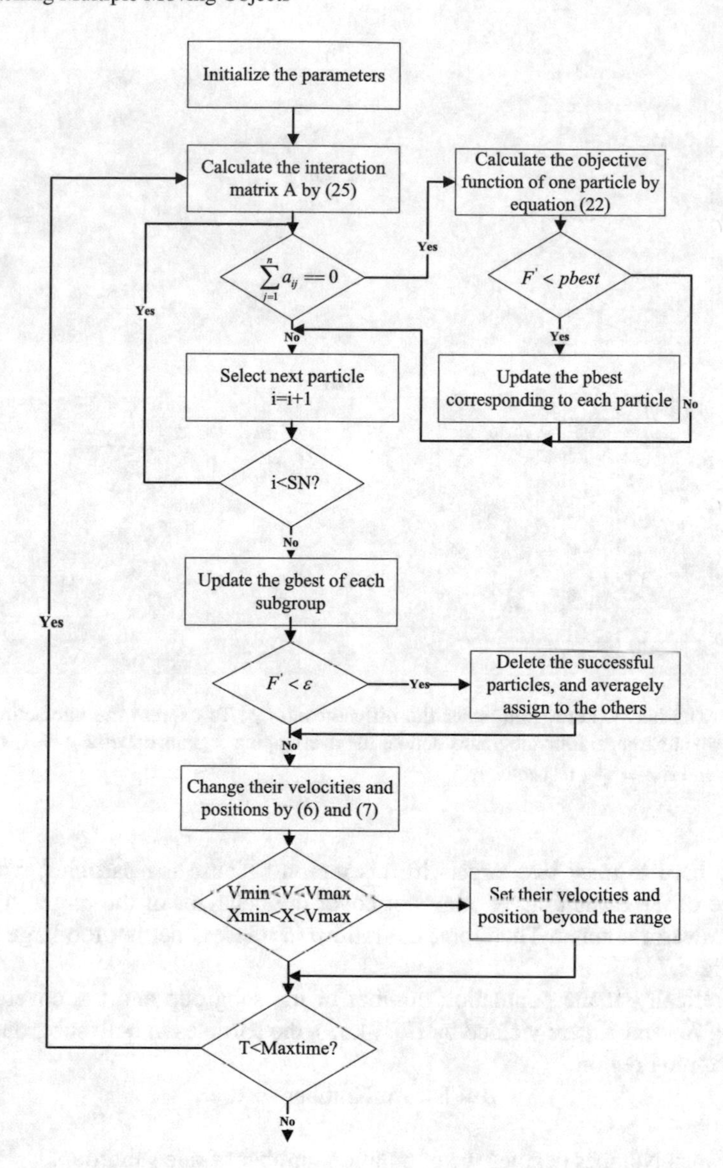

Fig. 4.23 The flow chart of locating multiple targets

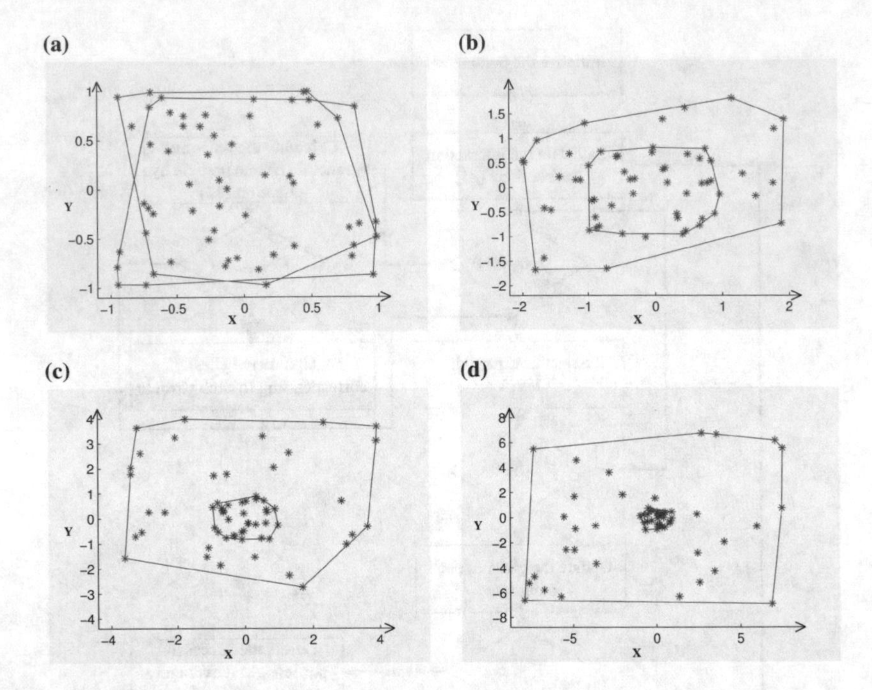

Fig. 4.24 The interaction region under the different ratio ρ. To express the interaction region between two subgroups, four subgraphs denote the overlapping regions under **a** $\rho = 1$, **b** $\rho = \frac{1}{4}$, **c** $\rho = \frac{1}{16}$, and **d** $\rho = \frac{1}{64.}$, respectively.

it is also hard to track two targets in this region because the particles, which are in charge of the distant target, may not cover the behavior of the target under the constant swarm number. Therefore, the ratio ρ is selected neither too large nor too small (Fig. 4.24).

Theoretically, if the population number of the subgroup and the covered radii including $\mathcal{R}_1$ and $\mathcal{R}_2$ are yielded by Eq. (4.53), the particles in both subgroups have no overlapping region.

$$\rho \times \text{SwarmNumber} < 1 \tag{4.53}$$

where SwarmNumber denotes the population number of one subgroup.

4.5.4.2 Algorithm of Tracking Multiple Targets

To successfully track multiple targets in the noisy environment, the random swarm optimization which is inspired by the essence of particle swarm optimization method is summarized as follows. In addition, the flow chart of this algorithm is shown in Fig. 4.25.

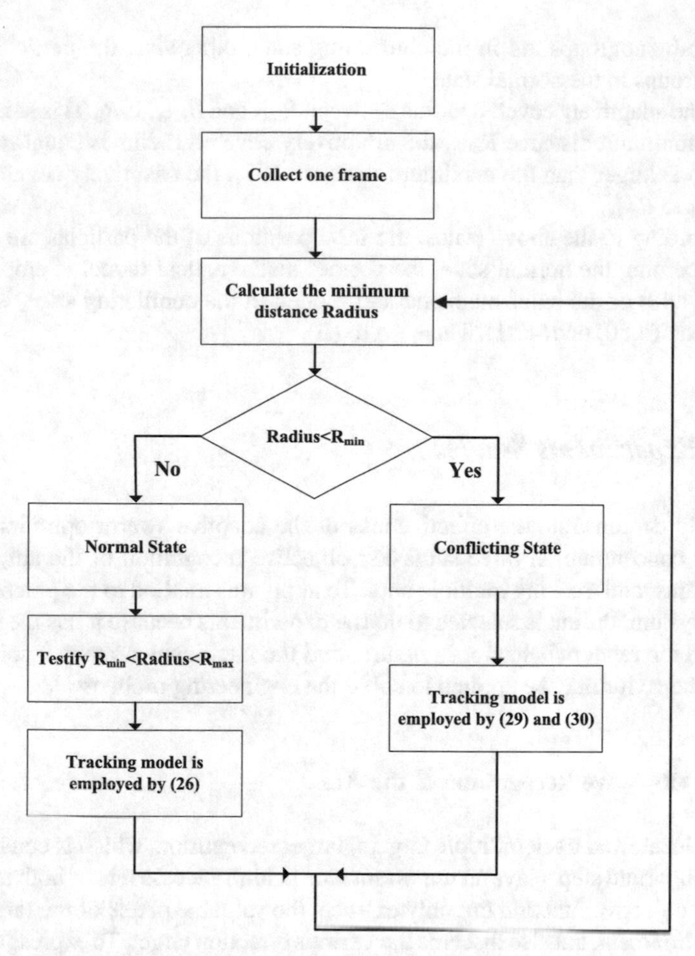

Fig. 4.25 The flow chart of the proposed tracking method. To understand the algorithm from the graphical point, the brief flow chart of tracking multiple targets is plotted in the above figure

The method of tracking multiple targets in the noisy environments is employed by the random swarm optimization, which is inspired by the essence of particle swarm optimization method.

(i) Initialize the subgroup's number, the particle's number of each subgroup, the minimum and maximum covered radius of each subgroup, etc.
(ii) Collect one graph from the camera.
(iii) To guide in setting the adaptively covered radius, $R_0(t)$ in each subgroup is equivalent to the smallest distances among the center positions of each subgroup.
(iv) Select the normal state using Eq. (4.48) or the conflicting state using Eqs. (4.50) and (4.51). When $R_0(t)$ is smaller than the minimum radius $R_{\min}$, the particles

in both subgroups are in the conflicting state, otherwise, the particles in two subgroups in the normal state.

(v) Set the adaptively covered radius between $R_{\min}$ and $R_{\max}$. If $R_0(t)$ is smaller than the minimum distance $R_{\min}$, the adaptively covered radius is equal to $R_{\min}$. If $R_0(t)$ is larger than the maximum distance $R_{\max}$, the adaptively covered radius is set to $R_{\max}$.

(vi) According to the above states, the new positions of the particles are updated. Concerning the normal state, the former mathematical model is employed by Eq. (4.48) or the latter mathematical model, in the conflicting state, is utilized by Eqs. (4.50) and (4.51). Then, go to (ii).

4.5.5 Experiments Studies

In order to demonstrate the effectiveness of the adaptive swarm optimization, we primarily concentrate on three subtasks: objective recognition of the ant, locating multiple ants, and tracking multiple ants. To apply this method to the practical complicated system, the ant is selected to do the experiments because it has the eccentric shape and the random behavior in nature, and the intelligent strategy inspired from the ant's behavior may be applied to solve the engineering problem.

4.5.5.1 Objective Recognition of the Ant

To better locate and track multiple targets, target recognition, which is considered as the first important step, plays an important role in high success rate of both tasks. The proposed objective function not only extracts the valuable pixels of the target in the noisy environment, but also discerns the various direction target. To express the effectiveness of the improved objective function, this subsection primarily concentrates on sensitive analysis regarding the coefficient_1 and the target's rotated angle.

First, coefficient_1 essentially controls to extract the valuable information of the target's window. Large coefficient_1 can get many valid pixels as well as much unrelated pixels from the target's window, while small coefficient_1 gets only the most valuable pixels from the target's window, however, too small coefficient_1 may not extract enough valid pixels the noisy environment. To select the suitable coefficient_1 from 0.05 to 0.50, the valuable information is extracted from the target's window in the following figures, where the window size of the ant is 50×50 and the rotated angle of the ant is selected to 45.

By changing the parameter coefficient_1, different objective fitness may be obtained. From extensive simulations, it can be seen that when coefficient_1 is selected 0.20 or 0.25, the main body of the ant can be efficiently extracted to achieve ant recognition. More exactly, small coefficient_1, corresponding to 0.05, 0.10, 0.15, and 0.20, can get the main body from the ant, while large coefficient_1, setting to 0.40 and 0.45, extracts several noisy signal in the ant's window. Meanwhile, the fitness under the

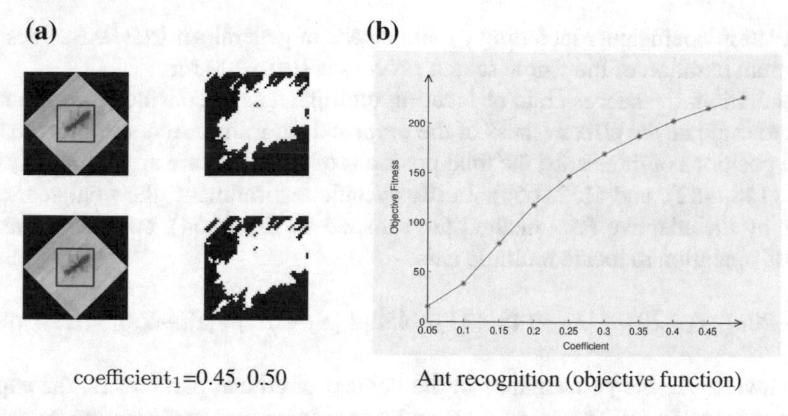

Fig. 4.26 Ant recognition ($\text{coefficient}_{1.} = 0.45, 0.50$) and objective fitness

fine objective function increases from 24 to 234. When the coefficient_1 is too large, the noisy signal plays an important role in the fitness value, describing in Fig. 4.26.

Second, the fine objective function can recognize the ant which consists of the differently rotated angle state. With extensive simulations, when the rotated angle modifies from 0 to 360°, the proposed objective function can successfully extract the main body and the valid position from the ant's window. According to the objective fitness, every state of the differently rotated angle has the closely objective fitness, corresponding to approximate 118.6°. Therefore, this method can extract the useful information from the target with the different directional angles and has the robust capability of object recognition.

Result 1 *From the experimental results, it has been notable that the fine objective function can successfully identify the ant under the noisy environment. The* coefficient_1 *can be selected by neither too large nor too small, and this objective function under the above parameter settings can extract the main body shape of the ant. Moreover, to identify the rotated ants of various rotation angles, the proposed objective function can be demonstrated by the experimental results.*

4.5.5.2 Locating Multiple Targets

For the sake of locating three ants in the first frame from the camera, the parameters in the adaptive particle swarm optimization are set to analysis the performance according to the success rate. The subgroup number is selected to 3 and the population number of each subgroup is set to 10. In terms of the colors of the ant, the pixel's RGB is set to 70, 70, 70, respectively. The central threshold distance of each subgroup is 50 pixels and the size of the ant's window is 20 × 20. Concerning the crucial parameter in the improved objective function, coefficient_1 setting to 0.3 can extract the valid information from the ant's window in the noisy environment. The inertia weight linearly decreases from 0.90 to 0.30 in the whole search process, while acceleration coefficients including c_1 and c_2 are in general set to 2.0. Besides, the maximum iteration of the whole search process is set to 200 runs.

In addition, the success rate of locating multiple ants is considered as the main index to evaluate the effectiveness of the proposed algorithm. According to the final actual positions of three ants, the final positions of three ants are approximately (20, 204), (148, 452), and (152, 656). In the specific environment, the final searching points by the adaptive *PSO* method are satisfied by Eq. (4.54), considering as the success operation to locate multiple ants.

$$|X_1-20|+|Y_1-204|+|X_2-148|+|Y_2-452|+|X_3-152|+|Y_3-656| < 120 \quad (4.54)$$

To investigate the performance in the context of crucial parameters, the experiments concentrate on the success rates under four components: the setting maximum velocity of the particles, the size of the target's window, the threshold radius of the central position in each subgroup, and the maximum iteration of the whole search process.

The particle's maximum velocity $V_{\max}$ representing the maximum search ability of the particle is crucial for the success rate of locating the ants. If the maximum velocity is set to too large, the particles will easily go beyond the setting regions and are initialized by the random value, resulting in low efficiency of locating multiple ants. If the maximum velocity is too small, the particles easily converge into the local optimum in the former search process.

To guide the setting the maximum velocity $V_{\max}$, several velocities are listed by

$$V_{\max 1} = \lceil 0.10 \times (Length\ Width) \rceil = \lceil 0.10 \times (425\ 688) \rceil = [43\ \ 69] \quad (4.55)$$

$$V_{\max 2} = \lceil 0.15 \times (Length\ Width) \rceil = \lceil 0.15 \times (425\ 688) \rceil = [64\ 104] \quad (4.56)$$

$$V_{\max 3} = \lceil 0.20 \times (Length\ Width) \rceil = \lceil 0.20 \times (425\ 688) \rceil = [85\ 138] \quad (4.57)$$

where $\lceil \bullet \rceil$ is the operator of getting the ceiling value. According to the experiments, large $V_{\max}$ can have the powerful global search ability, while small $V_{\max}$ may converge into the local optimum, resulting in premature convergence.

According to Table 4.6, which illustrates the success rates under 100 independent runs when the maximum velocity of the particles is [43 69] and the range of inertia weight is from 0.95 to 0.35 or 0.30. Under these settings, the best success rate of locating multiple ants is 87 % success rate. Roughly speaking, the particles may converge into the local optimum under the relatively small velocity, mainly leading to low efficiency of locating multiple ants. Because of possibly converging into the local optimum, the inertia weight range under Vmax_1 is difficultly selected to achieve the high and stable success rate.

Furthermore, the size of the target's window does not greatly play an important role in the success rate in following experiments, since the fine objective function may extract the valid pixels from the target's window. If the square length of the target's window is too large, the algorithm will have the high probability to search for the

Table 4.6 The success rates of locating the ants when $V_{max\,1.}$ is set to [43 69]

Rate	0.95	0.90	0.85	0.80	0.75	0.70	0.65	0.60	0.55	0.50
0.40	85	85	84	81	83	75	76	82	84	77
0.35	82	79	79	72	78	79	81	78	71	78
0.30	87	83	79	78	78	81	77	83	78	76
0.25	79	83	79	72	80	68	73	79	84	77
0.20	77	79	84	80	75	80	87	73	75	76
0.15	76	86	84	86	73	77	73	75	77	82
0.10	76	82	82	79	76	81	81	76	74	68
0.05	70	82	82	77	81	76	75	80	79	76
0.00	84	70	74	74	85	78	80	78	76	78

The horizon denotes the initial weight from 0.95 to 0.50, and the vertical denotes the final initial weight from 0.40 to 0.00

ants because of reducing multiple local optima of objective function. However, the computational time under large size of the ant's window is too large. If the square length of the target's window is too small, the algorithm can quickly locate the positions of the ants, but the optimization surface should have many local optima and it may suffer from premature convergence, giving rise to the low searching efficiency. To investigate the performance by the different size of the target's window, the square length of the ant changes from 1 to 25 pixels to compare the success rates.

With respect to Fig. 4.27, when the target's window is relative small, the success rate is approximately 80 % success rate. With the increase of the target's window, the success rate slowly increases, nearly equaling to 90 % success rate. More specifically, the best success rate of locating the multiple ants is 98 % success rate when the square length of the target is 25 pixels.

Remark 1 The window size does not greatly affect the success rate in the above experiments. After considering the surface of objective function and the computational time, the window size of the ant is selected to 20 × 20.

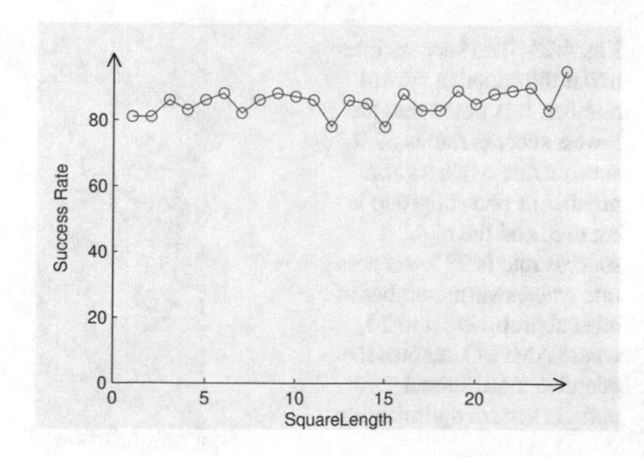

Fig. 4.27 The success results under the adaptive window size. According to the success results of locating the ants, the square length of the ant does not greatly have an influence on the result

Obviously, the population number of each subgroup plays an important role in ant identification. First, many particles in each subgroup obviously result in the high success rate, however, many particles may suffer from the high computational cost and do not apply to the practical system. Second, few particles may not result in the high success rate of locating multiple ants.

As remarked above in Fig. 4.28, the best success rate of locating multiple ants is 98 % when the whole population number is 60, while the worst success rate is 57 % when the whole population number is 15. When the whole population number is less than 33, the success rates quickly shoot up. When the population number is larger than 33, the success rates slowly increase from 33 particles to 60 particles.

Result 2 *The more the population in one subgroup, the higher the success searching rate will be. In the experiments, the best population number is selected to 30 particles, achieving the quick convergence speed and the high success rate of locating the ants.*

To locate all ants by the smallest subgroup number, the adaptive particle swarm optimization can cooperate and compete with each other. To express the competitive mechanism, the threshold region of the central point of each subgroup is crucial for the success rate. If the threshold region of each subgroup is set to too large, the efficiency of searching the ants is too low because the particles should be initialized by the random value when the particles go into the threshold region of another subgroup. If the threshold region of each subgroup is too small, the particles in the different subgroups may converge into the same target and the subgroups do not repel each other, leading to low efficiency. Based on the success rates in Fig. 4.29, the threshold region of each subgroup can be calculated from 5 to 80 pixels.

According to Fig. 4.29, when the threshold region of every subgroup is smaller than 40 pixels, the success rate quickly shoots up, because the small threshold results in converging into one target among many subgroups. When the threshold region of each subgroup is approximately 50 pixels, the success rate is about 85 % success rate, having the stable success rate. With the increase of the threshold region of the target, the success rate should be low because the overlapping area among the subgroups increases.

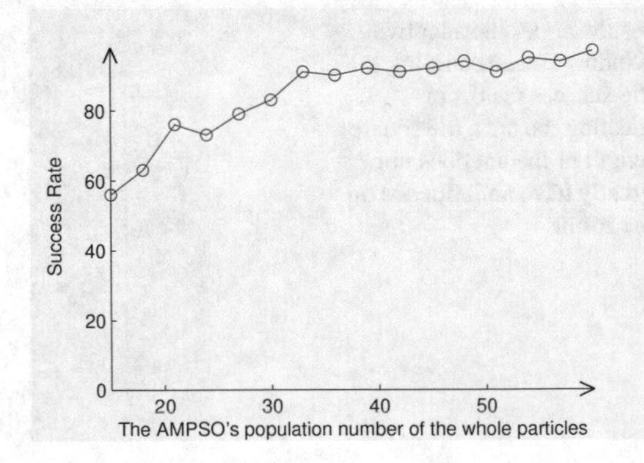

Fig. 4.28 The success rates under the adaptive swarm number. It is noted that the lowest success rate is 57 % success rate when swarm number of one subgroup is set to 5, and the highest success rate is 98 % success rate when swarm number of one subgroup is set to 20, where AMPSO denotes the adaptive multi-model particle swarm optimization

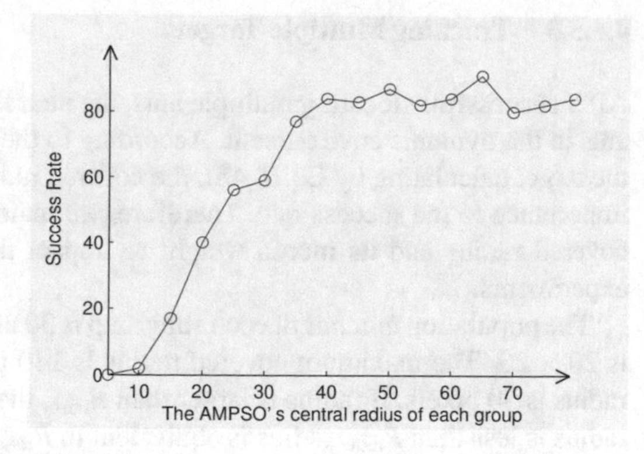

Fig. 4.29 The success rates under the central threshold radius of each subgroup. It can be clearly seen that the best success rate is 91 % when the square length of the target is 65. Then, the success rate shoots up when the central radius changes from 0 to 80 pixels

With respect to the generation runs of the search process, it is notable that the large generation can give rise to the high efficiency. However, the large iterations may suffer from a high computation load. Therefore, selecting the fitting iterations plays an important role in the practical real-time system.

As shown in Fig. 4.30, more generation times result in better success rate. The success rate reaches the highest point at 96 % success rate when the iteration time is approximately 350. More precisely, the success rate sharply rises from 30 to 150 iterations, corresponding to 20 % success rate to 82 % success rate. After the step 150, the success rate slowly increases from 150 to 400 iterations.

Remark 2 As mentioned before, the threshold region of the central point in the subgroup is selected to 40 pixels, while the best generation time is approximately selected to 300.

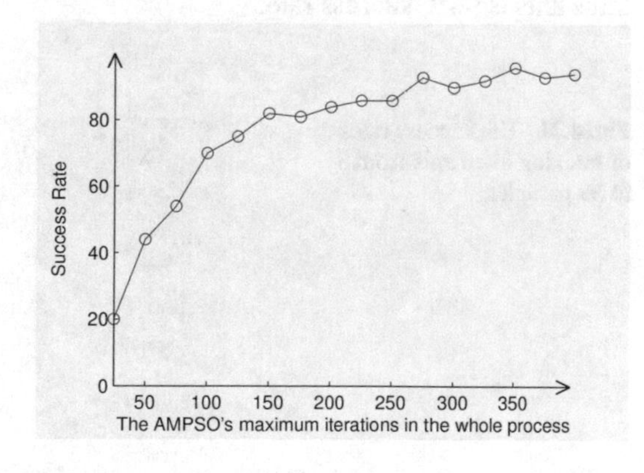

Fig. 4.30 The success rates under the maximum generations. Large iteration time in the whole process leads to good success result of locating multiple ants, while small iteration time results in low success rate

4.5.5.3 Tracking Multiple Targets

After successfully locating multiple ants, the next important task is to track multiple ants in the dynamic environment. According to the mathematical model of tracking the target calculating by Eq. (4.48), the covered radius as well as inertia weight is of importance to the success rate. Therefore, we mainly concentrate on the adaptively covered radius and its inertia weight to impact the success rate in the following experiments.

The population number of each subgroup is 30 and the size of the target's window is 20×20. The maximum covered radius is 140 pixels and the minimum covered radius is 50 pixels. If radius is larger than $R_{\max}$, the adaptive radius is set to $R_{\max}$. If radius is less than $R_{\min}$, radius is equivalent to $R_{\min}$.

Moreover, coefficient_2 in the tracking mathematical model influences the direction behavior by the last best position of the particle. If coefficient_2 is set to large value, it has the advantage over tracking the targets which do not have the random behavior. If coefficient_2 is too small, it is helpful for tracking the random behavior target. Owing to the random behavior of the ant, we consider coefficient_2 as 0 in the following experiments.

In the tracking swarm optimization, the population number of each subgroup mainly affects the success rate of tracking three ants. If the population number is set to too small, the computational time is too little but few particles do not track the ant's behavior. If the population number is too large, the particles can track the behaviors of the ants but the computational time is too long.

As shown from Fig. 4.31, when the population number of each subgroup increases, the success rate also increases. More precisely, the success rate quickly shoots up from 3 particles to 15 particles, corresponding to 4 % success rate to 85 % success rate. Setting to 33 particles of each subgroup, the highest success rate of tracking three ants is 98 % success rate.

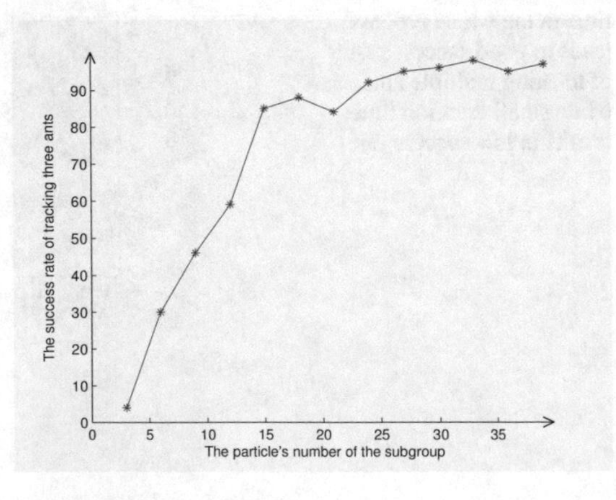

Fig. 4.31 The success rates of tracking three ants from 3 to 39 particles

Remark 3 Comparing the results in the experiments, the best population number is 24 under the trade-off between computational time and the efficiency.

Furthermore, we are concerned with the maximum covered radius of the subgroup. If the maximum covered radius is too small, the particles cannot track the ants and give rise to low efficiency. If the maximum setting radius is too large, the interaction region among the subgroups usually appears, leading to low success rate of tracking the targets.

Therefore, it can be clearly seen from Fig. 4.32 that the best parameter of the maximum covered radius is set to 110 pixels, resulting in 98 % success rate. When the maximum covered radius is less than 60 pixels, the success rate is equivalent to 0 success rate, resulting from the closeness tracking process between the second ant and the third ant. From the graphical view point, the whole process of tracking multiple ants should be illustrated by the following graphs. When this parameter is set to 70 to 90 pixels, the success rate quickly shoots up.

Result 3 *With the increase of the maximum covered radius, the success rate of tracking the ants decreases since the particles in the general process have many interaction steps and are initialized by the random value.*

To better demonstrate the efficiency of the adaptive swarm optimization from the graphical view, the success tracking process is recorded to express the adaptive radius of each subgroup and successfully address the tracking problem of two adjacent ants.

At the beginning of the tracking task, each subgroup track its own target by the adaptive swarm optimization. The red point in Fig. 4.33 is one particle in the first subgroup and the red square is the best position of the first subgroup. In addition, the red lines are the convex lines of the first subgroup. The green points are the particles in the second subgroup and the blue points represent the particles in the third subgroup.

The first frame graph in the tracking process is plotted in Fig. 4.33. In this simulation, there are three ants, whose trajectories are shown in Fig. 4.34. The second

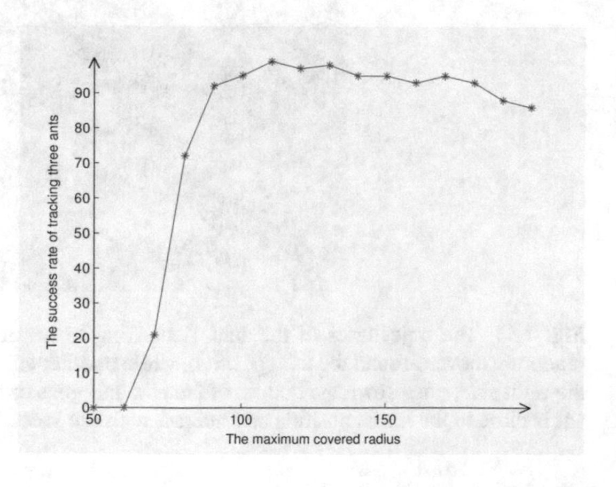

Fig. 4.32 The success rates of tracking the ants under the maximum covered radius from 50 to 200 pixels

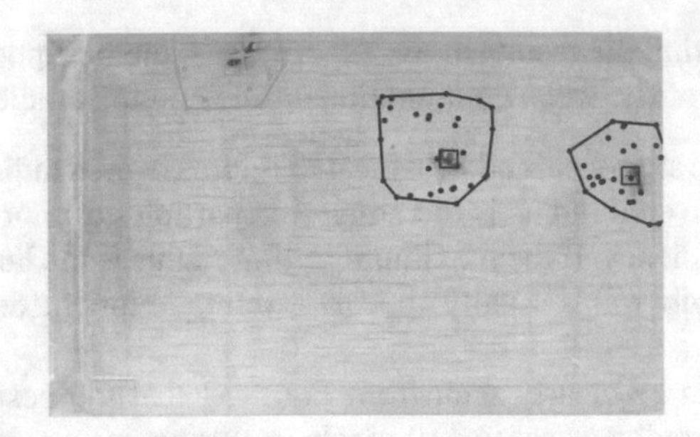

Fig. 4.33 The first graph of tracking three ants (t = 1)

ant moves from the left to the right, closing to the first ant and then leaving the first ant. During this process, initially the first subgroup and the second subgroup have no interaction region, later they go closer and interact with each other with adaptively covered radius, and finally they got separated.

During the tracking problem by swarm optimization, it is hard to track the closeness ants by the general tracking swarm optimization because the particles in two subgroups search for the identical ant, and finally it greatly influences the success rate of tracking three ants. To address this problem, one subgroup is in charge of the inner region and another subgroup is in charge of the outer region of the central position of both subgroups. The proposed tracking method can address the problem of the conflicting ants successfully tracking the closeness ants. In this subprocess, the important factor is to design the switch condition between the normal state and the conflicting state. Eventually, the subgroups in the normal state can easily track three ants, because the minimum distances of all ants is so large that the ants do not escape from the covered region of each subgroup.

According to the tracking trajectories of three ants in Fig. 4.34, the tracking swarm optimization can record the positions of three ants. The first ant randomly moves

Fig. 4.34 The trajectories of the ants. According to the actual trajectories of the ants, one ant randomly moves around the final position, while another ant clockwise moves from the left side to the right side, goes down the bottom of the box and goes up the final point. Additionally, the third ant is close to the initial position and interact with the second ant during the process

around the original position. Additionally, the second ant first moves the left side to the right side in the box, and then gets close to the third ant, finally, it goes down along the side of the box. The third ant also moves around the original position.

Result 4 *By employing the adaptive swarm optimization, the proposed method can successfully track three ants in the dynamic environment.*

According to the trajectories of the ants, the minimum distances among the subgroups play an important role on the adaptive radius of each subgroup. In the early tracking place, three ants are close to each other, corresponding to large distance among the subgroups. In the middle tracking place, the second and third ants get in touch with each other, leading to small distance among the subgroups. Under this situation, the mathematical equations should be utilized in Eqs. (4.50) and (4.51), coping with tracking the closeness ants. In the final tracking place, three ants have the large distance among them. To provide the guideline to set the adaptively covered radius of each subgroup or control the switch condition between the normal and conflicting states, the minimum distances among the subgroups are plotted by Fig. 4.35, to set the adaptively covered radius.

After tracking several ants in the experiments, the interesting intelligence phenomena of the ants' behavior are observed and may be helpful for the novel intelligence algorithm. When the ants are in the unknown environment, they first communicate with each other by their antenna, and then one large ant may detect the environment in the practical surrounding, while another large ant protects the small ants. Then, the detected ant comes back to the group and stays with others ants. After a while, the identical phenomenon is executed again by again. The phenomenon implies that the ants have the powerful cooperative ability and labor division in nature. Particularly, the experience from the ants' behavior can be applied to one troop in danger. For example, one small troop is encircled by another army, so the commander can select the excellent solider to get the help from the others and select another outstanding soldier to help the injured soldiers and monitor the environment avoiding the assault from the enemy. If the solider sending the message does not achieve the goal, the

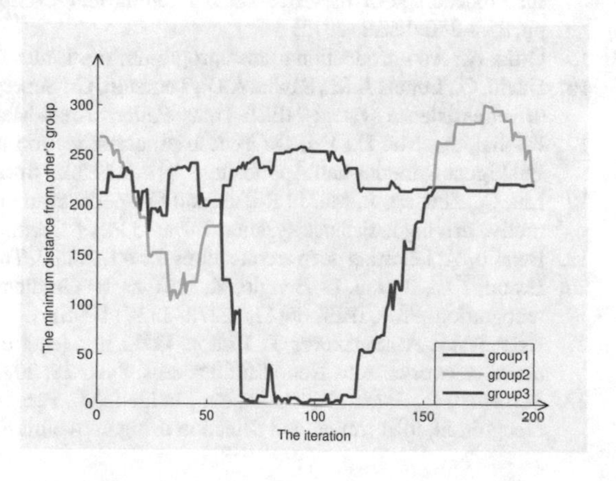

Fig. 4.35 The minimum radius among the ants in the whole process

commander also arranges another excellent solider to send the message to get the help from others.

References

1. Zitova, B., Flusser, J.: Image registration methods: a survey. Image Vis. Comput. **21**(11), 977–1000 (2003)
2. Reddy, B.S., Chatterji, B.N.: An fft-based technique for translation, rotation, and scale-invariant image registration. IEEE Trans. Image Process. **5**(8), 1266–1271 (1996)
3. Chui, H., Rangarajan, A.: A new point matching algorithm for non-rigid registration. Comput. Vis. Image Underst. **89**(2), 114–141 (2003)
4. Sotak, G., Boyer, K.L.: The laplacian-of-gaussian kernel: a formal analysis and design procedure for fast, accurate convolution and full-frame output. Comput. Vis., Graph., Image Process. **48**(2), 147–189 (1989)
5. Chen, B.Y.: An explicit formula of hessian determinants of composite functions and its applications. Kragujevac J Math **36**(1), 27–39 (2012)
6. Maini, R., Aggarwal, H.: Study and comparison of various image edge detection techniques. Int. J Image Process. (IJIP) **3**(1), 1–11 (2009)
7. Ramasubramanian, V., Paliwal, K.K.: Fast k-dimensional tree algorithms for nearest neighbor search with application to vector quantization encoding. IEEE Trans. Signal Process. **40**(3), 518–531 (1992)
8. Szeliski, R., Shum, H.-Y.: Creating full view panoramic image mosaics and environment maps. In: Proceedings of the 24th Annual Conference on Computer Graphics And Interactive Techniques, pp. 251–258. ACM Press/Addison-Wesley Publishing Co. (1997)
9. Introducing json. http://json.org/
10. Rapidjson: main page. http://miloyip.github.io/rapidjson/
11. Xiao, J., Hays, J., Ehinger, K.A., Oliva, A., Torralba, A.: Sun database: large-scale scene recognition from abbey to zoo. In: Proceedings of the 2010 IEEE Conference on Computer Vision and Pattern Recognition (CVPR), pp. 3485–3492 (2010)
12. Tighe, J., Lazebnik, S.: Scalable nonparametric image parsing with superpixels. Springer, Berlin (2010)
13. Bell, S., Upchurch, P., Snavely, N., Bala, K.: Opensurfaces: a richly annotated catalog of surface appearance. ACM Trans. Graph. (TOG) **32**(4), 111 (2013)
14. Song, S., Xiao, J.: Tracking revisited using RGBD camera: unified benchmark and baselines. In: Proceedings of the 2013 IEEE International Conference on Computer Vision (ICCV), pp. 233–240. IEEE (2013)
15. Ortiz, A.: An introduction to metaprogramming. Linux J. **158**, 6 (2007)
16. David, G., Lopez, A.M., Sappa, A.D., Thorsten, G.: Survey of pedestrian detection for advanced driver assistance systems. IEEE Trans. Pattern Anal. Mach. Intell. **32**(7), 1239–1258 (2010)
17. Khellal, A., Ma, H., Fei, Q.: Pedestrian classification and detection in far infrared images. Intelligent Robotics and Applications, pp. 511–522. Springer, Switzerland (2015)
18. Liu, Q., Zhuang, J., Ma, J.: Robust and fast pedestrian detection method for far-infrared automotive driving assistance systems. Infrared Phys. Technol. **60**(5), 288–299 (2013)
19. Bengio, Y.: Learning deep architectures for AI. Found. Trends Mach. Learn. **2**(1), 1–127 (2009)
20. Lecun, Y.L., Bottou, L., Bengio, Y., Haffner, P.: Gradient-based learning applied to document recognition. Proc. IEEE **86**(11), 2278–2324 (1998)
21. Krizhevsky, A., Sutskever, I., Hinton, G.E.: Imagenet classification with deep convolutional neural networks. Adv. Neural Inf. Process. Syst. **25**, 2012 (2012)
22. Sermanet, P., Eigen, D., Zhang, X., Mathieu, M., Fergus, R., Lecun, Y.: Overfeat: integrated recognition, localization and detection using convolutional networks. Eprint Arxiv (2013)

23. Simonyan, K., Zisserman, A.: Very deep convolutional networks for large-scale image recognition. Eprint Arxiv (2014)
24. Zeiler, M.D., Fergus, R.: Visualizing and Understanding Convolutional Networks. Lecture Notes in Computer Science, vol. 8689, pp. 818–833. Springer, Switzerland (2013)
25. Powers, D.M.: Evaluation: from precision, recall and F-measure to ROC, informedness, markedness and correlation. J. Mach. Learn. Technol. **2**, 37–63 (2011)
26. Liu, J., Ren, X., Ma, H.: Adaptive swarm optimization for locating and tracking multiple targets. Appl. Soft Comput. **12**(11), 3656–3670 (2012)
27. Eberhart, R., Kennedy, J.: A new optimizer using particle swarm theory. In: Proceedings of the Sixth International Symposium on Micro Machine and Human Science MHS '95, pp. 39–43 (1995)
28. Zhang, J.R., Zhang, J., Lok, T.M., Lyu, M.R.: A hybrid particle swarm optimization-back-propagation algorithm for feedforward neural network training. Appl. Math. Comput. **185**(2), 1026–1037 (2007)
29. Juang, C.F.: A hybri of genetic algorithm and particle swarm optimization for recurrent network design. IEEE Trans. Syst. Man Cybern. Part B-Cybern. **34**(2), 997–1006 (2004)
30. Meissner, M., Schmuker, M., Schneider, G.: Optimized particle swarm optimization (OPSO) and its application to artificial neural network training. BMC Bioinf. **7**, 125 (2006)
31. Coello, C.A.C., Pulido, G.T., Lechuga, M.S.: Handling multiple objectives with particle swarm optimization. IEEE Trans. Evol. Comput. **8**(3), 256–279 (2004)
32. Kao, Y.T., Zahara, E.: A hybrid genetic algorithm and particle swarm optimization for multi-modal functions. Appl. Soft Comput. **8**(2), 849–857 (2008)
33. Brits, R., Engelbrecht, A.P., van den Bergh, F.: Locating multiple optima using particle swarm optimization. Appl. Math. Comput. **189**(2), 1859–1883 (2007)
34. Liang, J.J., Qin, A.K., Suganthan, P.N., Baskar, S.: Comprehensive learning particle swarm optimizer for global optimization of multimodal functions. IEEE Trans. Evol. Comput. **10**(3), 281–295 (2006)
35. Jin, Y., Branke, H.: Evolutionary optimization in uncertain environments - a survey. IEEE Trans. Evol. Comput. **9**(3), 303–317 (2005)
36. Blackwell, T., Branke, J.: Multi-swarm optimization in dynamic environments. Appl. Evol. Comput. **3005**, 489–500 (2004)
37. Gaing, Z.L.: A particle swarm optimization approach for optimum design of PID controller in AVR system. IEEE Trans. Energy Convers. **19**(2), 384–391 (2004)
38. Abido, M.A.: Optimal design of power-system stabilizers using particle swarm optimization. IEEE Trans. Energy Convers. **17**(3), 406–413 (2002)
39. Esmin, A.A.A., Lambert-Torres, G., de Souza, A.C.Z.: A hybrid particle swarm optimization applied to loss power minimization. IEEE Trans. Power Syst. **20**(2), 859–866 (2005)
40. Yoshida, H., Kawata, K., Fukuyama, Y., Takayama, S., Nakanishi, Y.: A particle swarm optimization for reactive power and voltage control considering voltage security assessment. IEEE Trans. Power Syst. **15**(4), 1232–1239 (2000)
41. Ciuprina, G., Ioan, D., Munteanu, I.: Use of intelligent-particle swarm optimization in electromagnetics. IEEE Trans. Magn. **38**(2), 1037–1040 (2002)
42. Robinson, J., Rahmat-Samii, Y.: Particle swarm optimization in electromagnetics. IEEE Trans. Antennas Propag. **52**(2), 397–407 (2004)
43. De Falco, I., Della Cioppa, A., Tarantino, E.: Facing classification problems with particle swarm optimization. Appl. Soft Comput. **7**(3), 652–658 (2007)
44. Huang, C.L., Dun, J.F.: A distributed pso-svm hybrid system with feature selection and parameter optimization. Appl. Soft Comput. **8**(4), 1381–1391 (2008)
45. Shi, Y., Eberhart, R.: A modified particle swarm optimizer. In: Proceedings of the IEEE World Congress on Computational Intelligence, pp. 69–73 (1998)
46. Poli, R.: Mean and variance of the sampling distribution of particle swarm optimizers during stagnation. IEEE Trans. Evol. Comput. **13**(4), 712–721 (2009)

47. Clerc, M., Kennedy, J.: The particle swarm - explosion, stability, and convergence in a multidimensional complex space. IEEE Trans. Evol. Comput. **6**(1), 58–73 (2002)
48. Trelea, I.C.: The particle swarm optimization algorithm: convergence analysis and parameter selection. Inf. Process. Lett. **85**(6), 317–325 (2003)
49. van den Bergh, F., Engelbrecht, A.P.: A study of particle swarm optimization particle trajectories. Inf. Sci. **176**(8), 937–971 (2006)

Chapter 5
Visual Servoing Control of Robot Manipulator

Abstract Vision sensors are particularly useful for robots since they mimic our human sense of vision and allow for noncontact measurement of the environment. In this chapter, we first give a brief introduction of visual servoing and the applications of visual sensors. Then, a human–robot cooperation method is presented. The human operator is in charge of the main operation and robot autonomy is gradually added to support the execution of the operator's intent. Next, a stereo camera-based tracking control is developed on a bimanual robot. Stereo imaging and range sensors are utilized to attain eye-to-hand and eye-in-hand servoing. A decision mechanism is proposed to divide the joint space for efficient operability of a two-arm manipulator.

5.1 Introduction of Visual Servoing

In recent years, the presence of robotic systems more or less evolved is common in our daily life. Humans and robots can perform tasks together and their relation became more important than a basic remote control to realize a task. The human–robot interaction is a large research area, which has attracted more and more interests in the research community. Among human–robot interaction techniques, visual interaction is a major topic. Vision sensor is useful since it mimics the human sense of vision and allows for noncontact measurement of the environment. The human–robot interaction methods based on vision allows users just to perform in the front of the sensor devices without wearing or operating any control devices to achieve their control purposes. These sensor devices are convenient for use. The studies of this kind of interaction mostly use virtual environment. They provide a suitable environment of tests and studies decreasing the dangers of a direct interaction between a human and a robot. This virtual interpretation of the reality became an important step in the integration process of the vision application in the daily life.

Sense of vision is one of the most essential markers of intelligence and contributes immensely to the interactions with our environment. When applied to robotics, its machine replication serves as the basic communication link between time and space. Hence, for an anthropomorphic robot to operate in human surroundings, the importance of ocular feedback is next to none. Strides made in robotic and machine vision

© Science Press and Springer Science+Business Media Singapore 2016

C. Yang et al., *Advanced Technologies in Modern Robotic Applications*,
DOI 10.1007/978-981-10-0830-6_5

are of paramount importance for engineering, manufacturing, and design processes as predicted in [1]. Along with localization and mapping, the advancements in 3D Vision-Guided Robots (VGRs) have led to accurate noncontact geometrical measurements and reduction of workspace ambiguity. Visual servoing (VS) is defined as the use of visual feedback mechanisms for the kinematic control of a robot. Based on the positioning of the camera on the link and control techniques, VS ramifies into several types. Eye-in-hand and eye-to-hand VS are represented according to the position of the camera on the robotic manipulator. In [2], researches have manifested the preliminary essentials of visual servoing taxonomy, position and image based and methods using feature extraction and image correlation. Being attached on the robot arm, eye-in-hand VS provides a narrower field of view as compared to eye-to-hand servoing which, the former compensates with higher precision tracking. Several hybrid systems have been developed in order to embed the assets of both into a single control mechanism. A combination of force sensor, CCD cameras, and laser projector is used by Wen-Chung Chang et al. for accurate drilling on a 3D surface of unknown pose [3]. Ren C. Luo et al. used eye-to-hand for pose estimation and eye-in-hand for object grasping in a hybrid control loop [3]. But, the use of a visual negative feedback loop for locating objects in a nonindustrial scenario would result in delay due to low sensor sampling time. The procedure described uses a similar approach that subsumes stereo image processing for pose estimation and the range data from IR sensors near the end effectors for precision gripping. Several controllers use direct visual servoing or a dual loop system [4]. Such mechanisms are robust but come at the price of reduced performance and speed variables. The drawbacks of time delay between image acquisition and its relation to latency in robot controller is solved by the use of an open loop-based VS for computing the reference input of the joint controller from the targets projection once [5].

As introduced in Sect. 1.2 Microsoft Kinect is a vision sensor widely used in video games, robotics, image processing, human interfaces, etc. Figure 5.1 shows the structure of a Kinect sensor, consisting of an infrared (IR) projector, an IR camera, and a color camera. The RGB and depth camera can offer color image and depth of each pixel. Besides, there are some projects available for Kinect: OpenNI, OpenK-

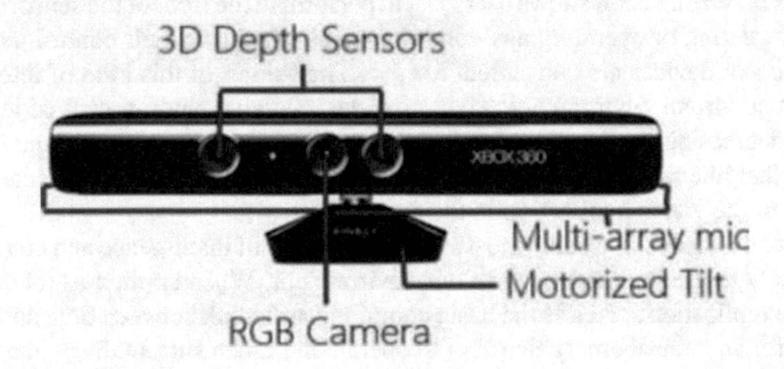

Fig. 5.1 The Kinect sensor

inect, and Microsoft Kinect for Windows. Some of the libraries which use these projects as their backend implementations are OpenCV, Unity3D, RGB-Demo, etc. The OpenKinect is an open source project enabling Kinect to be used with Linux, Mac, and Windows. The library provides wrappers to several languages such as Python, C++, C#, and Java.

The Kinect for Windows SDK offers several capabilities like seated skeleton recognition, skeleton tracking, facial tracking, and speech recognition [6].

OpenNI (Open Natural Interaction) [7] defines an API for writing applications utilizing Natural Interaction. The main purpose is to enable communication with visual, audio sensors, and audio perception middleware. Next, we will introduce a number of applications of the Kinect sensor.

- **Filtering**: Both the RGB image and the depth data obtained by Kinect have noises. There are two main methods to resolve the problems. The first approach is using Gaussian (or an average, or a median) filter, the other approach is temporal filtering [8]. The Gaussian filter is simple, but it will change low-frequency structure. Temporal filtering approach can maintain low-frequency information and removes high-frequency noise. However, the two methods cannot guarantee hole filling. An important approach to filter data was presented by Newcombe et al. [9]. In their work, the authors presented a parallelizable technique by using GPUs.
- **3D Reconstruction**: 3D reconstruction has motivated many researches in recent years. The two main contributions of the work by Newcombe et al. [9] are dense surface mapping and camera tracking. The sensor tracking to model uses Iterative Closest Point (ICP) algorithm [10], to calculate the 6 degrees of freedom of the camera [11]. Based on ICP algorithm, some 3D reconstruction projects are created.
- **Augmented Reality**: The depth data can also be used to help the creation of systems of Augmented Reality (AR). Using the depth data, it is possible to construct a marker-free application to AR using the Kinect Fusion system.
- **Image Processing**: Image segmentation is an important problem of image processing. Adding the depth channel, this problem will be simpler since we can define the segment boundary like the edges on depth data. However, the depth data has problems of noise and holes. A method based on the depth data has been proposed in [12]. It shows the input data (RGB color and depth), the priors processing (Depth Range Estimation), and the output (segmented image).
- **Interation**: The main motivation on Kinect creation was to interact with Xbox 360 through natural gestures. Using the Kinect, we can capture the skeleton of any person stood in front of it. With the skeleton, we can identify human's gestures. The algorithm using Kinect platform was presented in [13]. Based on this work, many human interface games have been developed.

- **Object Recognition**: With the depth and RGB data acquired by the Kinect, a lot of works were developed to help the object recognition task. Aniruddha Sinha and Kingshuk Chakravarty presented a gait based person identification system using 3D human pose modeling for any arbitrary walking pattern in any unrestricted indoor environment, using Microsoft Kinect sensor. They modeled the gait pattern with a spatiotemporal set of key poses and sub-poses [14]. Spinello and Arras introduced a method to detect people in RGB-D data. In this work, they combine Histogram of Oriented Depths and Histogram of Oriented Gradients to create a probabilistic function to detect people.

5.2 Kinect Sensor Based Visual Servoing for Human–Robot Cooperation

Recently, may efforts have been made to enable safe human–robot cooperation using vision interaction. In particular, Microsoft's Kinect sensor, which provides wealth of information like depth, which a normal video camera fails to provide. In addition, RGB-D data from the Kinect sensor can be used to generate a Skeleton model of humans with semantic matching of 15 body parts. Since human activities are a collection of how different body parts move across each time period, these information can be used to detect human activities. The existing human–robot interaction systems enhance the stability and reduce cost, but also have some limitations, such as comfort, accuracy and success rate. One solution to enhance the accuracy and success rate is to bridge the gap between full human control and full robot autonomy, the human operator is in charge of the main operation and robot autonomy is gradually added to support the execution of the operator's intent. Base on this idea we aim at enabling intuitive interaction between human and robot, in the context of an service scenario, where the two can collaborate to realize the task accurately.

In this section, we will explore how a human–robot team can work together effectively, and present the cooperation system environment as well as details of the technique developed to accomplish the human–robot cooperation for the robotic application [15].

5.2.1 System Architecture

The objective of the human–robot cooperation system is to enable a robot to pick up a cup on the table effectively with the direction of human. In this experiment, the robot itself has no any a priori information of the position of the cup and the robot is not programmed in advance for achieving the task of picking up the cup. In other words, we aim to make the robot capable to achieve different unprogrammed jobs under the guidance of the operator, while the operator needs only to deliver his or her intent via rough motion intent. The required operations can be summarized as following:

- For the human: to direct the robot move arm to the nearby of the cup on the table.
- For the robot: to detect the cup and adjust the position of the gripper, then, grasp the cup.

The proposed cooperation system is composed of a Laptop, one Microsoft Kinect Sensor and the Virtual Robot Experiment Platform (V-REP) software which have been introduced in Chap. 1. The Microsoft Kinect Sensor is used to detect and track human movement, these video data gathered by Kinect with 640×480 pixels at 30 Hz are streamed to a laptop using a USB 2.0 connection. These data as the raw control signal are analyzed and then conversed into control command of each joint of the robot by the remote OpenNI API, sent to the V-REP over serial port communication.

The service robot shown in Fig. 5.2 is Asti robot in V-REP. The Asti robot can be also replaced with other robot such as the humanoid Baxter robot. Figure 5.2 shows the structure of the cooperation system based on Kinect. The skeleton data extracted by Kinect sensor is the raw input signal of the system. Then the input signal is classified and processed to generate the control signal. The control object is the Asti robot's arms in V-REP.

5.2.2 *Experimental Equipments*

5.2.2.1 Microsoft Kinect Sensor

As introduced in Sect. 1.2, the Kinect sensor provides RGB video stream at a frame rate of 30 frames per second as well as a monochrome intensity encoded depth map, both in VGA resolution (640×480 pixels). The video offers 8-bits resolution while the depth data is represented in 11-bits in units of millimeters measured from the

Fig. 5.2 The structure of the human–robot cooperation system. Reprinted from Ref. [15], with permission of Springer

camera. The depth sensor is the most novel. It operates by projecting a known IR pattern onto the environment and measuring the returned IR pattern (using an IR camera). The received pattern is compared to the known projected pattern and the differences are used to extract object depth. In this experiment, the Kinect sensor is used for visual servoing control.

5.2.2.2 OpenNI (Open Natural Interaction)

OpenNI (Open Natural Interaction) is a multilanguage, cross-platform framework that defines APIs for writing applications utilizing Natural Interaction. OpenNI Application Programming Interfaces(API) are composed of a set of interfaces for writing NI applications. The OpenNI standard API enables the NI application developers to track real-life (3D) scenes by utilizing data types that are calculated from the input of a sensor. OpenNI is an open source API that is publicly available.

5.2.2.3 V-REP (Virtual Robot Experimentation Platform)

As mentioned in Sect. 1.4, the robot simulator V-REP, with integrated development environment, is based on a distributed control architecture: each object/model can be individually controlled via an embedded script, a plugin, a ROS node, a remote API client, or a custom solution. This makes V-REP very versatile and ideal for multirobot applications.

5.2.3 Implementation with V-REP

5.2.3.1 Control Signal Acquisition

The control signal of the teleoperation system is from skeleton data tracked by Kinect and the first feature can be extracted as joint data. For each joint, we have three main information. The first information is the index of the joints. Each joint has a unique index value. The second information is the positions of each joint in x, y, and z coordinates. These three coordinates are expressed in meters. The x, y, and z axes are the body axes of the depth sensor. This is a right-handed coordinate system that places the sensor array at the origin point with the positive z axis extending in the direction in which the sensor array points. The positive y axis extends upward, and the positive x axis extends to the left (with respect to the sensor array). The three coordinates for joint position are presented in Fig. 5.3.

The last information is the status of the joint. If Kinect is able to track the joint, it sets the status of this joint 'tracked'. In the case if the joint cannot be tracked, the algorithm tries to infer the joint position from other joints. If possible, then the status of this joint is inferred. Otherwise, the status of the joint is nontracked.

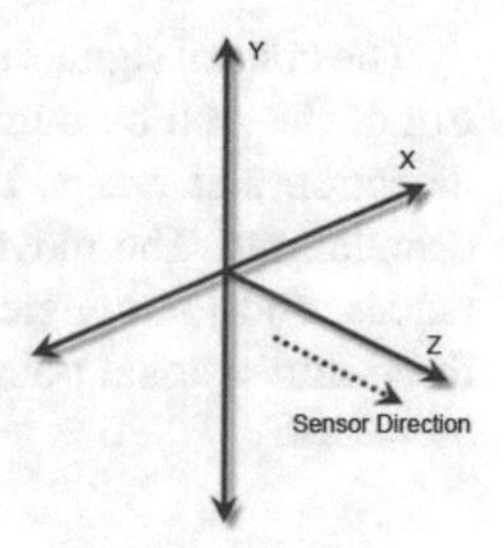

Fig. 5.3 Three coordinates for joint position. Reprinted from Ref. [15], with permission of Springer

Since, each joint has 3 values of 3 coordinates, a skeleton consist of 20 joints. Hence, the feature vector has 60 dimensions. The most important joints (A, B, I, J, K, L, M, N) in the teleoperation system are shown in Fig. 5.4. The programming structure of skeletal tracking as follows:

- Enable Skeletal Tracking.
- Activate User Tracking.
- Access Skeletal Tracking Information.
- Access Joint Information and Draw a Skeleton.

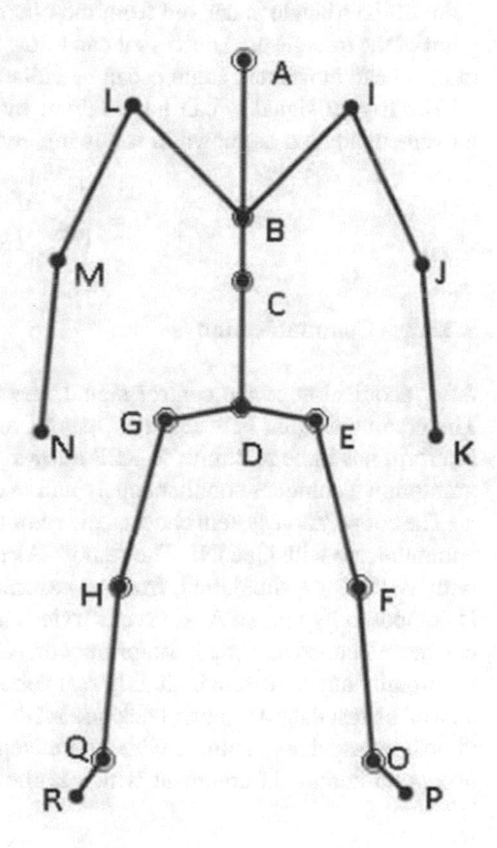

Fig. 5.4 Important joints. Reprinted from Ref. [15], with permission of Springer

The control signals include two parts: One part is the signal to accomplish the control of the Asti by human motion, and the other part is the signal trigger automatic detection and grasp. The trigger signal is the one-binary generated by movement calculations. The movement calculations are based on two or more locations, distances, and joint angles to calculate the distance between two points for two- and three-dimensional points, refer to the following equations:

$$d_2 = \sqrt{(x_2 - x_1)^2 + (y_2 - y_1)^2} \tag{5.1}$$

$$d_3 = \sqrt{(x_2 - x_1)^2 + (y_2 - y_1)^2 + (z_2 - z_1)^2} \tag{5.2}$$

where (x_1, y_1) and (x_2, y_2) are points in 2D space, d_2 is the distance between these two points , (x_1, y_1, z_1) and (x_2, y_2, z_2) are points in 3D space, and d_3 is the distance between these two points. The law of cosines can help to calculate the angle between the joints. The maximum computable angle is 180°. When calculating the angle between the joints, an additional point is needed to determine 180° $\sim$ 360°. From the skeleton tracking data, we can draw a triangle using joints A and B. The third point of the triangle is derived from the other two points. If the coordinates of each point of the triangle are known, we can know the length of each side. The magnitude of user head movement angle α can be calculated by applying the law of Cosines.

The trigger signal s_t can be obtained by applying coding process to the head movement angle α as shown in following equation:

$$s_t = \begin{cases} 1 & \text{if } \alpha \geq 5^\circ \\ 0 & \text{if } \alpha < 5^\circ \end{cases}$$

5.2.3.2 Communication

After acquisition of the control signal, the next step is to send signal to V-REP. The communication between the OpenNI and the Virtual Robot Experimentation Platform has to be real time. V-REP allows the user to choose among various programming techniques simultaneously and even symbiotically.

The cooperation system chooses the remote API client method to make V-REP to communicate with OpenNI. The remote API interface in V-REP allows interacting with V-REP or a simulation, from an external entity via socket communication. It is composed by remote API server services and remote API clients. The client side can be embedded as a small footprint code (C/C++, Python, Java, MATLAB, Urbi) in virtually any hardware including real robots, and allows remote function calling, as well as fast data streaming back and forth. On the client side, functions are called almost as regular functions, with two exceptions however: remote API functions accept an additional argument which is the operation mode, and return the same error code. The operation mode allows calling functions as blocking (will wait until the server replies), or nonblocking (will read streamed commands from a buffer, or start/stop a streaming service on the server side). The ease of use of the remote API, its availability on all platforms, and its small footprint, make it an interesting alternative for the teleoperation system.

In the teleoperation system, the client side is the OpenNI application named 'Kinect Server' which is used for tracking skeleton data of human and sending the raw data to V-REP. With remote APIs in V-REP, it is convenient to send the skeleton data to V-REP by one free serial port set via C++ codes.

On the other hand, to enable the remote OpenNI API on V-REP side, the first step is to make sure the remote OpenNI API plugin (v_repExtRemoteApi.dll) was successfully loaded at V-REP start-up. The remote OpenNI API plugin start as 'Kinect Server' communicating via a free port set in List. 1 before. Figure 5.5 illustrates how incoming remote OpenNI API commands are handled on the server side. The temporary remote API server service was chosen to start the 'Kinect Server' from within a script at simulation start. After the simulation started, the child script will detect the free serial port set in 'Kinect Server' and receive skeleton data from 'Kinect Server'.

5.2.3.3 Robot Arm Control

To remotely control the robot, the first required development was the arm control. Offsets were applied to define the workspace of the virtual arm. They provided a suitable workspace environment where the remote control of the Asti Robot arm was possible. Then the choice concerning was decided. As a mirror, the arms movements were matched to the Asti Robot Simulator arms, as the user moved its own arms. The

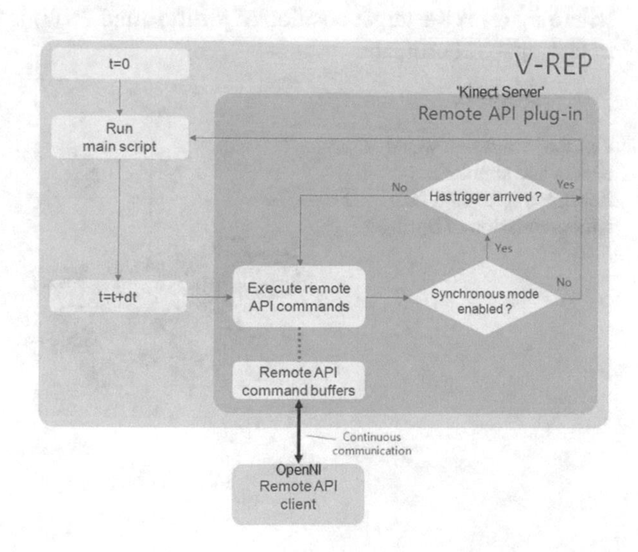

Fig. 5.5 The V-REP side remote API command handling. Reprinted from Ref. [15], with permission of Springer

Cartesian systems used for the Asti Robot Simulator and the Human in Real World are shown in Fig. 5.6.

The control of the arms was performed by sending positions of shoulders, elbows, and hands with a command via a free serial port to the Asti Robot Simulator. Then it moved its joints according to its inverse kinematics in order to reach the desired positions. In order to get the target position in simulator, the coordinate transformation is performed from the real-world space coordinates to the task space coordinates.

More specifically, the OpenNI tracker detects the position of the following set of joints in the 3D space $G = \{g_{i,\ i \in [1,I]}\}$. The position of joint g_i is implied by vector $P_{i0}(t) = [x\ y\ z]^T$, where t denotes the frame for which the joint position is located and the origin of the orthogonal XYZ coordinate system is placed at the center of the Kinect sensor. The task space is also a 3-dimensional space, the position of joint g_i is implied by vector $P_{i1}(t)$ in the task space. We obtain the vector $P_{i1}(t)$ by applying the coordinate transformation rule

$$P_{i1}(t) = T_s R_y P_{i0}(t) \tag{5.3}$$

$$R_y = \begin{bmatrix} 0 & 0 & -1 \\ 0 & 1 & 0 \\ -1 & 0 & 0 \end{bmatrix}$$

where $T_s = 0.7$ is the scaling factor between the task space and the real-world space. R_y is the rotation matrices for rotation of $-\frac{\pi}{2}$ about the y-axes. The translation transformation to the Asti Robot target can be written in matrix form as

$$P_{i2}(t) = P_{i1}(t) + P_A(t) \tag{5.4}$$

where $P_{i2}(t)$ is the target position of joint g_i, and $P_A(t)$ is the position of Asti Robot in task space coordinate.

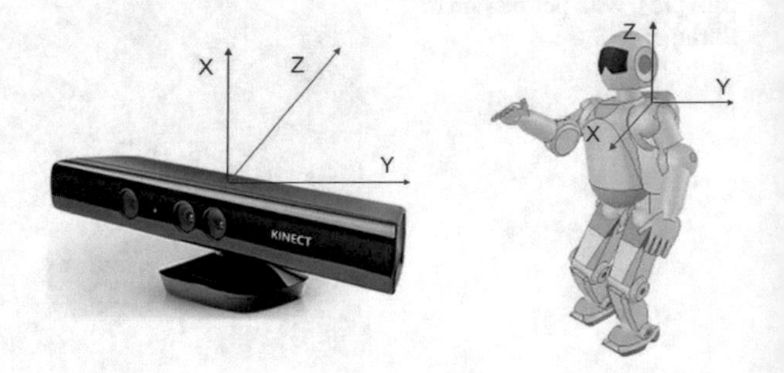

Fig. 5.6 Cartesian world of Kinect and simulator. Reprinted from Ref. [15], with permission of Springer

5.2.3.4 Inverse Kinematics Calculation Module Configuration

The problem of Inverse Kinematics can be seen as the one finding the joint values corresponding to some specific position and/or orientation of a given body element (generally the end effector). More generally, it is a transformation from the task space coordinates into the joint space coordinates. For a serial manipulator, for instance, the problem would be to find the value of all joints in the manipulator given the position (and/or orientation) of the end effector.

V-REP uses IK groups as shown in Fig. 5.7 to solve inverse kinematics tasks. An IK group contains one or more IK elements:

- *IK groups*: IK groups group one or more IK elements. To solve the kinematics of a simple kinematic chain, one IK group containing one IK element is needed. The IK group defines the overall solving properties (like what solving algorithm to use, etc.) for one or more IK elements.
- *IK elements*: IK elements specify simple kinematic chains. One IK element represents one kinematic chain. A kinematic chain is a linkage containing a base, several links, several joints, a tip, and a target.

There are many methods of modeling and solving inverse kinematics problems [16]. The most flexible of these methods typically rely on iterative optimization to seek out an approximate solution, due to the difficulty of inverting the forward kinematics equation and the possibility of an empty solution space. The core idea behind several of these methods is to model the forward kinematics equation using a Taylor series expansion, which can be simpler to invert and solve than the original system.

The Jacobian inverse technique is a simple yet effective way of implementing inverse kinematics. The model of the Asti Robot arms is shown in Fig. 5.8. The forward kinematics system space is the *span* of the joint angles with 3 dimensions. The IK system lives in a 2D space. Hence, the position function can be viewed as a mapping $p(x) : \mathcal{R}^3 \rightarrow \mathcal{R}^3$. Let $p_0 = p(x_0)$ give the initial position of the system. The Jacobian inverse technique computes iteratively an estimate of Δx that minimizes the error given by $\| p(x_0 + \Delta x) - p_1 \|$. For small Δx-vectors, the series expansion of the position function gives:

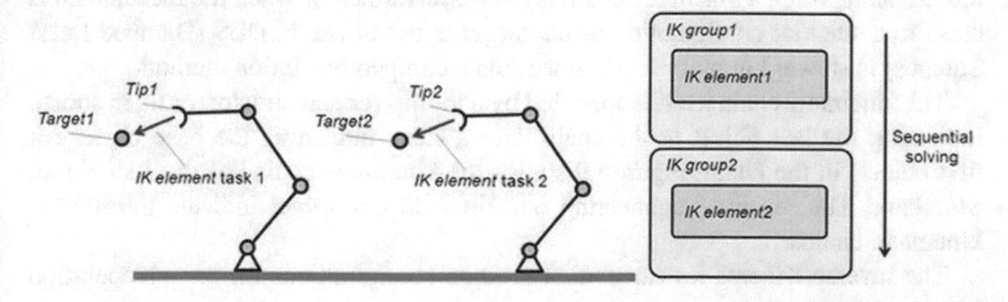

Fig. 5.7 Two separate IK chains. Reprinted from Ref. [15], with permission of Springer

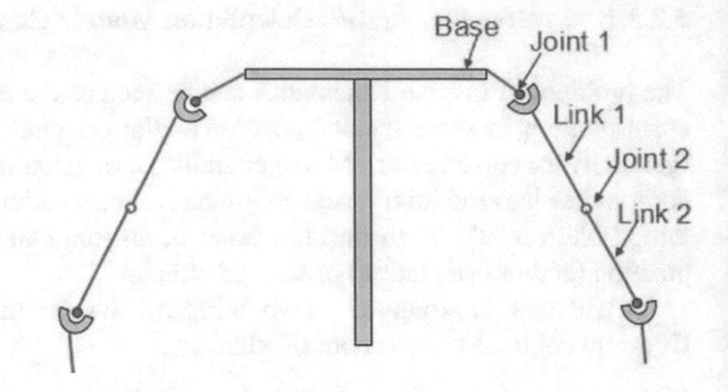

Fig. 5.8 The arms model of Asti Robot. Reprinted from Ref. [15], with permission of Springer

$$p(x_1) \approx p(x_0) + J_p(\hat{x_0})\Delta x \tag{5.5}$$

where $J_p(x_0)$ is the (3×3) Jacobian matrix of the position function at x_0. Note that the (i, k)th entry of the Jacobian matrix can be determined numerically:

$$\frac{\partial p_i}{\partial x_k} \approx \frac{p_i(x_{0,k} + h) - p_i(x_0)}{h} \tag{5.6}$$

where $p_i(x)$ gives the ith component of the position function, $x_{0,k} + h$ is simply x_0 with a small delta added to its kth component, and h is a reasonably small positive value. Taking the Moore–Penrose pseudoinverse of the Jacobian (computable using a singular value decomposition) and rearranging the terms results in

$$\Delta x \approx J_p^+(x_0)\Delta p \tag{5.7}$$

where $\Delta p = p(x_0 + \Delta x) - p(x_0)$. And in order to improve the estimation for Δx, the following algorithm (known as the Newton–Raphson method) was used:

$$\Delta x_{k+1} = J_p^+(x_k)\Delta p_k \tag{5.8}$$

There are two kinds of methods for the specified IK group resolution in V-REP. Pseudo Inverse is the fastest method but can be unstable when the target and tip lie too far apart, when a kinematic chain is over-constrained or when the mechanism is close to a singular configuration or the target is out of reach. DLS (Damped Least Squares) is slower but more stable since it is a damped resolution method.

The kinematic chain itself is specified by a tooltip (or end effector, or tip in short), indicating the last object in the chain, and a base, indicating the base object (or first object) in the chain. Figure 5.9 shows the kinematic chains of the Asti Robot Simulator. The red lines connecting one tip with one target indicate the inverse kinematic chains.

The Inverse Kinematics calculation module configurations for the teleoperation system as follows:

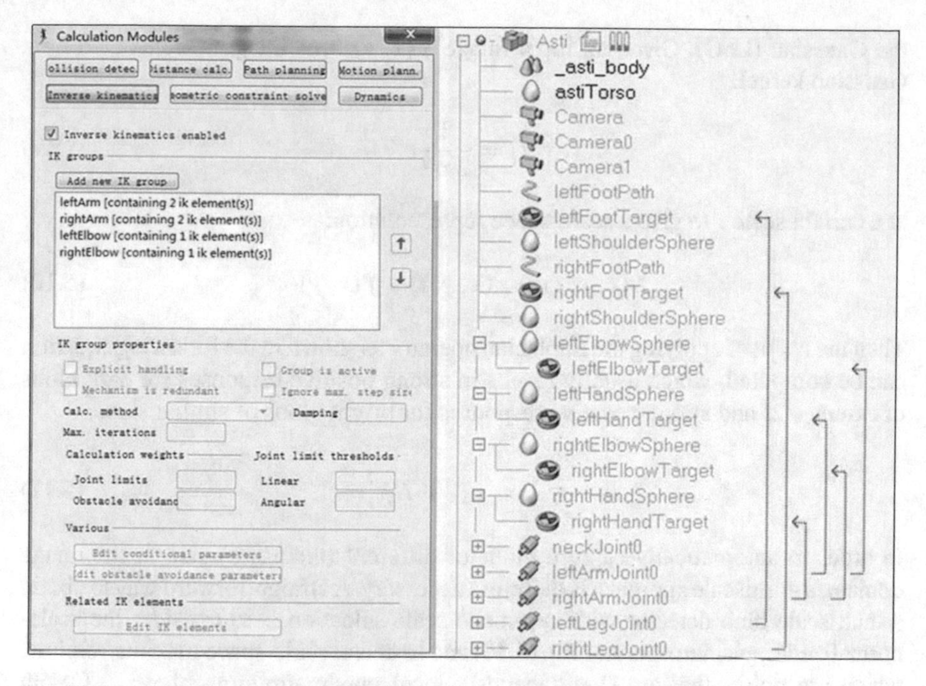

Fig. 5.9 The IK groups in the system. Reprinted from Ref. [15], with permission of Springer

- Set the bases of all the IK chains as Asti and specify their individual tip.
- Link the tip dummy to a target dummy to specify a target for every IK chain to follow.
- Set the Calc. Method as DLS.
- Set the IK Elements are constrained (X, Y, Z).
- Set the Max. Iterations as 6.
- Set the Joint limits and the Obstacle avoidance.

5.2.3.5 Visual Servoing System

In order to implement the function that the Asti robot can detect the object and grasp it correctly while receives the trigger signal, the color-based visual servoing system [17] was applied to Asti robot. There are fully integrated cameras on head and each wrist for visualizing end-effector interactions.

In V-REP, a vision sensor normally produces two images at each simulation pass: a color image and a depth map. Our approach to detect the target is directly based upon the physical characteristics of color reflection. There are two main steps in detecting the target. At first, select the main color of the target in the image by setting the RGB threshold value. Then applying the Blob detector to detect the target region [18]. In our vision servoing system, the Blob detector is based on the Laplacian of the Gaussian (LoG). Given an input image $f(x, y)$, this image is convolved by a Gaussian kernel:

$$g(x, y, t) = \frac{1}{2\pi t^2} e^{-\frac{x^2+y^2}{2t^2}} \tag{5.9}$$

at a certain scale t to give a scale-space representation:

$$L(x, y; t) = g(x, y, t) * f(x, y) \tag{5.10}$$

Then the result of applying the Laplacian operator as shown in the following equation can be computed, which usually results in strong positive responses for dark blobs of extent $\sqrt{2t}$ and strong negative responses for bright blobs of similar size.

$$\overset{2}{\nabla} L = L_{xx} + L_{yy} \tag{5.11}$$

In order to automatically capture blobs of different (unknown) size in the image domain, a multiscale approach is therefore necessary. A straightforward way to obtain a multiscale blob detector with automatic scale selection is to consider the scale-normalized Laplacian operator Eq. (5.10) and to detect scale-space maxima/minima, which are points that are simultaneously local maxima/minima of $\nabla^2_{norm} L$ with respect to both space and scale [19].

$$\underset{norm}{\overset{2}{\nabla}} L(x, y; t) = t(L_{xx} + L_{yy}) \tag{5.12}$$

Thus, given a discrete two-dimensional input image $f(x, y)$, a three-dimensional discrete scale-space volume $L(x, y, t)$ is computed and a point is regarded as a bright (dark) blob if the value at this point is greater (smaller) than the value in all its 26 neighbors. Thus, simultaneous selection of interest points $(\hat{x}, \hat{y})$ and scales $\hat{t}$ is performed according to

$$(\hat{x}, \hat{y}; \hat{t}) = \text{argmaxminlocal}_{(x,y;t)} \left(\underset{norm}{\overset{2}{\nabla}} L(x, y; t) \right) \tag{5.13}$$

After target position detection, the Asti robot will adjust the gripper position automatically. When the gripper arrives at the desired position, it will grasp the target object. For autonomous position adjustment, it is common to use a loop that controls the adjustment trajectory of the gripper. Thus, for controller design, we consider the target position $u = \left[x_{x0} \; x_{y0} \right]^T$ as the control input for the dynamics adjustment system. The state equation of the controller is:

$$\begin{cases} x(k+1) = \begin{bmatrix} 2 & 0 \\ 0 & 2 \end{bmatrix} x(k) + \begin{bmatrix} -1 \\ -1 \end{bmatrix} u \\ y(k) = x(k) \end{cases} \tag{5.14}$$

where the position of the gripper y is the output, $x(k) = \left[x_x(k) \; x_y(k) \right]^T$ is the state vector.

5.2.4 Experiment Studies

Our goal is to explore whether a human–robot team can accomplish the service task together effectively using the proposed cooperation method. The first idea was building the experiment environment in V-REP. The experimental environment includes one Asti robot, one table, and two cups. Two cups with different colors were used to test the accuracy of Asti target autonomous detection.

To examine the validity of the control of the robot arm with motion, we design four motions to observe the motion of subject and 3D robot simulator. As shown in Fig. 5.10, the first one shows the control of the robot simulator whose arms swing back and forth, and the second one is the procedure of the robots bending its elbow joint. In each subfigure of Fig. 5.10, the left side shows the motion of the subject detected by the Kinect based on the skeleton tracking technology, while the right side displays the motion of the simulated robot Asti. From the figures, we find that the simulated robot Asti can track the motion of subject's arm effectively.

In order to verify the accuracy of the visual servoing system, we do a series of experiments as follows. We move the Asti robot arm to different positions near the cups to test whether the robot can detect the target and adjust the gripper position

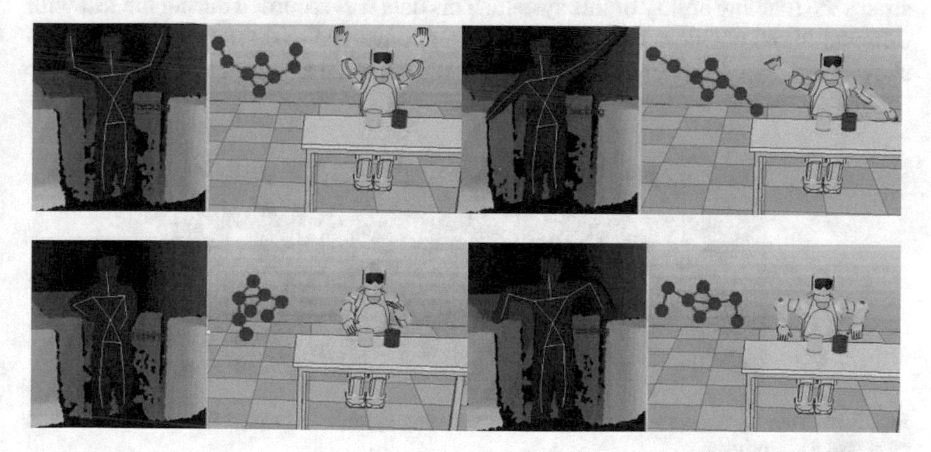

Fig. 5.10 Simulated robot movements. Reprinted from Ref. [15], with permission of Springer

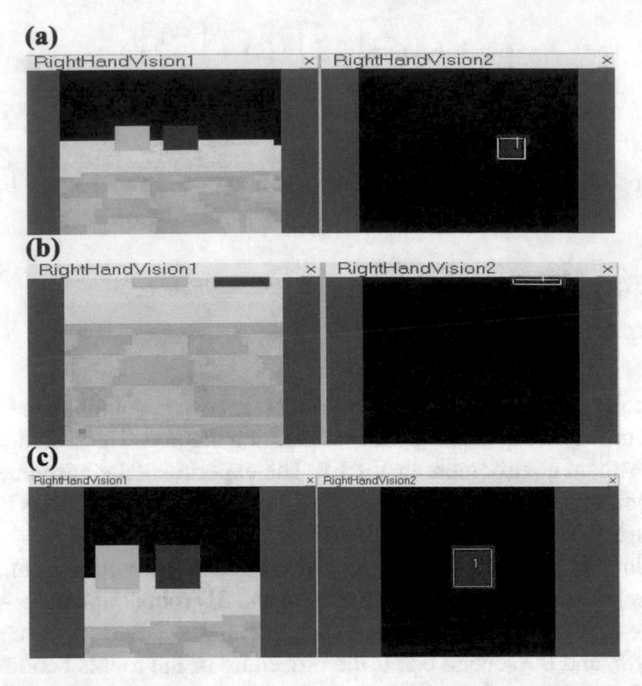

Fig. 5.11 The experiment results of visual servoing system. **a** and **b** are in different initial position; **c** is the adjustment result of **a** and **b**. Reprinted from Ref. [15], with permission of Springer

accurately. The left side of a and b in Fig. 5.11 display raw image of the hand camera in two different initial position, the right side of Fig. 5.11a,b display the detection results. The subfigure c is the autonomous adjustment result of a and b.

Figure 5.12 shows the autonomous adjustment error in the Z direction and demonstrates the tracking ability of this system. This data was recorded during the test with different initial position. Notice that the error converged to 0 after 10 s. Figure 5.13 shows the autonomous adjustment velocity, the maximal adjustment time is 10 s. We

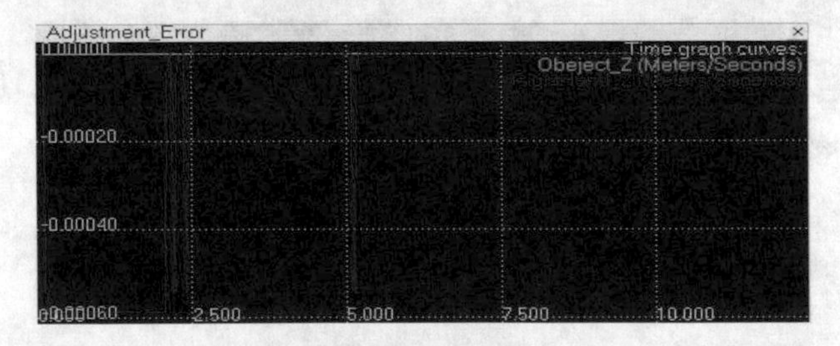

Fig. 5.12 The adjustment error of the visual servoing system. Reprinted from Ref. [15], with permission of Springer

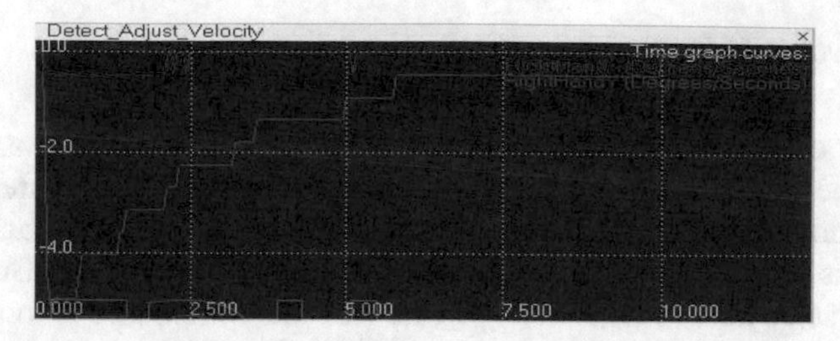

Fig. 5.13 The detection and adjustment velocity of the visual servoing system. Reprinted from Ref. [15], with permission of Springer

can find that the robot can detect the target position and adjust the gripper to the desired position accurately.

In this section, we have put forward one novel human–robot cooperation method which falls in between full human control and full robot autonomy. In order to verify the effectiveness of the method by experiments, an experiment environment based on the Kinect sensor and V-REP was built up. In the experiments, the Kinect sensor was used to obtain the skeleton framework information of the human, then, the data is classified and transformed to the robot as the motion control signal. Moreover, the Blob detector was applied to detect the target and applied the closed loop control to implement autonomous adjustment in the visual servoing system. It can be found that the human–robot team can work effectively through the experiments.

5.3 Visual Servoing Control Using Stereo Camera

In order to control a closely coupled two-arm manipulator system via eye-to-hand servoing, in this section, a hybrid visual servoing methodology is developed based on Baxter robot and a Point Grey Bumblebee2 stereo camera, which could be used to obtain the 3D point cloud of a specific-colored object [20]. The objective is to achieve the manipulation by detecting the world coordinates of the object in the Cartesian plane and target range data from an IR detector embedded on the gripper.

Manipulator workspace boundary estimation is essential for improvization of robotic algorithms and optimization of its overall design and analysis. In this section, an optimal control distribution is realized by workspace estimation of the manipulators and decentralization of control based on a constrained joint space. This is achieved by tracing the object location in the robot coordinate system in a convex hull of both the arms. While basing on the previous works [21] and bolstering the Monte Carlo algorithm with Delaunay Triangulation (DT) and Convex Hull boundary generation, a novel approach to incorporate a constraint joint space between the two manipulator arms is implemented.

5.3.1 System Integration

The system constitutes of a client–server UDP network that integrates the system components and confers parallel processing. It consists of a Baxter robot (see Sect. 1.1) and a Bumblebee2 stereo camera (see Sect. 1.2). The machine vision processing is done in the client computer connected with a Point Grey® Bumblebee2 stereo camera with IEEE-1394 Firewire connection. The robotic control is shared between the client and server systems. The client is responsible for the object pose and world coordinate extraction and constrained joint space control for left and right arm. Whereas, the trajectory generation and joint speed control is formulated by the server computer which has a direct communication with the robot. Range data from the robot-end effector is then received by the server for implementing precision gripping. The system architecture is described in Fig. 5.14. As mentioned in Sect. 1.2, the Point Grey Bumblebee2 stereo camera is a 2 sensor progressive scan CCD camera with fixed alignment between the sensors. The resolution-speed trade-off has to be managed concisely as an increased frame speed gives a smooth robot trajectory whereas enhances the processing time. And an increased resolution provides a denser, more accurate point cloud for feature extraction but with increased latency. Recall in Sect. 1.1, the Baxter robot is a semi-humanoid robot with two 7DOF arms installed on the left and right mounts, respectively, on a movable pedestal. Several attributes such as the presence of a CCD camera and an IR range sensor at the end effector, make Baxter the definable option. The DH parameters of Baxter are adopted for implementing pos-to-pos inverse kinematics control.

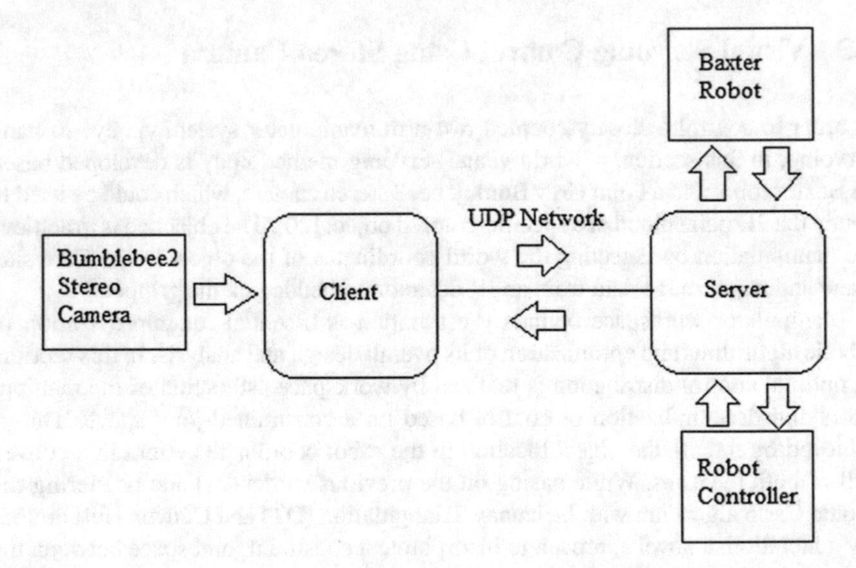

Fig. 5.14 System Architecture. Reprinted from Ref. [20], with permission of Springer

Table 5.1 Nomenclature

Symbol	Meaning
X, Y, Z	World coordinates of the point
x, y	Image plane coordinates of the point
f	Focal length of the camera lens
s	Scaling factor
c_x, c_y	Projection displacement parameter
T	Translational matrix
H	Homography matrix
A	Intrinsic matrix
r_1, r_2, r_3	Spatial rotational matrices
x^1	Column value of right image pixel
x^r	Column value of right image pixel
D	Depth
B	Baseline
d	Disparity
q	Projection matrix
$X/W, Y/W, Z/W$	3D world coordinate
W	World Frame Projector
T_x	Translational vector

5.3.2 *Preprocessing*

5.3.2.1 Camera Calibration

Camera calibration is necessary as the use of lenses introduces nonlinearities and deviates from the simple pin-hole model such as lens distortion namely radial and tangential distortion. It provides the camera geometry and distortion model which in turn define the intrinsic parameters of the camera. Each of the two cameras is calibrated by using a flexible planar reference object calibration technique described in [22]. The camera parameters namely intrinsic, extrinsic, and distortion are evaluated by the use of a 2D checker-board pattern. 3D reference models are avoided due to computation complexity and high cost of precise calibration objects. 20 checker-board images were fed to the calibrator algorithm encompassing differential angles in the projection space. This provides enough values to estimate the camera geometry parameters namely, 4 intrinsic (f_x, f_y, c_x, c_y), 5 distortion (radial -k_1, k_2, k_3) (2 tangential-p_1, p_2) and the camera extrinsics (rotation-Ψ, Ξ, Θ),(translation-T_x, T_y, T_z). The point 'Q' on the object plane is related with the image plane point 'q' by the following equations. r_1, r_2, and r_3 are the rotational matrices which represent the axis rotation between image and world in the X, Y, and Z axes, respectively. A summary to the definition of above mentioned notations is provided in Table 5.1.

$$\begin{bmatrix} x \\ y \\ 1 \end{bmatrix} = s.A \cdot [r_1\, r_2\, r_3\, T] \begin{bmatrix} X \\ Y \\ 0 \\ 1 \end{bmatrix} \tag{5.15}$$

$$x = f \cdot (X/Z) + c_x \tag{5.16}$$

$$y = f \cdot (Y/Z) + c_y \tag{5.17}$$

$$q = s \cdot H.Q \tag{5.18}$$

The camera parameters are finally obtained using Closed-Form solution and Maximum Likelihood estimation [23]. Further the distortion is nullified from the image points by introducing the following equations:

$$x_{corrected} = x(1 + k_1 r^2 + k_2 r^4 + k_3 r^6) \tag{5.19}$$

$$y_{corrected} = y(1 + k_1 r^2 + k_2 r^4 + k_3 r^6) \tag{5.20}$$

5.3.2.2 Coordinate Transformation

Both the camera and the robot must align in the same coordinate system in order to implement the algorithm. The coordinate transformation of the detected feature points from the Bumblebee2 coordinates to Baxter coordinates can be achieved by Eq. (5.21).

$$[x_1\; y_1\; z_1\; 1]^T = \mathbf{T}[X_1\; Y_1\; Z_1\; 1] \tag{5.21}$$

The homogenous transformation matrix is obtained by measuring the coordinates four noncollinear points from the robot coordinate and the Bumblebee2 coordinate. Let, the four points on the either coordinate systems be (x_1, y_1, z_1), (x_2, y_2, z_2), (x_3, y_3, z_3), (x_4, y_4, z_4) and (X_1, Y_1, Z_1), (X_2, Y_2, Z_2), (X_3, Y_3, Z_3), (X_4, Y_4, Z_4) respectively. The transformation matrix from Bumblebee2 to Baxter coordinate can be calculated by Eq. (5.22).

$$T = \begin{bmatrix} x_1 & x_2 & x_3 & x_4 \\ y_1 & y_2 & y_3 & y_4 \\ z_1 & z_2 & z_3 & z_4 \\ 1 & 1 & 1 & 1 \end{bmatrix} \begin{bmatrix} X_1 & X_2 & X_3 & X_4 \\ Y_1 & Y_2 & Y_3 & Y_4 \\ Z_1 & Z_2 & Z_3 & Z_4 \\ 1 & 1 & 1 & 1 \end{bmatrix}^{-1} \tag{5.22}$$

5.3.3 Algorithm Implementation

5.3.3.1 3D Reconstruction

The images captured by the Bumblebee2 stereo camera in active ambient lighting are shown in Fig. 5.15. Both the images are calibrated using the camera intrinsics and are corrected for distortion. Subsequently, the undistorted images are stereo rectified in order to align the epipolar lines of both the projection planes and ensure the presence of similar pixels in a specified row of the image. The images obtained are then frontal parallel and are ready for correspondence estimate. The essential and the fundamental matrix are calculated by using Epipolar geometry [24]. Essential matrix is a 3×3 matrix with 5 parameters; two for translation and three for the rotation values between the camera projection planes. The fundamental matrix on the other hand, represents the pixel relations between the two images and has seven parameters, two for each epipole and three for homography that relates the two image planes. Bouguets algorithm [25] is then implemented to align the epipolar lines and shift the epipoles to infinity. Figure 5.16 depicts the results of stereo rectification where the red and cyan colors represent the left and right images with row-aligned pixels. Stereo correspondence is a method of matching pixels with similar texture across two coplanar image planes. The distance between the columns of these perfectly matched pixels is called disparity (d).

$$d = x^1 - x^r \tag{5.23}$$

Block matching is implemented for evaluating the correspondence between the images. Block sizes of 15 pixel window are used to find the matches by the use of SAD (sum of absolute differences). The disparity range is kept low, [0 40] in order to match the indoor low texture difference and taking into account computational speed. Semi Global method is used to force the disparity values to the neighboring pixels for a more comprehensive result [26]. The disparity output is shown in Fig. 5.17. Disparity is inversely proportional to the depth of the pixel and is related by the

Fig. 5.15 Images captured from Bumblebee2. Reprinted from Ref. [20], with permission of Springer

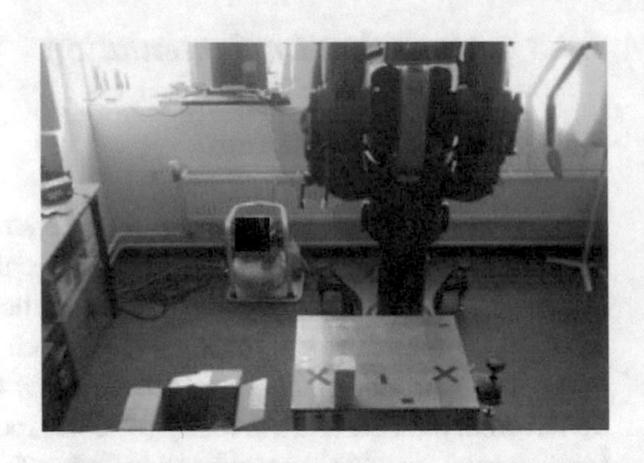

Fig. 5.16 Rectified stereo Images. Reprinted from Ref. [20], with permission of Springer

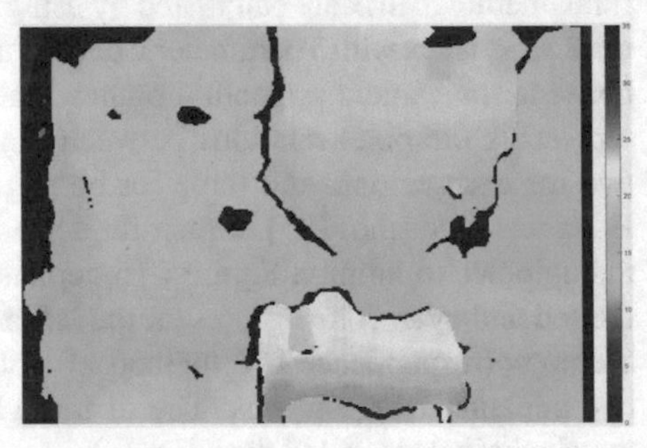

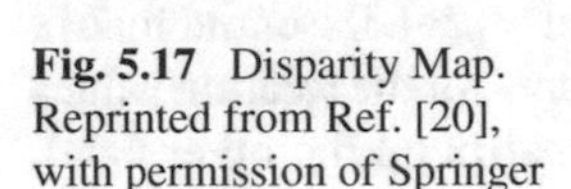

Fig. 5.17 Disparity Map. Reprinted from Ref. [20], with permission of Springer

Triangulation equation Eq. (5.24). Triangulation refers to the estimation of depth of an object by visualizing its location from two different known points.

$$D = T.f/d \tag{5.24}$$

The reconstruction of the image in the Cartesian coordinates is obtained by the use of projection matrix evaluated using Bouguet's algorithm Eq. (5.25). From the 2D homogenous point and its associated disparity d, the point can be projected into 3D using Eq. (5.28). Q is the reprojection matrix constituting the camera intrinsic parameters. Threedimensional coordinates are obtained when screen pixel coordinates are multiplied with Q. Figure 5.18 depicts the 3D reconstruction of the robot's workspace.

$$P\begin{bmatrix} X \\ Y \\ Z \\ W \end{bmatrix} = \begin{bmatrix} x \\ y \\ d \\ 1 \end{bmatrix} \tag{5.25}$$

$$\begin{bmatrix} x \\ y \\ d \\ 1 \end{bmatrix} = Q\begin{bmatrix} X \\ Y \\ Z \\ W \end{bmatrix} \tag{5.26}$$

Fig. 5.18 3D reconstruction of the image. Reprinted from Ref. [20], with permission of Springer

$$Q = \begin{bmatrix} q_1 1 \; q_1 2 \; q_1 3 \; 14 \\ q_2 1 \; q_2 2 \; q_2 3 \; 24 \\ q_3 1 \; q_3 2 \; q_3 3 \; 34 \\ q_4 1 \; q_4 2 \; q_4 3 \; 44 \end{bmatrix} = \begin{bmatrix} 1\,0 & 0 & -c_x \\ 0\,1 & 0 & -c_y \\ 0\,0 & 1 & f \\ 0\,0 & -1/T_x & (c_x - c_x')/T_x \end{bmatrix}^{-1} \tag{5.27}$$

5.3.3.2 Object Detection

Color-based segmentation is used in order to separate a single color object from the background. The image is converted into L*a*b* color space and the Euclidean distance between red-green and yellow-blue opponent components of the object and 'a' and 'b' matrices calculated. The minimum value gives the most accurate estimate of the object. Further, the corners of the object are calculated by Harris corner detector and the centroid calculated by intersection of the diagonals. The depth value of the centroid is then extracted from reconstructed point cloud of the task space. Figure 5.19 shows the calculated robot coordinates of the centroid of the object after coordinate transformation.

5.3.3.3 Workspace Estimation

The DH parameters of Baxter have been obtained in Sect. 2.1. The link lengths and the joint rotation limits of all the 7 DOFs are given as 1(0.27 m, −97.5°, 97.5°), 2(0.364 m, −123°, 60°), 3(0.069 m, −175°, 175°), 4(0.375 m, −2.865°, 150°), 5(0.01 m, −175.27°, 175.27°), 6(0.28 m, −90°, 120°), and 7(−175.27°, 175.27°). The DH parameters applied to forward kinematics gives the homogenous transformation matrices of the arm. The transformation matrix can be used to obtain the position of the end effector based on specific joint angle values. Monte Carlo method of random sampling is applied to the task space to estimate the workspace of both the arms. Homogenous radial distribution is used to generate 6000 points in the joint space for each arm separately. The joint angle values are chosen randomly and for-

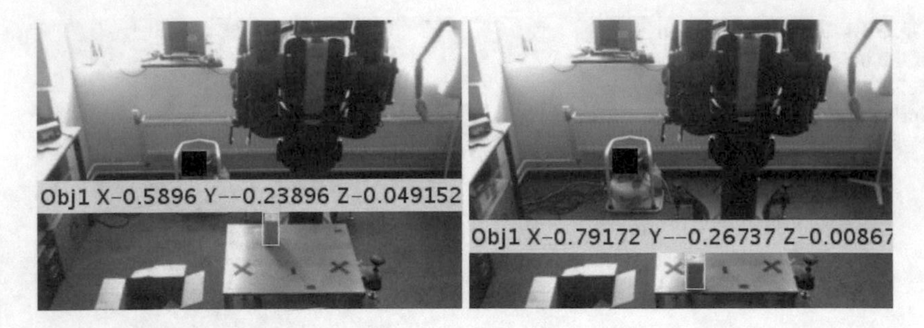

Fig. 5.19 Object detection results. Reprinted from Ref. [20], with permission of Springer

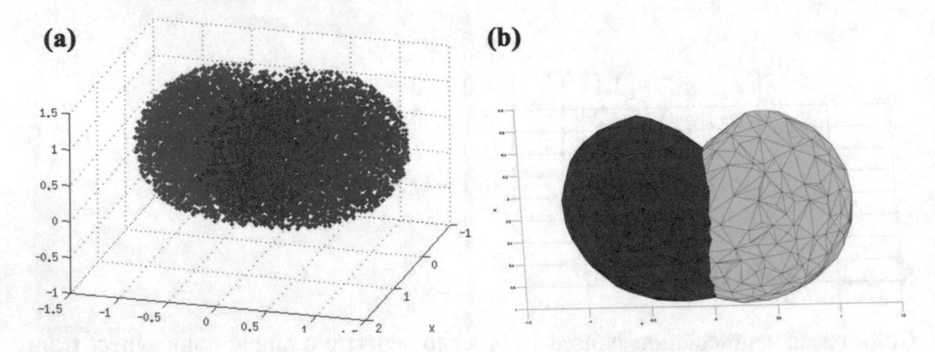

Fig. 5.20 Workspace Estimation. Reprinted from Ref. [20], with permission of Springer

ward kinematics is implemented to evaluate the end-effector points, creating a point cloud of the workspace. The generated point cloud of the workspace for both the manipulators is shown in Fig. 5.20.

Further, Delaunay triangulation is applied to the points in the 3D space to generate a set of points with a circumcircle without any points in its interior. This facilitates the creation of a convex hull of the joint space.

5.3.3.4 Arm Switching Mechanism

The obtained convex hull is used to constrain the workspace and decentralize the control between the left and right arms for efficient maneuvrability. The depth of the object is mapped to the convex hull to expedite the process of decision making as a point cloud matching technique would hinder the processing speed. This is done by determining the presence of the points respective coordinates in the convex hull projection on the three Cartesian planes. Hence, if (X_1, Y_1, Z_1) is the point representing the object, its existence in the 3D hull is detected by following five steps:

1. Check if the XY plane projection of the hull contains the point (X_1, Y_1), YZ plane contains point (Y_1, Z_1) and XZ plane contains point (X_1,Z_1) using Ray Casting Algorithm [24].
2. Obtain the presence decision in both the arm workspaces.
3. If the point is present in both of the manipulator workspace, give the control priority to the arm with the smallest Euclidean distance from manipulator.
4. Assign the control based on the detection of the point in left or right workspaces.
5. If the point lies outside both the workspaces, stop arm movement to avoid singularity.

The above workflow ensures an efficient methodology to reach the object in the robot workspace. It mimics the human intuition of using the nearest possible arm for grabbing in order to avoid the use of excessive body movements and minimize the use of energy.

Arm switching algorithm is based on the detection of the depth world coordinates in the workspace projection of the left- and right-arm manipulators. The presence of the respective coordinates of the depth is checked in all of the three polygons as shown in Figs. 5.21 and 5.22. If the existence of a point in all the polygons of a specific arm hull is confirmed, that arm is activated and assigned the coordinate for tracking of the object.

5.3.3.5 Inverse Kinematics

Once the object coordinates and the arm control have been specified, the joint angles have to be calculated for the 7DOF robotic arm. The approach taken for solving the inverse kinematics problem is to find a closed-form analytical solution for the equation Eq. (5.28). Resolved-rate motion control is applied to obtain the joint velocity vector from the arm Jacobian and end-effector velocity.

$$\dot{x} = J(q)\dot{q} \tag{5.28}$$

$$q = J^{\dagger}(q)\dot{x} \tag{5.29}$$

$$\dot{q} = KJ^{T}(q)e \tag{5.30}$$

The task space vector is replaced by Eq. (5.29), where the error is between the desired and the actual task trajectories. The use of transpose of Jacobian instead of inverse, ensures avoidance of kinematics singularity, and establishes a stable closed loop system. The detailed proof of the solution is shown in [27].

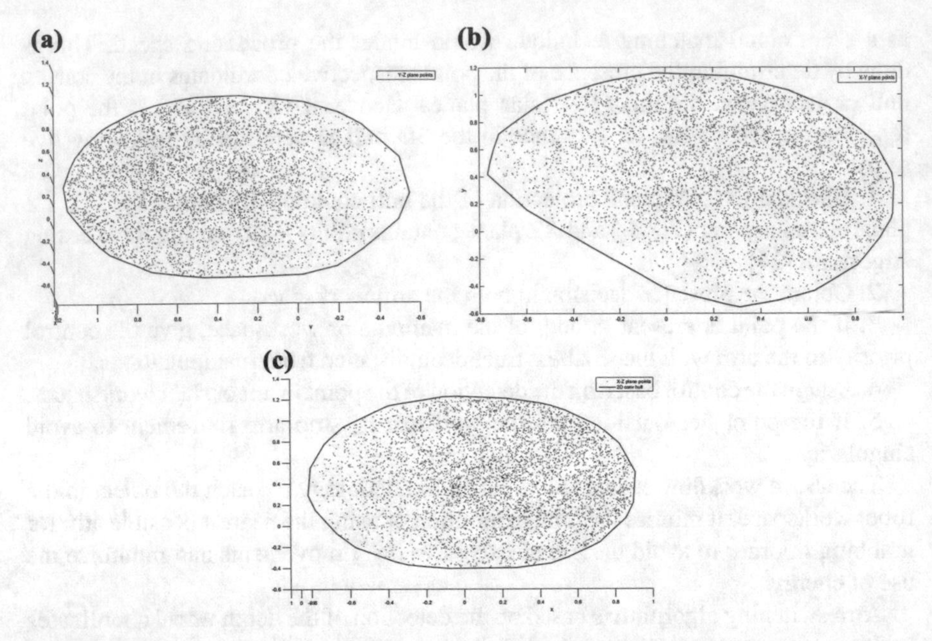

Fig. 5.21 Projections of the convex hull on the coordinate axes. (Left Arm) The *red points* depict the Cartesian coordinates of 3D point cloud on the three axes planes and the *blue lines* represent the convex hull enclosing the outer points. **a** XY plane projection. **b** YZ plane projection. **c** XZ plane projection

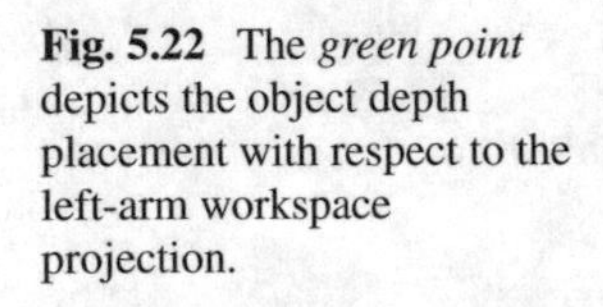

Fig. 5.22 The *green point* depicts the object depth placement with respect to the left-arm workspace projection.

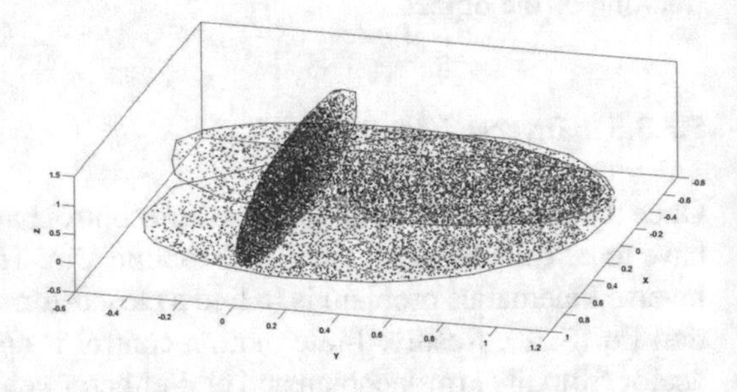

5.3.3.6 Precision Object Gripping

Stereo Vision provides a wider angle of view serving the eye-to-hand servoing system whereas for meticulous sensing of the object, an IR range sensor provides the feedback instead of an eye-in-hand system. This drastically reduces the processing time and provides satisfactory accuracy.

The range sensor used, provides 16-bit digital data and checks for the location of object along the Z and Y axes. The camera and robot X axes are perfectly aligned and hence the real-time error is quiet minimal. But, the error in other axes needs to be neutralized, partially due to the lack of visibility of some portions of the object and objects with ambiguous dimensions. A low pass filter is applied to the range data in order to remove the occasional anomalies.

After reaching 5 cm behind the desired world coordinate, the end-effector proceeds along the *Z* axis till the range value reaches below a threshold of 5 cm. This indicates the presence of an obstacle in the perpendicular *Y* axis. Further, the arm proceeds along the *Y* axis till a manageable distance for gripping is reached (2 cm). The threshold values can be modified depending on the dimensions of the gripper.

5.3.4 *Results*

5.3.4.1 Object Tracking and Following

In order to put the arm switching and object detection algorithm into scrutiny, Baxter is programmed to follow a green-colored object in its task space. The experiment was successful as the manipulator switched control when the object was transferred from left arm workspace to the right arm workspace. The latency of the algorithm implementation was evaluated which came out to be 400 ms. 20 frames per second were processed per second using computer vision resulting in a smooth trajectory while tracking operation.

5.3.4.2 Object Gripping and Placing

Visual servoing and object precision gripping was tested by devising a simple task where Baxter had to localize the object in the workspace and place it from one position to another. The algorithm involved using stereo vision to reach 10 cms behind the object and checking for the object using IR range detector with forward movements of 0.5 cm denominations. When the range crosses a certain threshold value, the object is gripped and placed in a predefined position.

References

1. Corby Jr., N.R.: Machine vision for robotics. IEEE Trans. Ind. Electron. **3**, 282–291 (1983)
2. Hutchinson, S., Hager, G.D., Corke, P., et al.: A tutorial on visual servo control. IEEE Trans. Robot. Autom. **12**(5), 651–670 (1996)
3. Chang, W.-C., Shao, C.-K.: Hybrid eye-to-hand and eye-in-hand visual servoing for autonomous robotic manipulation. In: Proceedings of the 2010 SICE Annual Conference, pp. 415–422. IEEE, (2010)
4. Gao, C., Li, F., Xu, X.H.: A vision open-loop visual servoing. In: Proceedings of the 2006 International Conference on Machine Learning and Cybernetics, pp. 699–703. IEEE (2006)
5. Lee, M.-F.R., Chiu, F.H.S.: A hybrid visual servo control system for the autonomous mobile robot. In: Proceedings of the 2013 IEEE/SICE International Symposium on System Integration (SII), pp. 31–36. IEEE (2013)

6. Fabian, J., Young, T., Peyton Jones, J.C., Clayton, G.M.: Integrating the microsoft kinect with simulink: real-time object tracking example. IEEE/ASME Trans. Mechatron. **19**(1), 249–257 (2014)
7. Cruz, L., Lucio, D., Velho, L.: Kinect and rgbd images: Challenges and applications. In: Proceedings of the 2012 25th SIBGRAPI Conference on Graphics, Patterns and Images Tutorials (SIBGRAPI-T), pp. 36–49. IEEE (2012)
8. Gonzalez, R.C.: Digital Image Processing. Pearson Education India (2009)
9. Newcombe, R.A., Izadi, S., Hilliges, O., Molyneaux, D., Kim, D., Davison, A.J., Kohi, P., Shotton, J., Hodges, S., Fitzgibbon, A.: Kinectfusion: real-time dense surface mapping and tracking. In: Proceedings of the 2011 10th IEEE International Symposium on Mixed and Augmented Reality (ISMAR), pp. 127–136. IEEE (2011)
10. Rusinkiewicz, S., Levoy, M.: Efficient variants of the icp algorithm. In: Proceedings of the 2001 Third International Conference on 3-D Digital Imaging and Modeling, pp. 145–152. IEEE (2001)
11. Pomerleau, F., Magnenat, S., Colas, F., Liu, M., Siegwart, R.: Tracking a depth camera: parameter exploration for fast icp. In: Proceedings of the 2011 International Conference on Intelligent Robots and Systems (IROS) IEEE/RSJ, pp. 3824–3829. IEEE (2011)
12. http://www.impa.br/faprada/courses/procImagenes/
13. Shotton, J., Sharp, T., Kipman, A., Fitzgibbon, A., Finocchio, M., Blake, A., Cook, M., Moore, R.: Real-time human pose recognition in parts from single depth images. Commun. ACM **56**(1), 116–124 (2013)
14. Sinha, A., Chakravarty, K.: Pose based person identification using kinect. In: Proceedings of the 2013 IEEE International Conference on Systems, Man, and Cybernetics (SMC), pp. 497–503. IEEE (2013)
15. Ma, H., Wang, H., Fu, M., Yang, C.: One new human-robot cooperation method based on kinect sensor and visual-servoing. Intelligent Robotics and Applications, pp. 523–534. Springer, Cham (2015)
16. Sun, Z., He, D., Zhang, W.: A systematic approach to inverse kinematics of hybrid actuation robots. In: Proceedings of the 2012 IEEE/ASME International Conference on Advanced Intelligent Mechatronics (AIM), pp. 300–305. IEEE (2012)
17. Hojaij, A., Zelek, J., Asmar, D.: A two phase rgb-d visual servoing controller. In: Proccedings of the 2014 IEEE/RSJ International Conference on Intelligent Robots and Systems (IROS 2014), pp. 785–790. IEEE (2014)
18. Espiau, B., Chaumette, F., Rives, P.: A new approach to visual servoing in robotics. IEEE Trans. Robot. Autom. **8**(3), 313–326 (1992)
19. Mikolajczyk, K.K., Schmid, C.: Scale and affine invariant interest point detectors. Int. J. Comput. Vis. **60**(1), 63–86 (2004)
20. Yang, C., Amarjyoti, S., Wang, X., Li, Z., Ma, H., Su, C.-Y.: Visual servoing control of baxter robot arms with obstacle avoidance using kinematic redundancy. Intelligent Robotics and Applications, pp. 568–580. Springer, Cham (2015)
21. Rastegar, J., Perel, D.: Generation of manipulator workspace boundary geometry using the monte carlo method and interactive computer graphics. J. Mech. Des. **112**(3), 452–454 (1990)
22. Zhang, Z.: A flexible new technique for camera calibration. IEEE Trans. Pattern Anal. Mach. Intell. **22**(11), 1330–1334 (2000)
23. Su, H., He, B.: A simple rectification method of stereo image pairs with calibrated cameras. In: Proceedings of the 2010 2nd International Conference on Information Engineering and Computer Science (ICIECS), pp. 1–4 (2010)
24. Bradski, G., Kaehler, A.: Learning OpenCV: Computer vision with the OpenCV library. O'Reilly Media, Inc., Sebastopol (2008)
25. Hong, L., Kaufman, A.: Accelerated ray-casting for curvilinear volumes. In: Proceedings of the Visualization'98, pp. 247–253. IEEE (1998)

26. Hirschmuller, H.: Accurate and efficient stereo processing by semi-global matching and mutual information. In: Proceedings of the 2005 IEEE Computer Society Conference on Computer Vision and Pattern Recognition, CVPR 2005, pp. 807–814 (2005)
27. Sciavicco, L., Siciliano, B.: Solving the inverse kinematic problem for robotic manipulators. RoManSy 6, pp. 107–114. Springer, Heidelberg (1987)

Chapter 6
Robot Teleoperation Technologies

Abstract This chapter gives an introduction of robot teleoperation and detailed analysis for up-to-date robot teleoperation technologies. The teleoperation based on body motion tracking is first introduced, using a Kinect sensor to control a robot with both vector approach and inverse kinematics approach. Fuzzy inference based adaptive control is then employed in teleoperation, such that the telerobot is able to adapt similarly as the practical case that our humans are able to adapt to other collaborators in a cooperative task. Next, the haptic interaction is implemented using a 3D joystick connected with a virtual robot created by the iCub Simulator, which works with YARP interface and simulates the real iCub robot. Finally, a teleoperation using position-position command strategy is employed to control a slave robot arm to move according to the action of the master side. A simple yet effective haptic rendering algorithm is designed for haptic feedback interaction.

6.1 Teleoperation Using Body Motion Tracking

6.1.1 Introduction of Robot Teleoperation

Robot teleoperation technologies have been widely used in various industrial and medical fields. Using teleoperation, a robot can carry out dangerous work like transportation of radioactive waste, also some elaborate tasks such as remote surgical operation, stoke patients rehabilitation, etc. Specifically, robot can be a useful therapy in neurological rehabilitation for people who suffer from the neurological injuries leading to immobility of upper limb [1]. However, unstructured environments remain a challenge for machine intelligence, which is not feasible for robot to deal with the versatile environment autonomously. The semi-autonomous or human-in-loop method is thus preferred, and plays an important role for robot manipulator to tackle the unknown circumstance.

Teleoperation or "remote control" indicates control, operation, and manipulation of a remote machine, typically a robot arm or mobile robot. A master–slave system is the most popular used teleoperation system and mainly consists of two parts: a master acting as the controller which is responsible for interaction with human

© Science Press and Springer Science+Business Media Singapore 2016

C. Yang et al., *Advanced Technologies in Modern Robotic Applications*,
DOI 10.1007/978-981-10-0830-6_6

and command the slave, and a slave which is commanded by the master to actually complete the task. In general, necessary feedback information sampled by the slave in a distance is transformed to the master, such as position, orientation, force, and torque. So that the human operator can command the manipulator according to this feedback information in real time in the distance. Because the slave system moves with the movement of the master system, the teleoperation manipulation has the advantage of high intelligence. This is the main reason why the teleoperation system such as telesurgeon system has become extremely popular in recent years.

Historically, the typical input device such as keyboard and joystick were widely used in teleoperation system. Through the advance of technology, haptic input devices are being more used in the past decades. The major advantage of such devices is to provide operator tactile feeling of the remote environment through force feedback and thus the immersion into the virtual reality of the local operation platform [2]. It is reported that introduction of force feedback into teleoperated systems can facilitate to reduce the energy consumption, the task completion time and the magnitude of errors [3–5]. Several teleoperation systems have been reported in the literature. For example, a self-assisted rehabilitation system for the upper limbs based on Virtual Reality is developed in [6]. In [7], a SensAble haptic device was used to command a mobile robot. The authors developed a preliminary teleoperation system based on YARP platform and iCub robot via hand gesture recognition [8]. In [9], a micro hand consisting of two rotational fingers is proposed.

In this section, the Kinect sensor is employed to implement the motion control of the Baxter robot, while two different methods using vector approach and inverse kinematics approach have been designed to achieve a given task as well [10].

6.1.2 Construction of Teleoperation System

Kinect Sensor

Recall Sect. 1.2, the Microsoft Kinect Xbox has been used in the visual-servoing system. It has also been widely used in human motion 3D tracking such as body language, gestures, etc., via its depth camera [11]. Human body can be simplified by chopsticks representing position and posture of human in 3D space. In human teleoperated system, Kinect maps human joints' positions and velocities to robot side, which enable human–robot interaction smoothly in real time. Compared with conventional motion tracking system based on RGB data which requires complicated programming and worn by users and high expense, Kinect can be embedded in the control system via open source drivers, e.g., SDK, openNI. Furthermore, it has low cost and need not wearing, making teleoperation tasks naturally and not subject to site constraints. Thus, Kinect is an ideal sensor to be used in the teleoperation system. The Kinect is used in the proposed teleoperation system.

Kinect Develop Software

Many a software is available to interface Kinect with PC, e.g., Libfreenect by OpenKinect, OpenNI, Microsoft Kinect for windows SDK. The choice of software is very important and should be on the basis that: (i) capability of extracting skeletal data; (ii) compatibility with multiple platforms such as Windows and Linux; (iii) good documentation; and (iv) simplicity for fast verification of algorithms. After proper comparison, processing software which satisfies all the requirements is employed. Processing is built on Java so its functionality is very similar to Java. The useful functions of processing that are employed in this work are detailed as below:

PVector: A class to describe a 2D or 3D vector, specifically a Euclidean (also known as geometric) vector. A vector is an entity that has both magnitude and direction. The data type, however, stores the components of the vector (x, y for 2D, and x, y, z for 3D). The magnitude and direction can be accessed via the methods mag () and heading (). See [12] for more details.

pushMatrix() and popMatrix(): Push the current transformation matrix onto the matrix stack. The pushMatrix() function saves the current coordinate system to the stack and popMatrix() restores the prior coordinate system. The functions pushMatrix() and popMatrix() are used in conjunction with the other transformation functions and may be embedded to control the scope of the transformations [13].

Baxter Research Robot

The experiment platform used here is the Baxter robot. Recall Sect. 1.1, the Baxter robot consists of a torso, 2-DOF head and two 7-DOF arms with integrated cameras, sonar, torque sensors and encoders. The setup of the Baxter robot in this experiment is depicted in Fig. 6.1.

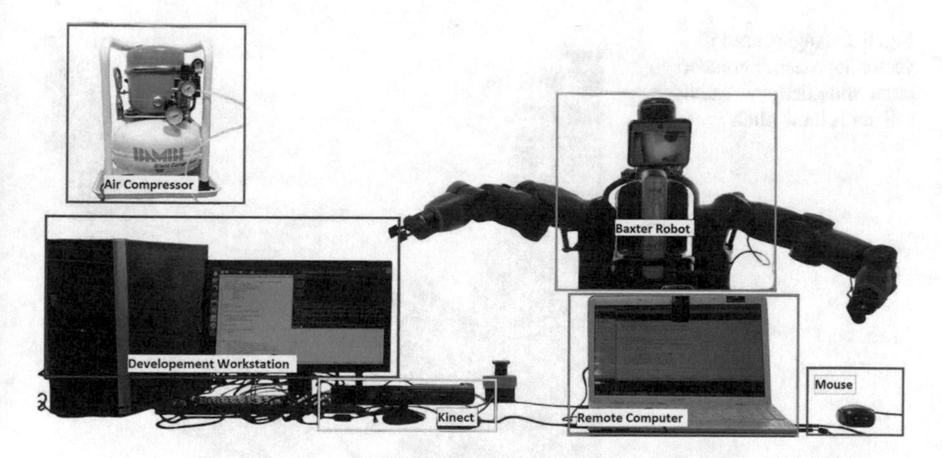

Fig. 6.1 Environment of teleoperation control with Kinect

6.1.3 Design Principles

6.1.3.1 Vector Approach

Kinect can detect all the coordinates of the joints of human body and can return its location coordinates. These coordinates can be converted into vectors and the respective angles of the joints can be calculated.

In this method the Cartesian coordinates of the body joints are extracted from Kinect and the respective angles made by limbs are calculated. They are then sent to Python code that controls Baxter, after mapping them according to our requirement. First, the four angles including shoulder pitch, shoulder yaw, shoulder roll, and elbow pitch, as shown in Fig. 6.2, are calculated from the limb position coordinates that are obtained from the Kinect.

The principle of angle calculation using vector approach is shown in the Fig. 6.3. The bold Line OC and CD represent left upper arm, left lower arm of a human respectively. Bold line BO is a line from left hip to left shoulder, and AO is line from right shoulder to left shoulder. The directed segment BX+, BY+, BZ+ represents the axis of frame in Cartesian space of Kinect, and the point B is the origin of the frame.

Calculation of Shoulder Pitch and Elbow Pitch: As shown in Fig. 6.3, the shoulder pitch angle ($\angle BOC$) is calculated from the angle between two vectors $\overline{OB}$ and $\overline{OC}$. The calculation can be solved by using the positions of three joints namely, hip (point B), shoulder (point O) and elbow (point C). Pass the three points to the angleOf() function, And this function returns the angle that can be directly sent to Baxter. In a similar fashion, elbow pitch ($\angle OCD$, the angle between $\overline{OC}$ and $\overline{CD}$) can be calculated by passing hand, elbow and shoulder points into the angleOf()

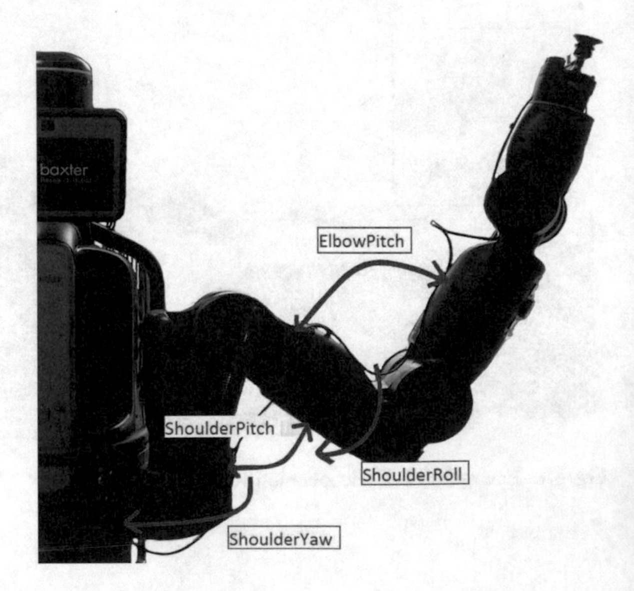

Fig. 6.2 Angles used in vector approach: Shoulder pitch, shoulder yaw, shoulder roll, and elbow pitch

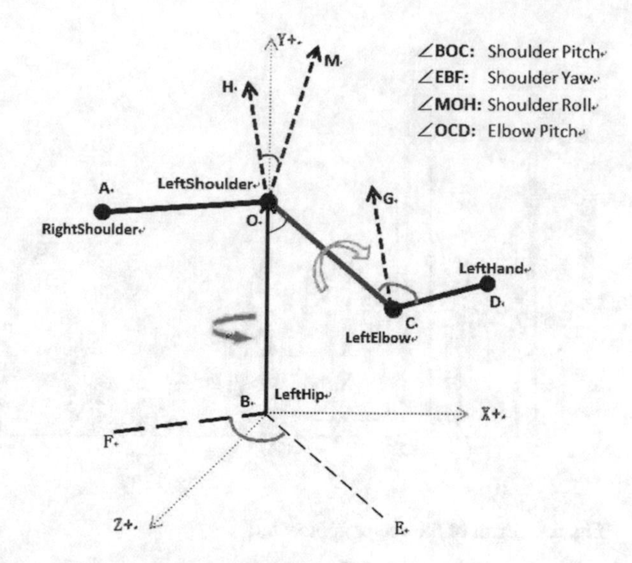

Fig. 6.3 The illustration of angle calculation in vector approach

function. In fact, any angle between two vector in the processing software can be calculated by using the angleOf() function via this method.

Calculation of Shoulder Yaw: As shown in Fig. 6.3, the shoulder yaw angle ($\angle EBF$) is calculated in a similar way by using both shoulder point (point A, O) and elbow point (point C), that is to say, using the vectors $\overline{OC}$ and $\overline{OA}$. But here these two vectors ($\overline{OC}$ and $\overline{OA}$) are projected into the XZ plane to get the vectors $\overline{BE}$ and $\overline{BF}$. Then shoulder yaw ($\angle EBF$, the angle between $\overline{BE}$ and $\overline{BF}$) is calculated by using angleOf() function.

Calculation of Shoulder Roll: Among the angle calculations, shoulder roll is the most challenging task. As the calculation is not intuitive and all the points obtained are in 2D plane the same method used for the previous angle calculations cannot be applied.

The idea here is to find the angle made by the elbow-hand vector in plane perpendicular to the shoulder-elbow vector and passing through the shoulder joint. The reference vector must be stable with respect to the body. So this reference vector is calculated by taking cross product between shoulder-shoulder vector and shoulder-elbow vector.

In this case, the *normal line* got from cross-product of two vectors can be used. First, the vector $\overline{OM}$ can be got by calculating the cross-product of vectors $\overline{OC}$ and $\overline{OA}$. The vector $\overline{OM}$ is perpendicular to plane decided by the vector $\overline{OC}$ and $\overline{OA}$, which is the *normal line* vector of this plane. Obviously, vector $\overline{OM}$ is perpendicular to vector $\overline{OM}$ (left upper arm). In this way, the *normal line* vector $\overline{CG}$ can be calculated from cross-product of vector $\overline{OC}$ and $\overline{CD}$, which is also perpendicular to vector $\overline{OM}$. Then, translating vector $\overline{CG}$ along the vector $\overline{CO}$ to point O can get a vector $\overline{OH}$. The angle between vectors $\overline{OH}$ and $\overline{OA}$, $\angle EBF$, is the shoulder roll angle.

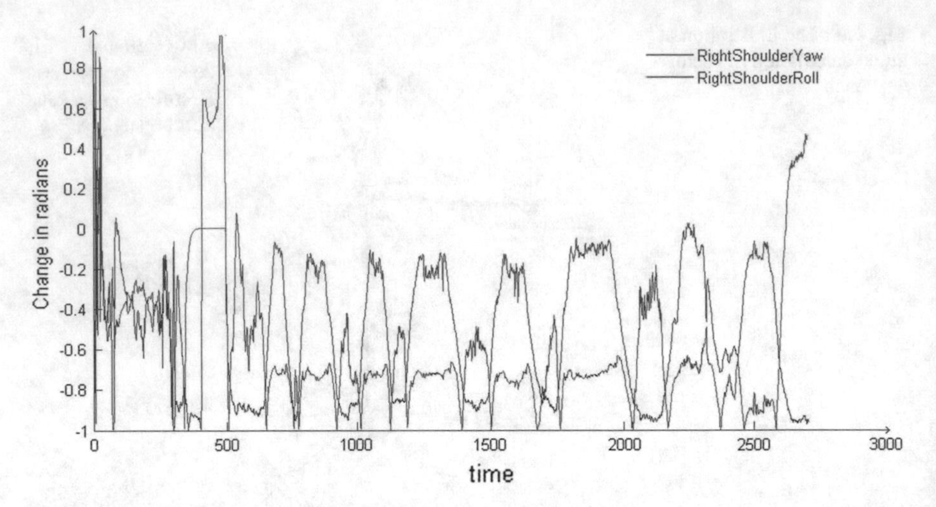

Fig. 6.4 Error of the vector approach

The orientation data sent by the Kinect can be extracted by using PMatrix3D instance in processing software. The orientation matrix is given into the PMatrix3D variable. Then, the present coordinate system is pushed and saved into the stack. The coordinate system is then moved to the shoulder joint and the orientation matrix is used to orient the transformed coordinate axis. All the calculations in this function from here will be calculated in this transformed coordinate system.

After the calculation of roll angle, the original coordinate system is regained by popping the matrix from the stack. The right shoulder roll is also calculated in a similar way. A small change has to be made in the direction of the vectors.

A little error correction is made to the shoulder roll, because the function used to find the roll is not completely accurate. The shoulder roll is observed to be changing with respect to the shoulder yaw. So when the data of shoulder roll and shoulder yaw is plotted, the relationship can be observed in Fig. 6.4. From trial-and-error method the error is partially corrected by the equation shown below:

$$leftShoulderRoll = -leftShoulderRoll - leftShoulderYaw/2 - 0.6 \quad (6.1)$$

Now these returned angles are sent to the Python code in the Baxter development work station using UDP communication. The function shown above will send the data packets through the server object created earlier. Now that all the angles are sent to the Python code, and Python code only has to use these angles to control Baxter robot.

6.1.3.2 Inverse Kinematics

Inverse kinematics is calculation of joint angles from the end-effector coordinate location using kinematic equations and the constraints such as length and angle

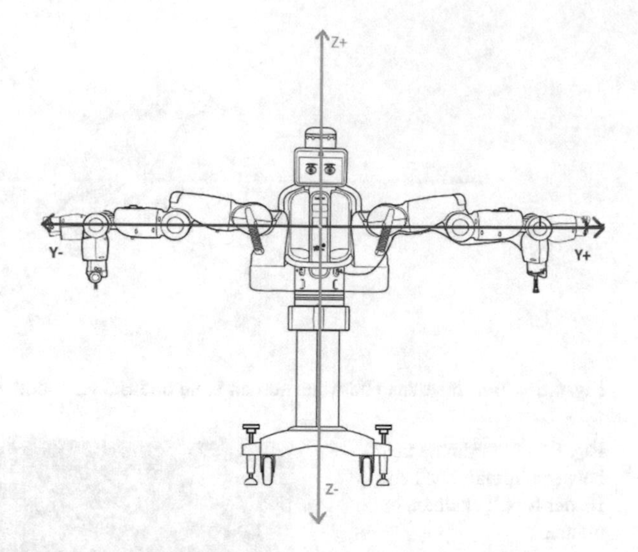

Fig. 6.5 Coordinate map between human hand and Baxter robot: *Front view* of robot

constraints of the robot. Kinect gives the coordinate location of the hand. These coordinate locations can be converted to the required joint angles of the Baxter robot.

Coordinates Extracted: First the coordinates of the hand joints are extracted. These coordinates are used for controlling the end effector by inverse kinematics. Then the elbow coordinates are extracted. These coordinates together with hand coordinates can be used to find the length of the joints.

Mapping human hands with Baxter hands: Human hands are not of the same size of those of Baxter robot. So it is essential to map human hands with Baxter hands for the inverse kinematics method to work correctly. As mentioned earlier the length of the hands can be found by calculating the magnitude of the hand-elbow vector. The Baxter limb joint is found to be 0.47 in its coordinate system.

Analyzing Baxter coordinate system: The Baxter coordinate system convention (see Figs. 6.5 and 6.6) is not same as Kinect coordinate system convention (see Fig. 6.7). So the coordinate axis must be mapped according to the respective conventions.

$$\begin{aligned} X_{Baxter} &= -Z_{Kinect} \\ Y_{Baxter} &= X_{Kinect} \\ Z_{Baxter} &= Y_{Kinect} \end{aligned} \tag{6.2}$$

After mapping the coordinates can be sent to the Python code to control Baxter. An error value of 2.2 is added to the Z coordinate in order to compensate for the error.

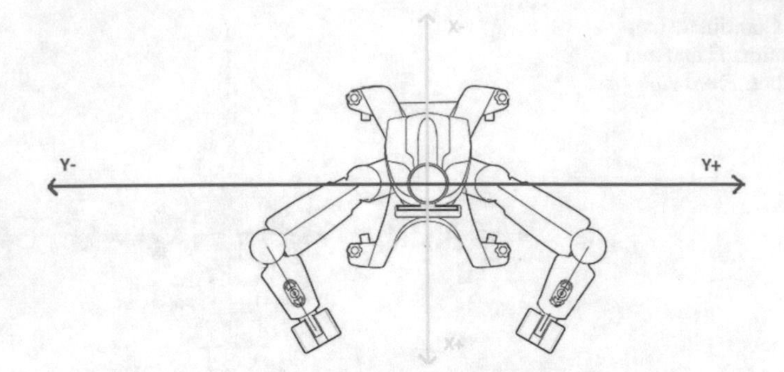

Fig. 6.6 Coordinate map between human hand and Baxter robot: *Top view* of robot

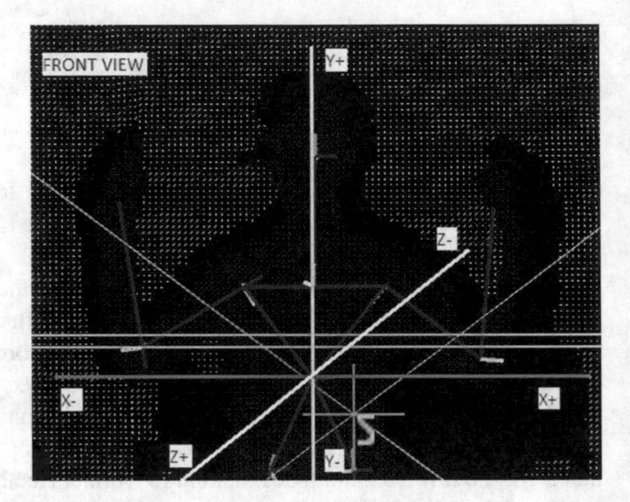

Fig. 6.7 Coordinate map between human hand and Baxter robot: skeleton of human

6.1.4 Experiment Study

6.1.4.1 Results of Vector Approach Method

When every step is done a little error is observed in the received angles and the required angles. So the received angles and the required angles are mapped with the map command in processing software. It is observed that when the required angle is −0.6, but the received angle is −0.1. When the required angle is 0.6, the received angle is −0.9. So these values are mapped accordingly. Similarly all the angles are corrected according to the required angles. In the end the Baxter works as expected but by using this method only 4 of 7 joints are controlled. In order to control the remaining joints, extra data of the palm and fingers is required, which is not available from the Kinect. This problem can be solved in the inverse kinematics approach.

6.1.4.2 Results of Inverse Kinematics Method

Processing is able to extract the coordinates from the Kinect and mouse click events from the mouse. The network protocol is able to send and the development workstation successfully receives the values sent by the processing. The Python code cannot solve all the positions sent by the processing. Baxter responds to all the solved positions at the same time avoiding collision between hands. The grippers and suction were able to respond to the mouse clicks. A new user may find it little difficult as the forward and backward movement of robot limbs is a mirror image to the human hand motion, whereas the sideward motion of the limbs will follow the human hands.

In this section, Kinect sensor is used to control robots with both vector approach and inverse kinematics approach. Both the methods of vector approach and inverse kinematics have its own advantages and disadvantages. The vector approach can exactly follow human arm but only four of the seven joints can be controlled. The inverse kinematics method can be used in pick and place industrial applications but cannot exactly replicate the human motion. Alongside advantages of the proposed control methods, there are some disadvantages as well. The PSI pose has to be made by every user in order for the Kinect to detect the user perfectly. Due to some technical reasons the program does not support too tall or too short people. Some of the data that is sent by the processing software to development work station is faulty and needs to be filtered. Filtered data may not have a joint solution in the inverse kinematics approach, such that sometimes lag/delay is observed in the robot motion.

6.2 Fuzzy Inference Based Adaptive Control for Teleoperation

In last section, the body motion tracking teleoperation using Kinect sensor is introduced. This section will design a control scheme for bimanual robot manipulation [14], in which the leading robot arm is directly manipulated by human operator through a haptic device and the following robot arm will automatically adjust its motion state to match the man's motion. This method combines intelligence of human operator and automatic function of a robot. Several different control algorithms have been designed and the results have shown that the robot can successfully complete the specified task by tracking the leading robot arm controlled by the human operator.

6.2.1 System Modeling and Problem Formulation

Recall Sect. 1.1, iCub robot is a humanoid robot which has 53 motors that move the head, arms, hands, waist, and legs. As is shown in Fig. 6.8, the task of the iCub robot is to move the object along the trajectory specified by a human operator. The leading (right) manipulator of iCub robot is controlled directly by position command set by human operator through Falcon.

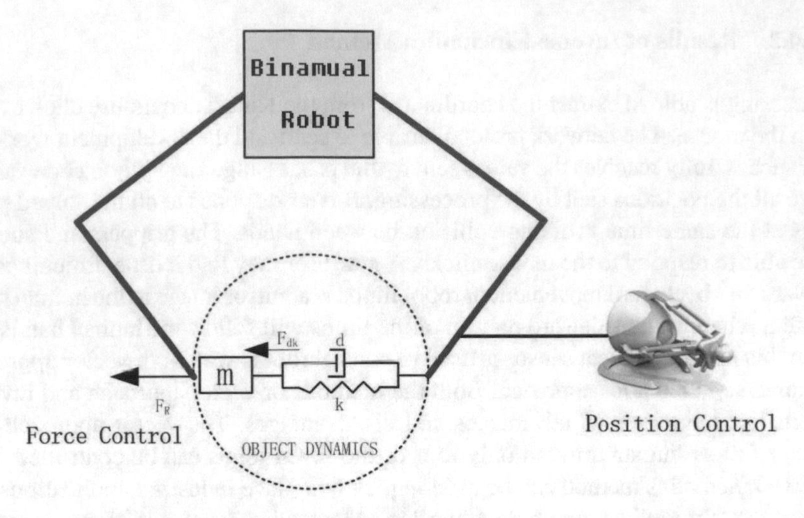

Fig. 6.8 Dynamics model of robot and object interaction

The object between the two hands of the robot is supposed to move synchronously with the leading (right) manipulator, equivalently, there is no relative motion between the object and the leading manipulator. The object dynamics is modeled as a combination of spring-mass-dampers, which is similar to [15] and the effect of gravity is not in consideration.

The following (left) manipulator will move autonomously under the designed controller. Because the following (left) manipulator of iCub robot is in touch with the object, when the object is moving, the force caused by deformation will affect the motion of the former. The force between the following manipulator and the object will be detected and added into the control law of the designed controller.

Once the distance between two hands of iCub robot is too large, the object will fall down on the ground. On the other hand, if the distance is too small, the object will be damaged. Ideally, the following manipulator could track the trajectory of the leading one while the distance between both hands maintains a constant, so the object can be carried naturally and smoothly. In order to simplify the process, we just consider the motion in one dimension. Some notations used in this section are listed in Table 6.1.

6.2.1.1 Robot Dynamic Equation

The dynamics of a single robot manipulator is set by Eq. (6.3):

$$M(q)\ddot{q} + C(q, \dot{q})\dot{q} + \tau_{int} + G(q) = \tau \tag{6.3}$$

where q denotes the joint position of the robot arm, $M(q) \in \mathbb{R}^{n\times n}$ is the symmetric bounded positive definite inertia matrix, and n is the degree of freedom (DOF) of the

Table 6.1 Notations

Symbol	Description
q^L q^R	Joint angles of left hand and right hand
q_d^L	Desired joint angles of left hand
X^L X^R	Actual Cartesian position of left hand and right hand
X_d^L	Expected Cartesian position of left hand
τ	Joint torque
F	Force in Cartesian space
e	Position error
d	Damping coefficient
k	Elasticity coefficient

robot arm; $C(q, \dot{q})\dot{q} \in \mathbb{R}^n$ denotes the Coriolis and centrifugal force; $G(q) \in \mathbb{R}^n$ is the gravitational force; $\tau \in \mathbb{R}^n$ is the vector of control input torque; and $\tau_{int} \in \mathbb{R}^n$ is the external force which represents the force applied to the end effector.

6.2.1.2 Robot Kinematic Model

Each arm of iCub robot comprises 16 joints, but we only study seven joints for shoulder, elbow, and wrist. Recall Sect. 2.1, DH notation method is applied to build kinematics of the Baxter robot. Similarly, the DH notation method is used to identify the robot kinematics model and the DH parameters of iCub robot are listed in Table 6.2 [16].

We employ the homogeneous transformation matrix T_i^{i-1} to represent the transformation from link coordinate frame $\{i\}$ to frame $\{i-1\}$ of the iCub robot as follows:

$$T_i^{i-1} = \begin{bmatrix} c\theta_i & -s\theta_i & 0 & \alpha_{i-1} \\ s\theta_i c\alpha_{i-1} & c\theta_i c\alpha_{i-1} & -s\alpha_{i-1} & -s\alpha_{i-1} d_i \\ s\theta_i s\alpha_{i-1} & c\theta_i s\alpha_{i-1} & c\alpha_{i-1} & -s\alpha_{i-1} d_i \\ 0 & 0 & 0 & 1 \end{bmatrix} \tag{6.4}$$

$$T_7^0 = T_1^0 T_2^1 T_3^2 T_4^3 T_5^4 T_6^5 T_7^6 \tag{6.5}$$

Therefore, when the seven joint angles $(\theta_1 \cdots \theta_7)$ are known, we can calculate the Cartesian position and orientation. To describe the relationship mentioned above, the forward kinematics is defined as follows:

$$X = fkine(q) \tag{6.6}$$

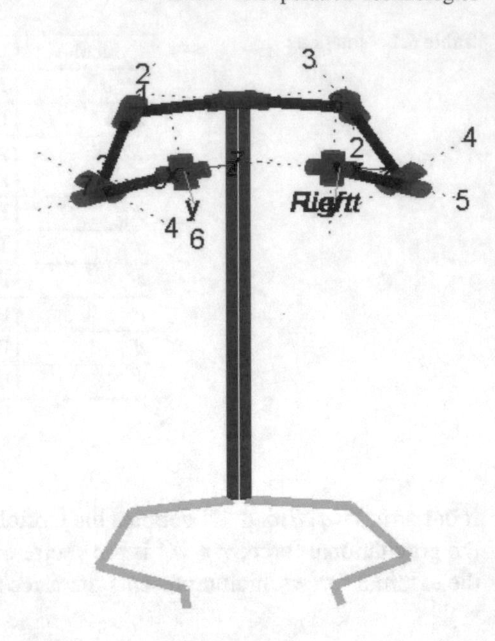

Fig. 6.9 Illustration of a simulated iCub robot

Table 6.2 DH parameters of an arm of iCub robot

Link i	a_i(mm)	d_{i+1}(mm)	α_i(rad)	θ_{i+1}(deg)
$i = 0$	0	107.74	$-\frac{\pi}{2}$	$90 + (5 \rightarrow 95)$
$i = 1$	0	0	$\frac{\pi}{2}$	$-90 + (0 \rightarrow 161)$
$i = 2$	15	152.28	$\frac{-\pi}{2}$	$75 + (-37 \rightarrow 100)$
$i = 3$	−15	0	$\frac{\pi}{2}$	$5.5 \rightarrow 106$
$i = 4$	0	137.3	$\frac{\pi}{2}$	$-90 + (-50 \rightarrow 50)$
$i = 5$	0	0	$\frac{\pi}{2}$	$90 + (10 \rightarrow -65)$
$i = 6$	62.5	−16	0	$-25 \rightarrow 25$

where $X \in \mathbb{R}^6$, which denotes the Cartesian position and orientation, $q \in \mathbb{R}^7$, which represent the seven angles in joint space.

In our experiment, human operator sends position commands to a virtual iCub robot (Fig. 6.9) through the joystick-Falcon to control its motion. Falcon is 3-DOF haptic device introduced in Sect. 1.2.

Seven joints constitutes iCub robot manipulator and every joint makes contributions to the Cartesian position of the end effector, therefore, there is a huge computational load to solve the inverse kinematic problem and improper solution of inverse kinematics would cause many troubles in position control. In order to solve this problem, we employ the method of velocity Jacobian [17]. Its mathematical description is as follows:

$$\dot{\hat{q}} = J_v(\hat{q})^{\dagger} \times \dot{X}$$
$$J_v = \begin{bmatrix} 1\,0\,0\,0\,0\,0\,0 \\ 0\,1\,0\,0\,0\,0\,0 \\ 0\,0\,1\,0\,0\,0\,0 \end{bmatrix} \times J \tag{6.7}$$

where q denotes the position of seven joints in joint space, $\hat{q}$ is the estimation of q, $J_v(\hat{q})^{\dagger}$ is the pseudo inverse of matrix $J_v(\hat{q})$, X denotes the end-effector position in Cartesian space whose dimension is three, J represents the Jacobian Matrix of iCub manipulator. Here, we suppose the initial state of the robot arm is $q(0)$, then the initial position is $X(0)$, which can be calculated according to the kinematics. After the transformation, J_v is a 3×7 matrix.

Assume

$$X(0) = fkine(q(0)) \tag{6.8}$$

Then we can get the solution of inverse kinematics through integration as follows:

$$\hat{q} = \int_{t_0}^{t_f} (J_v(\hat{q})^{\dagger} \times \dot{X})dt + q(0) \tag{6.9}$$

where $X(0)$ and $q(0)$ represent the initial position in Cartesian and joint space. In order to examine the precision of this method, we test it with Falcon joystick sending Cartesian position signals. In this experiment, the iCub's manipulator would move in y direction back and forth and we get the trajectory of both commands and solutions of inverse kinematics for comparison.

The result is shown in Fig. 6.10. We use the velocity Jacobian matrix of iCub robot to calculate the solution of joints, then we can get the corresponding Cartesian

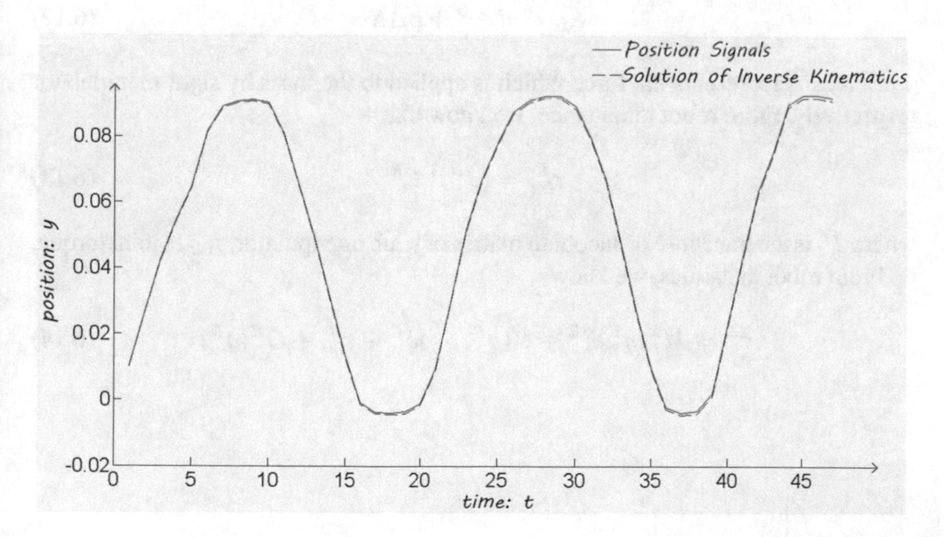

Fig. 6.10 Solution of inverse kinematics

solution—x, y and z with forward kinematics. Apparently, the error between the position signals transmitted by Falcon joystick and the solution of inverse kinematics is very small, i.e., the maximum of error is 1.02×10^{-3}. We repeat this experiment for 50 times and collect the experimental data. Analysis shows the average error is not larger than 1.1×10^{-3}. Thus the performance of the inverse kinematics algorithm mentioned above is satisfactory.

6.2.1.3 Object Dynamic Model

Just as shown in Fig. 6.8, the object is modeled as a combination of spring-mass-dampers. The object will produce slight deformation when it is handled by iCub robot's two manipulators [15]. This process can be expressed as follows:

$$\Delta X = 2r - (X^L - X^R) \tag{6.10}$$

where X^L represents the position of the left manipulator's end effector and X^R represents the right one's.

If $\Delta X \leq 0$, the distance between iCub robot's two manipulators is larger than the diameter of the object and the object is falling down; if $\Delta X > 0$, from Newton's Second Law of Motion, we can obtain that

$$m\dot{X}^R = F^R + F_{dk} \tag{6.11}$$

where F_{dk} is the resultant of the force from spring and the force from the damper, which can be expressed as follows:

$$F_{dk} = d\Delta\dot{X} + k\Delta X \tag{6.12}$$

Here, F^R represents the force which is applied to the mass by right manipulator respectively. From robot kinematics, we know that

$$\tau_{int}^R = (J^R)^T F^R \tag{6.13}$$

where J^T is the transpose of Jacobian matrix of iCub manipulator, τ_{int} is joint torque.

From robot dynamics, we know:

$$\tau^R = M^R(q^R)\ddot{q}^R + C(q^R, \dot{q}^R)\dot{q}^R + \tau_{int}^R + G^R(q^R) \tag{6.14}$$

6.2.2 *Fuzzy Inference Based Control*

6.2.2.1 Feedforward Compensation Controller

The PD control with feedforward compensation consists of a linear PD feedback plus a feedforward computation of the nominal robot dynamics along the desired joint position trajectory [18]. We apply the feedforward compensation and design the control law as follows:

$$\tau = M(q_d)\dot{q}_d + C(q_d, \dot{q}_d)\ddot{q}_d + G(q_d) + \tau_{int} + \tau_{pd} \tag{6.15}$$

where $\tau_{pd} = K_d\dot{e} + K_p e$, $e = q_d - q$, $\dot{e} = \dot{q}_d - \dot{q}$, and q_d is the desired angles of following $left$ manipulator.

In order to keep the distance between the leading manipulator and the following one, their velocity in Cartesian space should keep the same. We are able to obtain the velocity of the leading manipulator, v_d, which is also the desired velocity of the following manipulator. Via the Jacobian matrix, we know that

$$v_d = \dot{q}_d \times J(q) \tag{6.16}$$

Then, we can get

$$\dot{q}_d = v_d \times J^{\dagger}(q) \tag{6.17}$$

After integration, we can get the desired angles of the following manipulator q_d.

6.2.2.2 Fuzzy Feedforward Compensation Controller

In the feedforward compensation controller given in Eq. (6.15), we need to select suitable parameters, K_p and K_d and keep them constant while the control process. But when the external and internal environment change, the parameters may not be the best choice again. In this section we choose the fuzzy control to describe these parameters.

Fuzzyfication and Defuzzyfication

We select error $|e|$ in joint space and its derivative $|\dot{e}|$ as the input of fuzzy PD controller. The two crisp variables must be transformed to fuzzy variables in order to match the fuzzy rule described by the language, that is fuzzy subset. This process is called fuzzyfication.

We choose Gaussian curve as the membership function of the fuzzy subset, whose mathematical description is given by:

$$f(x, \sigma, c) = e^{\frac{-(x-c)^2}{2\sigma^2}} \tag{6.18}$$

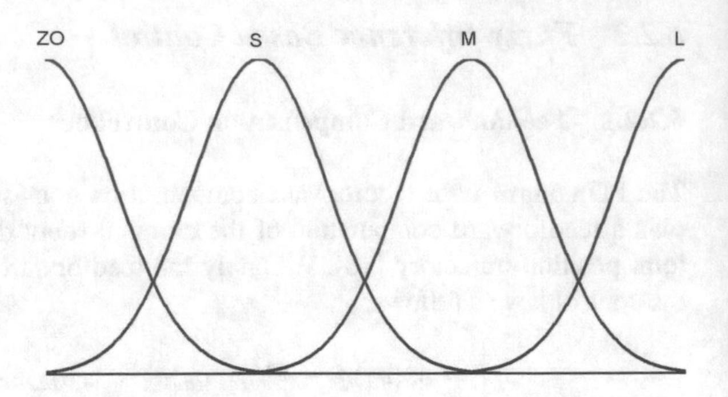

Fig. 6.11 Gaussian curve built in membership function

We can select the suitable parameters in Eq. 6.18, σ and c, to express the four fuzzy subsets in the fuzzy PD controller-A(ZO), A(S), A(M) and A(L), which represent the fuzzy variable is zero, small, medium, and large respectively, as shown in Fig. 6.11. We can express the fuzzy set by Zedeh Method [19]:

$$A = \frac{A(ZO)}{ZO} + \frac{A(S)}{S} + \frac{A(M)}{M} + \frac{A(L)}{L} \tag{6.19}$$

If we define A as the input fuzzy set, similarly, we can define set B as the output fuzzy set. B(ZO), B(S), B(M) and B(L) are the four fuzzy subsets. $f_A(u)$ and $f_B(y)$ are the membership functions of input and output. When one crisp input variable, such as u, needs to be fuzzificated, we can get its membership values(u_1, u_2, u_3, u_4) in four input fuzzy subsets-A(ZO), A(S), A(M) and A(L) through Eq. (6.18). According to Eq. (6.19), we can know the variable, u, can be expressed as the form of fuzzy set as follows:

$$u = [u_1 u_2 u_3 u_4] \tag{6.20}$$

After fuzzy logic inference, the output is a fuzzy set. We need to transform it to some representative value. There are a lot of methods to achieve defuzzyfication, such as centroid, bisector, and maximum. Here, we adopt centroid.

Suppose the output result is a fuzzy set, $Y = \{y_1, y_2, y_3, y_4\}$, the, f_B then the actual output can be expressed as follows:

$$u_{cen} = \frac{\int f_B(y) y dy}{\int f_B dy} \tag{6.21}$$

Fuzzy Rule

In Eq. (6.15), we know that

$$\tau_{pd} = K_d \dot{e} + K_p e \tag{6.22}$$

Therefore, Eq. (6.22) can be seen as a PD controller. Combining the traditional PD control algorithm and fuzzy control theory, this section designs a fuzzy PD controller

Table 6.3 Fuzzy rules for ΔK_p

$\dot{e}$	e			
	L	M	S	ZO
L	M	S	M	M
M	L	M	L	L
S	L	M	L	L
ZO	L	M	L	ZO

Table 6.4 Fuzzy rules for ΔK_d

$\dot{e}$	e			
	L	M	S	ZO
L	S	M	ZO	ZO
M	M	M	S	ZO
S	L	L	S	S
ZO	L	L	S	ZO

which can adjust the parameters of the PD controller online. The following is a simple introduction of the fuzzy rule.

We can conclude the relationship between error e, derivative of error $\dot{e}$ and the PD controller's parameters-K_p, K_i and K_d from experimental experience and theoretical analysis as follows:

- When $|e(t)|$ is large relatively, which represents L, in order to accelerate the response speed of the system, K_p should be large enough. K_p should not be too large, otherwise the system may be unstable; the parameter of K_d should be small to make it quicker.
- When $|e(t)|$ is a medium value, which represents M, K_p should be small which can restrain overshoot effectively; K_d should be appropriate.
- When $|e(t)|$is small relatively, which represents S, in order to get good steady-state performance, K_p should be large enough; K_d should be appropriate to avoid the vibration around the balance point.

Based on the relationship above, we can conclude the fuzzy rules shown in Tables 6.3 and 6.4.

In the tables above, L, M, S, and ZO represent the fuzzy variable is large, medium, small, and zero, respectively. K_{p0} and K_{d0} are the initial values of the PID controller parameters, ΔK_p and ΔK_d denote the modification of K_{p_0} and K_{d_0}, so the real-time parameters of PID controller, $\overline{K}_p$ and $\overline{K}_d$ are as follows:

$$\begin{aligned} \overline{K}_p &= K_{p_0} + \Delta K_p \\ \overline{K}_d &= K_{d_0} + \Delta K_d \end{aligned} \tag{6.23}$$

In the iCub manipulator's seven joint angles, all of them can affect the position of the end effector. In this experiment, we just want to control the end effector moving along a line such as in the y direction of reference coordinate system. There is no need to design controllers based on seven joints. Here, we choose three of them to finish the task. So we need three PID controllers and six fuzzy controllers.

6.2.3 Simulation Studies

The experiment is simulated on the computer. The motion model is simplified and only one dimension of the motion is considered which means the object is only moving in the y direction. Noise and disturbance is also neglected. In the process of simulation, one manipulator's Cartesian position is controlled by human operator through haptik device Falcon and its trace is arbitrary. While the robot manipulator is moving in the Cartesian space, the orientation is keeping a constant which means we just pay attention to the three joint angles that are related to the position of the end effector.

The main results of the simulation are shown in Fig. 6.12a, b. From the simulation results, we can see that both methods can achieve the goal. In Fig. 6.12a, b the Cartesian trajectories under different controllers are shown respectively. The blue one denotes the path that the left manipulator moves along Fig. 6.13 shows the error between left arm and right arm. Through calculation, we can get the mean of error under feedforward compensation controller is 0.117 and the variance is 0.0125; the mean of error under fuzzy feedforward compensation controller is 0.0409 and the variance is 0.0017. That means the tracking result under the fuzzy feedforward compensation controller has higher precision, less overshoots and its fluctuation is less than feedforward compensation controller. Figure 6.14a, b shows the tracking errors in joint space. Obviously, the fuzzy feedforward compensation controller performs better.

The feedforward compensation controller's structure is relatively simple and this algorithm can be calculated faster than the other. Because the fuzzy feedforward compensation controller can adaptively regulate parameters, it has a great advantage over the traditional feedforward compensation controller in the transient characteristics. Overall, fuzzy feedforward compensation controller's result is more satisfying.

In this work, we have designed one simple yet nontrivial experiment so as to model the practical case where human operators can adapt to other persons in the task of lifting a load together even if the following has no a priori knowledge to the leading's force profile. In our experiment, one robot arm plays the role of leading arm, and the other arm tries to adapt to the leader in the absence of the leader's exact mathematical model by using the feedback information through the object connected with the two arms.

From the preliminary experiments conducted, the merits of the introduced HRI model may be summarized as follows:

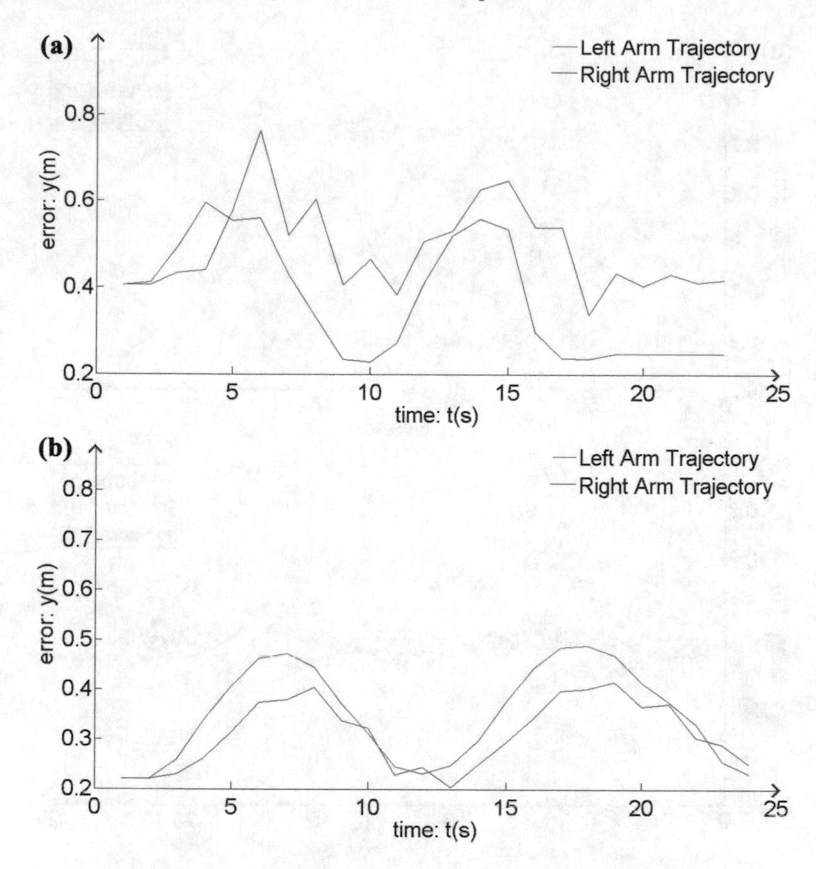

Fig. 6.12 Tracing performance. **a** Force feedback impedance controller. **b** Fuzzy PD controller

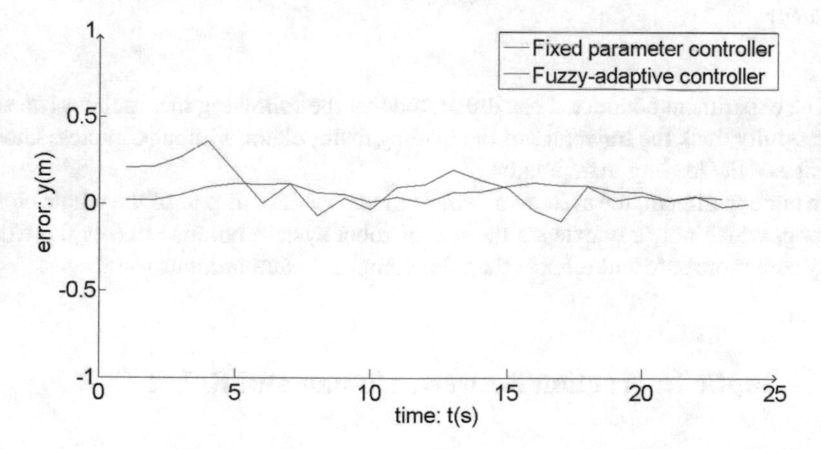

Fig. 6.13 Error comparison in Cartesian space

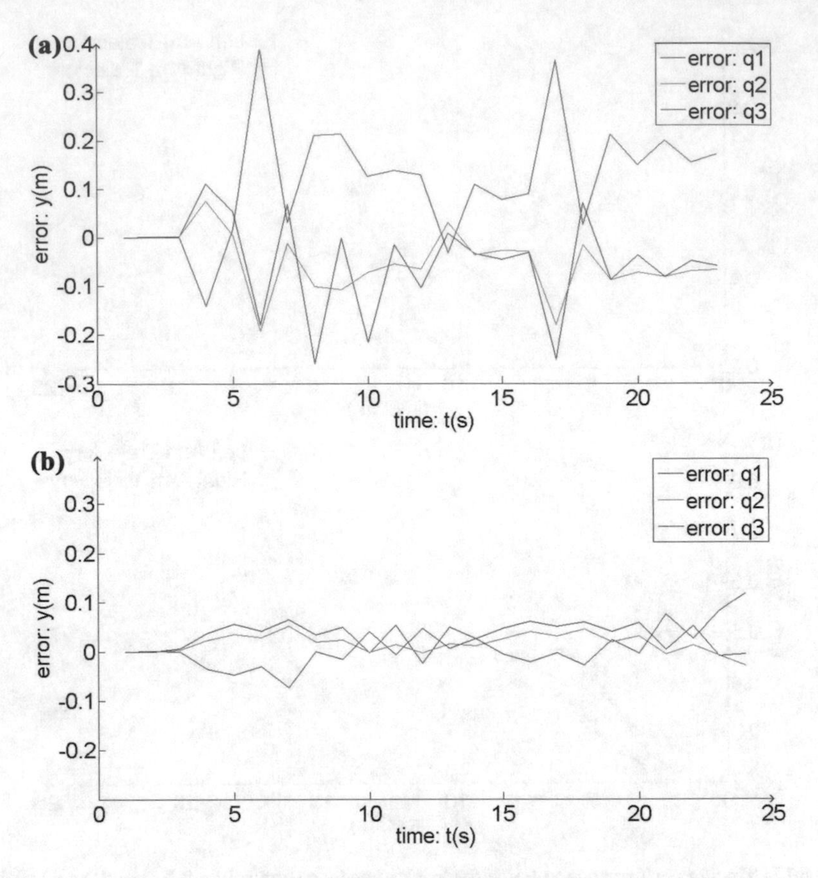

Fig. 6.14 Error comparison in joint space. **a** Force feedback impedance controller. **b** Fuzzy PD controller

- The experiment conducted has illustrated that the following manipulator can successfully track the trajectory of the leading manipulator without complete knowledge of the leading manipulator.
- In our experiment, the system includes human operator as part of the whole closed loop, which not only extends the use of robot system but also makes the whole system more safe and reliable than the completely autonomous robot.

6.3 Haptic Interaction Between Human and Robot

In the last two sections, robot teleoperations using body motion tracking and joystick manipulation were introduced. Devices such as joysticks are able to provide force feedback to the human operator, whom could be enabled to sense the robot in a haptic manner. As a matter of fact, haptic interaction is attracting more and more interests

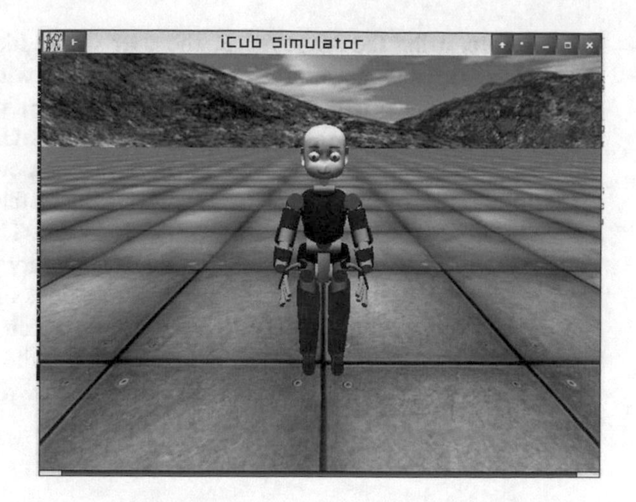

Fig. 6.15 iCub simulator 3D interface (screen snapshot)

in the research community. The concept of the haptic interaction is to feel by the tactile sense of robot environment. Haptic interaction can lead to many applications in different domains such as the rehabilitation or the body enhancement [20–22], the telemanipulation of device or the transfer of knowledge through robots [23]. In this section, we will show how to achieve haptic interaction between the human and a virtual iCub robot [24].

6.3.1 Tools Selection and System Description

6.3.1.1 Novint Falcon and iCub Simulator

Regarding the haptic interaction between human and virtual robot, several things have to be taken into account. A haptic interaction needs to be realized between two devices. So this kind of interaction needs a specific tool to transfer force feedbacks to the user. Several haptic devices are available nowadays, providing more or less the same capacities. Novint Falcon joystick introduced in Sect. 1.2 is used in many applications such as video games and ensures realistic communication capacities, and accurate force feedbacks giving the illusion of a tactile perception during a simulation. It provides a force-feedback up to 1kg allowing a good illusion of the tactile rendering.

The other part of the interaction is the virtual robot. The iCub Simulator (Fig. 6.15) [25], simulation of the real humanoid iCub [26], part of the European project Robot-Cub offers many advantages. First, iCub Simulator is an open source virtual robot, and it also provides an environment where the simulated iCub can already interact with. Moreover, its design follows the realistic version of the robot, so results on the simulator could be directly tested on the real robot.

Therefore, using the 3D joystick, the user would like to control the real iCub robot or the virtual counterpart, iCub Simulator, and with realistic force feedbacks sent to the user thanks to motors into it. In this part, we aim to resolve the problem of controlling a virtual iCub robot with a haptic device Novint Falcon by exploring two different approaches: one is based on CHAI3D, one powerful open-source haptics C/C++ library; the other one is based on MATLAB with the aid of HAPTIK, one open-source haptics library/toolbox for use with MATLAB. The former approach has the advantage of real-time, fine portability and extendability with C/C++ programming, however this approach may suffer from the difficulties of learning C/C++ language and implementing complex scientific algorithms which involve many matrix/vector operations or specific mathematical processes; while the latter the advantage of convenience in algorithmic research and integration with MATLAB, however this approach will be limited by restrictions of MATLAB.

6.3.1.2 Communication Architecture

The Novint Falcon communicates through the computer, data to a program based on libraries which will analyze the data, and will send it to the iCub Simulator. Then the iCub Simulator's data could be received in order to compute forces sent to the joystick to the motors allowing the tactile perception. Figure 6.16 illustrates the analysis of the working environment required to communicate. Later we will discuss the libraries used and the development platform in details.

The Falcon joystick is commonly installed on Windows. The iCub Simulator is also available on Windows or Linux with source files downloadable. Note that Falcon itself does not know anything on the iCub Simulator, hence the biggest problem of controlling the virtual iCub robot via the Falcon joystick is to resolve the communication between the joystick and the iCub Simulator.

Fortunately, the iCub robot and simulator are built on the top of one open-source framework, the so-called YARP (yet another robot platform), which allows data and command exchanging among any applications through the application programming interface of YARP.

Once the communication made, data received from the joystick have to be analyzed and translated in order to be understandable by the simulator. The data from

Fig. 6.16 Communication design

the motors of the Falcon provide only a 3D position and button values. Then a loop has to be made which will allow moving the robot arm with the joystick in real time. Parameters needed to be taken into account such as the fact that the Novint Falcon had its own coordinates system and the iCub robot too, the way of moving the arm to keep the maximum of fluidity and to allow the iCub robot to reach an area as large as possible. Some tests will finally have to be done in order to match correctly the human movement and the iCub one.

The way of communication, from the joystick to the simulator, had to be enhanced to receive data from the simulator as well. To create the haptic interaction, specific data from the virtual robot are needed in order to understand its position and its behavior into its environment. Then the application of force-feedbacks through the motors will allow the creation of the haptic interaction. The final result has to be as realistic as possible.

6.3.1.3 Falcon and iCub Structure

The Falcon device is a 3D joystick using a USB interface with commands sent from the computer by the firmware to provide perception. Motors used are three Mabuchi RS-555PH-15280 with motion monitored by a coaxial 4-state encoder with 320 lines per revolution. The reachable workspace allocated will allow a suitable working area where the robot will be able to move its arm. The Falcon device can be tested with the provided test software as shown in Fig. 6.17.

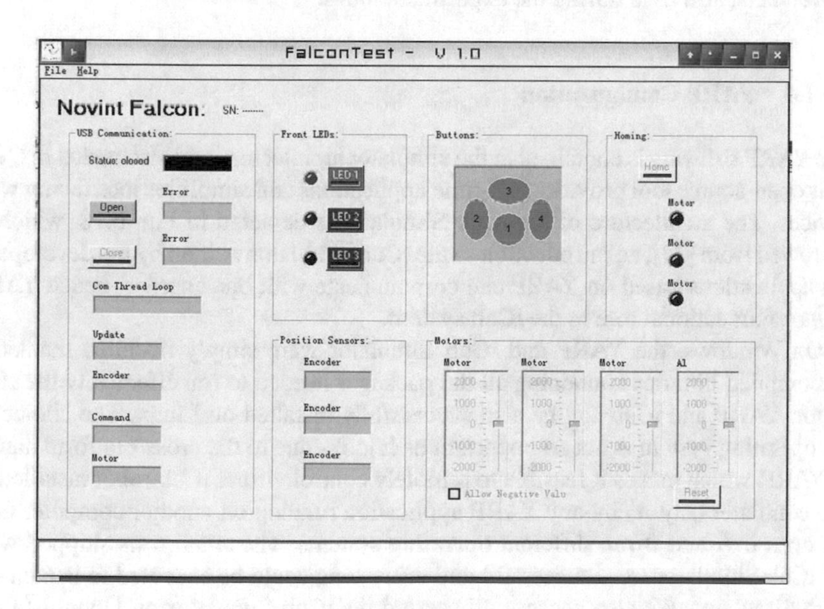

Fig. 6.17 Testing Novint Falcon (screen snapshot)

The iCub Simulator mostly coded in C++ is a 3D representation of the real iCub robot allowing movement of the joints, vision system, and pressure sensors on the hands. Its design using data from the real robot provides a virtual "clone" of the real robot. In the real and virtual iCub robot, there are 53 degrees of freedom which 12 for the legs, 3 for the torso, 32 for the arms and 6 for the head.

The iCub Simulator uses several external open-source libraries which can simplify the complex programming jobs occurred in the simulation of iCub robot. Following is one incomplete list of libraries depended:

- ODE (Open Dynamics Engine) libraries: used to simulate the body and the collision detection.
- ACE (Adaptive Communication Environment): used to implement high-performance and real-time communication services and applications across a range of OS platforms.
- GLUT (OpenGL utility toolkit): used to simplify programming with OpenGL 3D graphics rendering and user interface.
- GTKMM libraries: used to create the user interfaces for GTK+ and GNOME.
- IPOpt (Interior Point Optimizer): used to solve optimization problems.
- OpenCV libraries: used in the creation of real-time computer vision.
- QT3 libraries: generally used in the development of software with user interface.
- SDL (Simple DirectMedia Layer): used to provide low-level access to audio, keyboard, mouse, joystick, 3D hardware via OpenGL, and 2D video framebuffer.

In the virtual arm including the hand, 16 degrees of freedom ensure the left arm movements, arm used during the experimentations.

6.3.1.4 YARP Configuration

The YARP software is contained in the simulator architecture and also coded in C++. This open-source tool provides real-time applications and simplifies interfacing with devices. The architecture of the iCub Simulator is depicted in Fig. 6.18, which is borrowed from [27], an introduction to the iCub Simulator written by the developers. All applications based on YARP can communicate with one another; hence YARP plays a foundational role in the iCub system.

On Windows, the YARP and iCub Simulator were simply installed thanks to precompiled libraries containing all the packages needed to run effectively the simulator. YARP and iCub can be also successfully installed on Linux. The choice of the operating system is not so important as it looks due to the cross-platform nature of YARP, which makes it feasible to remotely control virtual iCub robot installed in one computer (say A) by any YARP application running on another computer (say B), even if A and B run different operating systems. The small tests shipped with the iCub Simulator were successful and everything could be executed to launch the iCub Simulator. We also successfully tested the iCub Simulator on Ubuntu Linux and AndLinux, among which the latter is one native Linux environment working cooperatively with Windows.

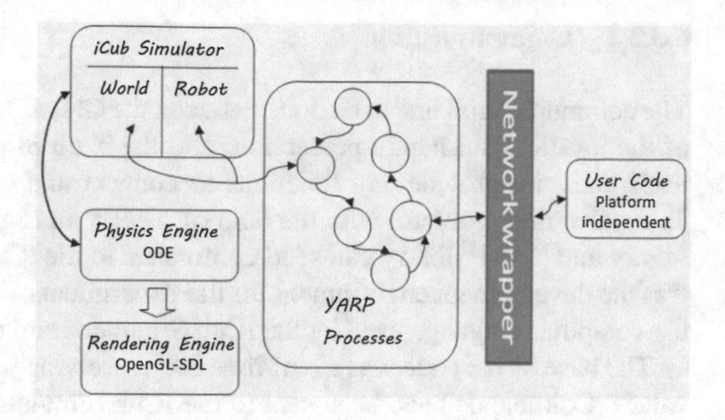

Fig. 6.18 Architecture of the iCub simulator

When the Falcon device and the iCub Simulator are installed on the same operation system of one computer, there is no special configuration for YARP needed. To make our studies more practical and useful, it would be desirable to install the Falcon device on one computer (say F) while to run the virtual iCub robot on another remote computer (say R). To make it possible to remotely control R via F, it is necessary to configure both F and R with the command-line command "yarp conf <ip>", where <ip> denotes the true IP address of computer R. Note that F and R should connect with each other (use ping command to test the connection), ideally in the same ethernet.

6.3.2 Implementation with CHAI3D

6.3.2.1 CHAI3D Library

The haptic interaction needs to compute forces with data received from the virtual environment, before the sending to the motors. A set of open-source C++ library functions called CHAI3D [28] allows the use of several haptic devices including the Novint Falcon in the Windows [32-bit] version. Moreover they offer haptic functions such as the computation of continuous forces sent to the device, simplifying the low level of the haptic interaction.

Finally the CHAI3D library is downloadable with examples which could be compiled and executed with Visual Studio 2010. These examples provided a good idea of the Novint capacities and the CHAI3D. Examples are smooth and the haptic interaction is really realistic. Even some bugs reported such as the modification of the button IDs during the example executions, the CHAI3D library is excellent for starting haptic development due to its easy-to-use programming interface and easy-to-understand example codes.

6.3.2.2 Communication

The communication had to be done between the CHAI3D library simplifying the use of the joystick and haptic perception, and the iCub Simulator. Note that CHAI3D itself does not provide any functions to connect and operate the iCub Simulator. To resolve this problem, with the help of YARP, a program based on the CHAI3D library and YARP library can send commands to the iCub Simulator. Visual Studio was the development environment for the experiments, coded in C++ because it was the computer language used by the iCub Simulator and the CHAI3D library used.

The base of the code was a real-time loop receiving joystick positions and button values. Commands have been sent to the iCub Simulator through the YARP while the real-time loop received data. The communication became possible and correct information had to be transferred.

6.3.2.3 iCub Arm Control

To remotely control the robot, the first required development was the arm control. Offsets were applied to define the workspace of the virtual arm. They provided a suitable workspace environment where the remote control of the iCub hand was possible.

Then the choice concerning was decided. As a mirror, the hand movements were matched to the iCub Simulator hand, as the user moved its own hand. The Cartesian systems used for the iCub Simulator and the Novint Falcon joystick are shown in Fig. 6.19.

The control of the arm was performed by sending positions with a command via YARP interface to the iCub Simulator. Then it moved its joints according to its kinematics in order to reach the desired position (Fig. 6.20).

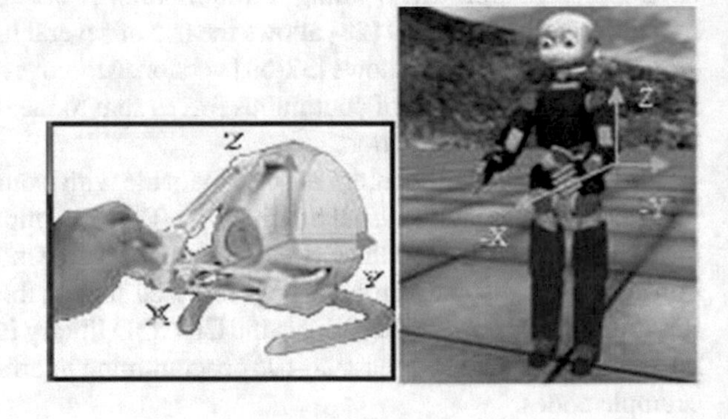

Fig. 6.19 Cartesian world of joystick and simulator

Fig. 6.20 The configured simulator interface in our simulations (screen snapshot)

Fig. 6.21 Representation of the magnet simulation (screen snapshot)

6.3.2.4 Simulations Implementation

The first idea was the use of a function called “magnet” which allows the robot to “glue” the shape to the hand, or more precisely, to keep a constant distance between the hand simulator and the shape. This function can be used if the iCub hand is not in running mode which means the hand sensors are off. This solution allowed the application of a small force-feedback when the object was grabbed. To be a bit more realistic, the distance between the shape and the simulator hand should be small in order to be able to grasp the object.

A second solution was tried with the sensors. Using the hand sensors providing 12 boolean values which allowed knowing when the hand touched something, the robot had to detect a table placed in front of him. Then, a retro-action into the joystick, equal to the inverse of the force applied by the human on the “table”, would be implemented to “feel” the table (Figs. 6.21 and 6.22).

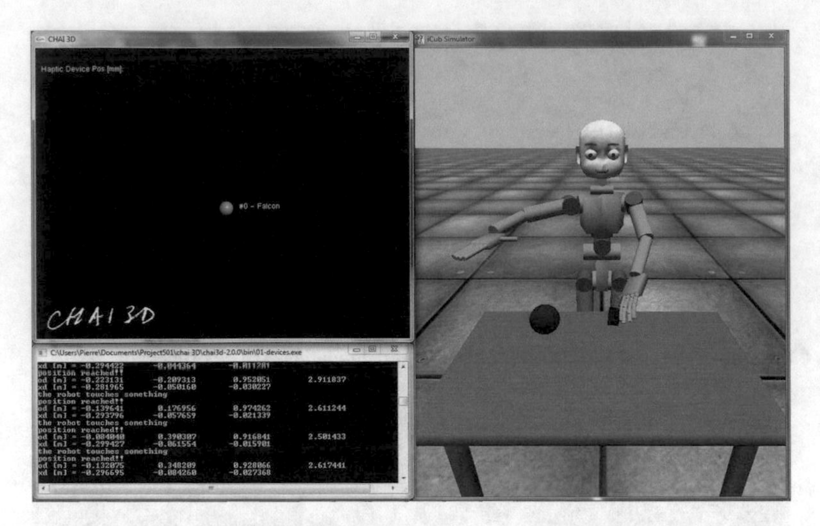

Fig. 6.22 Simulation using hand sensors values (screen snapshot)

6.3.2.5 Reducing Time Delay in Arm Control

By the above mentioned approaches, it can be seen that the virtual iCub robot can be successfully controlled by the Novint Falcon joystick, yet with some time delay if the joystick moves fast. This time delay may be caused by hardware issues and software issues, among which the former may be restricted by the performance of the computer or the Novint Falcon device while the latter may include the implementation details of our program or design defects of CHAI3D.

Forgetting the real-time interaction, the last solution tried was the programming of buffers which could save the robot hand positions thus the simulator could reach the points with its own pace. Therefore, a better accuracy of movement with less computer bugs was expected.

To realize this solution, two threads have been implemented, one running the real-time loop allowing the reception of the Novint Falcon joystick data, and the other running the sending of commands to simulator via the YARP. Tests where realized, and the Novint Falcon joystick could perform movements, and then executed by the iCub Simulator with its own speed.

6.3.3 Implementation with MATLAB

Besides the solution of using CHAI3D, we also tried the other solution based on MATLAB. To enable using haptic devices in MATLAB, we rely on one open-source library, HAPTIK, which provides also MATLAB bindings for scripting with MATLAB.

6.3.3.1 HAPTIK Library

The HAPTIK library is an open-source lightweight library with a component based architecture that acts as a hardware abstraction layer to provide constant access to HAPTIK devices. HAPTIK was developed with the aim of providing an easy but powerful low-level access to devices from different vendors [29]. Prior to HAPTIK, when focusing on hardware access, all current libraries present some drawbacks and fail to satisfy all the following needed requirements:

- Available devices are enumerated by the library, and applications can easily let the user choose one of them.
- Loading of hardware specific plugins is performed transparently at runtime, allowing the same executable to run on systems with different hardware configurations, different driver versions, or even without any device.
- The fallback loading scheme allows the same executable to use the most recent hardware and driver versions.

These requirements motivate the development of HAPTIK, which overcomes such limitations, achieving many advantages for the end user as well as the developers.

HAPTIK has a very simple API (Application Programming Interface). A small amount of code lines are enough to start using any haptic device. Devices operations such as enumeration, automatic default selection, device information querying and auto-recalibration are already built in and ready to use. HAPTIK is developed in C++ programming language but it can be used in MATLAB and Simulink since it provides MATLAB bindings. Unlike CHAI3D, HAPTIK does not contain graphic primitives, physics- related algorithms, or complex class hierarchies. It exposes instead a set of interfaces that allows the differences between devices to be hidden and thus making applications device-independent. The library has been built with a highly flexible infrastructure of dynamically loadable plugins.

With HAPTIK library, it is convenient to obtain 3D position of the Novint Falcon joystick via MATLAB codes shown in Listing 6.1.

Listing 6.1 MATLAB codes for reading position of Falcon joystick

```
1 function XYZ = callFalcon(a)
2 global h;
3 h = haptikdevice;
4 pos = read_position(h);
5 XYZ = pos;
6 end
```

6.3.3.2 YARP for MATLAB

To interact with the iCub Simulator via MATLAB, it is necessary to call YARP within MATLAB. Fortunately, YARP provides also MATLAB binding through JAVA with the help of SWIG (Simplified Wrapper and Interface Generator), which is an open-source software tool used to connect computer programs or libraries written in C or

C++ with scripting languages such as Python and other languages like Java. According to the instructions given in [30], after steps of installing SWIG, compiling YARP binding (a dynamic loading library jyarp.dll), compiling JAVA classes, and configuring MATLAB paths, it is ready to call YARP in MATLAB by running "LoadYarp" command in MATLAB. The codes shown in Listing 6.2 load YARP in MATLAB, initialize iCub Simulator, and set positions of left arm through YARP interface.

Listing 6.2 MATLAB codes for using YARP to initialize iCub Simulator

```
1  LoadYarp; % imports YARP and  connect to YARP network
2  options = yarp.Property;
3  options.put('device','remote_controlboard');
4  options.put('remote','/icubSim/left_arm');
5  options.put('local','/matlab/left');
6  robotDeviceLeft = yarp.PolyDriver(options);
7  if isequal(robotDeviceLeft.isValid,1)
8      disp '[success] robot available';
9  else
10     disp '[warning] robot NOT available, does it exist?';
11 end
12 posLeft = robotDeviceLeft.viewIPositionControl;
13 encLeft = robotDeviceLeft.viewIEncoders();
14 nDOF = encLeft.getAxes();
15 theta = [ -45 40 60 90 -45 -65 10 45 10 0 0 0 0 0 0 0 ]*180/pi;
16 home = yarp.DVector(nDOF);
17 encodersLeft = yarp.DVector(nDOF);
18 for k=0:15,
19         home.set(k,theta(k+11));
20 end;
21 posLeft.positionMove(home);
```

6.3.3.3 Robotics Toolbox

To control the arms of the virtual iCub robot by MATLAB, we use the Robotics Toolbox which has been introduced in Sect. 1.3.

With Robotics toolbox, we have created iCub left arm and right arm as manipulators. For easy implementation of control algorithms on the virtual iCub robot, a Simulink model, icubSIM.mdl, is created for simulations of iCub Simulator through the haptic device. Listing 6.3 shows some codes to create the iCub arms with Robotics Toolbox.

Listing 6.3 Creating robot arms with Robotics toolbox

```
1 % Create iCub arms
2 %                 \Theta d        a       \alpha   \sigma
3 Left(1)   = Link([ 0    0.10774   0       -pi/2   0 ], 'standard');
4 Left(2)   = Link([ 0    0         0        pi/2   0 ], 'standard');
5 Left(3)   = Link([ 0    0.15228   0       -pi/2   0 ], 'standard');
6 Left(4)   = Link([ 0    0        -0.015    pi/2   0 ], 'standard');
7 Left(5)   = Link([ 0    0.1373    0        pi/2   0 ], 'standard');
8 Left(6)   = Link([ 0    0         0        pi/2   0 ], 'standard');
```

```
9   Left(7)   = Link([ 0      -0.016      0.0625   0      0 ], 'standard');
10
11  for i = 1:7
12      Left(i).m = ml(i);
13      Left(i).I = Il(:,:,i);
14      Left(i).G = 1;
15      Left(i).r = rl(:,:,i);
16      Left(i).Jm = 00e-6;
17  end
18
19  Left(1).qlim = [-0.38 1.47];
20  Left(2).qlim = [-0.68 0.68];
21  Left(3).qlim = [-1.03 1.03];
22  Left(4).qlim = [-1.66 0.09];
23  Left(5).qlim = [ 0.00 2.81];
24  Left(6).qlim = [-0.65 1.75];
25  Left(7).qlim = [ 0.10 1.85];
26
27  robotArm1 = SerialLink(Left,'name', 'Left','base', ...
28      transl(0.0170686, -0.0060472, 0.1753)*trotx(pi/2)*troty(-15*pi/180));
```

6.3.3.4 Simulations

With the installed HAPTIK toolbox, Robotics Toolbox and YARP loaded in MATLAB, we can use the Novint Falcon device to control the virtual iCub robot. The MATLAB codes shown in Listing 6.4 outline the main loop of simulations, where functions `pos2ang` and `yarp_set_pos` should be implemented further according to forward and inverse kinematics of the robot arm model. Function `pos2ang` converts the end-point position to angles of links, and function `yarp set pos` sets the position of the virtual iCub robot arm through YARP interface. In this main loop, `robotArm1` was created by the Robotics Toolbox and `robotDeviceLeft` was initialized through the codes in Listing 6.2.

Listing 6.4 Main loop of simulation

```
1   mdl_icubsim; % Load iCub robot model
2   h = haptikdevice; % Get handle of haptic joystick
3   pos = read_position(h); % Read haptic position
4   tic; % Timer
5   while toc < 100; % Do this loop for 100 seconds
6       pos = read_position(h); % Read haptic position
7       qpos = pos2ang(pos,robotArm1); % Convert 3D position to angles of links
8       robotArm1.plot(qpos); % Plot the position read
9       drawnow(); % Plot the position now
10      yarp_set_pos(robotDeviceLeft,pos); % Set position of left arm via YARP
11  end;
12  close(h); % close the haptic device
13  clear h % clear variable
```

The experiments perform the communication and arm control via the Novint Falcon joystick using YARP and CHAI3D libraries. Even if the software solutions implemented need to be improved further, they allowed a better understanding of the haptic interaction with a virtual robot and highlighted relevant problems which could be solved in further works. Also, based on these techniques, more work can be done to introduce human intelligence by programming the adaptation law of the virtual robot for the purpose of integrating the haptic device and the remote robot.

6.4 Teleoperation Using Haptic Feedback

In last section, the haptic interaction based on the Falcon joystick and a virtual iCub robot was introduced. This section will present a haptic feedback based master–slave teleoperation simulation system which consists of a master with haptic feedback and a physical Baxter robot [31]. The haptic device Omni is employed as master to command slave, a 7-DOF robotic arm of Baxter robot. With haptic feedback, the teleoperation system can be manipulated bilaterally.

6.4.1 System Description

The master system mainly consists of the Omni haptic device, and the slave system mainly consists of Baxter robot which has been introduced in Sect. 1.1. On the 7-DOF robot arm, the force sensor is equipped in each joint. The position and force information of the robot hand are sampled and sent to the control computer. On the other hand, the equal force with the robot arm is exerted on the stylus of the SensAble haptic device. So the manipulator can operate the distant object according to the haptic feedback and visual feedback. Figure 6.23 shows the profile of the system. Figure 6.24 shows the structure of the system. The block diagram of the system and the information flow have also been described in this figure.

6.4.2 Workspace Mapping

When operating a kinematically dissimilar teleoperation system, the manipulator can only work within the limit of reachable workspace. Since workspace and region boundaries of master and slave are often quite different, evaluating whether or not a given location is reachable is thus a fundamental problem. To resolve this problem, several mapping approaches have been developed, which can map the master motion trajectory into reachable workspace of slave manipulator [32–34].

Analytical methods determine closed-form descriptions of workspace boundary but these methods are usually complicated by nonlinear equations and matrix inversion involved in manipulator kinematics. Numerical methods, on the other hand, are relatively simple and more flexible. Rastegar and Perel [35] introduced the

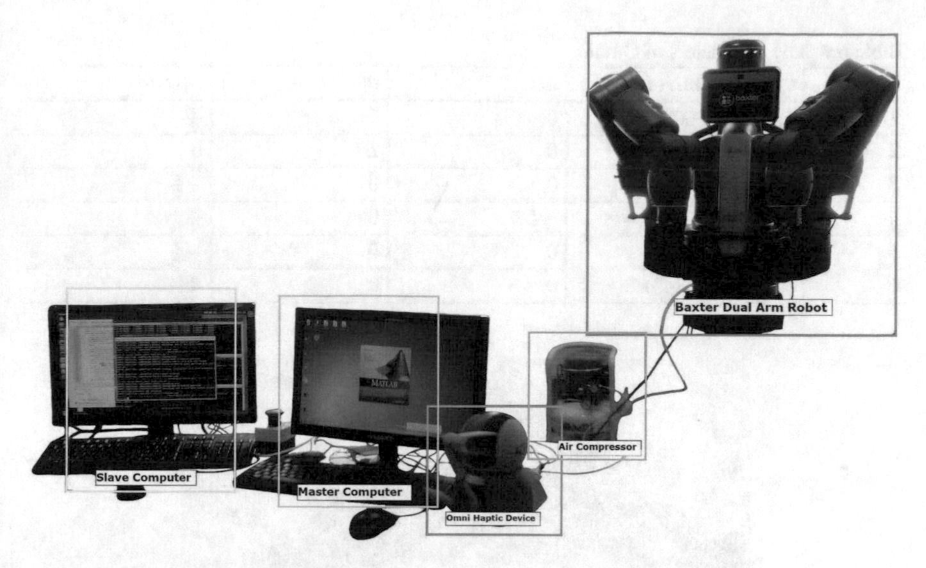

Fig. 6.23 The profile of teleoperation system

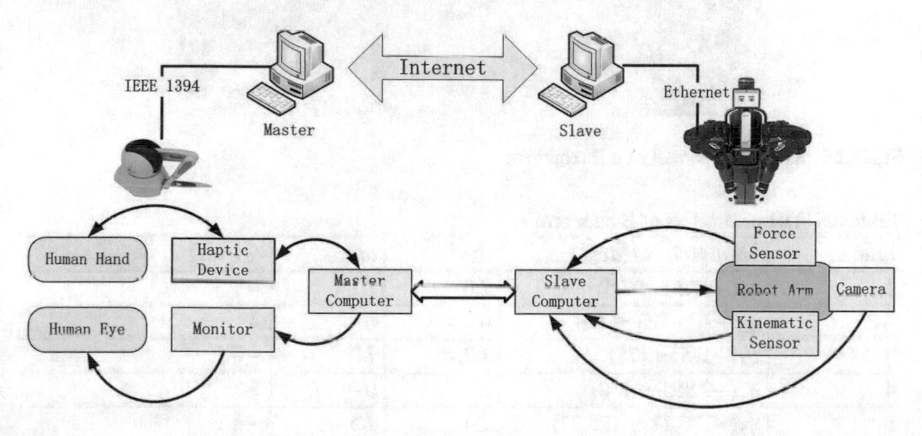

Fig. 6.24 The connection of and information in the system

Monte Carlo random sampling numerical method to generate the workspace boundary of some simple manipulators using only forward kinematics. The method is relatively simple to apply and by which we create workspace mapping model.

6.4.2.1 Forward Kinematics

Recall Sect. 2.1, forward kinematics describes the relationship between the joint angle of the serial manipulator and the position and orientation of its end effector.

Table 6.5 DH parameters of Omni

Link i	$\theta_i(anglelimit(deg))$	d_i	a_i	$\alpha_i(rad)$
1	$q_1(-60 \sim 60)$	0	0	$\frac{\pi}{2}$
2	$q_2(0 \sim 105)$	0	$L1$	0
3	$q_3(-100 \sim 100)$	0	0	$-\frac{\pi}{2}$
4	$q_4(-145 \sim 145)$	$-L2$	0	$\frac{\pi}{2}$
5	$q_5(-70 \sim 70)$	0	0	$-\frac{\pi}{2}$
6	$q_6(-145 \sim 145)$	0	0	$\frac{\pi}{2}$

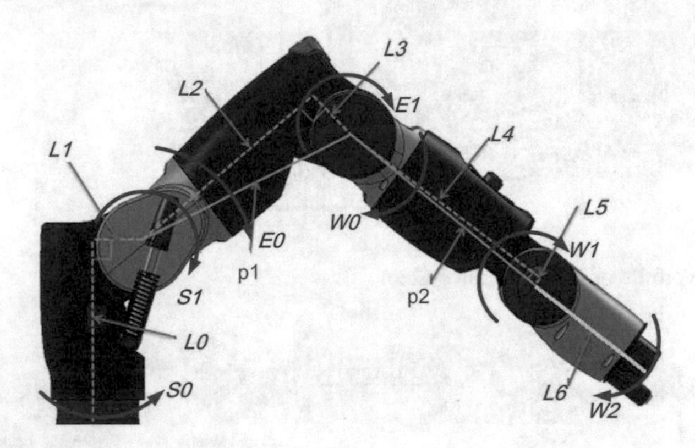

Fig. 6.25 Kinematic model of a Baxter arm

Table 6.6 DH parameters of Baxter arm

Link i	$\theta_i(anglelimit(deg))$	d_i	a_i	$\alpha_i(rad)$
1	$q_1(-97.5 \sim 97.5)$	$L0$	$L1$	$-\frac{\pi}{2}$
2	$q_2 + \frac{\pi}{2}(-123 \sim 60)$	0	0	$\frac{\pi}{2}$
3	$q_3(-175 \sim 175)$	$L2$,	$L3$	$-\frac{\pi}{2}$
4	$q_4(-2.865 \sim 150)$	0	0	$\frac{\pi}{2}$
5	$q_5(-175.27 \sim 175.27)$	$L4$	$L5$	$-\frac{\pi}{2}$
6	$q_6(-90 \sim 120)$	0	0	$\frac{\pi}{2}$
7	$q_7(-175.27 \sim 175.27)$	$L6$	0	0

The master omni device has a 6-DOF stylus and its DH table is shown in Table 6.5. where a_i are the link lengths, the twist angle α_i, the link offset d_i, and the joint angles that are variable θ_i. Here, $L1 = L2 = 133.35$ mm. The joints 4, 5, 6 have a spherical wrist thus the link lengths and link offset are all zero.

The slave, Baxter robot arm, has a 7-DOF manipulator and its kinematic model can be shown in the Fig. 6.25. Refer to Fig. 6.25. The DH table of Baxter arm is given in Table 6.6. Here, $L0 = 0.27$ m, $L1 = 0.069$ m, $L2 = 0.364$ m, $L3 = 0.069$ m, $L4 = 0.375$ m, $L5 = 0.01$ m, and $L6 = 0.28$ m.

When using DH notation method to model a manipulator, we note that the links of the manipulator are numbered from 1 to n, the base link is 1 and the outermost link or hand is n. A coordinate system is attached to each link for describing the relative arrangements among the various links. The coordinate system attached to the ith link is numbered i. The 4×4 transformation matrix relating $i+1$ coordinate system to i coordinate system is:

$$ {}^{i-1}A_i(\theta_i) = \begin{bmatrix} c\theta_i & -s\theta_i c\alpha_i & s\theta_i s\alpha_i & a_i c\theta_i \\ c\theta_i & c\theta_i c\alpha_i & -c\theta_i s\alpha_i & a_i s\theta_i \\ 0 & s\alpha_i & c\alpha_i & d_i \\ 0 & 0 & 0 & 1 \end{bmatrix} \tag{6.24} $$

where $s\theta_i = \sin\theta_i$, $c\theta_i = \cos\theta_i$, θ_i is the ith joint rotation angle and α_i is twist angle, a_i, d_i is length of link $i+1$ and offset distance at joint i. From DH table and by using Eq. (6.24), the homogeneous transformation matrices can be given by matrices multiplying as follows:

$$ {}^0A_n = {}^0A_1 {}^1A_2 \cdots {}^{n-1}A_n \tag{6.25} $$

The frame axis direction of the omni device is different from the Baxter, as illustrated in the Fig. 6.26. Thus the Cartesian coordinate of Omni, $[x_m' y_m' z_m']^T$, needs to be modified according to the equation below:

$$ A_o' = R_z\left(\frac{\pi}{2}\right) R_x\left(\frac{\pi}{2}\right) A_o R_y\left(\frac{\pi}{2}\right) R_z\left(\frac{\pi}{2}\right) \begin{bmatrix} 1 & 0 & 0 \\ 0 & -1 & 0 \\ 0 & 0 & -1 \end{bmatrix} \tag{6.26} $$

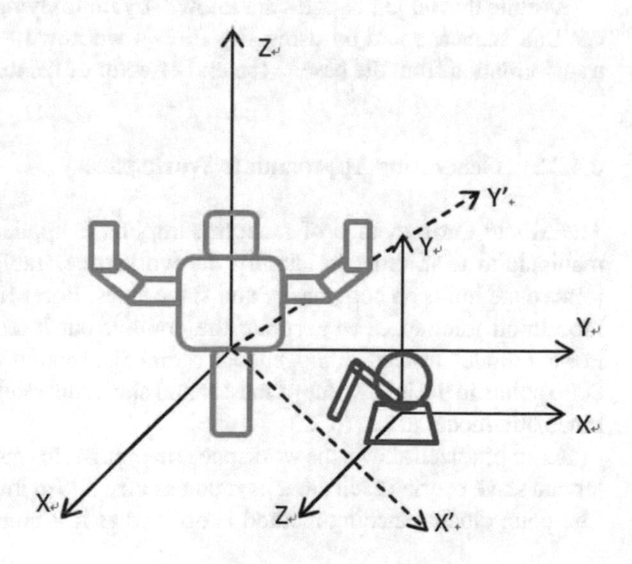

Fig. 6.26 Frame axis directions of the Omni and the Baxter

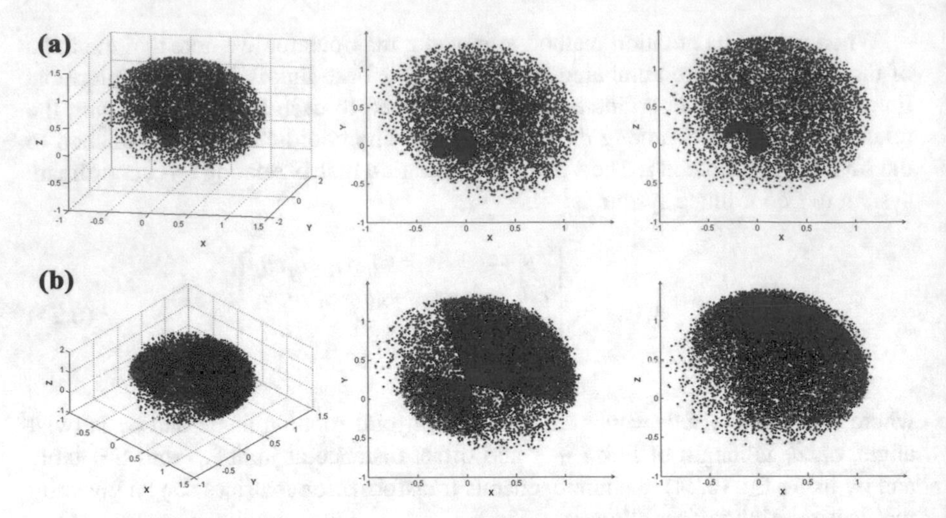

Fig. 6.27 Workspace mapping. The *red point* cloud is the workspace of slave and blue point cloud represents the master workspace. **a** Contour plot before workspace mapping. **b** Contour plot after workspace mapping

where A_o is the transform matrix of the Omni and $A_o{}'$ is the correspondent modified matrix.

Forward kinematic equation specifies the relation between Cartesian space and joint space can be represented by Eq. (6.27)

$$\overline{X_0} = ({}^0A_n)^T \overline{X_n} \tag{6.27}$$

Assume that all joint angles are known, by multiplying the transformation matrices in a sequence and by using Eq. (6.27), we now have the forward kinematics transformation from the base to the end effector of the manipulator.

6.4.2.2 Generating Approximate Workspace

The Monte Carlo method of random sampling is applied to the joint space of the manipulator to in order to identify the workspace. Tables 6.5 and 6.6 specify the joint rotate limits on both master and slave sides. Bart Milne et al. [3] use a distance loop in all joint space to generate the contour but it tends to be time exhausting. For our model, instead, homogeneous radial distribution was employed to generated 8000 points in the joint space of master and slave separately and by using the forward kinematic model in Eq. (6.27).

As emphasized above, the workspace mapping is to enable the workspaces of master and slave overlap each other as much as possible to improve the maneuverability. The point cloud matching method is utilized as it is convenient by considering the position of the end effector instead of its structure (Fig. 6.27). The mapping processing can be deduced as Eq. (6.28)

$$\begin{bmatrix} x_s \\ y_s \\ z_s \end{bmatrix} = \begin{bmatrix} \cos\delta & -\sin\delta & 0 \\ \sin\delta & \cos\delta & 0 \\ 0 & 0 & 1 \end{bmatrix} \times \left(\begin{bmatrix} S_x & 0 & 0 \\ 0 & S_y & 0 \\ 0 & 0 & S_z \end{bmatrix} \begin{bmatrix} x_m \\ y_m \\ z_m \end{bmatrix} + \begin{bmatrix} T_x \\ T_y \\ T_z \end{bmatrix} \right) \tag{6.28}$$

where $[x_s\, y_s\, z_s]^T$, $[x_m\, y_m\, z_m]^T$ are the Cartesian coordinates of end effector of the Baxter and the Omni respectively, δ is the revolute angle about Z-axis of the Baxter base frame, $[S_x S_y S_z]^T$ and $[T_x T_y T_z]^T$ are the scaling factor and translation about X, Y, Z axis. For the left arm of Baxter, the mapping parameters in equation can be calculated and given as

$$\delta = \frac{\pi}{4}, \quad \begin{bmatrix} S_x \\ S_y \\ S_z \end{bmatrix} = \begin{bmatrix} 0.041 \\ 0.040 \\ 0.041 \end{bmatrix}, \quad \begin{bmatrix} T_x \\ T_y \\ T_z \end{bmatrix} = \begin{bmatrix} 0.701 \\ 0.210 \\ 0.129 \end{bmatrix}$$

In Eq. (6.28), only the revolution about Z-axis is done. The master Omni joystick is placed on a horizontal platform and consequently the Z-axis of master is perpendicular to horizontal plane. Likewise, the slave, Baxter robot is adjusted carefully to make sure that the Z-axis is upward vertical and also perpendicular to horizontal plane. So, the Z-axis of master is parallel to Z-axis of slave. The revolution about X-axis and Y-axis are neglected due to a human-in-loop control approach is employed. That is to say, as long as the X-axis and Y-axis of master and slave are aligned approximately, operator can adjust the arm of robot according to the posture depicted on the master computer's screen instead of to control the slave directly using the accurate coordinate value.

6.4.3 Command Strategies

The strategy employed to control the slave in this section is by using the position-position scheme. In [36], Farkhatdinov et al. used a position-position command strategy to steer a mobile robot of two wheels going through a corridor with prescribed stopping points along the way, and a Phantom Premium 1.5A was employed in this strategy on the master side. Position-position strategy was also used in [37] for their master–slave setup, and input saturation compensation has also been investigated. In this section, we use the numerical inverse kinematic method to get the joint configuration and then send the joint angles to implement the control of the slave device.

6.4.3.1 Joint Space Position-Position Control

To implement position-position control strategy, a simple and easy approach is to map the joint angle of master to the slave directly, namely, applying teleoperation in the joint space. Because the slave Baxter arm is a redundant manipulator and has 7 joints, while the master Omni device only has six joints, so that a certain joint of the slave has to be restricted in this method. Consider the geometric structure similarity of the master and slave. We let the third joint to be fixed in direct angle mapping method, i.e., set angle of joint *E0* to zero, and match other joint one by one, then the slave can move according to the master. For position-position control strategy, the 6-DOF-6-DOF direct angle mapping method can be a choice. For example, Ahn et al. restricted 6-DOF master and 7-DOF slave to just 2 DOFs, such that the two axes for master and slave can be matched [37].

6.4.3.2 Task Space Position-Position Control

To make use of the advantage of 7-DOF slave's redundant manipulator, an inverse kinematic algorithm needs to be developed to implement the Cartesian position transforming from master to slave. In this section, it is the CLIK method.

If the DH parameters are known, the relation between task vector $\overline{x}$ and joint vector $\overline{q}$ can be represents [38] as

$$\overline{x} = f(\overline{q}) \tag{6.29}$$

where f is a continuous nonlinear function, whose structure and parameters are known. It is the fundamental importance for the teleoperation to solve the inverse kinematic problem, i.e., solving Eq. (6.29) with respect to time yields the mapping of the joint velocity vector $\dot{\overline{q}}$ into the end-effector task velocity vector $\dot{\overline{x}}$ [39],

$$\dot{\overline{x}} = J(\overline{q})\dot{\overline{q}} \tag{6.30}$$

where $J(\overline{q}) = \frac{\partial f}{\partial q}$ is Jacobian matrix. By use of the Jacobian matrix in Eq. (6.30), that is

$$\dot{\overline{q}} = J^{+}(\overline{q})\dot{\overline{x}} \tag{6.31}$$

where $J^{+}(\overline{q})$ is a pseudoinverse of Jacobian matrix, and can be computed as $J^{+} = J^{T}(JJ^{T})^{-1}$. The joint displacements $\overline{q}$ are then obtained by integrating Eq. (6.31) over time.

However, the method using Eq. (6.31) can be represented as

$$\dot{\overline{q}} = KJ^{T}(\overline{q})e \tag{6.32}$$

6.4.3.3 Haptic Rendering Algorithm

Our command strategies are based on the position-position scheme and it is hard for the human operator to know the position error between the human and the robot only through the robot motion simulation in the monitor. Therefore, the haptic rendering algorithm based on the error information should be added to the system. In the current setting, force-feedback is inversely proportional to position difference which is a simple haptic rendering algorithm based on Hook's Law. This haptic algorithm acts as an indicator that when the error is growing, the force-feedback will increase at the same time which in turn aids the operator to gain situation awareness and improve the overall accuracy of the motions.

The absolute value of the force-feedback can be calculated as

$$|f| = K_f\sqrt{(x_s - x_d)^2 + (y_s - y_d)^2 + (z_s - z_d)^2} \tag{6.33}$$

where K_f is the force-feedback scale factor, (x_s, y_s, z_s) is the actual Cartesian position of the slave end effector and (x_d, y_d, z_d) is desired value.

6.4.4 Experiment Studies

Based on the system profile, command strategy and corresponding algorithm mentioned above, two experiments are designed and carried out to verify the effectiveness of the proposed teleoperation system.

6.4.4.1 Free Space Moving

A free space moving is used to test the position-position command performance. The operator holds the Omni stylus firmly, and moves slowly in its workspace. First, it moves in the translation direction and then turns the stylus to the desired orientation. These two actions can be used to check the translation and rotational capability of the system. Figure 6.28 shows the results of the experiment.

In Fig. 6.28, the blue curve represents trajectory of master, Omni stylus. The green one represents trajectory of end-effector of slave, Baxter left arm. From the results, we can conclude that the position-position teleoperation control algorithm works well on the human input. The maximum position error between master and slave is no more than 0.1 m. Considering the error of workspace mapping, and the fact that the position control law is only a simple PD strategy, the results can be acceptable.

6.4.4.2 Haptic Feedback Test

Haptic feedback features the system with the force-feedback from the manipulator of slave to the master, by which operator can feel the environment and manipulate the system more accurately and instantly. When human moves the stylus of the slave, a force-feedback is exerted to the stylus continuously. We designed an experiment to check the force-feedback got from manipulator and exerted on the stylus when moving stylus slowly, the results are shown in Fig. 6.29.

Fig. 6.28 Robot moving in the free space

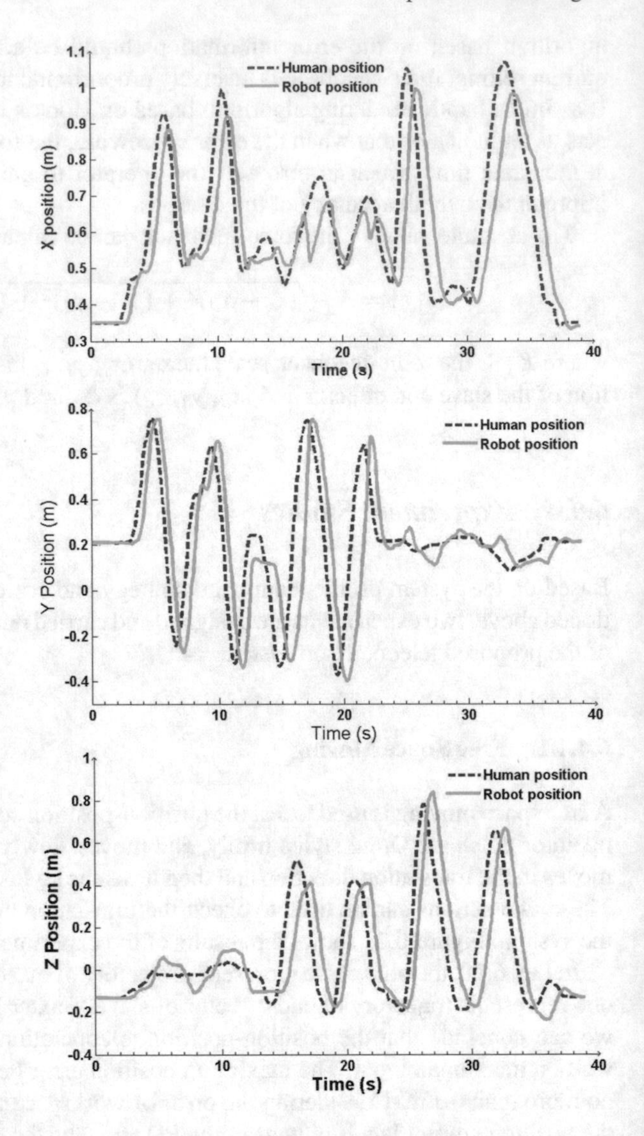

In Fig. 6.29, the dotted line represents the trajectory of stylus and the solid line represents the actual trajectory of manipulator. Several arrows represent the force-feedback. The force-feedback is to the direct ration of the position error and point to the manipulator. This force-feedback abides by the control strategy given in Eq. (6.33), with which the operator can know the error between the position of manipulator and the desired position.

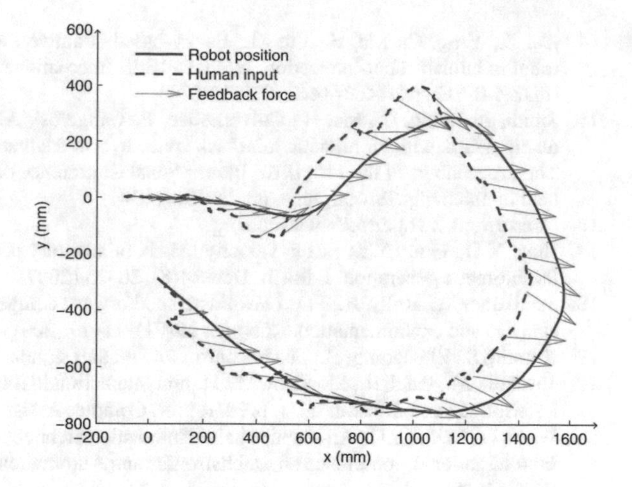

Fig. 6.29 The force-feedback when manipulator moving

References

1. Krebs, H.I., Hogan, N., Aisen, M.L., Volpe, B.T.: Robot-aided neurorehabilitation. IEEE Trans. Rehabil. Eng. **6**(1), 75–87 (1998)
2. P. Renon, C. Yang, H. Ma, and M. Fu, "Haptic interaction between human and virtual robot with chai3d and icub simulator," in *Proceedings of the 32nd Chinese Control Conference (CCC2013)*, (Xi'an), CCC, 2013
3. Burdea, G.C., Burdea, C., Burdea, C.: Force and Touch Feedback for Virtual Reality. Wiley, New York (1996)
4. Shimoga, K.B.: A survey of perceptual feedback issues in dexterous telemanipulation. II. Finger touch feedback. In: Virtual Reality Annual International Symposium, 1993, IEEE, pp. 271–279. IEEE (1993)
5. Stone, R.J.: Haptic feedback: A brief history from telepresence to virtual reality. Haptic Human-Computer Interaction, pp. 1–16. Springer, Berlin (2001)
6. Guo, S., Song, G., Song, Z.: Development of a self-assisted rehabilitation system for the upper limbs based on virtual reality. IEEE International Conference on Mechatronics (2007)
7. Wang, H., Liu, X.P.: Design of a novel mobile assistive robot with haptic interaction. In: IEEE International Conference on Virtual Environments Human-Computer Interfaces and Measurement Systems (VECIMS), pp. 115–120. IEEE (2012)
8. Li, C., Ma, H., Yang, C., Fu, M.: Teleoperation of a virtual icub robot under framework of parallel system via hand gesture recognition. In: IEEE International Conference on Fuzzy Systems (FUZZ-IEEE), pp. 1469–1474. IEEE (2014)
9. Inoue, K., Tanikawa, T., Arai, T.: Micro hand with two rotational fingers and manipulation of small objects by teleoperation. In: MHS 2008. International Symposium on Micro-NanoMechatronics and Human Science, pp. 97–102. IEEE (2008)
10. Reddivari, H., Yang, C., Ju, Z., Liang, P., Li, Z., Xu, B.: Teleoperation control of baxter robot using body motion tracking. In: 2014 International Conference on Multisensor Fusion and Information Integration for Intelligent Systems (MFI), pp. 1–6. IEEE (2014)
11. Machida, E., Cao, M., Murao, T., Hashi, H.: Human motion tracking of mobile robot with kinect 3d sensor. In: 2012 Proceedings of SICE Annual Conference (SICE), pp. 2207–2211. IEEE (2012)
12. http://processing.org/reference/PVector.html
13. http://www.processing.org/reference/pushMatrix_.html

14. Xu, Z., Yang, C., Ma, H., Fu, M.: Fuzzy-based adaptive motion control of a virtual iCub robot in human-robot-interaction. In: 2014 IEEE International Conference on Fuzzy Systems (FUZZ-IEEE), pp. 1463–1468. IEEE (2014)
15. Smith, A., Yang, C., Ma, H., Culverhouse, P., Cangelosi, A., Burdet, E.: Bimanual robotic manipulation with biomimetic joint/task space hybrid adaptation of force and impedance. In: The Proceedings of the 11th IEEE International Conference on Control and Automation to be held in Taichung, Taiwan, June, pp. 18–20 (2014)
16. Consortium, T.R.: http://www.icub.org
17. Bing, X.H., Qing, Y.Y.: Jacobi velocity matrix of arbitrary point in the robot mechanism and its automatic generation. J. Mach. Des. **24**(8), 20–25 (2007)
18. Santibañez, V., Kelly, R.: PD control with feedforward compensation for robot manipulators: analysis and experimentation. Robotica **19**(01), 11–19 (2001)
19. Xinmin, S., Zhengqing, H.: Fuzzy control and matlab simulation (2008)
20. Tavakoli, M., Patel, R., Moallem, M.: Haptic interaction in robot-assisted endoscopic surgery: a sensorized end-effector. Int. J. Med. Robot. Comput. Assist. Surg. **1**(2), 53–63 (2005)
21. Wang, D., Li, J., Li, C.: An adaptive haptic interaction architecture for knee rehabilitation robot,. In: International Conference on Mechatronics and Automation, 2009. ICMA 2009, pp. 84–89. IEEE (2009)
22. Gupta, A., O'Malley, M.K.: Design of a haptic arm exoskeleton for training and rehabilitation. IEEE/ASME Trans. Mechatron. **11**(3), 280–289 (2006)
23. Medina, J.R., Lawitzky, M., Mörtl, A., Lee, D., Hirche, S.: An experience-driven robotic assistant acquiring human knowledge to improve haptic cooperation. In: 2011 IEEE/RSJ International Conference on Intelligent Robots and Systems (IROS), pp. 2416–2422. IEEE (2011)
24. Renon, P., Yang, C., Ma, H., Cui, R.: Haptic interaction between human and virtual icub robot using novint falcon with chai3d and matlab. In: 2013 32nd Chinese Control Conference (CCC), pp. 6045–6050 (2013)
25. Simulator Web Site. Wiki for the icub simulator specifications from its installation to its use. http://www.eris.liralab.it/wiki/ (2012)
26. iCub robot Web Site. Official web site concerning the iCub projects around the europe. http://www.iCub.org/ (2012)
27. Martin, S., Hillier, N.: Characterisation of the novint falcon haptic device for application as a robot manipulator. In: Australasian Conference on Robotics and Automation (ACRA), pp. 291–292. Citeseer (2009)
28. CHAI3D Web Site. Set of open-source libraries providing an environment for haptic real time interactive simulations. http://www.chai3d.org/documentation.html (2012)
29. de Pascale, M., de Pascale, G., Prattichizzo, D., Barbagli, F.: The haptik library: a component based architecture for haptic devices access. In: Proceedings of EuroHaptics (2004)
30. http://wiki.icub.org/yarpdoc/yarp_swig.html#yarp_swig_matlab
31. Ju, Z., Yang, C., Li, Z., Cheng, L., Ma, H.: Teleoperation of humanoid baxter robot using haptic feedback. In: International Conference on Multisensor Fusion and Information Integration for Intelligent Systems (MFI), 2014, pp. 1–6 (2014)
32. Wang, H., Low, K.H., Gong, F., Wang, M.Y.: A virtual circle method for kinematic mapping human hand to a non-anthropomorphic robot. In: ICARCV 2004 8th Control, Automation, Robotics and Vision Conference, 2004, vol. 2, pp. 1297–1302. IEEE (2004)
33. Dubey, R.V., Everett, S., Pernalete, N., Manocha, K.A.: Teleoperation assistance through variable velocity mapping. IEEE Trans. Robot. Autom **17**(5), 761–766 (2001)
34. Pernalete, N., Yu, W., Dubey, R., Moreno, W.: Development of a robotic haptic interface to assist the performance of vocational tasks by people with disabilities. In: Proceedings. ICRA'02. IEEE International Conference on Robotics and Automation, vol. 2, pp. 1269–1274. IEEE (2002)
35. Rastegar, J., Perel, D.: Generation of manipulator workspace boundary geometry using the monte carlo method and interactive computer graphics. J. Mech. Des. **112**(3), 452–454 (1990)
36. Farkhatdinov, I., Ryu, J.-H.: Hybrid position-position and position-speed command strategy for the bilateral teleoperation of a mobile robot. In: International Conference on Control, Automation and Systems, 2007. ICCAS'07, pp. 2442–2447. IEEE (2007)

37. Ahn, S.H., Park, B.S., Yoon, J.S.: A teleoperation position control for 2-dof manipulators with control input saturation. In: Proceedings. ISIE 2001. IEEE International Symposium on Industrial Electronics, 2001, vol. 3, pp. 1520–1525. IEEE (2001)
38. Corke, P.: Robotics, Vision and Control: Fundamental Algorithms in MATLAB, vol. 73. Springer Science & Business Media, Sydney (2011)
39. Craig, J.J.: Introduction to Robotics, vol. 7. Addison-Wesley, Reading (1989)

Chapter 7
Obstacle Avoidance for Robot Manipulator

Abstract Kinematic redundancy allows a manipulator to have more joints than required and could make the robot more flexible in the obstacles avoidance. This chapter first introduces the concept of kinematic redundancy. Then, a human robot shared controller is developed for teleoperation of redundant manipulator by developing an improved obstacle avoidance strategy based on the joint space redundancy of the manipulator. Next, a self-identification method is described based on the 3D point cloud and the forward kinematic model of the robot. By implementing a space division method, the point cloud is segmented into several groups which represent the meaning of the points. A collision prediction algorithm is then employed to estimate the collision parameters in real-time. The experiment using the Kinect sensor and the Baxter robot has demonstrated the performance of the proposed algorithms.

7.1 Introduction of Kinematic Redundancy

A manipulator is termed kinematic redundancy if the number of degrees of freedom is higher than the number of task-space coordinates. In general, an object in the Cartesian space is located by three position parameters and three orientation parameters. In this case, a manipulator which has more than six DOFs is considered to have kinematics redundancy. A manipulator is intrinsically redundant when the dimension of the joint space is greater than the dimension of the task space.

Kinematic redundancy allows that a manipulator has more joints than required to give the end-effector a desired position and orientation. The additional degrees of freedom permit a dexterous motion of the end-effector even when an actuator is failed to mobilize, and also avoid kinematic singularities. The kinematic redundancy is also useful for obstacle avoidance of manipulators. The redundancy can be used to meet constraints on joint range availability and to obtain trajectories in the joint space which are collision-free in presence of obstacles along the motion.

The forward kinematics allows one to specify a unique relationship between the joint vector q and the Cartesian vector x. It could be formulated as

$$x = f(q) \tag{7.1}$$

© Science Press and Springer Science+Business Media Singapore 2016

C. Yang et al., *Advanced Technologies in Modern Robotic Applications*,
DOI 10.1007/978-981-10-0830-6_7

where f is a continuous nonlinear function.

In order to control a manipulator following a given reference trajectory, the conventional way is to transform the task-space reference trajectory to the joint space. Then the controller can be designed to track the reference in joint space. The solution of transforming the task-space coordinates into the joint space coordinates is called the inverse kinematics.

The most direct approach for solving the inverse kinematic problem is to obtain a closed-form solution of the inverse of Eq. (7.1) as $q = f^{-1}(x)$. But this solution cannot be directly applied to redundant manipulators since is no unique transformation from a task-space trajectory to a corresponding joint space trajectory, which means that the inverse solution of each x in Eq. (7.1) may have more than one solution and is hard to analytically express.

In order to solve the inverse kinematics problem, an alternative method is based on the relationship between joint velocities $\dot{q}$ and Cartesian velocities $\dot{x}$

$$\dot{x} = J(q)\dot{q} \tag{7.2}$$

where $J(q)$ is the Jacobian matrix describes the linear mapping from the joint velocity space to the end-effector velocity space.

For a redundant manipulator, the Jacobian matrix $J(q)$ is not invertible. To obtain the solution of the inverse kinematics, the inverses of the Jacobian is generalized as

$$\dot{q} = J^{\dagger}(q)\dot{x} \tag{7.3}$$

where $J^{\dagger}(q)$ is the pseudo-inverse of $J(q)$. It yields locally a minimum norm joint velocity vector in the space of least squares solutions.

The solution can be also obtained by

$$\dot{q} = J^{\dagger}\dot{x} + (I - J^{\dagger}J)z \tag{7.4}$$

where $(I - J^{\dagger}J)$ is the projection operator selecting the components of z which can be used for optimization purposes.

The joint displacements q are then obtained by integrating Eq. (7.3) over time.

This method using Eq. (7.3) is inherently open-loop and causes numerical drifts in the task-space unavoidably. In order to overcome this drawback, the CLIK (closed-loop inverse kinematics) algorithm is applied. The task-space vector $\dot{x}$ is replaced by $\dot{x} = Ke$ in (7.3), where $e = x_d - x$ describes the error between the desired task trajectory x_d and the actual task trajectory x. It can be computed from the current joint variables via the forward kinematics Eq. (7.1). Here K is a positive definite matrix, that shapes the error convergence. Considering the computational burden, the pseudo-inverse of the Jacobian in Eq. (7.3) can be replaced by the transposed Jacobian. Then Eq. (7.3) can be written as

$$\dot{q} = KJ^{T}(q)e \tag{7.5}$$

This solution may avoid the typical numerical instabilities, which occur at kinematic singularities, since no pseudo-inverse of the Jacobian matrix is required. The details and the steady state proof of this solution are shown in [1–3].

The solution is also called the Jacobian control method, which can be used to solve the minimization of actuator energy consumption and obstacle avoidance problems for redundant manipulators.

7.2 Shared Controlled Teleoperation with Obstacle Avoidance

In the past decades, tele-operated robots have been extensively studied because of their potential wide applications in special environment like space and deep ocean. Unlike the automatic controlled manipulator, which can deal with fixed task under simple environment, the tele-operated robots can deal with more unstructured environments and variable tasks. Considerable efforts have been made in the existing literature. Li et al. developed a preliminary teleoperation system based on YARP platform for iCub robot via hand gesture recognition [4]. Inoue et al. proposed a tele-operated micro hand which consists of two rotational fingers in [5]. In [6], a new control method has been introduced, which modified the model-based teleoperation system, to control the real robotic system Engineering Test Satellite VII manipulator. In [7], an Internet-based teleoperation system has been developed for a robot manipulator-Thermo CRS A465. In [8], Kofman presents a noncontacting vision-based method of robot teleoperation that allows a human operator to communicate simultaneous six-degree-of-freedom motion tasks to a robot manipulator by having the operator perform the three-dimensional human hand-arm motion that would naturally be used to complete an object manipulation task. In most of these studies, the feedback information is sent to the operator directly and the operator needs to consider the environment around the manipulator. The limited feedback information and the control flexibility limits the application range of the conventional teleoperation system, especially in the scenarios which require safe interaction with environment.

This section presents a shared control strategy to reduce the operation burden of human operator [9]. In this system, the human operator only needs to consider the motion of the end-effector of the manipulator, while the manipulator will avoid the obstacle by itself without sacrificing the end-effector performance. The environment information is not only sent to the operator, but also the automatic control module. Meanwhile, a dimension reduction method is employed to achieve a more efficient use of the manipulator redundancy.

This method enables the robot manipulator to achieve the goal of moving away from the obstacle, as well as to restore its original pose by designing an artificial parallel system by using the kinematic model of the telerobot manipulator. In the restoring task, the trajectory of each joint of the manipulator is controlled at the same time by implementing the proposed dimension reduction method. Therefore, the pose of the manipulator will not be permanently changed by the obstacle, which may cause the robot become singularity and cannot fulfill the remaining tasks. The stability of the closed-loop system is proved using Lyapunov direct method to ensure the asymptotic stability of the system to deal with the real-world tasks. Experiments are carried out on the Baxter robot to verify the effectiveness of the proposed method.

7.2.1 System Components

In the teleoperation system, the user teleoperates the manipulator by giving producing reference trajectory for the telerobot end-effector by using the Omni® haptic device, which is connected on the client computer. The manipulator is controlled by the server computer, which is also equipped with a Kinect® sensor to give the visual feedback of the manipulator and to detect the surrounding environment.

The SensAble Omni Haptic Device has been introduced in Sect. 1.2. The first three DOFs describe the position of the end-effector, while the last three form a gimbal and are related to its orientation. The Kinect® sensor (see Sect. 1.2) is used to achieve the obstacle detection. The obstacle stick is assumed to be of a single color such that the stick can be easily separated from the complex background. The HSV color space is used in the color segmentation, which can separate the points on the stick from the background. Then the projection of the separated points on X–Y, Y–Z, and X–Z planes are calculated and the lines on the planes are detected by Hough transform. Finally the 3D line is obtained from the projection lines.

The manipulator will follow the end-effector trajectory given by the user, and at the same time, avoid the obstacle detected by the Kinect® sensor automatically. The collision warning information will be sent to the user in the form of haptic feedback, as shown in Fig. 7.1.

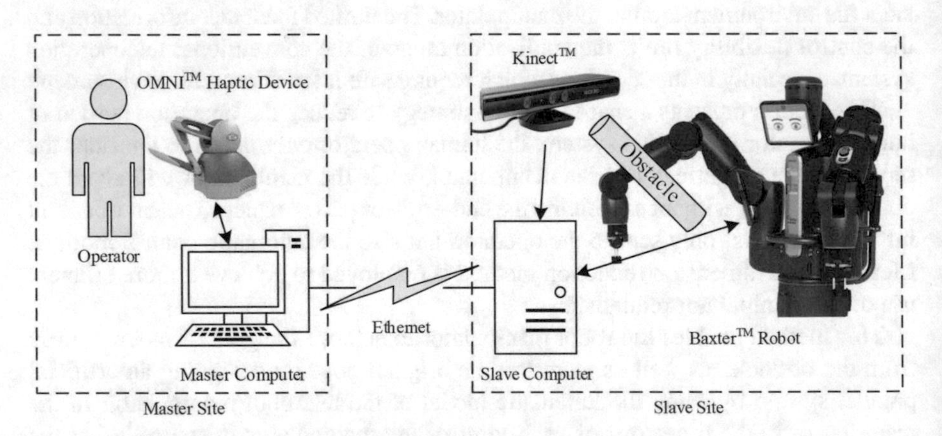

Fig. 7.1 The setup of the teleoperation system

7.2.2 Preprocessing

To process the shared controlled teleoperation and the obstacle avoidance of the robot, some preliminaries are introduced as follows.

7.2.2.1 Workspace Mapping

The workspace mapping aims to let the workspaces of master and slave overlap each other as much as possible, and to improve the maneuverability of the slave. The point cloud matching method is utilized, since it is convenient by considering the position of the end effectors instead of the structures of the different devices. The mapping process follows in [10] as below:

$$\begin{bmatrix} x_s \\ y_s \\ z_s \end{bmatrix} = \begin{bmatrix} \cos\delta & -\sin\delta & 0 \\ \sin\delta & \cos\delta & 0 \\ 0 & 0 & 1 \end{bmatrix} \times \left(\begin{bmatrix} S_x & 0 & 0 \\ 0 & S_y & 0 \\ 0 & 0 & S_z \end{bmatrix} \begin{bmatrix} x_m \\ y_m \\ z_m \end{bmatrix} + \begin{bmatrix} T_x \\ T_y \\ T_z \end{bmatrix} \right) \tag{7.6}$$

where $[x_s\, y_s\, z_s]^T$ and $[x_m\, y_m\, z_m]^T$ are the Cartesian coordinates of the end effectors of the manipulator and Omni respectively, δ is the revolution angle about the Z-axis of the manipulator base frame and $[S_x S_y S_z]^T$ and $[T_x T_y T_z]^T$ are the scaling factors and translations about the X, Y, and Z axis. The mapping parameters of Eq. (7.6) are given by

$$\begin{bmatrix} S_x \\ S_y \\ S_z \end{bmatrix} = \begin{bmatrix} 0.0041 \\ 0.0040 \\ 0.0041 \end{bmatrix}, \quad \begin{bmatrix} T_x \\ T_y \\ T_z \end{bmatrix} = \begin{bmatrix} 0.701 \\ 0.210 \\ 0.129 \end{bmatrix}, \quad \delta = \frac{\pi}{4} \tag{7.7}$$

7.2.2.2 Coordinate Transformation

The detected obstacle and the robot need to be placed into a unified coordinate system, in order to achieve obstacle avoidance. The coordinate transformation of the detected endpoints of the obstacle from the Kinect® coordinate to the robot coordinate can be obtained by Eq. (7.8).

$$\begin{aligned} [x_{l1}\ y_{l1}\ z_{l1}\ 1]^T &= \boldsymbol{T}\left[x'_{l1}\ y'_{l1}\ z'_{l1}\ 1\right]^T \\ [x_{l2}\ y_{l2}\ z_{l2}\ 1]^T &= \boldsymbol{T}\left[x'_{l2}\ y'_{l2}\ z'_{l2}\ 1\right]^T \end{aligned} \tag{7.8}$$

The transform matrix $\boldsymbol{T}$ can be obtained by measuring the coordinates of four noncollinear points under robot coordinate and the Kinect® coordinate. Assuming that we have four non-collinear points $\boldsymbol{s_1}$, $\boldsymbol{s_2}$, $\boldsymbol{s_3}$, and $\boldsymbol{s_4}$, the coordinates of the points on the robot coordinate X–Y–Z and the Kinect® coordinate X'–Y'–Z' are (x_1, y_1, z_1), (x_2, y_2, z_2), (x_3, y_3, z_3), (x_4, y_4, z_4), and (x'_1, y'_1, z'_1), (x'_2, y'_2, z'_2),

(x'_3, y'_3, z'_3), (x'_4, y'_4, z'_4), respectively. The transfer matrix $\boldsymbol{T}$ from the Kinect® coordinate to the robot coordinate can be calculated by Eq. (7.9).

$$\boldsymbol{T} = \begin{bmatrix} x_1 & x_2 & x_3 & x_4 \\ y_1 & y_2 & y_3 & y_4 \\ z_1 & z_2 & z_3 & z_4 \\ 1 & 1 & 1 & 1 \end{bmatrix} \begin{bmatrix} x'_1 & x'_2 & x'_3 & x'_4 \\ y'_1 & y'_2 & y'_3 & y'_4 \\ z'_1 & z'_2 & z'_3 & z'_4 \\ 1 & 1 & 1 & 1 \end{bmatrix}^{-1} \tag{7.9}$$

7.2.2.3 Identification of Collision Points

The collision points, $\boldsymbol{p_{cr}}$ and $\boldsymbol{p_{co}}$, are the two points either on the robot or on the obstacle, which covers the nearest distance between the robot and the obstacle. The forward kinematic of the robot is used to estimate the collision points. Each link of the robot can be seen as a segment in 3-D space. The coordinates of the endpoints of the segments, i.e. the coordinates of the joints, in Cartesian space can be obtained by Eq. (7.10).

$$^{n}X_o = {}^{0}A_1 {}^{1}A_2 \ldots {}^{n-1}A_n X_n, \tag{7.10}$$

where $X = [x, y, z, 1]^T$ is an augmented vector of Cartesian coordinate; and $^{j-1}A_j$ is the link homogeneous transform matrix, the definition of which can be found in [11].

The problem of estimating the collision points can be seen as the problem of searching the nearest point between the obstacle segment $\boldsymbol{q_{l1}} - \boldsymbol{q_{l2}}$ and the segments that stand for the robot links. First, the distance between segment $\boldsymbol{q_{l1}} - \boldsymbol{q_{l2}}$ and the ith segment of the robot links $[x_i, y_i, z_i] - [x_{i+1}, y_{i+1}, z_{i+1}]$ can be calculated by 3-D geometry, which is denoted by d_i; the nearest points on the robot link and the obstacle are denoted by $\boldsymbol{p_{cri}}$ and $\boldsymbol{p_{coi}}$, respectively. Then we have the collision point $\boldsymbol{p_{cr}} = \boldsymbol{p_{cri_{\min}}}$ and $\boldsymbol{p_{co}} = \boldsymbol{p_{coi_{\min}}}$ and the distance $d = d_{i_{\min}}$ where $i_{\min} = \underset{i \in \{0,1,\ldots,n\}}{\arg\min}\, d_i$.

7.2.3 Obstacle Avoidance Strategy

Consider one manipulator arm of the robot. Denote the joint velocities as $\dot{\boldsymbol{\theta}}$, then the end-effector velocity can be given by Eq. (7.11).

$$\dot{\boldsymbol{x}} = \boldsymbol{J}\dot{\boldsymbol{\theta}}, \tag{7.11}$$

where $\dot{\boldsymbol{x}}$ is the end-effector velocity and $\boldsymbol{J}$ is the Jacobian matrix of the manipulator.

The teleoperation of the manipulator can be seen as to control the end-effector of the manipulator following the reference Cartesian space trajectory given by the master device. The numerical inverse kinematic method is used to calculate the joint velocities. If the dimension of $\dot{\boldsymbol{\theta}}$ is larger than the dimension of $\dot{\boldsymbol{x}}$, i.e., the DOF (degree of freedom) of the manipulator arm is more than the desired end-effector velocity component, the manipulator will become kinematically redundant, which can be used to achieve secondary goals, e.g., obstacle avoidance, in addition to the end-effector trajectory following task.

7.2.3.1 Obstacle Avoiding

For the inverse kinematic problem of the kinematically redundant manipulator, infinite number of solutions can be find. The general solution is given by Eq. (7.12).

$$\dot{\boldsymbol{\theta}} = \boldsymbol{J}^{\dagger}\dot{\boldsymbol{x}} + (\boldsymbol{I} - \boldsymbol{J}^{\dagger}\boldsymbol{J})\boldsymbol{z}, \tag{7.12}$$

where $\boldsymbol{J}^{\dagger} = \boldsymbol{J}^T(\boldsymbol{J}\boldsymbol{J}^T)^{-1}$ is the pseudo-inverse of $\boldsymbol{J}$ and $\boldsymbol{z}$ is an arbitrary vector, which can be used to achieve the obstacle avoidance [12].

When the obstacle is near the manipulator arm, assume that the collision point on the manipulator arm is $\boldsymbol{p_{cr}}$, and the desired velocity under Cartesian space to move away from the obstacle is $\dot{\boldsymbol{x}}_o$, as shown in Fig. 7.2. The manipulator needs to satisfy Eqs. (7.13) and (7.14).

$$\dot{\boldsymbol{x}}_e = \boldsymbol{J_e}\dot{\boldsymbol{\theta}} \tag{7.13}$$

$$\dot{\boldsymbol{x}}_o = \boldsymbol{J_o}\dot{\boldsymbol{\theta}}, \tag{7.14}$$

where $\dot{\boldsymbol{x}}_e$ is the reference end-effector velocity; $\boldsymbol{J_e}$ is the end-effector Jacobian matrix; and $\boldsymbol{J_o}$ is the Jacobian matrix of the collision point $\boldsymbol{p_{cr}}$, and $\boldsymbol{J_o}$ can be obtained by using zero column vectors to replace the column vectors in J_e which correspond to distal joints beyond the collision point $\boldsymbol{p_{cr}}$.

However, if the collision point is found close to the base of the manipulator, the rank of the Jacobian matrix $\boldsymbol{J_o}$ might be less than the dimension of $\dot{\boldsymbol{x}}_o$. For example, when the collision is found between the elbow and the shoulder, the rank of $\boldsymbol{J_o}$ will be 2, yet the dimension of $\dot{\boldsymbol{x}}_o$ is 3. This will result in that the manipulator does not have enough DOF to achieve the avoidance velocity $\dot{\boldsymbol{x}}_o$. In other words, the inverse kinematics problem of the manipulator will be over defined.

One solution is to reduce the dimension of $\dot{\boldsymbol{x}}_o$. In fact, to avoid the obstacle, the avoidance velocity do not need to be exactly along the direction away from the obstacle. The only requirement is that the projection of the avoidance velocity equals to $\dot{\boldsymbol{x}}_o$. Denote the normal plane of $\dot{\boldsymbol{x}}_o$ as $\boldsymbol{N_o}$, then there exist at least a velocity vector $\dot{\boldsymbol{x}}'_o$ with end tip touching on $\boldsymbol{N_o}$ meet the requirement, as shown in Fig. 7.3, and satisfy

$$\dot{\boldsymbol{x}}_o^T(\dot{\boldsymbol{x}}'_o - \dot{\boldsymbol{x}}_o) = 0. \tag{7.15}$$

This velocity vector also satisfies $\dot{\boldsymbol{x}}'_o = \boldsymbol{J_o}\dot{\boldsymbol{\theta}}$, such that we have

$$\dot{\boldsymbol{x}}_o^T\dot{\boldsymbol{x}}_o = \dot{\boldsymbol{x}}_o^T\boldsymbol{J_o}\dot{\boldsymbol{\theta}}. \tag{7.16}$$

The dimension of $\dot{\boldsymbol{x}}_o^T \dot{\boldsymbol{x}}_o$ is 1 as it is a scalar, and the row rank of $\dot{\boldsymbol{x}}_o^T \boldsymbol{J}_o$ is equal to or more than 1. Hence the manipulator is no longer over defined by using z_o' instead of z_o.

The obstacle avoidance velocity strategy is designed to let the collision point on the manipulator $\boldsymbol{p}_{cr}$ move along the direction away from the obstacle and the closer the obstacle is, the faster $\boldsymbol{p}_{cr}$ should move. Thus the obstacle avoidance velocity $\dot{\boldsymbol{x}}_o$ is designed as

$$\dot{\boldsymbol{x}}_o = \begin{cases} \boldsymbol{0}, d \geq d_o \\ \dfrac{d_o - d}{d_o - d_c} v_{\max} \dfrac{\boldsymbol{p}_{cr} - \boldsymbol{p}_{co}}{d}, d_c < d < d_o \\ v_{\max} \dfrac{\boldsymbol{p}_{cr} - \boldsymbol{p}_{co}}{d}, d \leq d_c \end{cases}, \tag{7.17}$$

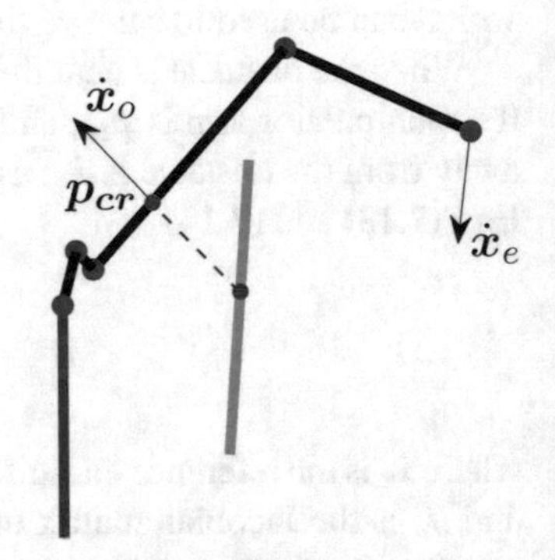

Fig. 7.2 The manipulator moves away from the obstacle

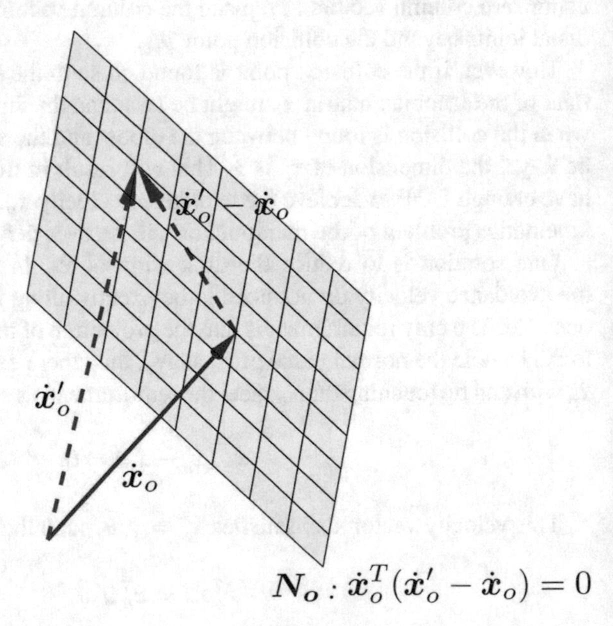

Fig. 7.3 The normal plane of $\dot{\boldsymbol{x}}_o$

where $d = \|\boldsymbol{p_{cr}} - \boldsymbol{p_{co}}\|$ is the distance between the obstacle and the manipulator; $v_{\max}$ is the maximum obstacle avoidance velocity; d_o is the distance threshold that the manipulator starts to avoid the obstacle; d_c is the minimum acceptable distance that the manipulator will avoid at the maximum speed.

7.2.3.2 Pose Restoration

On the other hand, in order to eliminate the influence of the obstructing when the obstacle has been removed, the manipulator is expected to restore its original state. To achieve that, an artificial parallel system of the manipulator is designed in the controller in real time to simulate its pose without the influence of the obstacle, as shown in Fig. 7.4, where the dashed black line indicates the parallel system. The kinematic model of parallel system is designed following our previous work in [11] using the Robotics Toolbox [13]. The control strategy of this parallel system is simply a classic inverse kinematic problem and the solution can be obtained in a different way to achieve different optimization objective. In this case, the pseudo-inverse is used. The control strategy of the parallel system is given by Eq. (7.18).

$$\dot{\boldsymbol{\theta}}_r = \boldsymbol{J}_e^{\dagger}\dot{\boldsymbol{x}}_e, \tag{7.18}$$

where $\dot{\boldsymbol{\theta}}_r$ is the joint velocities of the parallel system.

The motion of the manipulator is expected to follow as much as possible the parallel system. To achieve that, each of the joints on the real manipulator are controlled to follow the corresponding joints on the parallel system, as shown in Fig. 7.4. Three joints, shoulder, elbow, and wrist, need to be considered. Thus, they may also encounter the over-defined problem, and the joints cannot move toward the corresponding joints on the parallel system along a straight trajectory at the same time. Thus, the dimension reduction method can also be applied here. The manipulator needs to satisfy Eqs. (7.13) and (7.19) when the obstacle is absent.

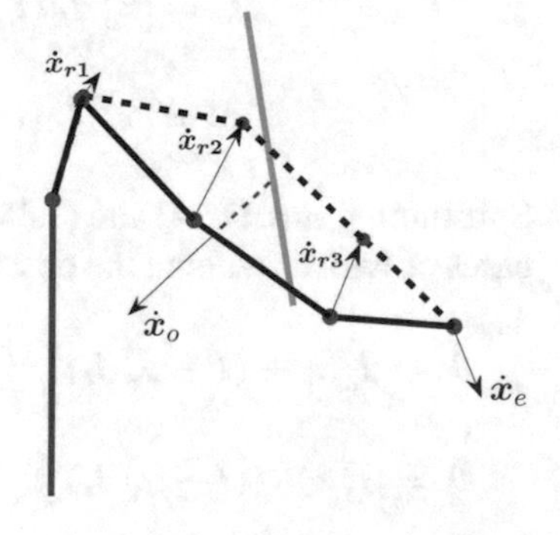

Fig. 7.4 The artificial parallel system built using Baxter®.kinematics model. The *solid black line* indicates the real manipulator. The *dashed black line* indicates the manipulator in the artificial parallel system

$$\begin{bmatrix} \dot{x}_{r1} \\ \dot{x}_{r2} \\ \dot{x}_{r3} \end{bmatrix}^T \begin{bmatrix} \dot{x}_{r1} \\ \dot{x}_{r2} \\ \dot{x}_{r3} \end{bmatrix} = \begin{bmatrix} \dot{x}_{r1} \\ \dot{x}_{r2} \\ \dot{x}_{r3} \end{bmatrix}^T \begin{bmatrix} J_{r1} \\ J_{r2} \\ J_{r3} \end{bmatrix} \dot{\theta}, \tag{7.19}$$

where J_{r1}, J_{r2}, and J_{r3} are the Jacobian matrices of the three joints, $\dot{x}_{r1}$, $\dot{x}_{r2}$, and $\dot{x}_{r3}$ are the desired joint velocities moving toward the corresponding joints on the parallel system. Define $\dot{x}_r = \left[\dot{x}_{r1}\, \dot{x}_{r2}\, \dot{x}_{r3}\right]^T$ and $J_r = \left[J_{r1}\, J_{r2}\, J_{r3}\right]^T$. Then Eq. (7.19) can be written as Eq. (7.20).

$$\dot{x}_r^T \dot{x}_r = \dot{x}_r^T J_r \dot{\theta} \tag{7.20}$$

The restoring velocity $\dot{x}_r$ is designed as a closed-loop system as

$$\dot{x}_r = K_r e_r, \tag{7.21}$$

where K_r is a symmetric positive definite matrix and $e_r = [e_{r1}\, e_{r2}\, e_{r3}]^T$ is the position errors of the joints between the parallel system and the real system.

7.2.3.3 Control Law

Substituting the general solution Eq. (7.12) into Eqs. (7.16) and (7.19) respectively, we have Eqs. (7.22) and (7.23).

$$\dot{x}_o^T J_o J_e^{\dagger} \dot{x}_e + \dot{x}_o^T J_o \left(I - J_e^{\dagger} J_e\right) z_o = \dot{x}_o^T \dot{x}_o \tag{7.22}$$

$$\dot{x}_r^T J_r J_e^{\dagger} \dot{x}_e + \dot{x}_r^T J_r \left(I - J_e^{\dagger} J_e\right) z_r = \dot{x}_r^T \dot{x}_r \tag{7.23}$$

The homogeneous solution can be obtained from Eqs. (7.22) and (7.23) by solving z_o and z_r as Eqs. (7.24) and (7.25).

$$z_o = \left[\dot{x}_o^T J_o \left(I - J_e^{\dagger} J_e\right)\right]^{\dagger} \left(\dot{x}_o^T \dot{x}_o - \dot{x}_o^T J_o J_e^{\dagger} \dot{x}_e\right) \tag{7.24}$$

$$z_r = \left[\dot{x}_r^T J_r \left(I - J_e^{\dagger} J_e\right)\right]^{\dagger} \left(\dot{x}_r^T \dot{x}_r - \dot{x}_r^T J_r J_e^{\dagger} \dot{x}_e\right) \tag{7.25}$$

Substituting Eqs. (7.24) and (7.25) into Eq. (7.12), the homogeneous solution for the cases of with or without the obstacle, as Eqs. (7.26) and (7.27).

$$\dot{\theta}_o = J_e^{\dagger} \dot{x}_e + (I - J_e^{\dagger} J_e) \left[\dot{x}_o^T J_o \left(I - J_e^{\dagger} J_e\right)\right]^{\dagger} \left(\dot{x}_o^T \dot{x}_o - \dot{x}_o^T J_o J_e^{\dagger} \dot{x}_e\right) \tag{7.26}$$

$$\dot{\theta}_r = J_e^{\dagger} \dot{x}_e + (I - J_e^{\dagger} J_e) \left[\dot{x}_r^T J_r \left(I - J_e^{\dagger} J_e\right)\right]^{\dagger} \left(\dot{x}_r^T \dot{x}_r - \dot{x}_r^T J_r J_e^{\dagger} \dot{x}_e\right) \tag{7.27}$$

To avoid abrupt switch between these two cases, the weighted sum of Eqs. (7.26) and (7.27) is used as the solution for both cases, as Eq. (7.28).

$$\dot{\boldsymbol{\theta}} = \boldsymbol{J}_e^{\dagger}\dot{\boldsymbol{x}}_e + \left(\boldsymbol{I} - \boldsymbol{J}_e^{\dagger}\boldsymbol{J}_e\right)\left[\alpha \boldsymbol{z}_o + (1-\alpha)\boldsymbol{z}_r\right], \tag{7.28}$$

where $\alpha \in [0, 1]$ is the weighting factor indicating whether the obstacle is near the manipulator and is used to switch the obstacle avoidance mode and the restoring mode, which is given by Eq. (7.29)

$$\alpha = \begin{cases} 0, d \geq d_o \\ \dfrac{d_o - d}{d_o - d_r}, d_r < d < d_o \\ 1, d \leq d_r \end{cases}, \tag{7.29}$$

where d_r is the distance threshold that the manipulator starts to restore its original pose.

As the desired end-effector trajectory is always given in a set of position $\boldsymbol{x}$ in Cartesian space, a closed-loop controller is designed as Eq. (7.30).

$$\dot{\boldsymbol{x}}_e = \dot{\boldsymbol{x}}_d + \boldsymbol{K}_e\boldsymbol{e}_x, \tag{7.30}$$

where $\boldsymbol{x}_d$ is the desired end-effector position given by the operator, $\boldsymbol{e}_x = \boldsymbol{x}_d - \boldsymbol{x}_e$ is the end-effector position error between the desired end-effector position and the actual position, which can be obtained by the forward kinematics; and $\boldsymbol{K}_e$ is a symmetric positive definite matrix. The control strategy can be obtained by substituting Eq. (7.30) into Eq. (7.28), as Eq. (7.31).

$$\dot{\boldsymbol{\theta}}_d = \boldsymbol{J}_e^{\dagger}\left(\dot{\boldsymbol{x}}_d + \boldsymbol{K}_e\boldsymbol{e}_x\right) + \left(\boldsymbol{I} - \boldsymbol{J}_e^{\dagger}\boldsymbol{J}_e\right)\left[\alpha \boldsymbol{z}_o + (1-\alpha)\boldsymbol{z}_r\right] \tag{7.31}$$

To show that the closed-loop system is asymptotically stable, the Lyapunov direct method is utilized.

Proof Choose a positive definite Lyapunov function as

$$V(\boldsymbol{e}_x) = \frac{1}{2}\boldsymbol{e}_x^T\boldsymbol{K}_e\boldsymbol{e}_x. \tag{7.32}$$

Differentiating Eq. (7.32) we have

$$\dot{V} = \boldsymbol{e}_x^T\boldsymbol{K}_e\dot{\boldsymbol{x}}_d - \boldsymbol{e}_x^T\boldsymbol{K}_e\dot{\boldsymbol{x}}_e. \tag{7.33}$$

Substituting Eq. (7.13) into it, we have

$$\dot{V} = \boldsymbol{e}_x^T\boldsymbol{K}_e\dot{\boldsymbol{x}}_d - \boldsymbol{e}_x^T\boldsymbol{K}_e\boldsymbol{J}_e\dot{\boldsymbol{\theta}}. \tag{7.34}$$

With the control law given by Eq. (7.31), the above equation becomes

$$\dot{V} = e_x^T K_e \dot{x}_d - e_x^T K_e J_e \left[J_e^\dagger (\dot{x}_d + K_e e_x) + \left(I - J_e^\dagger J_e\right) [\alpha z_o + (1-\alpha) z_r] \right]. \tag{7.35}$$

It can be further simplified as

$$\dot{V} = -e_x^T K J_e J_e^\dagger K e_x \leq 0 \tag{7.36}$$

such that the closed-loop system is asymptotic stable.

7.2.4 Experiment Studies

Three groups of experiments are designed to show the obstacle avoidance with Baxter® robot. Each group includes two comparative experiments, with and without the obstacle.

Static End-effector Position and Dynamic Obstacle

In the first group of comparative experiments, the end-effector position is maintained as static and held by the operator, while the position of the obstacle is changed dynamically, as shown in Fig. 7.5. The end-effector position error is shown in Fig. 7.6 and the joints angle is shown in Fig. 7.7. As can be seen, when the obstacle moves near the manipulator, the elbow moves downward and avoids the obstacle. At the same time, the end-effector position was barely affected by the obstacle. After the obstacle has been removed, the pose of the manipulator restores to its original pose.

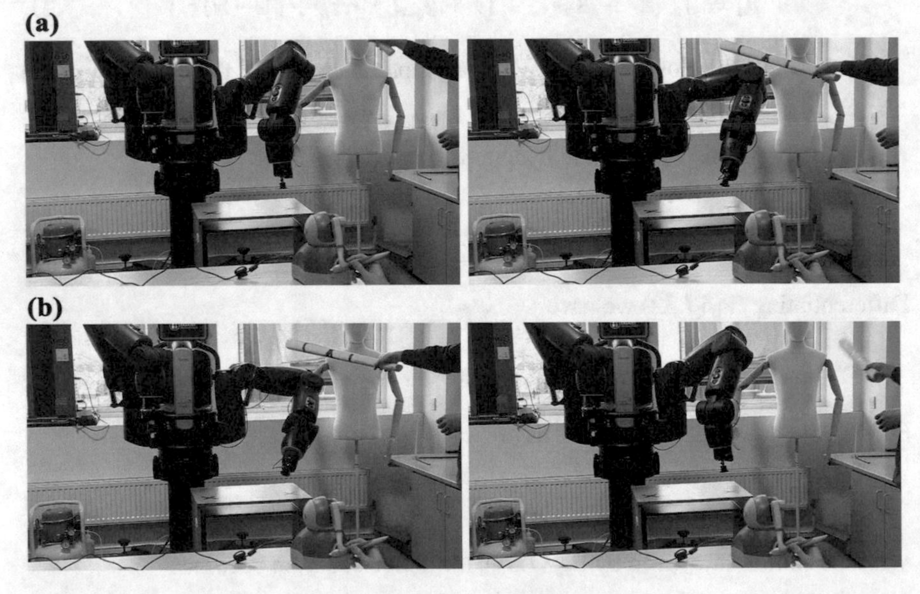

Fig. 7.5 Video frames of the first group of experiment, with static end-effector position and dynamic obstacle. **a** $t = 36$ s. **b** $t = 38$ s. **c** $t = 39$ s. **d** $t = 43$ s

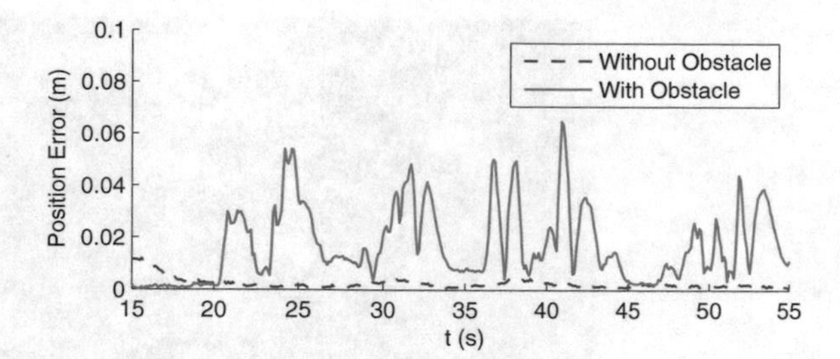

Fig. 7.6 The end-effector position error with static end-effector position and dynamic obstacle

Dynamic End-effector Position and Static Obstacle

In the second group of comparative experiments, the operator controls the manipulator's end-effector position moving along a predetermined trajectory from initial point A, via points B and C and finally returning to point A, as shown in Fig. 7.8. And the obstacle is placed statically near the predetermined trajectory, as shown in Fig. 7.9. The end-effector position trajectory is shown in Fig. 7.10 and the joints angle is shown in Fig. 7.11.

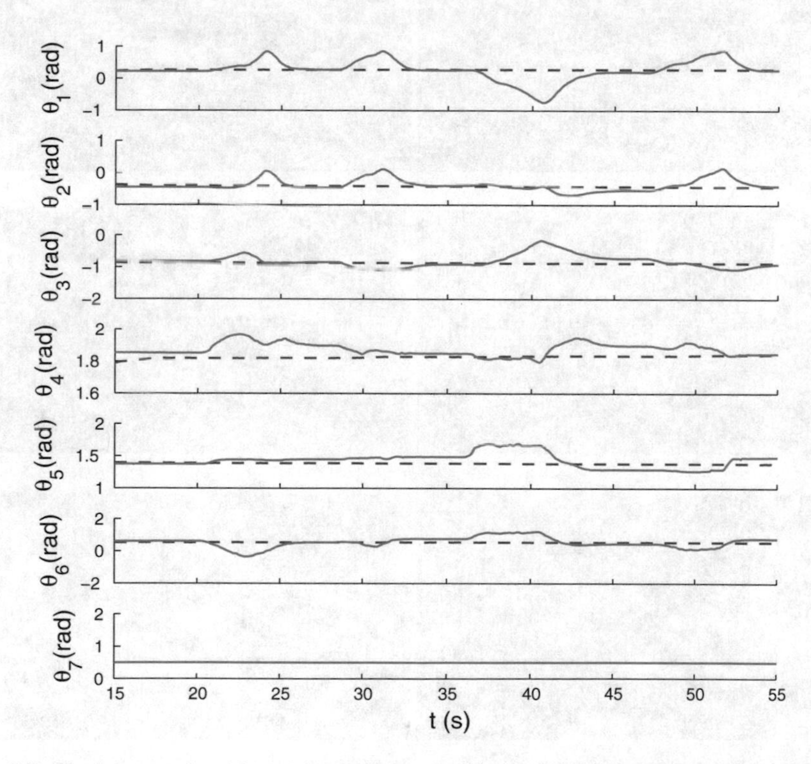

Fig. 7.7 The joints angle with static end-effector position and dynamic obstacle. The *solid line* indicates the joints angle with obstacle, while the *dashed line* indicates the joints angle without the obstacle

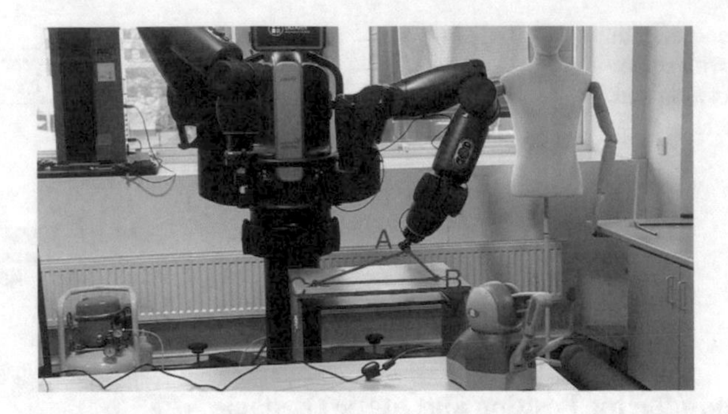

Fig. 7.8 The predetermined trajectory. Point A is the initial point, points B and C are the way points

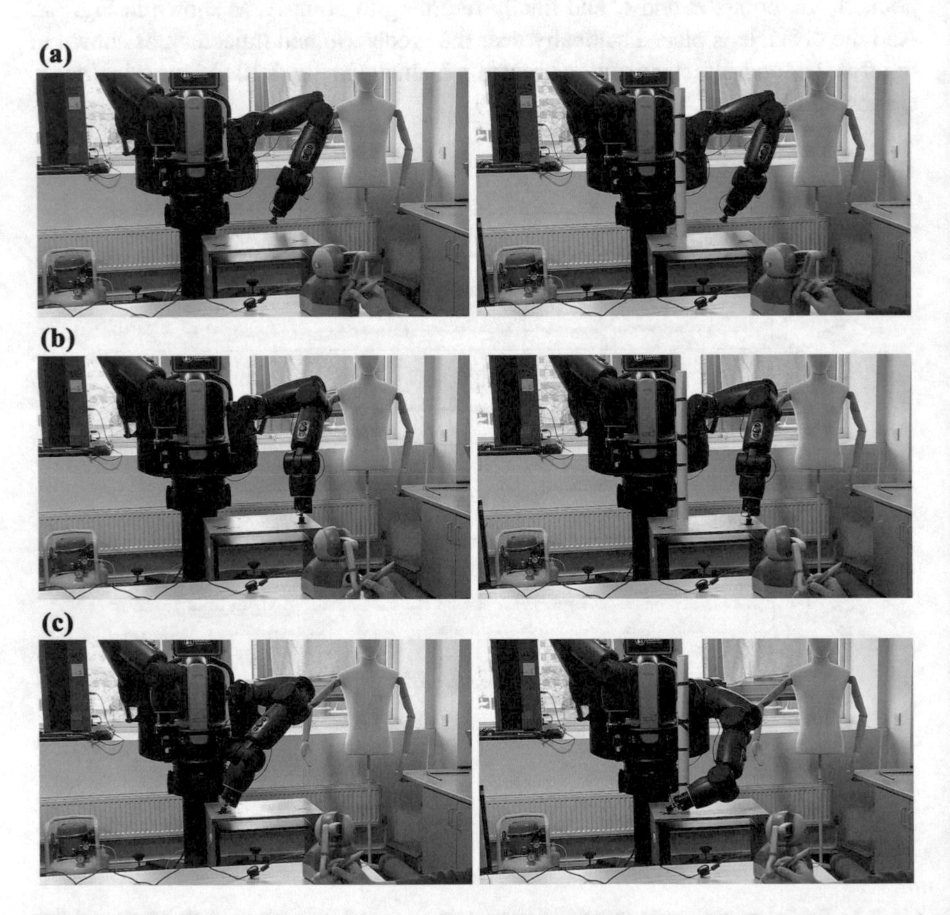

Fig. 7.9 Video frames of the second group of experiment, with dynamic end-effector position and static obstacle. **a** Without obstacle. $t = 7$ s. **b** Without obstacle. $t = 14$ s. **c** Without obstacle. $t = 14$ s. **d** Without obstacle. $t = 21$ s. **e** Without obstacle. $t = 19$ s. **f** Without obstacle. $t = 27$ s

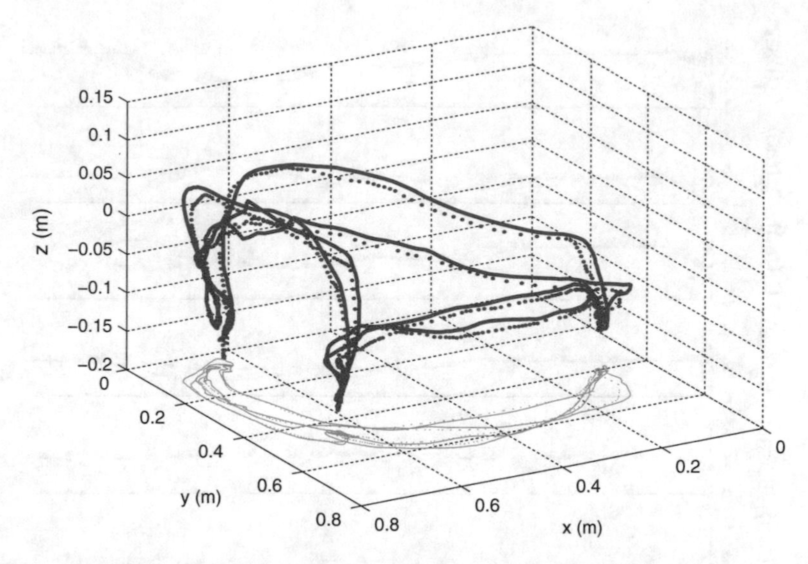

Fig. 7.10 The end-effector position trajectory with dynamic end-effector position and static obstacle. The *blue dashed line* and the *red solid line* indicate the desired and actual trajectory, respectively, without the obstacle while the *red lines* indicate the trajectory with the obstacle

As we can see, the manipulator can change its pose in order to reach point C without touching the obstacle. Although the joints angles change a lot when the manipulator is avoiding the obstacle, the end-effector position still follows the desired trajectory.

Dynamic End-effector Position and Dynamic Obstacle

In the third group, the manipulator's end-effector position is also controlled by the operator along the predetermined trajectory, while the obstacle is dynamically moved around the manipulator, as shown in Fig. 7.13. The end-effector position trajectory is shown in Fig. 7.12 and the joints angle is shown in Fig. 7.14. As can be seen, after the manipulator completed the obstacle avoidance task, its pose restored to its original pose.

7.3 Robot Self-Identification for Obstacle Avoidance

In recent years, the application fields of robotics increase rapidly. Different from traditional working environment, such as on the production line, robots are meeting more complex and dynamic environments in their recent tasks, as well as cooperating with human. As a prerequisite of collision avoidance and motion planning, environmental perception will play an important role to guarantee the safety of the human co-worker, the surrounding facilities and the robot itself, and also to enable the robots to complete their tasks more intelligently.

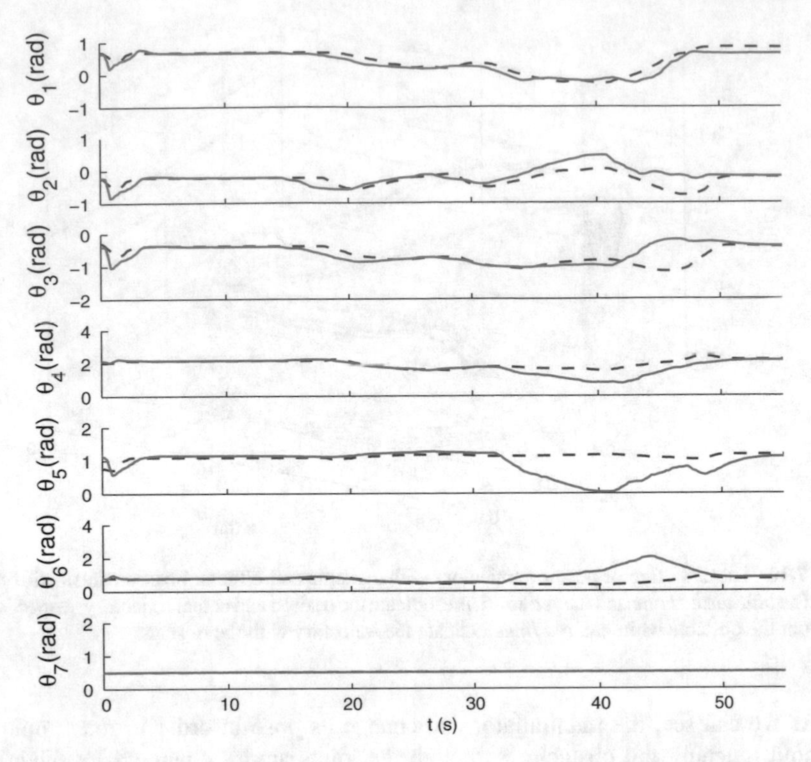

Fig. 7.11 The joints angle with dynamic end-effector position and static obstacle. The *solid line* indicates the joints angle with obstacle, while the *dashed line* indicates the joints angle without the obstacle

Various types of sensors have been used in previous studies of the environmental perception. Most of the sensors provide 3D point cloud of the manipulator and the surrounding environment. In [14], Natarajan used a 3D Time-Of-Flight (TOF) camera as a proximity sensor to identify the position of the obstacle near the elbow of the manipulator. In [15] Danial introduced a manipulator robot surface following algorithm using a 3D model of vehicle body panels acquired by a network of rapid but low-resolution RGB-D sensors. In [16], a fast normal computation algorithm for depth image was proposed, which is able to use normal deviation along eight directions to extract key points for segmenting points into objects on manipulation support plane in an unstructured table top scene.

In most of the previous studies, the manipulator is simply detected and deleted from the point cloud based on the 3D model. While the inevitable calibration error and the simplified 3D model of the manipulator may cost incompletely deleting, and the remaining points near the manipulator will cause great trouble to the subsequent motion planning or obstacle avoidance method. The most commonly used sensors include the 3D TOF camera, the stereo cameras and the Kinect® sensor. The noise in the depth channel of the raw data is also an unavoidable challenge.

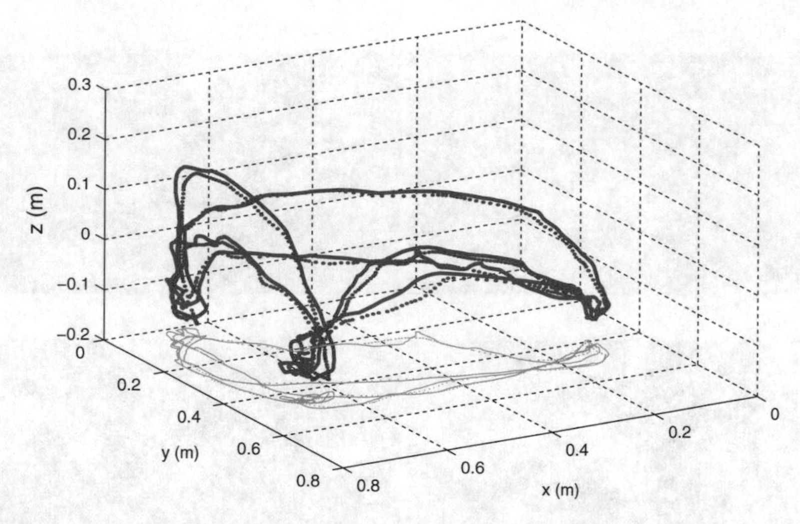

Fig. 7.12 The end-effector position trajectory with dynamic end-effector position and dynamic obstacle

In this section, the robot skeleton is first matched with the point cloud by calibration. The point cloud is then segmented into five groups which represent the meaning of the points using space division. The collision prediction method is proposed based on the point groups to achieve the real-time collision avoidance feature [17]. A first-order filter is designed to deal with the noise in the point cloud which can provide a smooth estimation of the collision parameters. Finally, a shared control system equipped with the proposed environment perception method is designed using the Kinect® sensor, the Omni® haptic device, and the Baxter® robot, which can achieve remote control and real-time autonomous collision avoidance.

7.3.1 Kinect® Sensor and 3D Point Cloud

As a low-cost typical RGB-D sensor, the Microsoft Kinect® sensor can generate a detailed point cloud with around 300,000 points at 30 Hz, and is widely used for the point cloud capture [15, 18–22]. In [19], Pan presents a new collision and distance query algorithms, which can efficiently handle large amounts of point cloud sensor data received at real-time rates. Sukmanee presents a method to distinguish between a manipulator and its surroundings using a depth sensor in [20], where the iterative least square (ILS) and iterative closest point (ICP) algorithms are used for coordinate calibration and the matching between the manipulator model and point cloud.

Unlike normal webcams, which generate images by collecting light that is reflected by the objects in its field of view and digitalizing the optical signals,

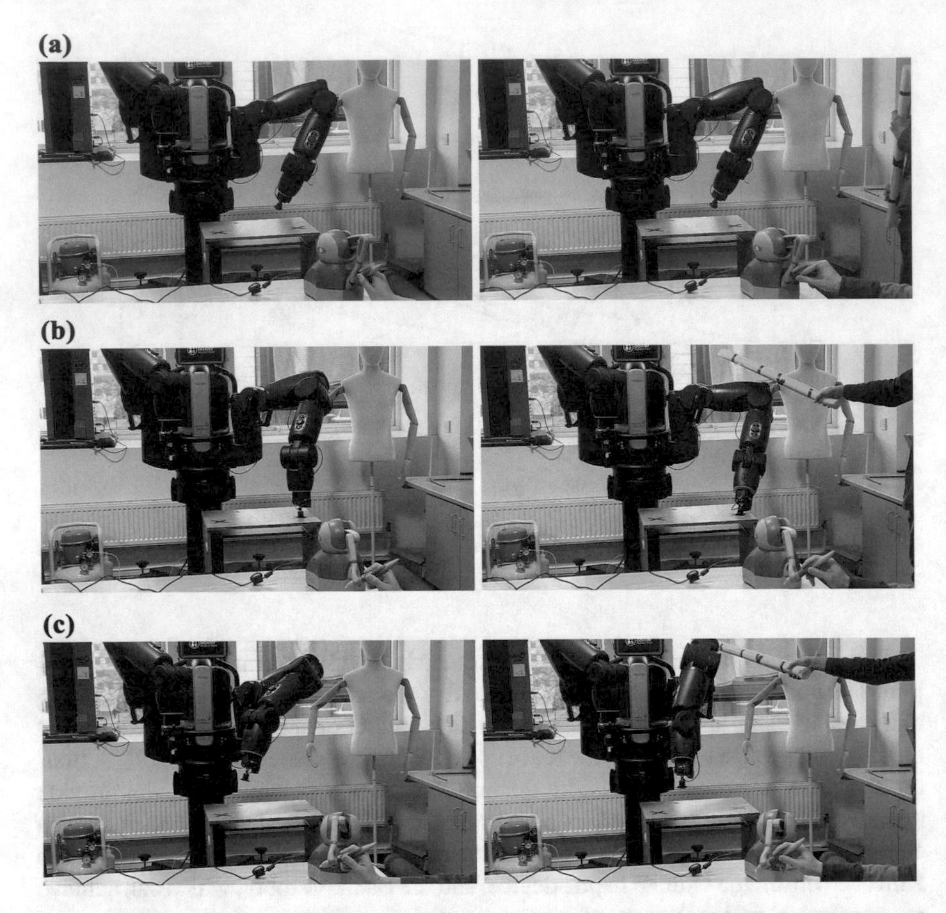

Fig. 7.13 Video frames of the third group of experiment, with dynamic end-effector position and dynamic obstacle. **a** Without obstacle. $t = 8$ s. **b** With obstacle. $t = 7$ s. **c** Without obstacle. $t = 14$ s. **d** Without obstacle. $t = 14$ s. **e** Without obstacle. $t = 18$ s. **f** Without obstacle. $t = 20$ s

Kinect®, a RGB-D camera, can also measure the distance between the objects and itself. A colored 3D point cloud can be generated from both of the RGB image and the depth image. In this way, the Kinect® sensor is capable of detecting the environment in front of it.

The raw data obtained from the Kinect® sensor includes a RGB image and a depth image, as shown in Fig. 7.15. A 3D point cloud in Cartesian coordinate is generated from these two images. Each pixel in the depth image corresponds to a point in the 3D point cloud. Denote the set of points in the 3D point cloud as $\boldsymbol{P} = \{\boldsymbol{p}_i\,(x_i, y_i, z_i)\,|i = 0, 1, \ldots, n\}$, where n is the total number of pixels in the depth image. The coordinate of the ith point $\boldsymbol{p}_i$ is calculated by Eq. (7.37) [21].

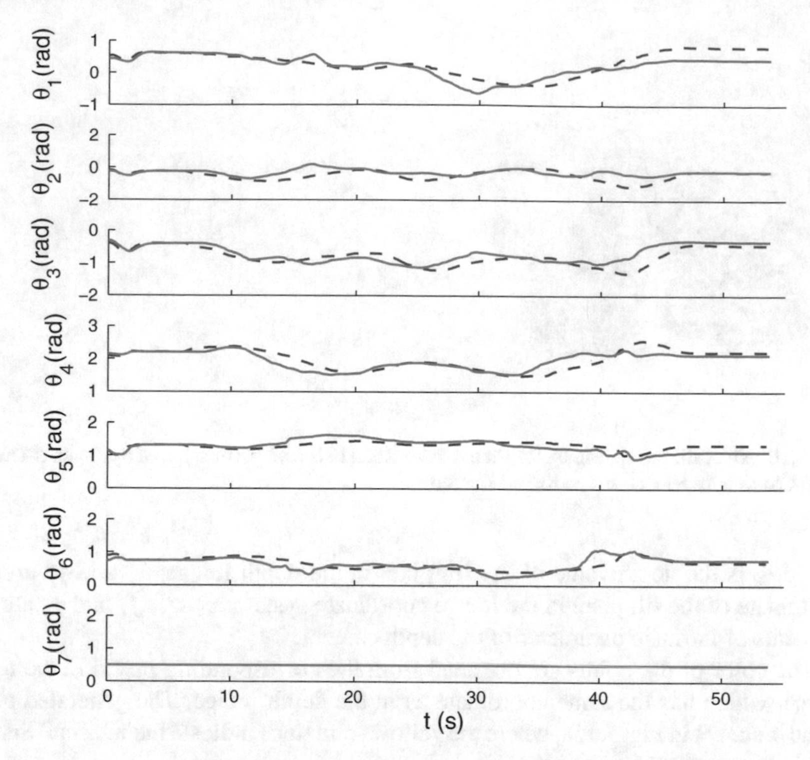

Fig. 7.14 The joints angle with dynamic end-effector position and dynamic obstacle. The *solid line* indicates the joints angle with obstacle, while the *dashed line* indicates the joints angle without the obstacle

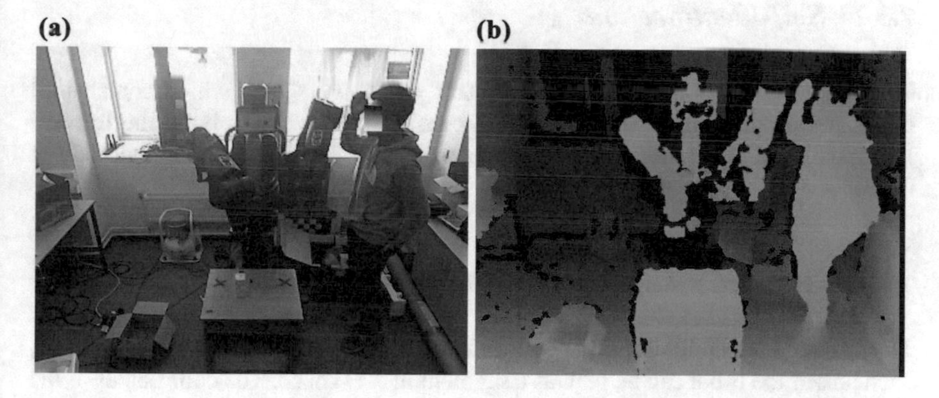

Fig. 7.15 The raw data obtained from the Kinect®.sensor. **a** The RGB image. **b** The depth image

$$\begin{aligned} x_i &= d_i(x_i^c - c_x)/f_x \\ y_i &= d_i(y_i^c - c_y)/f_y \\ z_i &= d_i, \end{aligned} \tag{7.37}$$

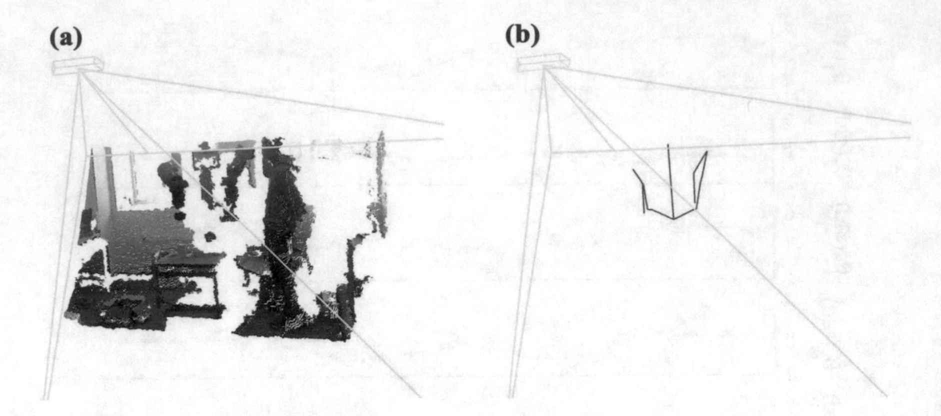

Fig. 7.16 The calibration results. Reprinted from Ref. [17], with permission of Springer. **a** The 3D Point Cloud with Kinect. **b** The robot skeleton

where d_i is the depth value of the ith pixel in the depth image; x_i^c and y_i^c are the coordinate of the ith pixel in the image coordinate system; c_x, c_y, f_x and f_y are the elements of intrinsic parameter of the depth camera.

The color of the points are obtained from the corresponding pixels in the RGB image, which has the same coordinate as in the depth image. The generated point cloud is shown in Fig. 7.16a, where the yellow solid lines indicate the Kinect® Sensor and its viewing angle.

7.3.2 Self-Identification

Self-identification is used to obtain a set of points $\boldsymbol{P_r} \subseteq \boldsymbol{P}$, which contains all the points on the surface of the robot. It consists of two steps: build the forward kinematics and calibration; generate the boundary of the skeleton neighborhood of the robot.

7.3.2.1 Forward Kinematics and Calibration

The forward kinematics is used to get the skeleton of the manipulator. In the skeleton, each link of the robot can be seen as a segment in 3-D space. The coordinates of the endpoints of the segments, i.e. the coordinates of the joints, in Cartesian space can be obtained by Eq. (7.38).

$$\boldsymbol{X}_i' = {}^{i-1}\boldsymbol{A}_i \ \ldots \ {}^{1}\boldsymbol{A}_2{}^{0}\boldsymbol{A}_1\boldsymbol{X}_0', \quad i = 1, 2, \ldots, n \tag{7.38}$$

where $\boldsymbol{X}_i' = [x_i', y_i', z_i', 1]^T$ is an augmented vector of relative Cartesian coordinate of the ith joint in the robot coordinate system; $\boldsymbol{X}_0'$ is the augmented coordinate of the base of the manipulator in the robot coordinate system; and ${}^{i-1}\boldsymbol{A}_i$ is the link homogeneous transform matrix, which can be found in Chap. 2, the kinematics modeling of the Baxter® robot.

The robot skeleton and the point cloud need to be placed into a unified coordinate system, in order to achieve self-identification of the manipulator. The coordinate transformation of the robot skeleton from the robot coordinate to the point cloud coordinate can be obtained by Eq. (7.39).

$$\boldsymbol{X}_i = \boldsymbol{T}\boldsymbol{X}'_i, \tag{7.39}$$

where $\boldsymbol{X}_i = [x_i, y_i, z_i, 1]^T$ is the coordinate in the point cloud coordinate system. The transfer matrix $\boldsymbol{T}$ from the Kinect® coordinate to the robot coordinate can be calculated by Eq. (7.9). The transformed robot skeleton is shown in Fig. 7.16b.

7.3.2.2 Neighborhood of the Robot Skeleton

The points on the robot must be in a neighborhood of the robot skeleton. To find the points $\boldsymbol{P}_r$, one solution is to calculate the distance between each point $\boldsymbol{p} \in \boldsymbol{P}$ and the robot skeleton. As the robot has as many as 19 links, it is foreseeable that the complexity of the algorithm will be very high. Alternatively, the boundary of the skeleton neighborhood can be generated first. To enable this, several points on each link of the robot are calculated by Eq. (7.40).

$$\boldsymbol{x}_k = \boldsymbol{X}_i + \frac{j}{m}(\boldsymbol{X}_{i+1} - \boldsymbol{X}_i), \; j = 0, 1, \ldots, m, \tag{7.40}$$

where $k = im + j$, $\boldsymbol{x}_k$ is the jth interpolation points on the ith link, and m is the amount of the points that is interpolated on each link. The spheres around the interpolation points with predefined radius is defined as the neighborhood of the interpolation points, and the union of all the spheres can be approximated as the neighborhood of the robot skeleton, as shown in Fig. 7.17, where the green spheres indicate the neighborhood of the robot skeleton with $m = 5$. The predefined radius is set such that the neighborhood can exactly contain the robot. Finally, all the points inside the neighborhood of the robot skeleton are considered as the points on the robot and is contained in $\boldsymbol{P}_r$, as shown in Fig. 7.18, where the green points indicate the set of points $\boldsymbol{P}_r$.

7.3.3 *Collision Predication*

The points in $\boldsymbol{P}$ can be divided into five categories, namely, the points on the robot $\boldsymbol{P}_r$, the points seen as the obstacle for left arm and right arm $\boldsymbol{P}_{ol} \subseteq \boldsymbol{P}$ and $\boldsymbol{P}_{or} \subseteq \boldsymbol{P}$, the points seen as the object to be operated $\boldsymbol{P}_d \subseteq \boldsymbol{P}$, and the other points. In order to get the points sets $\boldsymbol{P}_{ol}$, $\boldsymbol{P}_{or}$ and $\boldsymbol{P}_d$, the operating space is dynamically divided

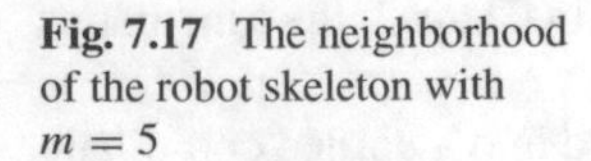

Fig. 7.17 The neighborhood of the robot skeleton with $m = 5$

into five regions, as shown in Fig. 7.19a, the green robot region as mentioned before, the blue obstacle region for left arm, the red obstacle region for right arm, the cyan operating object region, and the other space other than these regions. The operating object region is defined as the neighborhood of the end-effector positions with a radius r_d, as the cyan spheres in Fig. 7.19a. The obstacle regions are defined as the neighborhood of the movable links of the manipulators with a radius r_o.

The points inside the regions are included in the corresponding points sets respectively, as shown in Fig. 7.18b, the blue points indicate the obstacle points for left arm $\boldsymbol{P_{ol}}$ and the cyan points indicates the points $\boldsymbol{P_d}$ on the object to be operated. Because there are no obstacle near the right arm in Fig. 7.18b,

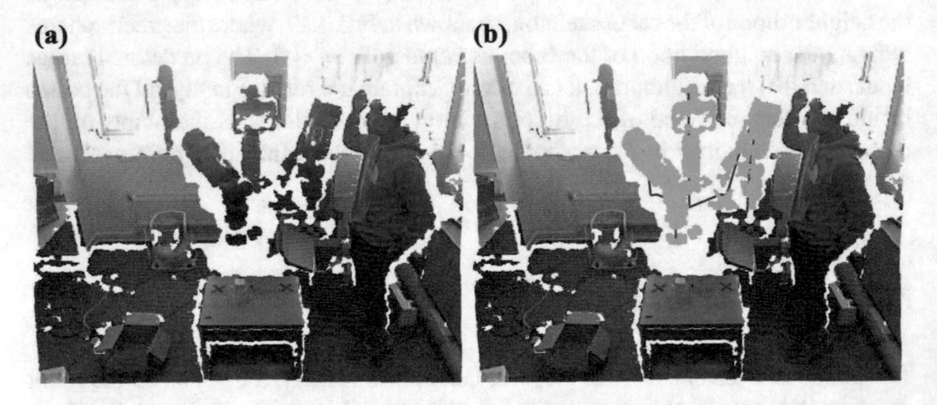

Fig. 7.18 Self-identification result. **a** The raw point cloud $\boldsymbol{P}$. **b** The *green points* indicate the set of points $\boldsymbol{P_r}$

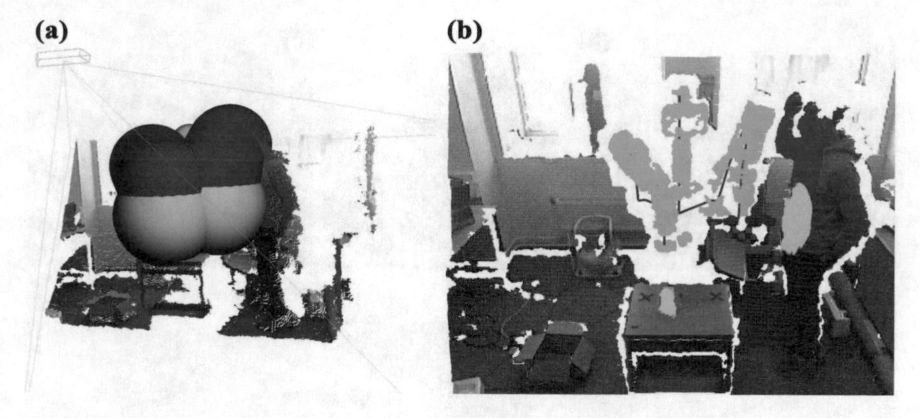

Fig. 7.19 Space division. **a** The space division regions. **b** The points in different regions

It should be noted that the points $\boldsymbol{P_d}$ on the object to be operated is found near the end-effector positions of the manipulators and is excluded from the obstacle points $\boldsymbol{P_{ol}}$ and $\boldsymbol{P_{or}}$. This is because that there will definitely be a collision between the operating object and the end-effector when the robot is manipulating the object. For instance, when the manipulator is grabbing the object with a gripper, the object will intrude into the gripper. However, these kinds of "collision" are intentional and should not be avoided.

For each manipulator, the collision parameters include the collision points on the robot and the obstacle, and the collision vector $\boldsymbol{v_c}$. The collision points, $\boldsymbol{p_{cr}} \in \{\boldsymbol{x_k} | k = 1, 2, \ldots, nm\}$ and $\boldsymbol{p_{co}} \in \boldsymbol{P_o}$, are the two points either on the robot or on the obstacle, which covers the nearest distance between the robot and the obstacle, as shown in Fig. 7.20. The noise on the depth image from the Kinect® sensor will transfer into the point cloud, and will lead to frequent changes in $\boldsymbol{p_{co}}$. A first-order filter is designed to obtain a stable and smooth changing collision point $\boldsymbol{p^*_{co}}$, which is given by Eq. (7.41).

$$\boldsymbol{p^*_{co}}(t) = \frac{k_f \boldsymbol{p^*_{co}}(t-1) + \boldsymbol{p_{co}}(t)}{k_f + 1}, \tag{7.41}$$

Fig. 7.20 Obstacle detection

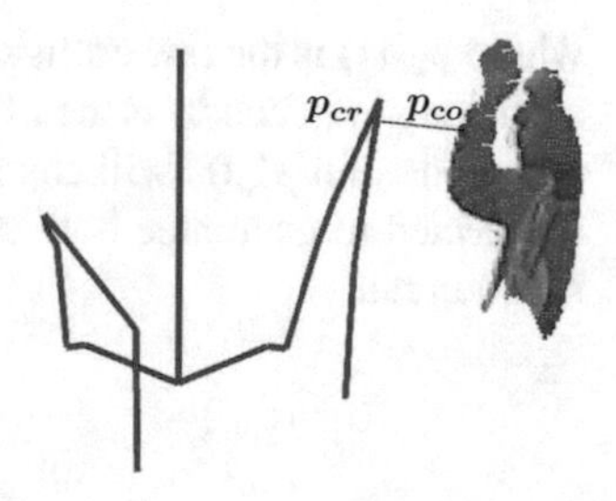

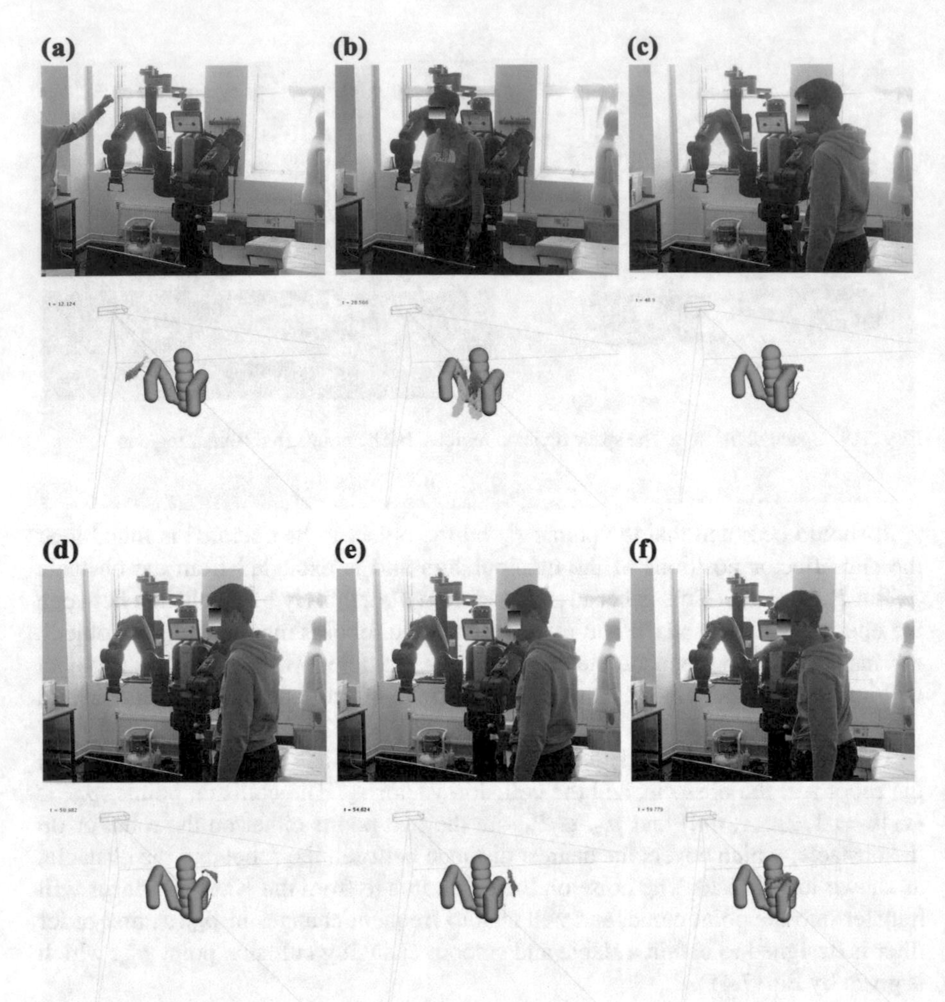

Fig. 7.21 Experiment results. **a** $t = 12.124$ s. **b** $t = 28.505$ s. **c** $t = 48.900$ s. **d** $t = 50.982$ s. **e** $t = 54.624$ s. **f** $t = 59.779$ s

where $\boldsymbol{p_{co}}(t)$ is the raw collision point on the obstacle at time t; $\boldsymbol{p_{co}^{*}}(t)$ is the filtered collision point; and k_f is the filtering strength parameter. With a larger k_f, the filtered collision point $\boldsymbol{p_{co}^{*}}(t)$ will change more smoothly yet with larger time delay. Thus k_f is selected to guarantee both the control smoothness of the robot and a satisfactory reaction rate.

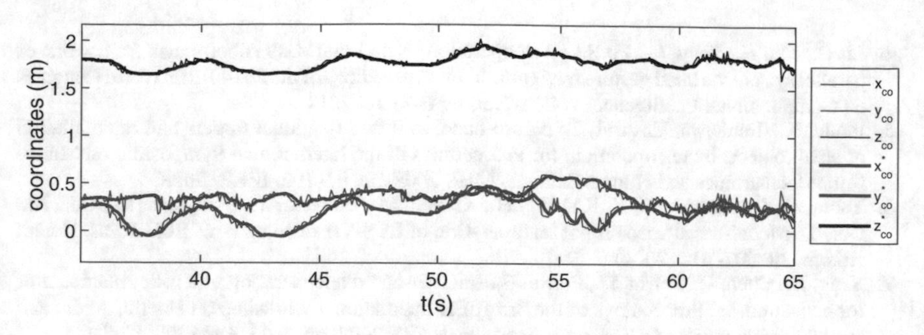

Fig. 7.22 The coordinates of $\boldsymbol{p}_{co}$ and $\boldsymbol{p}^*_{co}$

7.3.4 Experiments Studies

A shared control system is designed to verify the performance of the proposed self-identification and collision prediction method. In the system, the user teleoperates the Baxter® robot by giving the reference end-effector trajectory using the Omni® haptic device, which is connected on the client computer. The robot is controlled by the server computer, which is also equipped with a Kinect® sensor to give the RGB-D image of the manipulator and the surrounding environment.

The manipulator will follow the end-effector trajectory given by the user, and at the same time, avoid the obstacle automatically. The collision warning information will be sent to the used in the form of haptic feedback.

The experiment results are shown in Fig. 7.21. In each of the subfigures, the surrounding environment is shown upper and the detection result is shown lower, where the red and blue dots indicate the collision points, $\boldsymbol{p}^*_{cr}$ for the left and right arm, respectively. Notes that in Fig. 7.21b, the coworker is standing between the two arms, thus two collision points can be found on different side of the human body for different manipulator arm.

The coordinates of $\boldsymbol{p}_{co}$ and $\boldsymbol{p}^*_{co}$ for the left arm are shown in Fig. 7.22, where the thin lines indicate the coordinates of $\boldsymbol{p}_{co}$ and the thick lines indicate the coordinates of $\boldsymbol{p}^*_{co}$. As can be seen, the line of $\boldsymbol{p}^*_{co}$ is much smoother than $\boldsymbol{p}_{co}$ thanks to the filter and is ready for the following obstacle avoidance algorithm.

References

1. Sciavicco, L., Siciliano, B.: Solving the inverse kinematic problem for robotic manipulators. RoManSy 6, pp. 107–114. Springer, Cham (1987)
2. Slotine, J.-J., Yoerger, D.: A rule-based inverse kinematics algorithm for redundant manipulators. Int. J. Robot. Autom. **2**(2), 86–89 (1987)
3. Sciavicco, L., Siciliano, B.: A solution algorithm to the inverse kinematic problem for redundant manipulators. IEEE J. Robot. Autom. **4**(4), 403–410 (1988)

4. Li, C., Ma, H., Yang, C., Fu, M.: Teleoperation of a virtual iCub robot under framework of parallel system via hand gesture recognition. In: Proceedings of the 2014 IEEE World Congress on Computational Intelligence, WCCI. Beijing 6–11 Jul 2014
5. Inoue, K., Tanikawa, T., Arai, T.: Micro hand with two rotational fingers and manipulation of small objects by teleoperation. In: Proceedings of the International Symposium on Micro-NanoMechatronics and Human Science, MHS 2008, pp. 97–102. IEEE (2008)
6. Yoon, W.-K., Goshozono, T., Kawabe, H., Kinami, M., Tsumaki, Y., Uchiyama, M., Oda, M., Doi, T.: Model-based space robot teleoperation of ETS-VII manipulator. IEEE Trans. Robot. Autom. **20**, 602–612 (2004)
7. Yang, X., Chen, Q., Petriu, D., Petriu, E.: Internet-based teleoperation of a robot manipulator for education. In: Proceedings of the 3rd IEEE International Workshop on Haptic, Audio and Visual Environments and Their Applications, HAVE 2004, pp. 7–11. Oct (2004)
8. Kofman, J., Wu, X., Luu, T., Verma, S.: Teleoperation of a robot manipulator using a vision-based human-robot interface. IEEE Trans. Ind. Electron. **52**, 1206–1219 (2005)
9. Wang, X., Yang, C., Ma, H., Cheng, L.: Shared control for teleoperation enhanced by autonomous obstacle avoidance of robot manipulator. In: Proceedings of the 2015 IEEE/RSJ International Conference on Intelligent Robots and Systems (IROS), pp. 4575–4580 (2015)
10. Ju, Z., Yang, C., Li, Z., Cheng, L., Ma, H.: Teleoperation of humanoid baxter robot using haptic feedback. In: Proceedings of the 2014 International Conference on Multisensor Fusion and Information Integration for Intelligent Systems (MFI), pp. 1–6 (2014)
11. Ju, Z., Yang, C., Ma, H.: Kinematics modeling and experimental verification of baxter robot. In: Proceedings of the 2014 33rd Chinese Control Conference (CCC), pp. 8518–8523. IEEE (2014)
12. Maciejewski, A.A., Klein, C.A.: Obstacle avoidance for kinematically redundant manipulators in dynamically varying environments. Int. J. Robot. Res. **4**(3), 109–117 (1985)
13. Corke, P.: Robotics, Vision and Control: Fundamental Algorithms in MATLAB, vol. 73. Springer Science & Business Media, Sydney (2011)
14. Natarajan, S., Vogt, A., Kirchner, F.: Dynamic collision avoidance for an anthropomorphic manipulator using a 3D TOF camera. In: Proceedings of the 2010 41st International Symposium on Robotics (ISR) and 2010 6th German Conference on Robotics (ROBOTIK), pp. 1–7 (2010)
15. Nakhaeinia, D., Fareh, R., Payeur, P., Laganiere, R.: Trajectory planning for surface following with a manipulator under RGB-D visual guidance. In: Proceedings of the 2013 IEEE International Symposium on Safety, Security, and Rescue Robotics (SSRR), pp. 1–6 (2013)
16. Ma, S., Zhou, C., Zhang, L., Hong, W., Tian, Y.: Depth image denoising and key points extraction for manipulation plane detection. In: Proceedings of the 2014 11th World Congress on Intelligent Control and Automation (WCICA), pp. 3315–3320 (2014)
17. Wang, X., Yang, C., Ju, Z., Ma, H., Fu, M.: Robot manipulator self-identification for surrounding obstacle detection. Multimed. Tools Appl. pp. 1–26 (2016)
18. Luo, R., Ko, M.-C., Chung, Y.-T., Chatila, R.: Repulsive reaction vector generator for whole-arm collision avoidance of 7-DoF redundant robot manipulator. In: Proceedings of the 2014 IEEE/ASME International Conference on Advanced Intelligent Mechatronics (AIM), pp. 1036–1041. July (2014)
19. Pan, J., Sucan, I., Chitta, S., Manocha, D.: Real-time collision detection and distance computation on point cloud sensor data. In: Proceedings of the 2013 IEEE International Conference on Robotics and Automation (ICRA), pp. 3593–3599. May (2013)
20. Sukmanee, W., Ruchanurucks, M., Rakprayoon, P.: Obstacle modeling for manipulator using iterative least square (ILS) and iterative closest point (ICP) base on kinect. In: Proceedings of the 2012 IEEE International Conference on Robotics and Biomimetics (ROBIO), pp. 672–676 (2012)
21. Thumbunpeng, P., Ruchanurucks, M., Khongma, A.: Surface area calculation using kinect's filtered point cloud with an application of burn care. In: Proceedings of the 2013 IEEE International Conference on Robotics and Biomimetics (ROBIO), pp. 2166–2169. Dec (2013)
22. Wang, B., Yang, C., Xie, Q.: Human-machine interfaces based on EMG and kinect applied to teleoperation of a mobile humanoid robot. In: Proceedings of the 2012 10th World Congress on Intelligent Control and Automation (WCICA), pp. 3903–3908. July (2012)

Chapter 8
Human–Robot Interaction Interface

Abstract Human–robot interaction is an advanced technology and plays an increasingly important role in robot applications. This chapter first gives a brief introduction to various human–robot interfaces and several technologies of human–robot interaction using visual sensors and electroencephalography (EEG) signals. Next, a hand gesture-based robot control system is developed using Leap Motion, with noise suppression, coordinate transformation, and inverse kinematics. Then, another hand gesture control, which is one of natural user interfaces, is then developed based on a parallel system. ANFIS and SVM algorithms are employed to realize the classification. We also investigate controlling the commercialized Spykee mobile robot using EEG signals transmitted by the Emotiv EPOC neuroheadset. The Emotiv headset is connected to the OpenViBE to control a virtual manipulator moving in 3D Cartesian space, using a P300 speller.

8.1 Introduction of Human–Robot Interfaces

Human–robot collaboration represents key characters of next generation of robots working without cages and sharing the same workspace with human community [1, 2]. It allows human interacts with robots physically with safety and compliance guaranteed. Therefore, human–robot collaboration has a variety of applications in home care [3], office, industry [4], agriculture [5], school educations, rehabilitation, and various human motor skill training scenarios.

Conventional industrial robots have a lot of advantages such as working with precision and speed, but relying on high stiffness, high precise parts and components, and complicated control systems. This makes them relatively expensive, inflexible, and dangerous for humans [6]. Moreover, due to these disadvantages, such robots are only suitable and efficient for simple and repeated tasks in structured environment. However, the factory production should be more flexible and smart to meet diversity and personalization of customer demand which tends to be mainstream in the future [7]. Recently, collaborative robots are being extensively investigated by more researchers and technology companies and may be expected to dominate robot fields in the next few years [8]. The collaborative robots are required to fulfill more

© Science Press and Springer Science+Business Media Singapore 2016

C. Yang et al., *Advanced Technologies in Modern Robotic Applications*,
DOI 10.1007/978-981-10-0830-6_8

dexterous and compliant tasks to reduce the human workload and improve the quality of products.

Generally, this requires expert modeling and programming skills that only a few people can implement such skill transfer, which discourages median or small companies to employ robots in the production line. Therefore, an intuitive and augmented approach to human–robot complicated skills transfer will make robot application more practical and remove the virtual cage between human and robot completely [9]. For example, service robotics needs to perfectly fit customers need in home or office scenarios, and such kind of robot needs to interact with different stages of people with different cognitive and knowledge backgrounds. On the other hand, as the new generation of robots, collaborative robots are expected to share the same workspace with our humans, safety is a key for collaborative robotics [6]. It is often achieved through complex control systems, but robots are rarely designed to be inherently safe. The economic way is to make robot arm to be human like and as compliant as human limb and impedance of the robots can be adapted to dynamic environment. Thus the studies on compliant control have received increasing attentions [10] in the tasks involving uncertain and dynamic environment, e.g., physical human–robot cooperation [11–13].

The human–robot collaboration technique is a key feature in distinguishing new generation of robots from conventional robots which usually work remotely and separately from human. Human–robot collaboration with intuitive human–robot interface will be an compliant and efficient way to extend human and robot workspace and applicable areas.

A major problem of human–robot collaboration is to build the contact between human and robot, to make the human–robot collaboration possible. Therefore, it is necessary to design a interface between human and robot and transfer the human operation to the robot.

For an effective human–robot collaboration with safety, compliance, and stability, it is crucial to enable robots to be capable of integrating interactive cognition and manipulation with humans, similar to the mechanisms involved in human–human interaction [14].

Inspired by human mechanical impedance adaptability to various environments [15], human motor control extraction via EMG and bio-mimetic learning controller [16], [17] are developed to achieve more natural and efficient performance, as they directly reflect impedance variation without disturbing motor control during the task. EMG signals represent muscle activation regulated by central neutral system (CNS), and can be used to estimate limb stiffness for compliant demonstration, though the precise estimation is still a challenging due to muscle fatigue, noise perturbation, and estimation model error [18, 19]. In [20], a EMG-based human–machine interface was designed for tele-impedance control though without force feedback or autonomous controller after transferring human impedance regulations to robot. Furthermore, human can change arm posture to adapt stiffness as well. Augmented HRC can be exploited by human arm motion [21] and stiffness profile recognition and imitation.

According to different hardware equipments employed in human motion recognition, human motion extraction methods can be divided into two categories:

- Contact method using wearable sensors. In [22, 23] inertial measurement unit (IMU) is used to collect human motion angles for human motion recognition and interaction with robot, while such methods require humans to wear specific devices, which may induce inconvenience in human–robot interaction.
- Noncontact technology. N. Date [24] proposed a vision-based inverse kinematics approach to estimate full-body motion of human model which has 23 DOFS via visual cues. Noncontact methods do not need to put any physical restrictions on humans and provide a natural way to capture human motion though may bring occlusion.

In addition, human gesture also plays a significant role in human–robot interaction. Gesture information acquisition is similar to human motion extraction. It can be departed into vision-based method and wearable sensor-based method. Starner et al. [25] and Starner et al. [26] achieved American and Arab sign language recognition, respectively, based on vision information. Liang [27] Starner and Pentland [28], Kim [29] recognized their sign languages using data collected by data glove. Although data glove has high precision, its high price and cumbersome wearing prevent its widespread use. As for the human motion analysis, most researchers use inverse kinematics method. 7-DOF human arm is a redundant manipulator, when given the hand position and attitude, there are infinite solutions for the inverse kinematics. Moradi [30] defined a redundancy circle and used a position on it as the redundancy parameter to solve this problem. C. Fang [31] proposed a three-level task-motion planning framework and using a human arm triangle space to solve the motion plan of human arm.

Next, we will show some state-of-the-art human–robot interface techniques based on hand gesture recognition and the EEG signals.

8.2 Hand Gesture-Based Robot Control Using Leap Motion

Gesture recognition samples the characteristic of hand action to reflect the user's requirement or intention. Hand gesture recognition provides us an innovative, natural, and user friendly way which is familiar to our human beings. With the latest advances in the fields of computer vision, image processing, and pattern recognition, vision-based hand gesture classification in real time becomes more and more feasible for human–computer interaction. Gesture tracking is an important function of gesture-based human–robot interaction. There are two methods for gesture tracking research: one relies on vision, and the other one is based on data glove. The former one can be affected by the light, complexion, and it is difficult to capture the details with the former approach. The latter one may depend more on the device [32]. Low-cost sensors, such as Kinect and the Xtion Pro, make it convenient to achieve gesture recognition. But these sensors concentrate more on body actions. The Leap Motion is small in size, and has high precision and low power dissipation. It can detect details of hand actions [33].

This section presents a hand gesture-based robot control system using the Leap Motion Sensor [34]. The system obtains gesture information using the Leap Motion, and with the aid of algorithms, it analyzes data and understands robot control signals from the gesture information.

8.2.1 Hardware and Software

8.2.1.1 Leap Motion Device

We use the Leap Motion sensor to obtain the gesture information. The Leap Motion sensor has been introduced in Sect. 1.2 and has two cameras inside, which can take photos from different directions to obtain hand actions information in 3D space. The software development with the Leap Motion should base on the official SDK and the driver. When it is powered on, the Leap Motion sends hand action information periodically.

8.2.1.2 V-REP

We have used the V-REP for robot modeling and simulations in Sect. 1.4. V-REP is a robot simulator with integrated development environment. Each object/model can be individually controlled via an embedded script, a plugin, a ROS node, a remote API client, or a custom solution [35].

8.2.1.3 UR10 Robot

UR10 robot is produced by Universal Robots. Unlike traditional industrial robot, UR10 is light and cheap. It weights 28 kg, and can load 10 kg [36]. Its working range is 1300 mm (see Fig. 8.1).

8.2.1.4 BarrettHand

As a dexterous robot hand with multiple fingers, the BarrettHand is programmable to grasp various objects of different sizes, shapes, and orientations. The BarrettHand is of low weight (980 g) as well as super compact base (25 mm) [37].

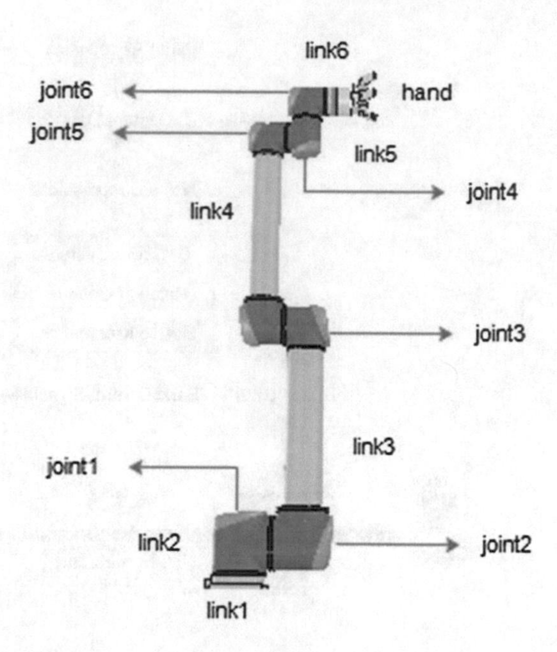

Fig. 8.1 UR10 & BarrettHand. Reprinted from Ref. [34], with permission of Springer

8.2.2 *Control System*

To achieve the goal of controlling robot with gesture, a client/server structured remote human–robot interface control software is developed [38]. First, the client sends a signal to the server to inform that the client is prepared to accept next action command. Then, the server gets the signal, and receives current frame information from the Leap Motion. The server finishes the work about noise suppression, coordinate transformation, and inverse kinematics, and packs the control signals sending to V-REP as well. Finally, V-REP runs the simulation. Figure 8.2 shows the process of the loop.

8.2.2.1 Noise Suppression

The data, including the position of palm, direction vector of finger, and normal vector of palm, will be handled by the server. The precision of the Leap Motion can be 0.01 mm in theory. Actually, there are some destabilizing factors which will affect the precision, such as shaking of hand, magnetocaloric effect, calculating, etc. We adapt a speed-based low-pass filter [39] to eliminate noise. The point of this method is changing the cut-off frequency of low-pass filter with the velocity of palm. The filter can be mathematically described by

$$\hat{X}_i = \alpha X_i + (1 - \alpha)\hat{X}_{i-1} \tag{8.1}$$

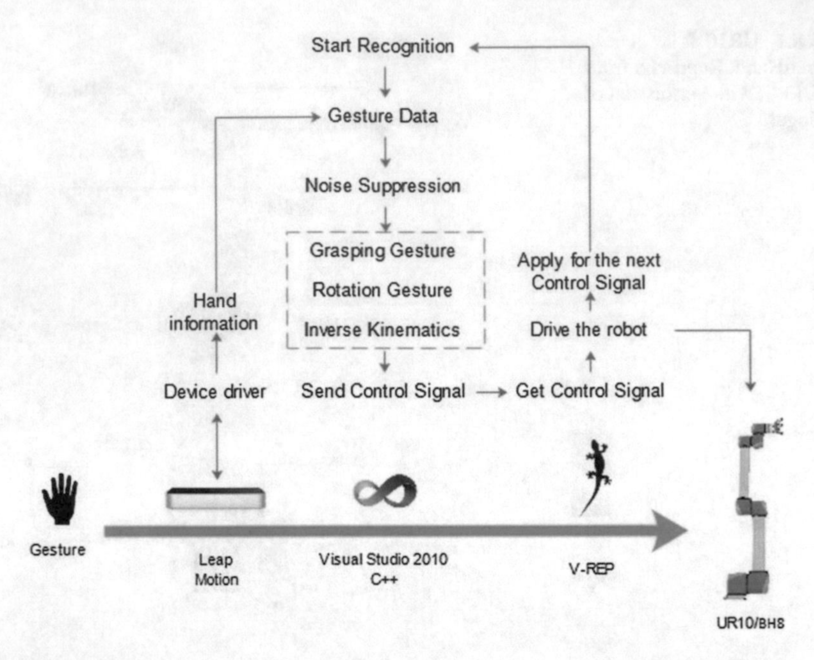

Fig. 8.2 Control system structure. Reprinted from Ref. [34], with permission of Springer

where X_i is a vector containing the coordinates and direction information given by Leap Motion, $\hat{X}_i$ is the vector after filter, and α is a factor which can be calculated by

$$\alpha_i = \frac{1}{1 + \tau_i / T_i} \tag{8.2}$$

$$\tau_i = \frac{1}{2\pi f_{ci}} \tag{8.3}$$

where T_i is the period of updating the data, τ_i is a time constant, f_{ci} is the cut-off frequency, which can be determined by

$$f_{ci} = f_{cmin} + \beta \mid \hat{V}_i \mid \tag{8.4}$$

where $\hat{V}_i$ is a derivative of $\hat{X}_i$, representing the velocity of palm. Based upon experience, make

$$f_{cmin} = 1\text{HZ}, \beta = 0.5 \tag{8.5}$$

8.2.2.2 Grasping Gesture

We add a hand on the terminal of robot, which can execute grasping task. We preset every joint of finger which can rotate between 0–120°. When the hand is open, joint rotation angle is defined as 0°. On the contrary, when the hand is closed, the angle is 120°. We use a coefficient μ to describe the level of grasping. With the API of the Leap Motion SDK, the parameter μ can be given by hand::grabStrength(), and the finger joint angle is 120 μ.

8.2.2.3 Rotation Gesture

To describe the rotation gesture mathematically, we build the coordinate system (see Fig. 8.3) for the hand, which is the same as Leap Motion. Therefore, the problem of hand rotation is equivalent to the problem of coordinate rotation. For example, the coordinate system for Leap Motion is named as frame A, and the coordinate system for hand is named as frame B. Starting with the frame B coincident with frame A, rotate frame B first about $\hat{Y}_B$ by an angle $\alpha(\alpha \in [0, 360°])$, then about $\hat{X}_B$ with an angle $\beta(\beta \in [-90°, 90°])$, and finally, about $\hat{Z}_B$ by an angle $\gamma(\gamma \in [0, 360°])$. We know the orientation of frame B relative to frame A. So if we can obtain the Euler angles of these rotation processes, we know how to control joints 4, 5, and 6 to reappear the rotation gesture.

Because all rotations occur about axes about frame B, the rotation matrix is

$$
{}^A_BR = R_Y(\alpha)R_X(\beta)R_Z(\gamma) = \begin{pmatrix} c\alpha & -s\alpha & 0 \\ s\alpha & c\alpha & 0 \\ 0 & 0 & 1 \end{pmatrix} \begin{pmatrix} c\beta & 0 & s\beta \\ 0 & 1 & 0 \\ -s\beta & 0 & c\beta \end{pmatrix} \begin{pmatrix} 1 & 0 & 0 \\ 0 & c\gamma & -s\gamma \\ 0 & s\gamma & c\gamma \end{pmatrix} \tag{8.6}
$$

$$
{}^A_BR = \begin{pmatrix} s\alpha s\beta s\gamma + c\alpha c\gamma & s\alpha s\beta c\gamma - c\beta s\gamma & s\alpha c\beta \\ c\beta s\gamma & c\beta c\gamma & -s\beta \\ c\alpha s\beta s\gamma - s\alpha c\gamma & c\alpha s\beta s\gamma + s\alpha s\gamma & c\alpha c\beta \end{pmatrix} \tag{8.7}
$$

where ${}^A_BR$ is a rotation matrix that specifies the relationship between coordinate systems A and B. And $c\alpha$ means cosine of angle α and $s\alpha$ means sine of angle α.

According to the definition of rotation matrix, we have

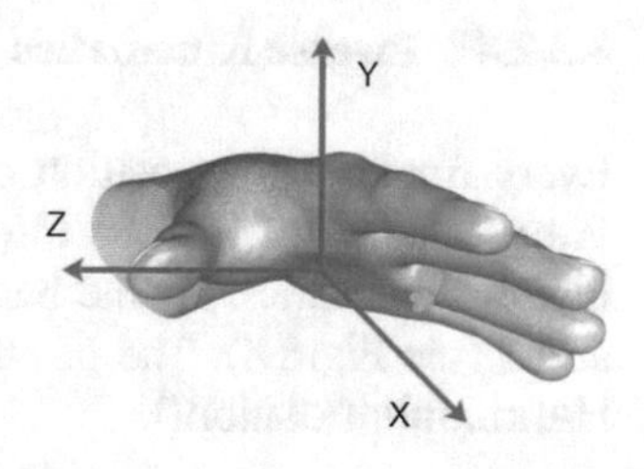

Fig. 8.3 Coordinate system for hand. Reprinted from Ref. [34], with permission of Springer

$$ {}^{A}_{B}R = \left({}^{A}\hat{X}_B \ {}^{A}\hat{Y}_B \ {}^{A}\hat{Z}_B \right) \tag{8.8} $$

where the unit vectors giving the principal directions of coordinate system B, when written in term of coordinate system A, are called ${}^{A}\hat{X}_B, {}^{A}\hat{Y}_B, {}^{A}\hat{Z}_B$. We can obtain the normal vector of hand with the function Hand::palmNormal(), and ${}^{A}\hat{Y}_B$ is in the opposite direction with the normal vector. We also can get the vector from palm to finger with the function Hand::direction(), and ${}^{A}\hat{Z}_B$ is in the opposite direction with it as well. What is more, we can obtain that

$$ {}^{A}\hat{X}_B = {}^{A}\hat{Y}_B \times {}^{A}\hat{Z}_B \tag{8.9} $$

when the coordinate system B employs a right-handed Cartesian coordinate system.

Let us assume that

$$ {}^{A}_{B}R = \begin{pmatrix} r_{11} & r_{12} & r_{13} \\ r_{21} & r_{22} & r_{23} \\ r_{31} & r_{32} & r_{33} \end{pmatrix} \tag{8.10} $$

Now, all the things have been prepared. First, we can get angle β which satisfies

$$ -s\beta = r_{23} \tag{8.11} $$

Then, we can get angle α which satisfies

$$ s\alpha c\beta = r_{13} \tag{8.12} $$

$$ c\alpha c\beta = r_{33} \tag{8.13} $$

Finally, we can get angle γ which satisfies

$$ c\beta s\gamma = r_{21} \tag{8.14} $$

$$ c\beta c\gamma = r_{22} \tag{8.15} $$

We control joint 4 to rotate angle β, make joint 5 to rotate angle α, and make joint 6 to rotate angle γ. Reappearing the rotation gesture is achieved.

8.2.2.4 Inverse Kinematics

Every frame of the position of the palm can be read when the hand is tracked. We use the palm position information to control joints 1, 2, and 3 of robot with inverse kinematics. At the beginning, we build coordinate system for joints 1, 2, and 3 (see Fig. 8.4). The position of black point, denoted by (x, y, z), is given by Hand::palmPosition().

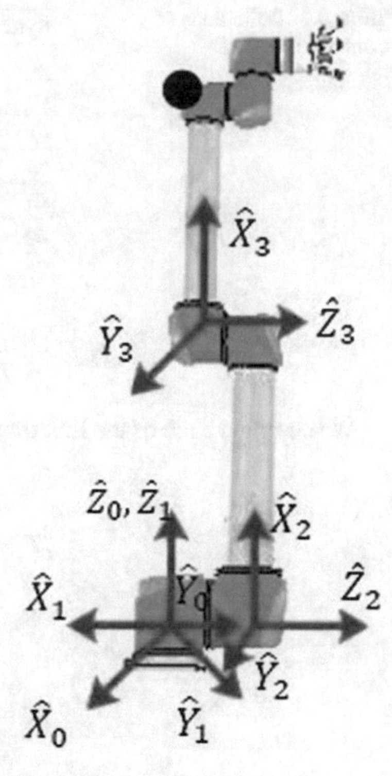

Fig. 8.4 Coordinate system for UR10. Reprinted from Ref. [34], with permission of Springer

Table 8.1 Link parameters for UR10

i	α_{i-1}	a_{i-1}	d_i	θ_i
1	0	0	0	θ_1
2	−90°	0	0	θ_2
3	0	L_1	0	θ_3

The coordinate system for robot is not coincident with that for the Leap Motion, and hence

$$x = palmPosition()[2]/150 \tag{8.16}$$

$$y = palmPosition()[0]/150 \tag{8.17}$$

$$z = palmPosition()[1]/150 - 0.4 \tag{8.18}$$

Table 8.1 shows the link parameters (Denavit–Hartenberg parameters), whose definitions are given in Table 8.2 [40], for UR10.

Table 8.2 Definitions of symbols

Symbol	Definition
a_i	The distance from $\hat{Z}_i$ to $\hat{Z}_{i+1}$ measured along $\hat{X}_i$
α_i	The angle from $\hat{Z}_i$ to $\hat{Z}_{i+1}$ measured along $\hat{X}_i$
d_i	The distance from $\hat{X}_{i-1}$ to $\hat{X}_i$ measured along $\hat{Z}_i$
θ_i	The angle from $\hat{X}_{i-1}$ to $\hat{X}_i$ measured along $\hat{Z}_i$
L_1	The length of link3
L_2	The length of link4

We compute each of the link transformations:

$$
{}^0_1T = \begin{pmatrix} c_1 & -s_1 & 0 & 0 \\ s_1 & c_1 & 0 & 0 \\ 0 & 0 & 1 & 0 \\ 0 & 0 & 0 & 1 \end{pmatrix} \tag{8.19}
$$

$$
{}^1_2T = \begin{pmatrix} c_2 & -s_2 & 0 & 0 \\ 0 & 0 & 1 & 0 \\ s_2 & c_2 & 0 & 0 \\ 0 & 0 & 0 & 1 \end{pmatrix} \tag{8.20}
$$

$$
{}^2_3T = \begin{pmatrix} c_3 & -s_3 & 0 & L_1 \\ s_3 & c_3 & 0 & 0 \\ 0 & 0 & 1 & 0 \\ 0 & 0 & 0 & 1 \end{pmatrix} \tag{8.21}
$$

where c_1 (or s_1) means cosine (or sine) of angle θ_1, and c_{12} means $\cos\theta_1 \cos\theta_2$.

Then,

$$
{}^0_3T = {}^0_1T \quad {}^1_2T \quad {}^2_3T \begin{pmatrix} c_{123} - c_{13}s_2 & -c_{12}s_3 - c_{13}s_2 & -s_1 & L_1c_{12} \\ s_1c_{23} - s_{123} & -s_{13}c_2 - s_{123} & c_1 & L_1s_1c_2 \\ s_2c_3 + c_2s_3 & -s_{23} + c_{23} & 0 & L_1s_2 \\ 0 & 0 & 0 & 1 \end{pmatrix} \tag{8.22}
$$

The position of black point relative to frame 3 is

$$
{}^3P = \begin{pmatrix} L_2 \\ 0 \\ 0 \end{pmatrix} \tag{8.23}
$$

The position of black point relative to frame 0 is

$$^{0}P = \begin{pmatrix} x \\ y \\ z \end{pmatrix} \tag{8.24}$$

Then,

$$\begin{pmatrix} ^{0}P \\ 1 \end{pmatrix} = ^{0}_{1}T^{1}_{2}T^{2}_{3}T \begin{pmatrix} ^{3}P \\ 1 \end{pmatrix} \tag{8.25}$$

From Eq. (8.25), we can get $^{0}_{3}T$, assuming

$$^{2}_{3}T = \begin{pmatrix} r_{11} & r_{12} & r_{13} & l_1 \\ r_{21} & r_{22} & r_{23} & l_2 \\ r_{31} & r_{32} & r_{33} & l_3 \\ 0 & 0 & 0 & 1 \end{pmatrix} \tag{8.26}$$

Now it is easy to obtain the value of θ_i $(i = 1, 2, 3)$.

8.2.3 Experiment and Result

Experiments are designed to test the system performance.The user first puts his hand upon the Leap Motion, and then can do the gestures such as translation, grasping, and rotation. The results of simulation shown in Fig. 8.5 demonstrate that the system can respond correctly and quickly to the gesture, which means that the system is efficient and practical.

Then, we test the accuracy of the system. The workspace of robot is a sphere with radius 1.4470 m. When the user conducts an action, the response time of system is limited in 0.1 s. The user can put his hand at any position upon Leap Motion. We get the position (29.7574, 175.155, 40.302 mm) instantly. In theory, the terminal position of robot should be (0.2687, 0.1983, 0.7677 m), and the real position is (0.2498, 0.3629, 0.7573 m). The open-loop error is 4.65 %. Do more experiments, the average open-loop error is limited in 5 %. That is the system has high precision.

Finally, we add a table into the scene, and put a cup on the table. We select five people to experience the system by grasping the cup with the Leap Motion (see Fig. 8.6). The testers' user experiences and feedbacks were recorded. Results demonstrate that people are satisfied to the system. Users also suggest that grasping gesture should be adapted to different shape things.

In this section, we have developed a hand gesture-based robot control system using the Leap Motion. The system contains noise suppression, coordinate transformation, and inverse kinematics, achieving the goal of controlling a robot with hand gesture. The system has advantages in terms of simple operation, high flexibility, and

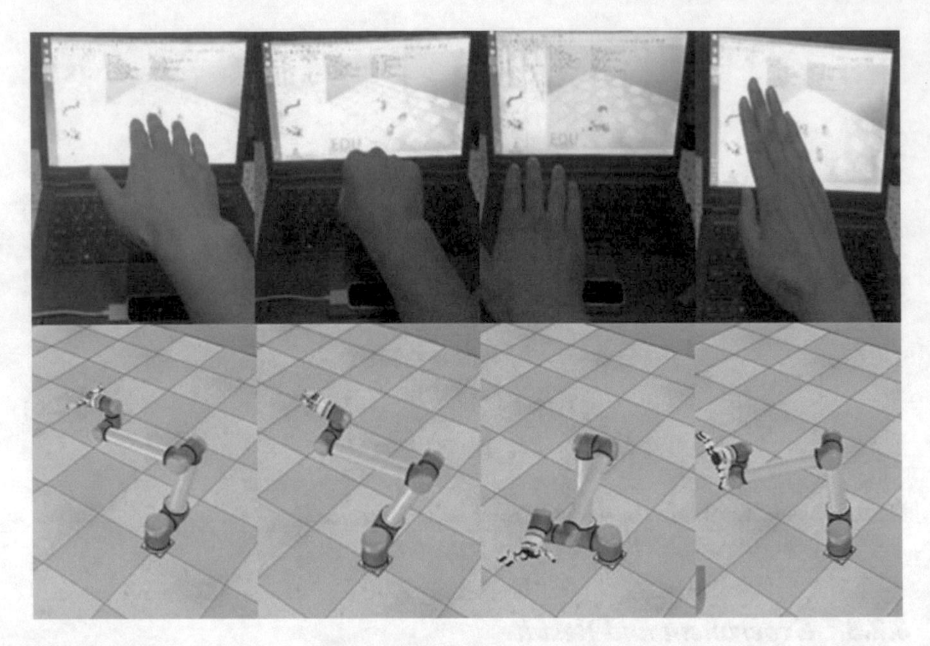

Fig. 8.5 Experiment result. Reprinted from Ref. [34], with permission of Springer

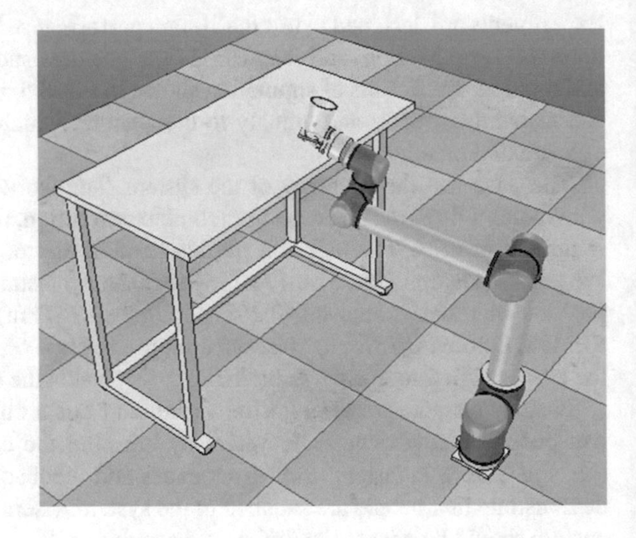

Fig. 8.6 Grasping the cup. Reprinted from Ref. [34], with permission of Springer

efficiency. This robot control system does not have any complex menu or buttons. This system considers more about user's experience in their daily life, so the control gesture desired will be natural and reasonable. It can be used in telemedicine, family nursing care, etc.

8.3 Hand Gesture-Based Control with Parallel System

In Sect. 8.2, we achieved the hand gesture control using the Leap Motion. The hand gesture recognition is also regarded as an innovative and user friendly way for humans to interact with robot.

In this section, we present a hand gesture-based robot control with a parallel system framework which can improve the performance of the teleoperation system [41]. Parallel system refers to the common system composed of a certain real natural system and a virtual artificial system. It is pointed out that parallel system plays an important role in the control and computer simulation system [42]. With the development of information technology and the increasing availability of the network, many problems involved in the management and control of complex systems will be solved by further investigation with the potential of artificial systems in the parallel systems. It helps artificial system to fully play its role in the management and control of actual system. In our telerobot control design, parallel system could help us to compensate for the effects caused by the uncertainty of complex system model.

8.3.1 Platform and Software

We carry our work based on the iCub Robot. As introduced in Sect. 6.2, the physical iCub robot is a one-meter tall humanoid robot, and it has also been used to study human's perception and cognition ability. As one simulator of real iCub robot, iCub simulator is an open-source virtual robot, and it also provides an environment where the simulated iCub can interact with [43]. Moreover, its design follows the realistic version of the robot, so the results on the simulator could be directly tested on the real robot. In this project, we use iCub simulator to achieve human–robot interaction and teleoperation.

Recall Sect. 6.3, Yet Another Robot Platform (YARP) [44] is a communication tool for the robot system which is composed of libraries, protocols, and tools. The module and the device separate without interfering each other. Through YARP, we can communicate with the robot. A dual arm iCub manipulator model built by Robotics Toolbox in MATLAB has been employed. The Robotics Toolbox provides many functions that are useful for the study and simulation of classical arm-type robotics, e.g., such things as kinematics, dynamics, and trajectory generation. An overview of the teleoperation platform is illustrated in Fig. 8.7.

8.3.2 Hand Gesture Recognition System Based on Vision for Controlling the iCub Simulator

For convenient and natural communication between human and robot, a lot of devices that can detect body motion, hand gestures, sound and speech, facial expression, and other aspects of human behaviors can be used. In order to interact with computer naturally, we adopt a hand gesture recognition system based on vision (Fig. 8.8).

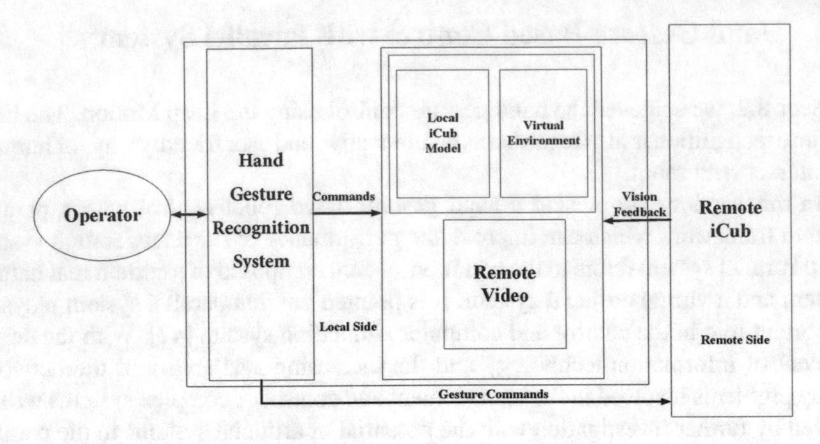

Fig. 8.7 The overall framework of the system

Hand gestures can be captured through a variety of sensors, e.g., "Data Gloves" that precisely record every digit's flex and abduction angles, and electromagnetic or optical position and orientation sensors for the wrist. While computer-vision-based interfaces offer unencumbered interaction, providing a number of notable advantages, e.g., it is nonintrusive, passive, and quite. Installed camera systems can perform other tasks in addition to hand gesture interfaces, sensing and processing

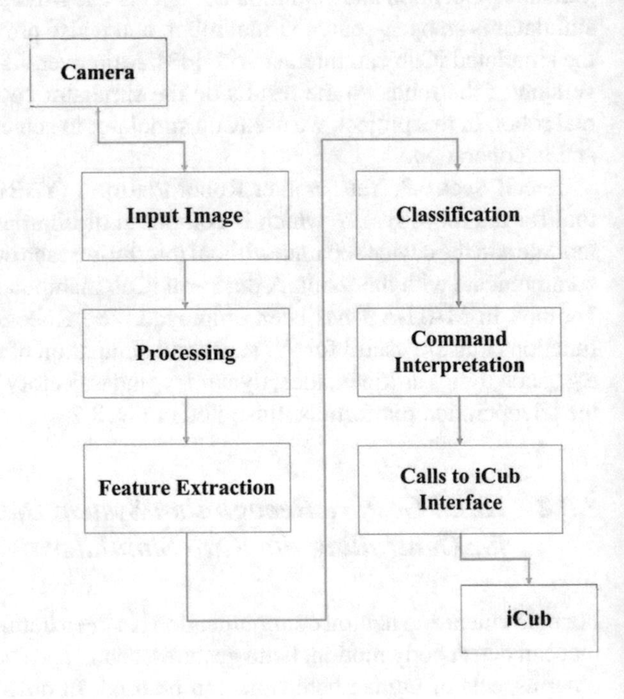

Fig. 8.8 The hand gesture recognition process

hardware is commercially available at low cost [45]. Here we design to control virtual iCub robot using iCub simulator through hand gesture recognition.

8.3.2.1 Processing

Image processing is necessary for feature extraction and classification. It consists of the following steps:

Filtering

This step is employed to reduce the noise gained in the acquisition process and to enhance the images quality. An eight-neighborhood filtering is applied to the image to reduce the noise.

Segmentation

The image is segmented into two regions: (i) hand gesture region and (ii) background region. In RGB color space images are more sensitive to different light conditions so we convert RGB images into YCbCr images. YCbCr is a kind of color space. Y is the brightness component, while Cb and Cr are blue and red concentration offset components. Skin color will not change with position, size, or the direction of hands. So skin color detection has great versatility in the gesture segmentation.

It is noted that color distribution is in concordance with the elliptical distribution on Cb'Cr' plane, while Cb'Cr' is obtained with nonlinear piecewise color transformation from CbCr [46] considering the Y component. Y component will affect the shape of skin color region. The proposed elliptical clustering model is obtained through trials, which is given by

$$\frac{(x-ecx)^2}{a^2} + \frac{(y-ecy)^2}{b^2} = 1 \tag{8.27}$$

$$\begin{bmatrix} x \\ y \end{bmatrix} = \begin{bmatrix} \cos\theta & \sin\theta \\ -\sin\theta & \cos\theta \end{bmatrix} \begin{bmatrix} Cb' - cb_0 \\ Cr' - cr_0 \end{bmatrix} \tag{8.28}$$

where $cb_0 = 109.38$, $cr_0 = 152.02$, $\theta = 2.53$, $ecx = 1.60$, $ecy = 2.41$, $a = 25.39$, $b = 14.03$ are computed from the skin cluster in the Cb'Cr' space.

The pixel belongs to the skin of the hand if $D(Cb, Cr)$ is 1.

$$D(Cb, Cr) = \begin{cases} 1 & \frac{(x-ecx)^2}{a^2} + \frac{(y-ecy)^2}{b^2} \le 1 \\ 0 & \text{others} \end{cases} \tag{8.29}$$

Boundary extraction

Boundary information is very important for feature extraction. Determining the border of the gesture region is the last step in the image processing phase. We use MATLAB command "bwboundaries" to extract contour of the quasi-binary image. Then we get binary images and extract the hand gesture contour as shown in Fig. 8.9.

Features extraction

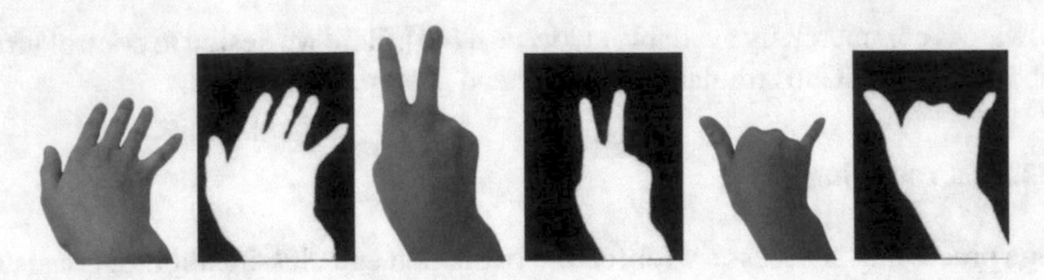

Fig. 8.9 Three gestures and their binary images

In order to improve the robustness of recognition, we use more than one feature vector to describe the gestures.

There are mainly three groups of features to be detailed in the followings: Hu moments, Fourier descriptor, and gesture regional feature.

(i) **Hu moments**: Hu introduced seven nonlinear functions defined on regular moments [47]. The moments are used to reflect the distribution of random variables in statistics. Extending to the mechanics, they are used as a quality distribution characterization. For the characteristics invariant to translation, scale, and rotation, Hu invariant moments are employed for the representation of hand gesture. The low-order moments mainly describe overall features of the image, such as area, spindle, and direction angle; high-order moments mainly describe details of the image, such as twist angle. We use M_1, M_2, M_3, M_4, M_5 [47].

(ii) **Fourier Descriptors**: Fourier descriptors [48] are the Fourier transform coefficients of the object boundary curve. We use Fourier descriptors to represent the boundary of the extracted binary hand as the set of complex numbers, $b_k = x_k + jy_k$; where $\{x_k, y_k\}$ are the coordinates of pixels on the boundary. This is resampled to a fixed length sequence, f_k, $k = 0, 1, \ldots$, for use with the discrete Fourier transform (DFT). Fourier descriptors will be influenced by scale, direction, and starting point of the shape. So it is necessary to normalize for Fourier descriptors. We use FFT to achieve fast Fourier transform of boundary points in MATLAB. Only the second to eleventh normalized coefficient is used.

(iii) **Gesture regional feature**: Regional characteristics generally include geometric features (such as area, center, and breakpoint), shape features, brightness characteristics (such as average brightness), and various texture statistical features. We calculate the ratio of actual area of hand to convex hull and the ratio of perimeter squared to area. Convex hull is a convex polygon formed by connecting the outermost points on plane which include all points. Boundary points and the convex figure of gesture "five" are drawn in the axis as shown in Fig. 8.10.

8.3.2.2 Recognition

In order to realize the classification and recognition, we adopt two methods: ANFIS and SVM.

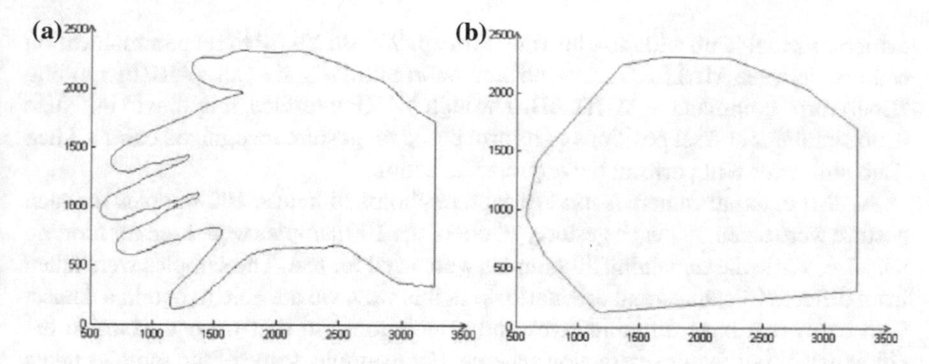

Fig. 8.10 **a**The contour of gesture five; **b** its convex hull

(i) **ANFIS**: The adopted Adaptive Neuro-Fuzzy Inference System (ANFIS) architecture is of the type that is functionally equivalent to the first-order Sugeno-type fuzzy inference system [49]. Fuzzy inference does not have its own learning mechanism; consequently, it is limited in practical use. Meanwhile, artificial neural network cannot express hazily. In fact it is similar to a black box lacking of transparency; therefore, it may not reflect human brain's functions such as abstract reasoning. Fuzzy inference based on adaptive neural network could combine fuzzy inference and artificial neural network and it preserves the advantages of two and overcomes the individual shortcomings. Adaptive neural fuzzy inference system provides a learning method to extract information from dataset for the process of fuzzy modeling. Learning in this method is very similar to the one using neural network. It can effectively calculate the best parameters of membership functions to make sure that the design of the Sugeno fuzzy inference system can simulate the desired or actual input–output relationship. We use the Fuzzy Logic Toolbox in MATLAB to build the adaptive fuzzy neural network. We create an initial fuzzy inference system, combine the hand gesture feature vectors with gesture labels according to a certain format to be the training data, and train the fuzzy inference system in ANFIS method to adjust the membership function parameters. The FIS model will approximate the given training data continuously.

(ii) **SVM**: Support vector machine is a popular machine learning method for classification, regression, and other learning tasks. LIBSVM is an integrated software for support vector classification [50]. It is developed by Lin Chih-Jen who is an associate professor of Taiwan University. It supports multi-class classification. Here we use the LIBSVM toolbox for gesture classification.

8.3.2.3 Experiment

The virtual iCub is controlled with the identified gestures with specific meaning. We move virtual iCub by changing the 16 joints' angels of the arms and the hands. For example, we extend our index finger. It is recognized in the picture captured by the camera. Virtual iCub will raise his right arm up. We use YARP to set communication channel between MATLAB and iCub simulator. Similarly, we call YARP by running "LoadYarp" command in MATLAB. Through YARP interface, it is able to initialize iCub simulator and set positions of its arm based on gesture recognized before. Then iCub simulator will perform the appropriate action.

A USB external camera is used to capture photos of hands. 100 samples for each gesture were taken. For each gesture, 70 out of the 100 samples were used for training purpose, while the remaining 30 samples were used for test. The samples were taken from different distances and orientations. In this way, we are able to obtain a dataset with cases that have different sizes and orientations, so that we can examine the efficiency of our feature extraction scheme. For example, some of the samples taken for the hand gesture of "five" are shown in Fig. 8.10. When the program is running, we shoot a picture every few seconds for testing. Dimension of boundary feature vector for each image is 17 including five Hu moments (see Fig. 8.11), ten Fourier coefficients, and two ratios about gesture regional features. As shown in Fig. 8.12, the iCub movements are represented by ten gestures.

Three examples of the obtained feature vectors are listed in Table 8.3, while results of hand gesture recognition are presented in Tables 8.4 and 8.5.

Commands of "one," "three," and "ten" mean to make iCub raise his right arm, raise his two arms, and make a fist as is shown in Fig. 8.13.

8.3.3 *Teleoperation Platform and Parallel System*

The teleoperation system consists of two parts: iCub simulator and iCub manipulator model built in MATLAB Robotics Toolbox. We use a virtual iCub. It is designed as same as the real one. Based on human–computer interaction, we accomplish controlling virtual iCub robot conveniently and interacting with the immersion environment naturally. To facilitate the subsequent research on network time delay problem and for better application, we introduce the concept of parallel system. We build iCub manipulator model in MATLAB locally and install iCub simulator in the remote side while implement the local control of iCub manipulator model and the remote control of virtual iCub using iCub simulator. In future work, the technology mentioned in this section will be used, so teleoperation system based on the theory of parallel

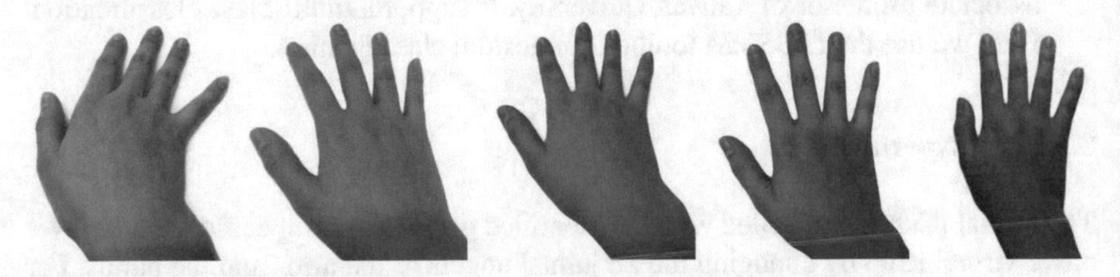

Fig. 8.11 Some of the samples taken for the gesture "five"

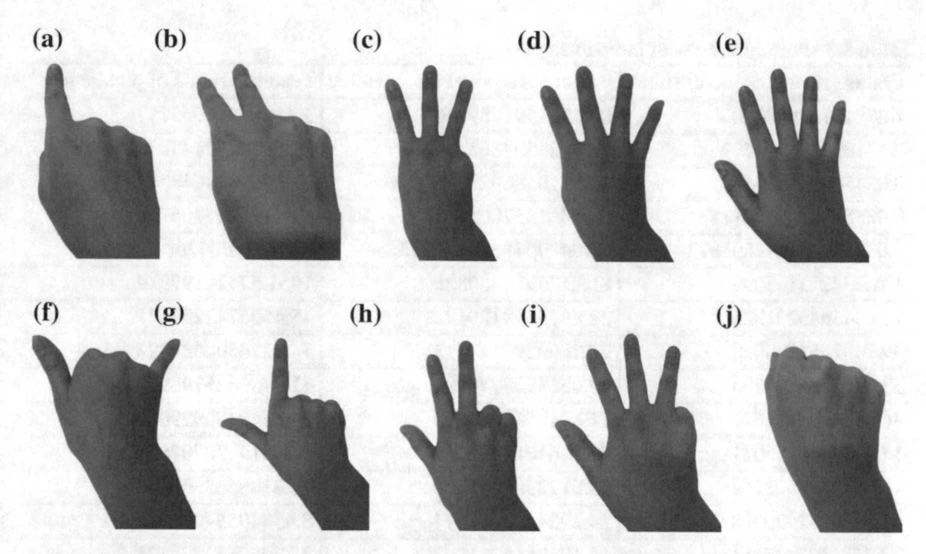

Fig. 8.12 The ten basic hand poses making up the system gestures. **a** raise his right arm up; **b** raise his left arm up; **c** raise both his arms up; **d** raise his right arm straight forward; **e** raise his left arm straight forward; **f** raise both his arms 45° upward; **g** arms swing up and down; **h** left arm bends at 90°; **i** right arm bends at 90°; **j** clenched fist

mode also brings a number of other problems. As there are some errors between the simulation model and real iCub movement, such as the virtual robot has touched objects but the real has none, and the real work environment is invisible, it is not so reliable for completely relying on the predictive simulation system that we must know the remote real situation. So we had better obtain remote video information and accomplish the integration of the models and the video.

8.3.3.1 Parallel System

In order to control the robot better, to make the robot evolve better, and to realize the remote operation, we employ the framework of parallel system mentioned above.

We adopt a parallel system framework based on virtual reality to optimize the performance of the robot. In virtual simulation environment, the operator can get feedback of the real-time operating results, and the remote robot repeats the simulation results after a certain delay, so the operator can effectively operate continuously facing the operating virtual simulation model with the influence of time delay. ICub simulator and the manipulator model form a collaborative intelligent simulation platform. On one hand, iCub manipulator model acts as the virtual model for the real robot while complete the parallel, online, and real-time simulation. On the other hand, iCub manipulator model is collaborative with virtual iCub robot, develop high-level decision-making ability, and extend and expand iCub robot functions.

Table 8.3 Feature vectors of three examples

Feature vectors of picture one	Feature vectors of picture three	Feature vectors of picture ten
1.27344128668080	0.768013917897862	1.15287474965257
0.341034945794739	0.205285988633656	0.231987711436551
0.0345098405446921	0.0551058692738506	0.0200987494896591
0.0885600310397111	0.100618520664892	0.0419426570679495
0.000681031472695372	0.00479048615623522	−3.63390033686653e-05
1013.78223135593	818.120925380823	1051.57474197919
103.045643221326	295.425599120192	45.4506742294877
49.7487851990050	78.2868290117613	74.7276500637026
33.7888356801995	40.6368968772091	41.7750625708967
46.5428205412630	119.211566543840	24.8569400929022
17.7746780455021	67.2151017492618	5.39515370707666
16.7520226903149	23.6122301399800	11.8188665303865
8.91967851920738	24.2924115274574	8.45940554065641
7.67213502927261	10.1949456741707	7.77960672239471
2.65334909384366	15.2443595339593	5.54551570458614
0.896527786365779	0.729194795728057	0.924981631713313
25.1125100024321	61.4839999790745	23.0492411591836

Table 8.4 Recognition result with SVM

Gesture	Train number	Test number	Successful cases	Recognition rate
1	70	30	30	100
2	70	30	29	96.67
3	70	30	30	100
4	70	30	28	93.33
5	70	30	30	100
6	70	30	30	100
7	70	30	30	100
8	70	30	29	96.67
9	70	30	30	100
10	70	30	30	100

The pattern that the real and the model work together intelligently to complete real-time simulation is different from the previous traditional one. Traditional simulation is offline, serial, open loop, but the coordinated intelligent simulation is dynamic, closed-loop, real-time, parallel, and online referring to Fig. 8.14.

The iCub robot has the ability of parallel evolution and intelligent coordination with the model platform.

Table 8.5 Recognition result with ANFIS

Gesture	Train number	Test number	Successful cases	Recognition rate
1	70	30	29	96.67
2	70	30	29	96.67
3	70	30	28	93.33
4	70	30	27	90
5	70	30	29	96.67
6	70	30	30	100
7	70	30	30	100
8	70	30	29	96.67
9	70	30	28	93.33
10	70	30	30	100

Fig. 8.13 **a** Raise his right arm; **b** raise his two arms; **c** make a fist

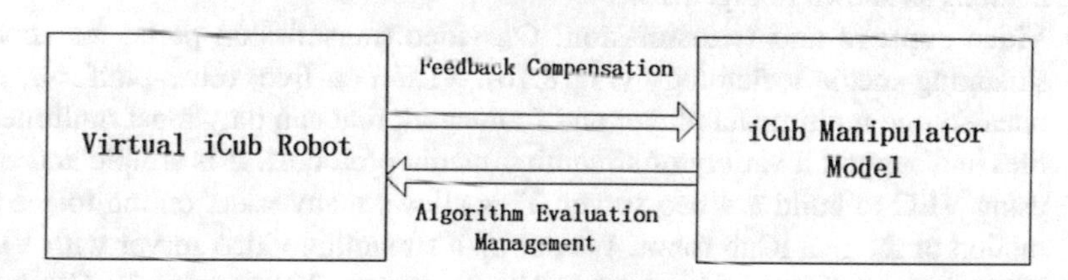

Fig. 8.14 The parallel system

8.3.3.2 iCub Manipulator Model in MATLAB

Based on the Robotics Toolbox [51], an iCub robot manipulator model is set up in MATLAB. We can use the model to study the specific mechanical arm control algorithm in future work. The Robotics Toolbox provides powerful support for a lot of classical mechanical arm model on dynamics, kinematics, and trajectory planning. We built the model of iCub manipulator in MATLAB on the local computer. Then we can use hand gesture recognition system to control it.

8.3.3.3 iCub Simulator

We communicate with the virtual robot via YARP. The created ports transfer information through a server called YARP server. Then we can realize the remote control of virtual iCub.

First, we must configure both A and B with the command line "yarp conf $< ip >$," where $< ip >$ denotes the true IP address of the computers. We should connect the local YARP output port a and the remote YARP input port b. Then we can control the remote robot through sending commands from port a to b.

For simplicity, we install a virtual machine on local machine. Connect the virtual and the local one. Then we can control iCub robot on local computer through sending commands from virtual machine.

In order to enable the operator to monitor the movement of the real robot far away, we need to feedback information of the remote side. Through the network, live video image information and the robot motion state information may be fed back to the operating end. The video monitor should not only provide information to the operator, but also be fused with the virtual model for later correction of the simulation model, and modify the virtual simulation environment, while we can also make the video of local hand gesture transported to remote side which will help to improve security.

There are mainly three steps: camera calibration, acquisition, and transmission of video information, overlay of video images and simulation graphics.

(i) **Camera calibration**: Camera calibration involves several coordinate transformations as shown in Fig. 8.15.
(ii) **Video capture and transmission**: On video transmission parts, we choose streaming media technology (Fig. 8.16). VLC is a free, cross-platform, and open-source multimedia player and framework that can play most multimedia files and support a variety of streaming media protocols. It is simple and easy using VLC to build a video server. This allows many users online to see the motion of the real iCub robot. We put up a streaming video server with VLC, transmitting real-time video captured by the camera. When using VLC to build streaming media server, we can choose the port, encoding mode, and the transmission protocols. We could put the VLC as video server in remote controller and transmit the real robot motion captured as real-time video streaming. We put the VLC as client in local place to play the real-time video streaming. On the local side, MATLAB can call VLC ActiveX control, so the real-time video

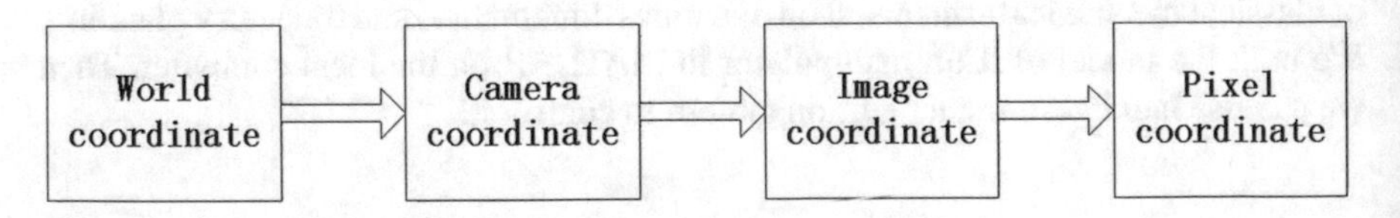

Fig. 8.15 The coordinate transformation

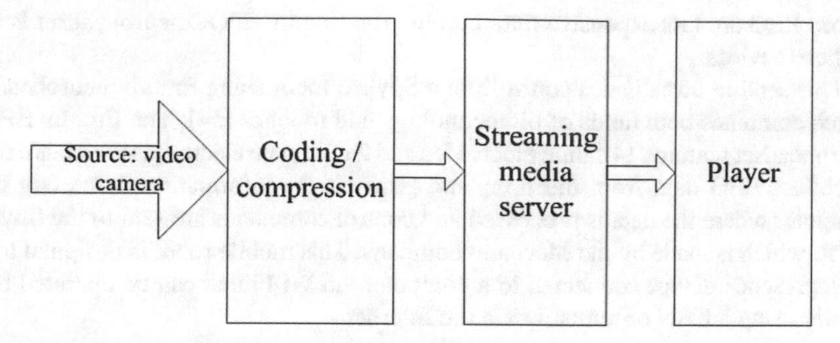

Fig. 8.16 Streaming media work diagram

can display in MATLAB window. VLC can also preserve the received real-time video streaming for facilitating the real-time processing.

(iii) **Overlay of video images and simulation graphics**: In the simulation environment, we calculate the calibration matrix of camera according to the camera calibration algorithm. Multiply the matrix with the left arm points of iCub [52]. Two-dimensional pattern can be obtained by perspective projection with the orthogonal projection onto the two-dimensional graphic image plane. Then iCub manipulator model is superimposed on the image with graphics and images overlap in the same window. Then we achieve the simulation graphics and image overlay. For ease of comparison and contrast, virtual simulation model is a wireframe model.

In this section, we have established a basic robot teleoperation platform with a hand gesture recognition system based on vision information. The features we select are beneficial to real-time control and can enhance robustness. We make it possible to control virtual iCub robot easily with gestures. The vision-based hand gesture recognition system has met the requirements including real-time performance, accuracy, and robustness. We also realize the virtual iCub teleoperation which may help to manipulate a robot remotely.

8.4 BCI Controlled Mobile Robot Using Emotiv Neuroheadset

The way robots communicate and interact with our human is important for the successful application of human–robot interaction. The brain electroencephalography (EEG) signals provide a possible way to connect the human mind to robots. The first report of a human EEG record was published in 1929 by Hans Berger [53] and after that scientists and psychologists have developed much knowledge of EEG especially in the neuroscience area. Nowadays, the price and quality of the different devices to record EEG are less expensive than before. The Emotiv EPOC neuroheadset is one of these devices.

This section investigates controlling a Spykee robot using Emotiv neuroheadset which combines both fields of biotechnology and robotics [54]. The Emotiv EPOC neuroheadset features 14 saline electrodes and 2 reference electrodes which are used to collect EEG data from the user, and transmit the information wirelessly to a computer where the data is processed and control commands are sent to the Spykee robot, which is made by the Meccano company. This mobile robot is designed to be a telepresence device connected to a computer via Wi-Fi that can be operated both locally using a LAN or remotely via the Internet.

8.4.1 EEG and Brain–Computer Interface (BCI) System

8.4.1.1 EEG Signal

The neuron is the fundamental cell in neuroscience. The neuronal activity comes from the neuron network in the brain. A neuron is composed by a cell body (also called soma), the axon, and the dendrites [55]. The soma contains the protein synthesis activity, and the axon is a cylinder used to transfer electrical impulse and protein. The dendrites are connected with the other neurons (axon and dendrites). The dendrites are the network of communication between nerve cells (neurons). The electrical activity of the brain is generated by action potentials (AP). An AP is an information transmitted through the axon. The mechanism of AP generation is based on the neuron's membrane potential [56]. The dendrites of the neuron receive a stimulus; at that moment the membrane potential is depolarized (it becomes more positive) and it reaches a maximum of $+30\,\mathrm{mV}$ [55]. Then the membrane is repolarized and it returns to the rest potential of $-70\,\mathrm{mV}$ (the potential can reach $-90\,\mathrm{mV}$ and it needs a hyperpolarization phase to return to the rest potential) [56].

According to [56], "An EEG signal is measurement of currents that flow during synaptic excitations of the dendrites of many pyramidal neurons in the cerebral cortex." The signal cannot register the activity of each neuron but it is a measurement of many neurons. Finally, an EEG signal is the measurement of the electrical field generated by the AP in the neuronal network. Electrodes are used to record the signals. The positions of the electrodes are important because they describe very different parts of the brain. The brain owns three main layers. Each layer can be simulated as impedance for EEG signals. The skull is the layer with the highest resistivity and this is the reason why the electrode's position is well described in the conventional electrode setting called 10–20 system (see Fig. 8.17 left) [57].

8.4.1.2 Brain–Computer Interface (BCI)

The goal of any BCI system is to record the brain activity to manage computer actions. A BCI can be described in six steps [58]: measure of cerebral activity, preprocessing,

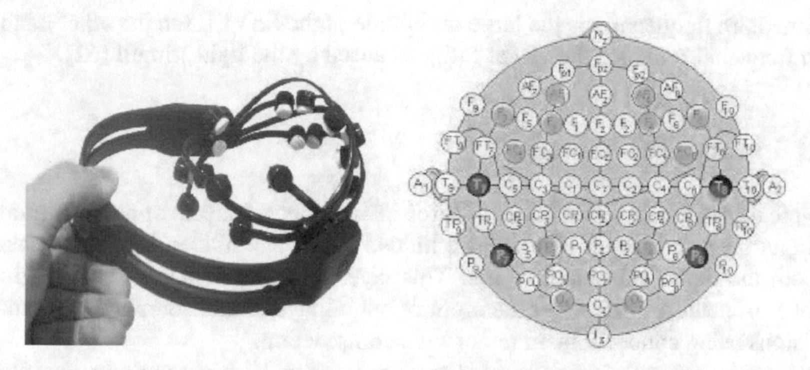

Fig. 8.17 The neuroheadset Emotiv EPOC (*left*) and the 10–20 system with the EPOC channels in *green* (*right*)

extraction of the data, classification, translation in commands (and application of the command), and visual feedback from the device. The experimental protocol follows these steps in the case of this work.

In order to increase the fidelity of extracted EEG data, a number of mental training methods could be exploited. In this work, the visualization and motor control methods will be employed. The visualization method of neural training is the method used by the Emotiv EPOC "Cognitiv Suite." This method involves the user visualizing the movement of a 3D cube in a three-dimensional environment and can be seen in Fig. 8.20. The software records 7 s of training data per action, assisting the user by moving the cube according to the desired action whilst training to help with visualization. This training can be repeated multiple times and the software presents the user with a skill rating for each trained action; the software can record up to four trained actions and neutral data. In the motor control method of mental training the subject first uses movements of their body, such as extending an arm, to stimulate neural activity in the frontoparietal zone. Then through repetition, eventually no longer physically makes the movement but still mentally performs the action [59], and this mental action is detected by the Emotiv EPOC neuroheadset. There are of course many other training methods to generate a certain EEG patterns, e.g., the user can close his eyes and think about a specific feeling. The user can think about left or right direction or he can solve an equation in his mind. The user can think about elementary subtractions by a fixed number ($20 - 3 = 17$, $17 - 3 = 14$) [60]. The objective is to produce patterns with significant particularities.

In addition, it is worth to mention another popular approach to use visual evoked potentials (VEPs) to simulate the human brain activity [61]. The brain is stimulated with a source of light. The light source is flickering with a frequency above 4 Hz [62]. The steady state visual evoked potential (SSVEP) is the cortical response to visual stimuli generated by a light source flickering. Most of the applications based on SSVEP use low (up to 12 Hz) and medium (from 12 to 30) frequencies. However, high frequencies (higher than 30 Hz) can be used as well. The main advantage of low and medium frequencies is the large amplitude of the SSVEP. On the other hand the high frequencies reduce the visual fatigue caused by the light stimuli [62].

8.4.1.3 Echo State Network

A typical neural network consists of three interconnected layers of neurons (nodes), an input layer, an output layer, and a hidden layer. The hidden layer is connected to both the input and output layers. This layer typically contains a vastly greater number of neurons than either the input or output layers. The neural network makes decisions using connectionism rather than computation.

When the network is first created the connections between neurons are all randomly weighted and the network must be taught what to do with input data; this is known as the learning phase. The network is given a stream of data and it is told that at which points in the data stream which output should be correctly assigned. The network then takes the data and adjusts the weights of connections accordingly. Connections that lead to incorrect results are weakened, whilst connections that lead to correct results are strengthened. In this way the network tries to minimize the cost function, this function is a measure of how far away from the optimal solution the current solution is. Via this method of supervised learning, a generic system is able to adapt to and identify patterns in a variety of situations.

The particular type of neural network used in this project is an echo state network (see Fig. 8.18). An ESN differs from the simple model of the neural network described in a few ways. Instead of having a simple hidden layer of neurons the ESN has a large dynamic reservoir of randomly interconnected neurons with a sparse level of connectivity, typically around 1 %. This reservoir of n neurons has an $n \times n$ matrix of connection weights; the maximum eigenvalue of this matrix must be 1. This means that when you divide all nonzero values into the matrix by the eigenvalue, the spectral radius will be 1. The spectral radius being 1 or less prevents the network from being an attractor network. An attractor network is not input driven, and in the state shown in Fig. 8.18 the interneuron connections amplify themselves so much that the input no longer matters. However, with a spectral radius of 1 it still may be possible for attractor states to appear within the network, so to be safe it is best to have the spectral radius less than 1. The spectral radius also affects how much of the past data the ESN takes into account when it assesses a data point, so in an effective network the spectral radius should be close to 1. In a successful ESN the system must decay. The dynamics of the reservoir map the input into a higher dimension allowing a nonlinear problem to be linearly solvable. Increasing the number of neurons in the reservoir increases the dimensionality of the system and allows more complex problems to be solved.

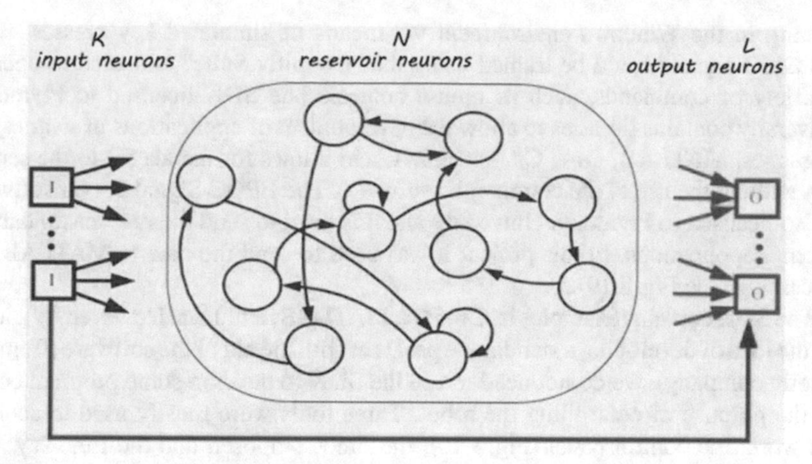

Fig. 8.18 A representation of an echo state network showing the reservoir of neurons

8.4.2 *Experimental System*

8.4.2.1 Emotiv EPOC Headset

The Emotiv EPOC headset (see Fig. 8.17 left) is a device created to record EEG signals equipped with 16 sensors. Data are recorded by 14 saline electrodes and two others sensors used as reference. The electrodes' positions are described using the 10–20 conventional system (see Fig. 8.17 right). The electrodes are mounted on a semi-flexible plastic structure which does not allow large moves to change the electrodes' positions. The advantage of this structure is the stability of the measure because the electrodes are always at the same position. On the other hand the positions cannot be freely modified by the user. The device has a 128 Hz sample rate. It preprocesses the data with a low-pass filter with a cut-off at 85 Hz, a high-pass filter with a cut-off at 0.16 Hz, and a notch filter at 50 and 60 Hz. The notch filter prevents any interference with the power line artifact.

The device combines the headset and a software development kit (SDK) to work with the encrypted data from the computer, and the software that comes with the device features a number of different packages. The control panel is used to inform the user of the connectivity of the electrodes and power levels of the battery, and it also has three different applications that demonstrate different uses of the headset. The "Expressive Suite" uses the headset electrodes to collect EMG data from facial muscles and allows the user to map facial expressions virtually. The "Affectiv Suite" extracts the users' emotional state and concentration levels based on gathered EEG data. Finally, the "Cognitiv Suite" (see Fig. 8.20) allows the user to control the motion of a three-dimensional cube using 13 actions, 6 motions along the 3D axes, 6 rotations around the axes, and 1 command to make the cube disappear.

The software can detect up to four different mental states in the same instance. The "EmoKey" tool allows the Emotiv software to interface with any other software running in the Windows environment via means of simulated key presses. EEG and EMG signals could be trained using the "Cognitiv Suite" and used to operate a variety of commands, such as mouse control. The SDK licensed to Plymouth University contains libraries to allow the development of applications in a variety of languages, MATLAB, C++, C#, and JAVA, and allows for the access to the sensor data without the use of the commercial software. The EPOC Signal Server software is also licensed to Plymouth University and it is used to send the raw sensor data to external applications. In this project it was used to send the data to MATLAB for processing (see Fig. 8.19).

The SDK contains examples in C++, C#, MATLAB, and Java. However, by adopting the idea of developing a standalone program with the aid of the software (from the Emotiv company), we do not need to use the SDK to develop some program codes for the purpose of controlling the robot. Three tools were mostly used to achieve this work: the control panel (Fig. 8.20), the EmoComposer, and the EmoKey. The control panel is the main tool of the Emotiv software. It allows the user to control a 3D cube with the user's brain. The EmoComposer can simulate a headset. This tool is useful in the case one does not have to prepare the headset and the user can control the detected state. The utilization of the three tools will be more described in the experimental system part.

8.4.2.2 Spykee Robot

Spykee robot (see Fig. 8.21) is a mobile robot made by the Meccano company. It can be controlled by a computer via Wi-Fi communication. The communication between the computer and the robot travels both across a local network or the Internet network. The user is able to control the robot from anywhere with a Wi-Fi-equipped computer. This robot is adapted for this work because it is easy to communicate with as most

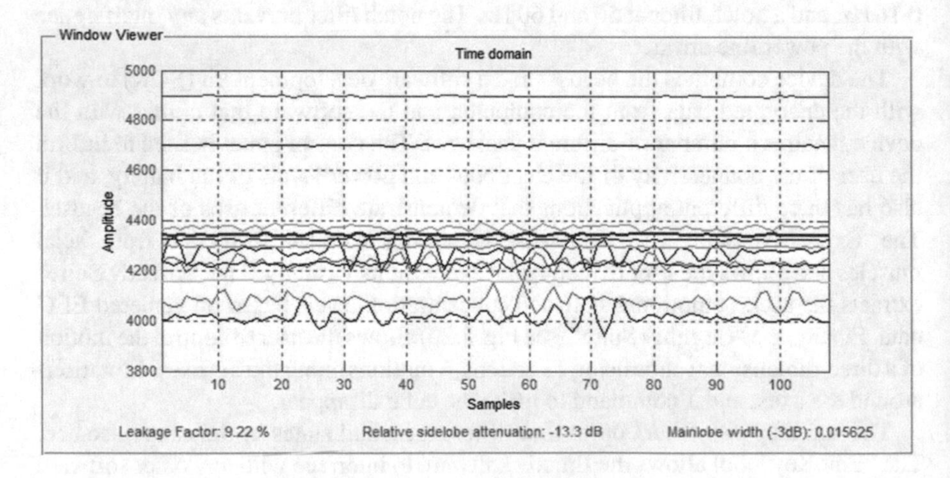

Fig. 8.19 The raw data displayed by MATLAB (screen snapshot)

Fig. 8.20 This is the cognitive control panel where a user can control a 3D cube using mind (screen snapshot)

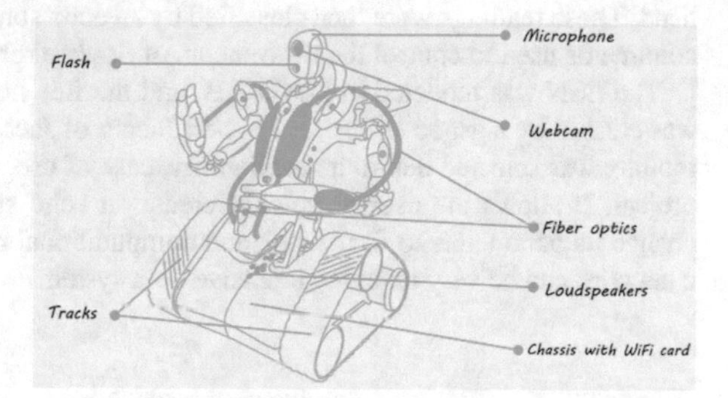

Fig. 8.21 The Spykee robot is equipped with two caterpillars to move, a webcam into the "head," a microphone, and loudspeakers

of the computer integrates Wi-Fi technology and the area covered by local access points is very good in urban zones.

In this section, we have developed a user interface by C# for controlling the Spykee robot. This interface is a part of the BCI because it sends the commands to the robot. The interface provides several functions. The video from the Spykee's camera is displayed into the interface; the user can see where the robot sees, and this is an important feedback of the robot's behavior. The interface allows the user to control the Spykee manually using four buttons; the robot can be connected and disconnected, too.

8.4.3 Training and Control Strategy

8.4.3.1 Training Process of ESN

The experimental system is based on four interfaces, three devices, and two phases. The three interfaces are the cognitiv control panel, the EmoKey tool, MATLAB and the Spykee's interface. The three devices are a computer, the Emotiv neuroheadset, and the Spykee robot. Two phases are needed to control the robot with the brain. The first phase is called the training phase, where the user has to generate training patterns before he/she can control the robot. The second phase is called the control phase where the user can control the robot freely with his mind in this phase.

As mentioned above, we have employed two different training methods in this work. In the first visualization method, the training was conducted in two directions at any given time, by visualizing one direction of movement for a 7 s training session; then switching to another direction for a session, mental fatigue was minimized. In the second motor control method, the test was conducted by taking EEG readings from a subject whilst performing physical actions, in this case tapping ones feet, throwing a punch with the right arm, and stroking a stuffed animal with the left hand. These readings were then classified by an echo state network and the resulting commands used to control the movements of Spykee robot.

The ESN was modeled in MATLAB, and the free noncommercial toolbox used was created by a group led by Dr Herbert Jaeger of Jacobs University Bremen. The toolbox was selected due to its availability, ease of use, and compatibility with this project. It allows the user to quickly create an echo state network and manually change its parameters to fit the task and computational resources, as the training of a network can be very resource intensive on a system.

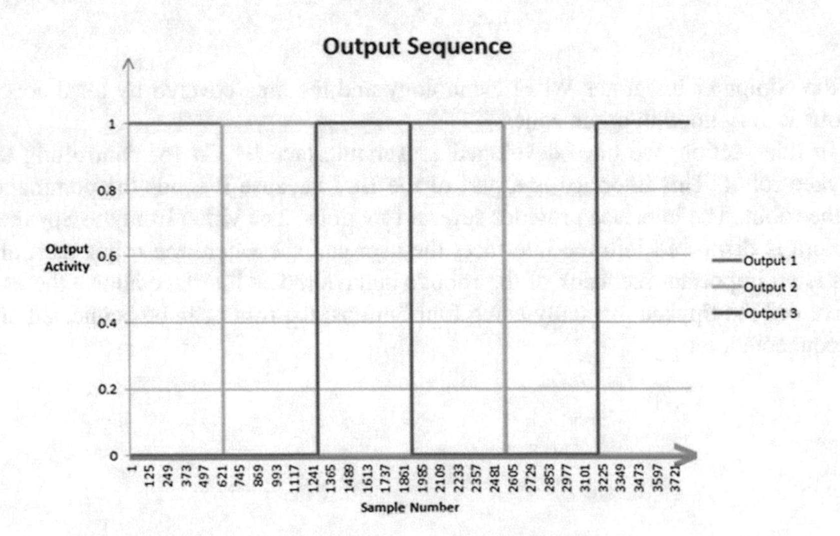

Fig. 8.22 The arrangement of 1 and 0 s in the output sequence array

In order to create, train, and test an ESN using this toolbox, the input data must be in the correct format. This test used 10 s of EEG data taken from the test subject using the Emotiv EPOC neuroheadset to create the first training set. During this time the subject would focus on one physical action (tapping feet, throwing a punch with the right arm, or stroking a stuffed animal with the left hand). This data was recorded within MATLAB using the EPOC Signal Server software to connect to the headset. This process was then repeated two more times, and each time the subject would focus on performing a different action. Data was sampled from all 14 electrode channels on the headset, and each channel required normalization independently of the others, so as to center each channel on zero. This was done by calculating the mean value of each channel for the 10 s sample period, and then subtracting that value from each of the data points in the sample. Once the data was centered on zero, its amplitude had to be reduced so that the majority of data lay between -1 and 1; this was done simply by dividing each data point on each channel by 30, which is the average amplitude of signals from the headset. Using the normalized data, a train/test input sequence had to be created; this sequence takes samples of data from each of the recorded EEG actions and concatenates them into a single sequence. This sequence was composed of the first five seconds of data from the first action, followed by the first five seconds of the second, and then third actions. This sequence was then repeated for the second five seconds of each of the actions. A fifteenth channel was then added, and this input channel had a constant value of 1 and was used to stabilize the ESN and improve the accuracy of the system. The reason for this sequence is that the ESN uses the "split_train_test" function with a value of 0.5 which splits the input sequence in half after the first five seconds of action 3. This first 15 s sample was used to train the ESN, and the second 15 s were used to test the network.

In order to train and test the network an output sequence also had to be created so that the network could be taught when the output was correct and when it was incorrect. The output sequence is an array of 1 and 0 s equal in length to the input sequence, with three fields one for each of the actions. The arrangement of 1 and 0 s within the array is as shown in Fig. 8.22, with only one output active at any given time. For the ESN to learn correctly this has to match the pattern of input activity, i.e., when action 1 is performed in the sample data, output 1 should be 1, and the other outputs 0.

After creating and training the ESN the test moved on to attempting to classify live data. The system took 2 s worth of data from the Emotiv EPOC neuroheadset and normalized it in the same manner as the training data, and then data was entered into the ESN for classification. Each 2 s sample from each electrode channel was normalized independently with respect to itself to avoid any possible sensor drift. The classification of data with a network of this size took about 1 s. Once the data had been classified, MATLAB displayed visual feedback for the user on screen in the form of text stating which of the three trained mental states the ESN had detected.

As the Spykee robot required four-directional control, the ESN program was expanded to be able to classify four mental states. This was done by simply adding another training state and another output state to the described three-way classification system.

8.4.3.2 Control Strategy

A user interface is developed to control the Spykee robot (see Fig. 8.23) and the control objective is to mentally manipulate four actions: moving forward/backward and turning left/right, such that the robot is able to move freely on the 2D floor. As in the training process, we have two different methods, visualization and motor control; in the control process, we also adopt these two methods.

Similar as in the training process, in the first control method using visualization provided by the "Cognitiv Suite," the extraction of the data is carried out by the Emotiv software using its own algorithm. During the extraction the software creates a characteristic arrow to describe each pattern previously recorded. The type of the arrow depends on the method used. There exist a lot of different methods to describe a signal: amplitude of waves, shape of the signal, and every representative of the pattern [63]. The classification step finds the state of mind corresponding with the characteristic arrow.

Using the EmoKey software (see Fig. 8.24), we map four keys on the keyboard, namely, W, S, A, and D, to the four actions of the robot, i.e., moving forwards, moving backwards, turning left, and turning right. Once a certain state of mind is detected, a command will be sent to the Spykee interface using the mapped keys.

In the second control method of motor control, we need to export the commands from MATLAB to the Spykee control window. This was done via the use of a free .mex file created by MATLABcentral user "Phillip" called TextInject. This file allows MATLAB to send a string of text to any X/Y coordinate in a target window. The target window in this instance being the Spykee control interface shown in Fig. 8.2, and the X/Y coordinates were those of the "command" text box. Once the ESN had classified a 2 s sample of data, TextInject could be used to send a command, based

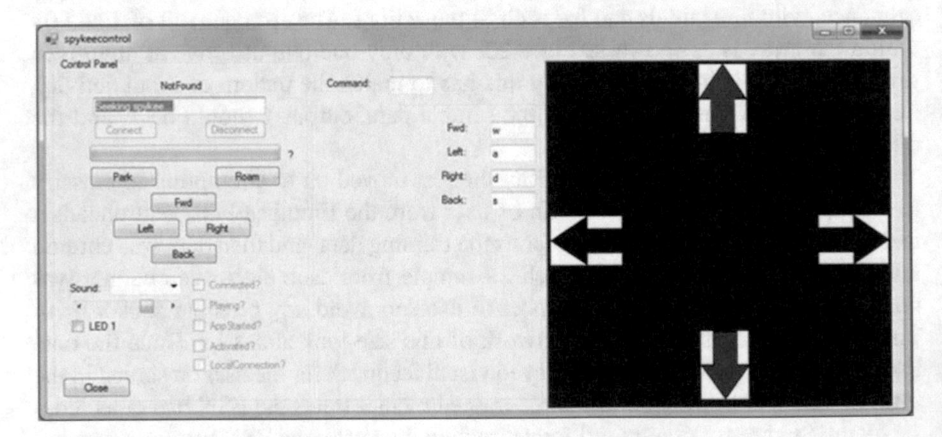

Fig. 8.23 Illustration of the Spykee interface, on which the video is displayed in the *black square*, and the *red arrows* inform the user to moving forward/backward while the *black arrows* indicating turning *left*/*right* (screen snapshot)

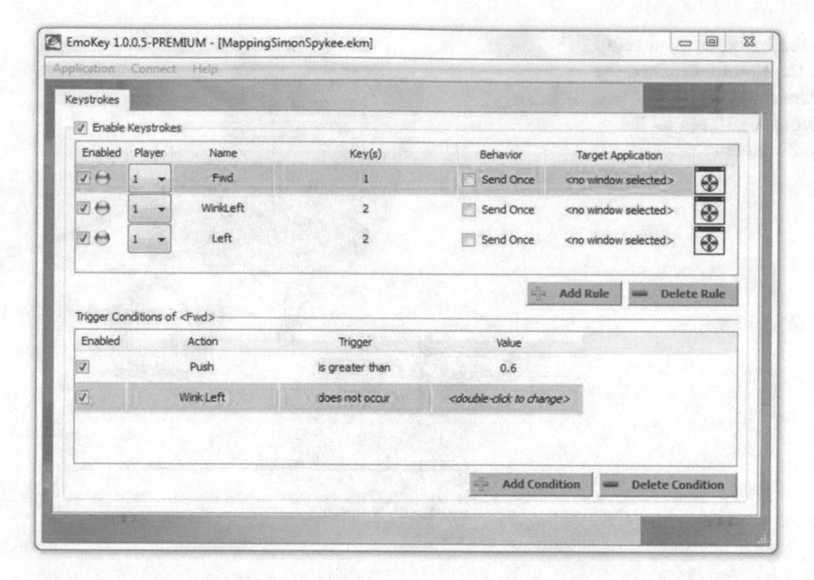

Fig. 8.24 The EmoKey tool allows the system to communicate between the Emotiv software and the Spykee robot (screen snapshot)

on what mental state was detected, to the Spykee control interface, thus controlling the actions of the robot.

8.4.4 Results and Discussions

In the experiment (Fig. 8.25), a system allowing for the four-way control of a Spykee robot using live EEG data taken from the Emotiv EPOC neuroheadset and classified by an ESN had been created. EEG samples were taken from three test subjects performing four physical actions of their choice each. An ESN was created and trained for each subject; the ESN was retrained until each classifier could be identified with at least 60 % accuracy, in two cases a sample had to be retaken to improve the quality of the classification. Whilst performing the physical action, 2 out of the 3 subjects were able to control the movement of the Spykee robot to a significant enough degree that the motion clearly was not random, i.e., piloting the robot is a certain direction on command. One of the two subjects able to move the robot at all was able to operate the robot with any significant level of control when only mentally performing the actions.

'There are also other ways to stimulate the cognitive system and the steady state visually evoked potential (SSVEP) method, which is reported to have high success rate [62], seems promising. The theory behind SSVEP is that, when a subject's retina is excited by a flashing visual stimulus within the frequency range of 3.5–75 Hz, a

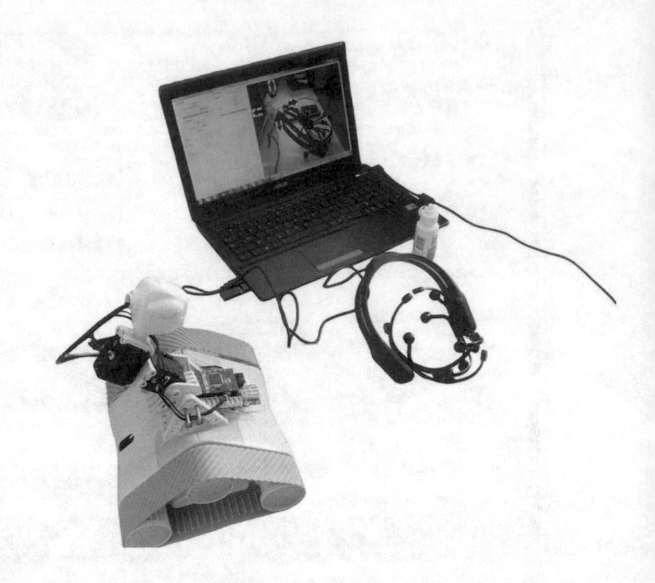

Fig. 8.25 Experiment setup with the Emotiv headset, the Spykee robot and the different interfaces on the laptop's screen

strong electrical signal is generated in the brain at the same frequency, or at a multiple thereof.

To test SSVEP method using Emotiv neuroheadset, we conducted an experiment using a small 4.5" screen flashing at 7 Hz to see if the Emotiv EPOC neuroheadset could detect the 7 Hz (alpha) EEG signals generated by the SSVEP signals in the brain [64]. The test was conducted with two people's results of the tests showed that the Emotiv EPOC neuroheadset was unable to pick up SSVEP signals, which was most likely due to the positioning of the electrodes on the device. SSVEP signals appear in the primary visual cortex, an area which the electrodes on the device do not cover. Electrodes on positions O1 and O2 are positioned over the occipital lobe, however, are not ideally positioned to detect the SSVEP signals. Other researches have also concluded that the Emotiv EPOC neuroheadset does not cover all the important locations in the scalp [65].

8.5 EEG Signal-Based Control of Robot Manipulator

In last section, we have used the EEG signals and BCI to control the mobile robot. BCI established by Jacques is a direct human–robot communication way by converting raw data acquired from human brain into forms of exploitable data vectors [66]. Current BCIs are designed to exploit four types of control signals, including sensor-motor rhythms, visual evoked potentials (VEPs), P300, and slow cortical potentials (SCPs). P300 control signals do not require extensive training, and are able to provide a large number of possible commands with a high information transfer rates and therefore a higher application speed. There are mainly three types of BCIs, namely, invasive, partially invasive, and noninvasive. Invasive and partially invasive (ElectroCorticoGraphy or ECoG) methods can offer better temporal resolution; however, there are certain short- or long-term risks because of the involved surgeries and many inserted electrodes (up to 256) [67].

In this section, we use the Emotiv EPOC headset and the P300 speller BCI to control a 7-DoF virtual robot manipulator [68]. It is totally noninvasive and nonintrusive, and contains only 14 electrodes. This system can operate independently and incorporated into larger platforms such as medicine, video game robotics, and so on. It provides a new way to use P300 to control a virtual manipulator in the Cartesian space via a noninvasive BCI.

8.5.1 Hardware and Software

OpenVIBE: OpenVIBE is a free and open-source software platform for the design, test, and use of BCIs. OpenVIBE can acquire, filter, process, classify, and visualize brain signals in real time. It enables also a convenient interaction with virtual display. OpenVIBE can be connected to Emotiv and retrieve directly the data collected.

V-REP: In the experiment, we use the V-REP to create a 7-DoF virtual robot, namely Manipulator DoF 7. Recall in Sect. 5.2, V-REP is a powerful 3D robot simulator with features and several extension mechanisms. The object or model can be commanded separately by a specific controller. Extensions and V-REP commands can be written in C++ or other languages [69].

P300 Wave: The P300 wave is a positive deflection which occurs when a user detects an expected and unpredictable stimulus. This deflection is detectable and can be exploited by a BCI to control devices [70]. It is an event-related potential (ERP), discovered by Sam Sutton et al. in 1965. The time required by the user to discriminate the target is approximately 250–500 ms after the apparition of the target.

8.5.2 Experimental Methodology

The EEG data is collected with the neuroheadset Emotiv EPOC. First of all, the connectivity of each electrode with the headset setup panel of the Emotiv software is checked. In order to get better connection, the electrodes with non-green indicators shown on the user interface can be moistened again or move a bit and push on the scalp. When eventually most circles of conductivity indicators become green it is ready to start the connection with OpenViBE.

8.5.2.1 Design of the BCI

P300 speller is a P300 spelling paradigm that depends on the visual evoked potentials. It is based on the so-called oddball paradigm, where low-probability target items are mixed with high-probability nontarget items [71]. It displays a grid of alphanumeric character as shown in Fig. 8.26.

The Speller application creates visual stimulation by flashing randomly the grid rows and columns. The user is instructed to focus on a particular target (a letter or number for instance) he/she wants to spell out. In order to actually selecting the letter, an algorithm is employed to decide which row and column contains the target letter. The user is asked to focus on this rare stimulus (target) and whenever he/she detects it, a wave peak is evoked. The intersection of the selected row and column determines the letter selected by the user. After some repetitions, a vote determines which letter the BCI system selects.

First of all, an acquisition scenario is developed online. In order to acquire some data for training classifiers that will detect the P300 brainwaves, both the P300 stimulus and regular EEG data are recorded. The collected data can be easily reused on other related scenarios [72]. A classifier to discriminate the P300 brainwaves from others is then trained in a second offline scenario, and finally applied on the developed online speller scenario [73].

Linear discriminant analysis (LDA) is employed in this scenario to discriminate whether the class is a P300 or not. LDA is a method to identify a linear combination of features which separates two or more classes of objects or events. Prior to the classification, the data is first preprocessed with 1–30 Hz filtering, decimated and segmented into epochs. Features are then extracted from the epochs of EEG signals to train the LDA classifier.

LDA algorithm assumes the covariance is the same for all classes; the aim is to seek a linear transformation of features such that the separation between the group means on the transformed scale is the best.

For K groups of features with means $(\mu_1, \mu_2, \ldots \mu_K)$, covariance $(\Sigma_1, \Sigma_2, \ldots, \Sigma_K)$, the discriminant function of the kth group is

$$\delta_k(x) = x' \Sigma^{-1} \mu_k - \frac{1}{2} \mu_k \Sigma^{-1} \mu_k + \log \pi_k \tag{8.30}$$

where π_k is the prior probability of class k,

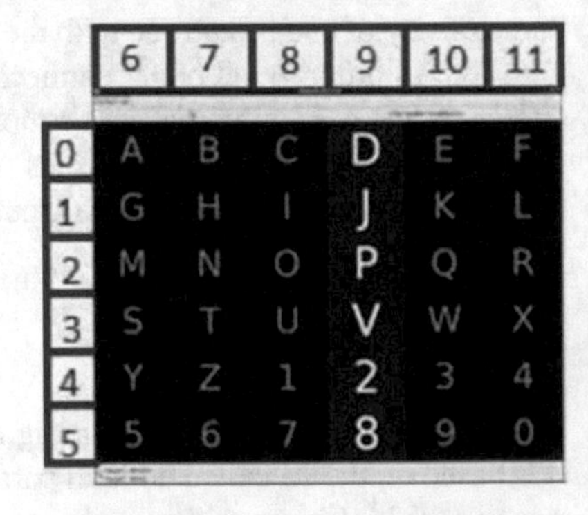

Fig. 8.26 The P300 speller grid with the ID of each row and channel

$$\sum_{k=1}^{K} \pi_k = 1$$

$\Sigma_k = \Sigma, (k = 1, 2, K)$. In practice, mean μ_k and covariance matrix Σk are not available beforehand, and training data are used to estimate them as follows:

$$\hat{\pi}_k = \frac{N_k}{N} \tag{8.31}$$

$$\hat{\mu}_k = \frac{1}{N_k} \sum_{g_i=k} x_i \tag{8.32}$$

$$\hat{\sum} = \frac{1}{N-K} \sum_{k=1}^{K} \sum_{g_i=k} (x_i - \hat{\mu}_k)(x_i - \hat{\mu}_k)' \tag{8.33}$$

where N_k is the training data features number in class k, and $g_i = k$ means the ith sample belongs to class k. The classifier generates a configuration file that will be used later on by another related classifier.

In another scenario a similar process is done to train the spatial filter called xDAWN. The signal space is reduced, and only the three dimensions of most significant for detecting a P300 are retained. It is found that the P300 is only visible in certain parts of the brain; therefore, the coefficients associated with these significant channels are high positive values, while with those irrelevant channels are 0. Coefficient is calculated and reused during the online phase.

In this last step, the speller is ready to be used in real time. This scenario is an association of the previous scenarios. The classifier is trained and employed in real time here. The goal is to detect the row and column which contains the target letter to be spelled. For this purpose, each column and row is associated with a stimulation. The stimulation is an impulsion sent from one box on the graphic programming interface to another, with two possible states, i.e., high or low. To train the classifier, each column and row is labeled as "target" or "nontarget." Among the twelve targets and nontarget classes emitted, two are voted and sent both to the speller visualization (in order to have a visual feedback of the letter spelled) and to the "analog VRPN server" [72]. This last box is not on the initial scenario provided by OpenViBE. It has been added to establish the communication with the virtual application V-REP. As this box can only receive analog value as input, a switch is used to associate each simulation with a specific channel and value. If a stimulation is not voted, the value of the corresponding channel will be NULL. Figure 8.27 shows the analog VRPN server used in the final scenario.

The end effector of the virtual robot manipulator is supposed to move in six different directions; thus, six letters need to be determined to control those directions. The main idea is to associate one letter with on direction in the Cartesian space. It should be noted that a target letter is a result of two stimuli, one from the row and one

Table 8.6 Letters associated with channel combinations and stimulus

Letters	Row stimulation	Column stimulation	Associated channels
Left L	OVTK_Stimulationid_ Label_02	OVTK_ Stimulationid_ Label_0C	Channel 1 Channel 11
Right R	OVTK_Stimulationid_ Label_03	OVTK_Stimulationid_ Label_0C	Channel 2 Channel 11
Backward B	OVTK_Stimulationid _ Label_01	OVTK_Stimulationid_ Label_08	Channel 0 Channel 7
Forward F	OVTK_Stimulationid_ Label_01	OVTK_Stimulationid_ Label_0C	Channel 0 Channel 11
Up U	OVTK_Stimulationid_ Label_04	OVTK_Stimulationid _ Label_09	Channel 3 Channel 8
Down D	OVTK_Stimulationid _ Label_01	OVTK_Stimulationid _ Label_0A	Channel 0 Channel

from the column. Hence, each direction or letter must be associated with a channel. As a result, 2×6 inputs are required.

There are six letters corresponding to six robot movements in the Cartesian spaces: $+X$, $-X$, $+Y$, $-Y$, $+Z$, and $-Z$. Letters are sent via a VRPN server on OpenViBE to a first client coded in C++. In order to be transmitted, a letter is associated with two different channels (one for the row and the other for the columns). Those data are transmitted via a C# socket to the second project, i.e., the V-REP client. In this one, V-REP remote API is used to translate incoming data into robot command which are finally sent to the V-REP server.

Table 8.6 summarizes the different letter associations, the name of stimulation retrieved from OpenVIBE, and the column and row dependent on the P300 speller grid.

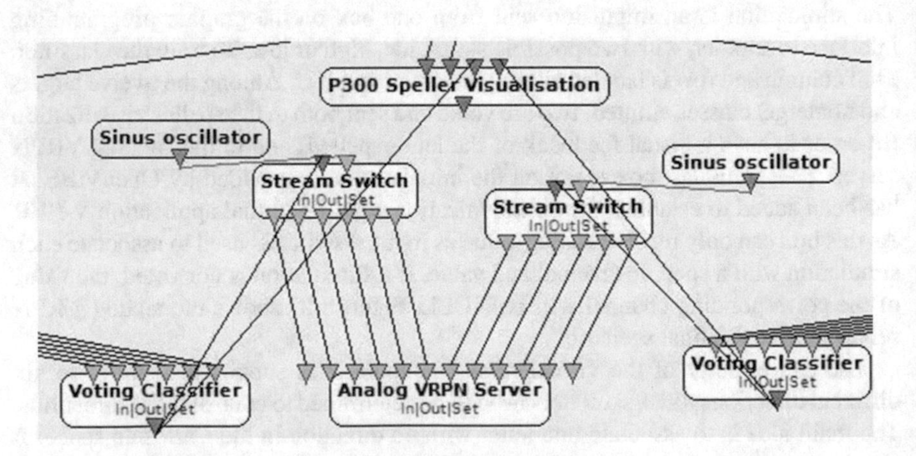

Fig. 8.27 Analog VRPN server employed in the final scenario

A C++ client is designed on a Visual Studio project. It is modified to only take into account the analog value. The 12 channels and their values are retrieved on this side. This communication is maintained via a while loop and allows real-time data exchange. In order to have the VRPN client in side and the V-REP in another, and allows them to communicate, a C# socket communication is established on both. Therefore, the received channels and their values will be sent through the C# client implemented in this Visual Studio project (Fig. 8.28).

8.5.2.2 V-REP Implementation

The second C++ client: In order to communicate with external applications, the best way is to use the V-REP remote API, since it can be easily used in a C++ project. Thus, a second C++ client is designed and the C# server is implemented on this second project in order to receive data (channels and values) sent by the first project. **The 7-DoF manipulator and remote V-REP C++ API**: Since a server is already integrated in a V-REP plugin, it only requires launching it on the V-REP installation folder. Once launched, several robots are directly accessible, and the 7-DoF virtual robot manipulator to be used must be selected and drag into the scene.

For moving the end effector of the manipulator it is necessary to move the object called "Target." Setting a new position in the Cartesian space of the "Target" will result in a movement of the manipulator from the current "tip" to the new "Target" position. All commands are sent in Cartesian space. According to the specific data received, the manipulator will be controlled to make corresponding (Fig. 8.29).

The following codes provide an example of how the end effector is controlled when a specific letter (such "u") is received.

Listing 8.1 A coding example of moving the end effector

```
1 if (Row = = 3 && Column = = 8){
2    position_copy[2]=position_copy[2]+0.01; //increment the  position of 0.01 on
         the Z \text{axis}.
3    simxSetObjectPosition(clientID, handles_copy[2], −1, position_copy,
         simx_opmode_oneshot); }//Set the \pg{\enlargethispage{12pt}}new object
         position
4 end if
```

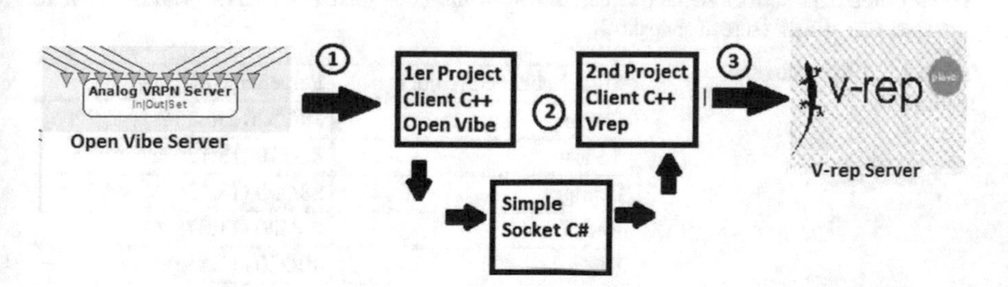

Fig. 8.28 Communication process between OpenVIBE and V-REP

As mentioned, remote API functions are used on a second C++ project. The remote APIs are V-REP functions used by external applications and in many different programming languages. V-REP libraries are shared on the C++ application in order to use those commands. Then according to the data received, the position of target will be modified accordingly.

8.5.2.3 Performance

On a total of 20 trials of data collection from Emoti EPOC headset, the numbers of green, yellow orange, and black circle indicators of conductivity obtained were recorded and can be seen in Table 8.7. Those results vary from a person to another according to his/her hair and scalp shape.

The P300 speller during training scenarios obtains an overall success rate between 75 and 80 %. However, results obtained during the online scenarios vary. In this study, three 24-year-old subjects participate in the experiments: a trained male, an untrained female, and an untrained male. They are asked to perform 30 trials each. The success rate results are summarized in Fig. 8.30. It is shown that for the trained subject, 25 out of the 30 trails are successful, and for the other two untrained subjects, an average number of 20 successful attempts are observed.

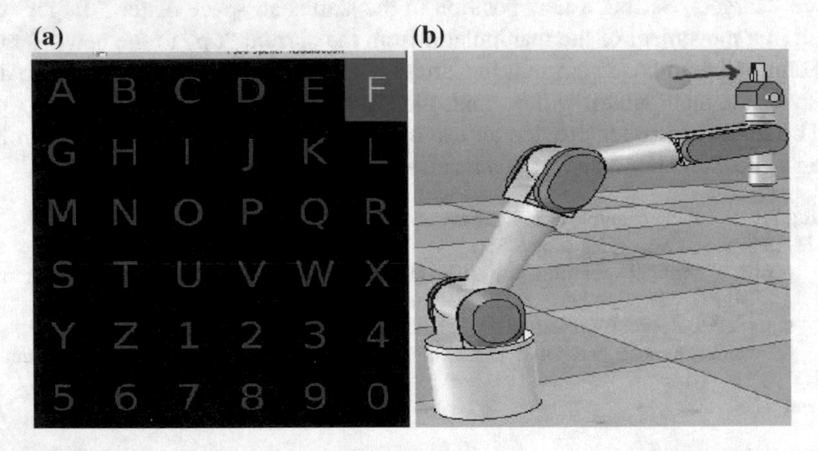

Fig. 8.29 An example of a letter selection on OpenVIBE and the corresponding movement on V-REP (screen snapshot). **a** Letter F selected on open vibe P300 speller (screen snapshot). **b** Forward movement on V-REP (screen snapshot)

Table 8.7 EEG collection results

Signal obtained (color)	Percentage of circle
Green	180/320 (56.25 %)
Yellow	42/320 (13.125 %)
Orange	38/320 (13.125 %)
Red	20/320 (11.875 %)
Black	40/320 (12.5 %)

There is a total of 50 correct letters selected over 90 trials. These data are transferred efficiently through the different sockets and then trigger a command each. If the right channel combination is received on the V-REP client, the associated command will be correctly activated. Figure 8.31 shows the final manipulator control results achieved, where the error probability in each section is indicated. It is found that over the 90 trails, for a total of 65 attempts, a correct robot motion is achieved, which delivers a 63 % success rate as a result. The final accuracy is then 65 results with a correct robot motion over 90 trials so 63 % of chance.

8.5.3 Discussion

As a first step, it is essential to collect signals with quality as good as possible in order to obtain accurate result afterward on OpenViBE. Many factors can lead to a bad connection. The headset could be not enough charged, and the electrodes can suffer of corrosion such that a white/green layer can be found on them. However, the results may vary from one person to another, for instance, people with more hairs are less likely to obtain good results.

Regarding the P300 speller results, the failures are mainly resulted from the signal quality and how good the subject is trained. The signal quality relates to the acquisition system as well as the computing power of the computer. Thus, it is important to obtain as many green circles as possible and turn off all unnecessary applications running. In addition, even if the P300 speller does not require an extensive training, a trained subject usually obtain better results since they are better focused. Other personal varieties also exist, such as some P300 waves are naturally cleaner than others.

On the other hand, since only six letters are used in this study, the 6 × 6 speller grid could be reduced to a 4 × 4 or 3 × 3 grid. However, considering the oddball paradigm conditions that when the matrix is getting smaller, the P300 waves could

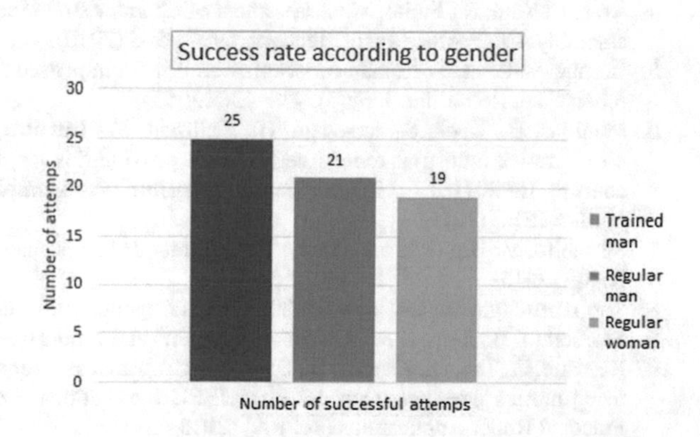

Fig. 8.30 Successful attempts obtained by three subjects in the P300 spelling experiments

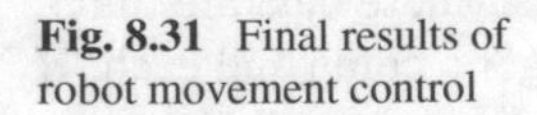

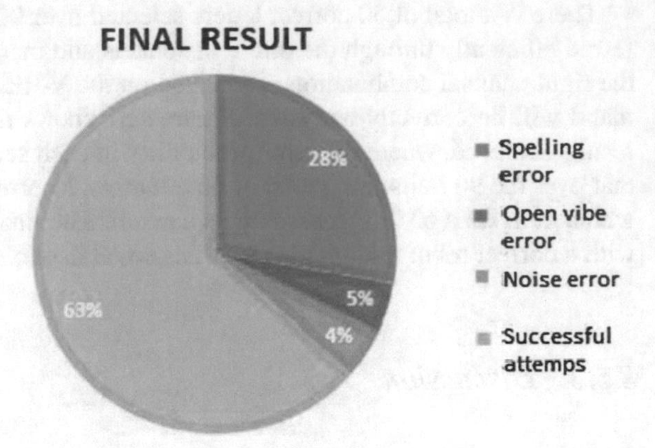

Fig. 8.31 Final results of robot movement control

be less significant, and we end up taking a 6×6 grid in the end. In addition, in future studies, the letters can be exchangeable by words or symbols.

Overall, the use of remote APIs and the communication system establishment is successful with a very low probability of error. However, the performance obtained with the P300 speller using Emotiv can still be improved. The implementation of a plug in rather than employing the Remote APIs could be considered in the future. In addition, controlling a virtual application via P300 speller offers more degree of freedom, and makes more commands possible in executing a single task.

References

1. Green, S.A., Billinghurst, M., Chen, X., Chase, G.: Human-robot collaboration: a literature review and augmented reality approach in design (2008)
2. Bauer, A., Wollherr, D., Buss, M.: Human-robot collaboration: a survey. Int. J. Humanoid Robot. **5**(01), 47–66 (2008)
3. Lu, J.-M., Hsu, Y.-L.: Contemporary Issues in Systems Science and Engineering. Telepresence robots for medical and homecare applications, p. 725. Wiley, New Jersey (2015)
4. Arai, T., Kato, R., Fujita, M.: Assessment of operator stress induced by robot collaboration in assembly. CIRP Ann. Manuf. Technol. **59**(1), 5–8 (2010)
5. Bechar, A., Edan, Y.: Human-robot collaboration for improved target recognition of agricultural robots. Ind. Robot: Int. J. **30**(5), 432–436 (2003)
6. Matthias, B., Kock, S., Jerregard, H., Kallman, M., Lundberg, I., Mellander, R.: Safety of collaborative industrial robots: certification possibilities for a collaborative assembly robot concept. In: 2011 IEEE International Symposium on Assembly and Manufacturing (ISAM), pp. 1–6 IEEE (2011)
7. Morabito, V.: Big Data and Analytics. Big data driven business models, pp. 65–80. Springer, Berlin (2015)
8. Top 10 emerging technologies of 2015. https://agenda.weforum.org. (2015)
9. Scassellati, B., Tsui, K.M.: Co-robots: Humans and robots operating as partners (2015)
10. Reardon, C., Tan, H., Kannan, B., Derose, L.: Towards safe robot-human collaboration systems using human pose detection. In: 2015 IEEE International Conference on Technologies for Practical Robot Applications (TePRA) (2015)

11. Kronander, K., Billard, A.: Learning compliant manipulation through kinesthetic and tactile human-robot interaction. IEEE Trans. Haptics **7**(3), 367–380 (2014)
12. Peternel, L., Petric, T., Oztop, E., Babic, J.: Teaching robots to cooperate with humans in dynamic manipulation tasks based on multi-modal human-in-the-loop approach. Autonomous Robots **36**(1–2), 123–136 (2014)
13. Rozo, L., Calinon, S., Caldwell, D.G., Jimnez, P., Torras, C.: Learning collaborative impedance-based robot behaviors. In: Association for the Advancement of Artificial Intelligence (2013)
14. Ganesh, G., Takagi, A., Osu, R., Yoshioka, T., Kawato, M., Burdet, E.: Two is better than one: physical interactions improve motor performance in humans. Sci. Rep. **4**(7484), 3824–3824 (2014)
15. Burdet, E., Osu, R., Franklin, D.W., Milner, T.E., Kawato, M.: The central nervous system stabilizes unstable dynamics by learning optimal impedance. Nature **414**(6862), 446–449 (2001). doi:10.1038/35106566
16. Burdet, E., Ganesh, G., Yang, C., Albu-Schaffer, A.: Interaction force, impedance and trajectory adaptation: by humans, for robots. Springer Tracts Adv. Robot. **79**, 331–345 (2010)
17. Yang, C., Ganesh, G., Haddadin, S., Parusel, S., Albu-Schäeffer, A., Burdet, E.: Human-like adaptation of force and impedance in stable and unstable interactions. In: IEEE Transactions on Robotics, vol. 27(5) (2011)
18. Ajoudani, A., Tsagarakis, N.G., Bicchi, A.: Tele-impedance: preliminary results on measuring and replicating human arm impedance in tele operated robots. In: 2011 IEEE International Conference on Robotics and Biomimetics (ROBIO), pp. 216–222 (2011)
19. Gradolewski, D., Tojza, P.M., Jaworski, J., Ambroziak, D., Redlarski, G., Krawczuk, M.: Arm EMG Wavelet-Based Denoising System. Springer International Publishing, Berlin (2015)
20. Ajoudani, A.: Tele-impedance: teleoperation with impedance regulation using a bodycmachine interface. Int. J. Robot. Res. **31**(13), 1642–1656 (2012)
21. Dragan, A.D., Bauman, S., Forlizzi, J., Srinivasa, S.S.: Effects of robot motion on human-robot collaboration. In: Proceedings of the Tenth Annual ACM/IEEE International Conference on Human-Robot Interaction, pp. 51–58. ACM (2015)
22. Saktaweekulkit, K., Maneewarn, T.: Motion classification using imu for human-robot interaction. In: 2010 International Conference on Control Automation and Systems (ICCAS), pp. 2295–2299. IEEE (2010)
23. Shi, G.Y., Zou, Y.X., Li, W.J., Jin, Y.F., Guan, P.: Towards multi-classification of human motions using micro imu and svm training process. In: Advanced Materials Research, vol. 60, pp. 189–193. Trans Tech Publication (2009)
24. Yoshimoto, H., Arita, D., Taniguchi et al., R.-I.: Real-time human motion sensing based on vision-based inverse kinematics for interactive applications. In: Proceedings of the 17th International Conference on Pattern Recognition, ICPR 2004, vol. 3, pp. 318–321. IEEE (2004)
25. Starner, T., Weaver, J., Pentland, A.: Real-time american sign language recognition using desk and wearable computer based video. IEEE Trans. Pattern Anal. Mach. Intell. **20**(12), 1371–1375 (1998)
26. Shanableh, T., Assaleh, K., Al-Rousan, M.: Spatio-temporal feature-extraction techniques for isolated gesture recognition in arabic sign language. IEEE Trans. Syst. Man Cybern. Part B: Cybern. **37**(3), 641–650 (2007)
27. Liang, R.-H., Ouhyoung, M.: A sign language recognition system using hidden markov model and context sensitive search. Proc. ACM Symp. Virtual Real. Softw. Technol. **96**, 59–66 (1996)
28. Starner, T., Pentland, A.: Real-time american sign language recognition from video using hidden markov models. In: Motion-Based Recognition, pp. 227–243. Springer (1997)
29. Kim, J., Mastnik, S., André, E.: Emg-based hand gesture recognition for realtime biosignal interfacing. In: Proceedings of the 13th international conference on Intelligent user interfaces, pp. 30–39. ACM (2008)
30. Moradi, H., Lee, S.: Joint limit analysis and elbow movement minimization for redundant manipulators using closed form method. In: Advances in Intelligent Computing, pp. 423–432. Springer (2005)

31. Fang, C., Ding, X.: A set of basic movement primitives for anthropomorphic arms. In: 2013 IEEE International Conference on Mechatronics and Automation (ICMA), pp. 639–644. IEEE (2013)
32. Pan, J.J., Xu, K.: Leap motion based 3D. gesture. CHINA. SCIENCEPAPER **10**(2), 207–212 (2015)
33. Jiang, Y.C.: Menacing motion-sensing technology, different leap motion. PC. Fan **11**, 32–33 (2013)
34. Chen, S., Ma, H., Yang, C., Fu, M.: Hand gesture based robot control system using leap motion. In: Intelligent Robotics and Applications, pp. 581–591. Springer (2015)
35. V-rep introduction. http://www.v-rep.eu/
36. UR10 introduction. http://news.cmol.com/2013/0530/33267.html
37. Barretthand introduction. http://wiki.ros.org/Robots/BarrettHand
38. Qian, K., Jie, N., Hong, Y.: Developing a gesture based remote human-robot interaction system using Kinect. Int. J. Smart Home **7**(4), 203–208 (2013)
39. Casiez, G., Roussel, N., Vogel, D.: 1 filter: a simple speed-based low-pass filter for noisy input in interactive systems. In: Proceedings of the 2012 ACM Annual Conference on Human Factors in Computing Systems, pp. 2527–2530. (Austin, TX, USA, 2012)
40. Craig, J.J.: Introduction to Rbotics: Mechanics and Control, 3rd edn. China Machine Press, Beijing (2006)
41. Li, C., Ma, H., Yang, C., Fu, M.: Teleoperation of a virtual icub robot under framework of parallel system via hand gesture recognition. In: 2014 IEEE International Conference on Fuzzy Systems (FUZZ-IEEE), pp. 1469–1474. IEEE (2014)
42. Wang, F.-Y.: Parallel system methods for management and control of complex systems. Control Decis. **19**, 485–489 (2004)
43. S.W.S. (2012), wiki for the icub simulator specifications from its installation to its use. http://www.eris.liralab.it/wiki/
44. Yet another robot platform. http://yarp0.sourceforge.net/
45. Wachs, J.P., Kölsch, M., Stern, H., Edan, Y.: Vision-based hand-gesture applications. Commun. ACM **54**(2), 60–71 (2011)
46. Hsu, R.-L., Abdel-Mottaleb, M., Jain, A.K.: Face detection in color images. Pattern Anal. Mach. Intell. **24**(5), 696–706 (2002)
47. Hu, M.-K.: Visual pattern recognition by moment invariants. IRE Trans. Inf. Theory **8**(2), 179–187 (1962)
48. Granlund, G.H.: Fourier preprocessing for hand print character recognition. IEEE Trans. Comput. **100**(2), 195–201 (1972)
49. Nedjah, N.e.: Adaptation of fuzzy inference system using neural learning, fuzzy system engineering: theory and practice. In: Studies in Fuzziness and Soft Computing, pp. 53–83 (2001)
50. Chang, C.-C., Lin, C.-J.: Libsvm: a library for support vector machines. ACM Trans. Intell. Syst. Technol. (TIST) **2**(3), 27 (2011)
51. Corke, P.I.: Robotics toolbox (2008)
52. Zhu, G.C., Wang, T.M., Chou, W.S., Cai, M.: Research on augmented reality based teleoperation system. Acta Simul. Syst. Sin. **5**, 021 (2004)
53. TF, C.: History and evolution of electroencephalographic instruments and techniques. J. Clin. Neurophysiol. **10**(4), 476–504 (1993)
54. Grude, S., Freeland, M., Yang, C., Ma, H.: Controlling mobile spykee robot using emotiv neuro headset. In: Control Conference (CCC), 2013 32nd Chinese, pp. 5927–5932. IEEE (2013)
55. Hoffmann, A.: Eeg signal processing and emotivs neuro headset, Hessen: sn (2010)
56. Sanei, S., Chambers, J.A.: EEG Signal Processing. Wiley, New Jerssey (2013)
57. Jasper, H.H.: The ten twenty electrode system of the international federation. Electroencephal. Clin. Neurophysiol. **10**, 371–375 (1958)
58. Lotte, F.: Les interfaces cerveau-ordinateur: Conception et utilisation en réalité virtuelle. Revue Technique et Science Informatiques **31**(3), 289–310 (2012)
59. Carmena, J.M., Lebedev, M.A., Crist, R.E., O'Doherty, J.E., Santucci, D.M., Dimitrov, D.F., Patil, P.G., Henriquez, C.S., Nicolelis, M.A.: Learning to control a brain-machine interface for reaching and grasping by primates. Plos Biol. **1**(2), 193–208 (2003)

60. Milln, J.D.R., Frdric, R., Josep, Mourio, A.W.G.: Non-invasive brain-actuated control of a mobile robot by human eeg. In: IEEE Transtition on Biomedical Engineering, Special Issue on Brain-Machine Interfaces, vol. 51, No. 6, pp. 1026–1033 (2004)
61. Wang, Y., Gao, X., Hong, B., Jia, C., Gao, S.: Brain-computer interfaces based on visual evoked potentials. IEEE Eng. Med. Biol. Mag. Q. Mag. Eng. Med. Biol. Soc. **27**(5), 64–71 (2008)
62. Diez, P.F., Mut, V.A., Perona, E.M.A., Leber, E.L.: Asynchronous bci control using high-frequency ssvep. J. Neuroeng. Rehabil. **8**(2), 642–650 (2011)
63. Bashashati, A., Fatourechi, M., Ward, R.K., Birch, G.E.: A survey of signal processing algorithms in brainccomputer interfaces based on electrical brain signals. J. Neural Eng. **4**(2), R32–R57 (2007)
64. Ding, J., Sperling, G., Srinivasan, R.: Attentional modulation of ssvep power depends on the network tagged by the flicker frequency. Cereb. Cortex **16**(7), 1016–1029 (2006)
65. Ekanayake, H.: P300 and emotiv epoc: Does emotiv epoc capture real eeg?. P300 and Emotiv EPOC: Does Emotiv EPOC capture real EEG? - ResearchGate (2011)
66. Vidal, J.-J.: Toward direct brain-computer communication. Ann. Rev. Biophys. Bioeng. **2**(1), 157–180 (1973)
67. Sullivan, T.J., Deiss, S.R., Jung, T.P., Cauwenberghs, G.: A brain-machine interface using dry-contact, low-noise eeg sensors. In: 2008 IEEE International Symposium on In Circuits and Systems. p. 1986–1989. ISCAS (2008)
68. Malki, A., Yang, C., Wang, N., Li, Z.: Mind guided motion control of robot manipulator using eeg signals. In: 2015 5th International Conference on Information Science and Technology (ICIST, 2015)
69. Coppeliarobotics, Vrep description. http://www.coppeliarobotics.com (2014)
70. Piccione, F., Priftis, K., Tonin, P., Vidale, D., Furlan, R., Cavinato, M., Merico, A., Piron, L.: Task and stimulation paradigm effects in a p300 brain computer interface exploitable in a virtual environment: A pilot study. Psychnol. J. **6**(1), 99–108 (2008)
71. Chen, W.D., Zhang, J.H., Zhang, J.C., Li, Y., Qi, Y., Su, Y., Wu, B., Zhang, S.M., Dai, J.H., Zheng, X.X.: A p300 based online brain-computer interface system for virtual hand control. J. Zhejiang Univ. Sci. C **11**(08), 587–597 (2010)
72. Renard, Y., Lotte, F., Gibert, G., Congedo, M., Maby, E., Delannoy, V., Bertrand, O., Cuyer, A.: Openvibe: an open-source software platform to design, test, and use brain-computer interfaces in real and virtual environments. Presence Teleop. Virtual Environ. **19**(1), 35–53 (2010)
73. Inria: P300: Old p300 speller. http://openvibe.inria.fr/openvibe p300-speller (2014)

Chapter 9
Indoor/Outdoor Robot Localization

Abstract Robot localization is essential for a wide range of applications, such as navigation, autonomous vehicle, intrusion detection, and so on. This chapter presents a number of localization techniques for both indoor and outdoor robots. The localization problem for an unknown static single target in wireless sensor network is investigated with least squares algorithm and Kalman filter. And an algorithm of passive radio frequency identification (RFID) indoor positioning is proposed based on interval Kalman filter, according to the geometric constraints of responding tags, combined with the target motion information. Next, the simultaneous localization and mapping algorithm (SLAM) for indoor positioning with the low-cost laser LIDAR—RPLIDAR—is investigated. Finally, for outdoor environments, we investigate two integration strategies to fuse inertial navigation system (INS) and vehicle motion sensor (VMS) outputs from their stand-alone configurations. The INS/VMS integration system is an entirely self-contained navigation system, and it is thus expected to benefit the GPS/INS when GPS signals are unavailable for long term.

9.1 Localization with Wireless Sensor Networks

Wireless sensor network localization is an essential problem that has attracted increasing attention due to wide requirements such as indoor navigation, autonomous vehicle, intrusion detection, and so on. With a priori knowledge of the positions of sensor nodes and their measurements to targets in the WSNs, i.e., posterior knowledge, such as distance and angle measurements, it is possible to estimate the position of targets through different algorithms. In this section, we present two localization approaches for one static target with WSNs base on least squares and Kalman filter, respectively [1].

9.1.1 Problem Formulation

Consider a planar wireless sensor network composed of N stationary sensor nodes and suppose that their position vector $W = [x_1, y_1, x_2, y_2, \ldots, x_N, y_N]^T$ is known.

© Science Press and Springer Science+Business Media Singapore 2016

C. Yang et al., *Advanced Technologies in Modern Robotic Applications*,
DOI 10.1007/978-981-10-0830-6_9

Each sensor may be able to detect some partial information of one target if the target is located in its sensing range or region; however, each sensor is not able to determine exact position of the target due to its low-cost, low power, limited computation ability, and measurement noise involved. The problem is how to determine the position of the target as exactly as possible. In order to get the position of the unknown target, assume that n nodes (indexed by $k_1, k_2, \ldots, k_n$) are available to detect the target and measure the information of the distance, the orientation angle, or both of them. Later, one would see that it is the great advantage of WSNs by combining plenty of noisy measurements to get a relatively exacter estimation of the target's position. In this section, we restrict ourselves to consider only static target rather than more complex cases of dynamic target; hence, without loss of generality, suppose that the active nodes which can detect the target are indexed by $1, 2, \ldots, n$, respectively. According to the range/angle information available, we only consider three typical cases of sensor measurements.

Case 1: The distance vector $L = [l_1, l_2, \ldots, l_n]^T$ between the nodes and the target can be only obtained, as shown in Fig. 9.1. Since l_i is a noisy distance measurement of node i it can be represented as

$$l_i = d_i + \varepsilon_i \tag{9.1}$$

$$d_i = \sqrt{(x - x_i)^2 + (y - y_i)^2} \tag{9.2}$$

where d_i is the true distance between node i and the target, (x_i, y_i) is the known location of node i, (x, y) is the position of the unknown target, and ε_i is the measurement noise of node i which is unrelated with other ε_j $(j \neq i)$ and the true distance d_i.

Case 2: The angle vector $\theta = [\theta_1, \theta_2, \ldots, \theta_n]^T$ is the only measurement, as shown in Fig. 9.2.

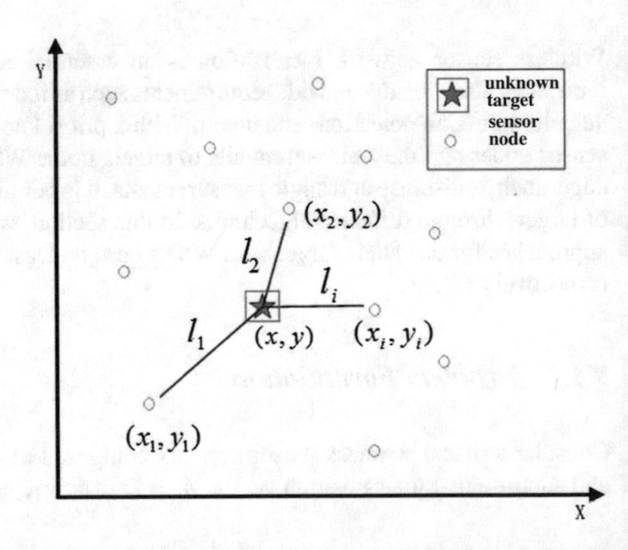

Fig. 9.1 Case 1: distance measurements only

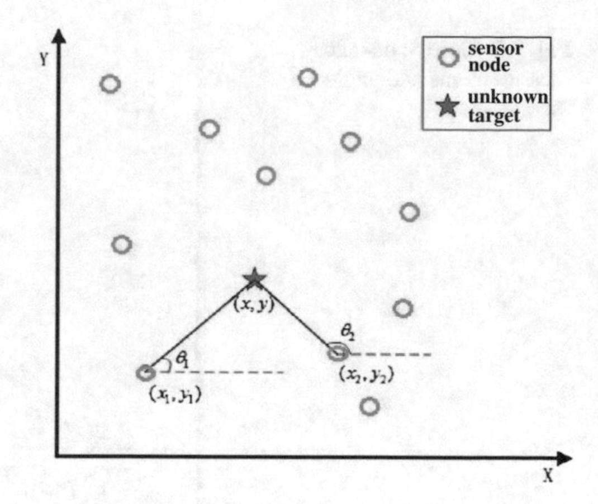

Fig. 9.2 Case 2: angle measurements only

For node i, the noisy measurement θ_i is

$$\theta_i = \varphi_i + \epsilon_i \tag{9.3}$$

$$\varphi_i = \arctan(y - y_i, x - x_i) \tag{9.4}$$

$$\arctan(y, x) = \begin{cases} \arctan\frac{y}{x} & x > 0 \\ \arctan\frac{y}{x} - \pi & x < 0, y \leq 0 \\ \arctan\frac{y}{x} + \pi & x < 0, y > 0 \\ \frac{\pi}{2} & x = 0, y > 0 \\ -\frac{\pi}{2} & x = 0, y < 0 \end{cases} \tag{9.5}$$

where φ_i is the true angle between the sensor i and the target, $\varphi_i \in [-\pi, \pi)$, and ϵ_i is the measurement noise of node i which is unrelated with each other and the true angle φ_i. Hereinafter, $\arctan(y, x)$ denotes the arctangent angle in the correct quadrant determined by the coordination (x, y).

Case 3: Both the distance measurement L and angle measurement θ between sensor nodes and the target are available, as shown in Fig. 9.3. For sensor node i, the coordinate of the target can be approximately represented as

$$\begin{cases} x_{ri} = x_i + l_i \cos\theta_i \\ y_{ri} = y_i + l_i \sin\theta_i \end{cases} \tag{9.6}$$

9.1.2 Algorithm Design

9.1.2.1 Least Squares Method

Distance model: The following equations can be obtained:

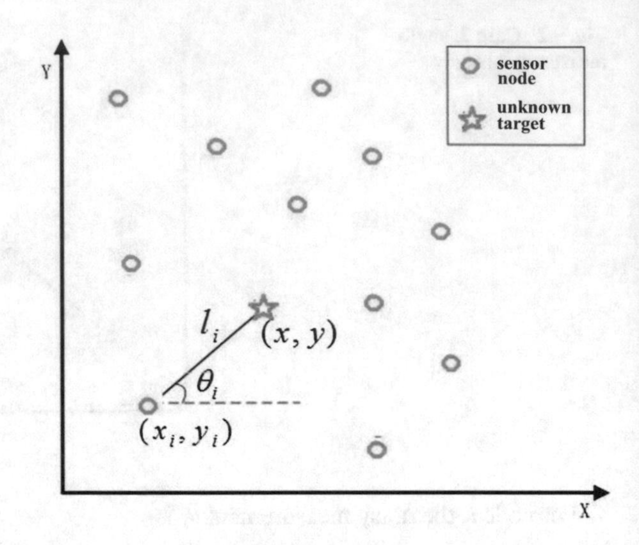

Fig. 9.3 Case 3: distance measurements and angle measurements

$$\begin{cases} (x - x_1)^2 + (y - y_1)^2 = (l_1 - \varepsilon_1)^2 \\ (x - x_2)^2 + (y - y_2)^2 = (l_2 - \varepsilon_2)^2 \\ \vdots \\ (x - x_n)^2 + (y - y_n)^2 = (l_n - \varepsilon_n)^2 \end{cases} \tag{9.7}$$

Subtracting the nth equation from each equation j ($j \in [1, 2, \ldots, n-1]$), it can be obtained as

$$2x\Delta x_j + 2y\Delta y_j = (l_j - \varepsilon_j)^2 - (l_n - \varepsilon_n)^2 - (x_j^2 + y_j^2) + (x_n^2 + y_n^2) \tag{9.8}$$

where

$$\Delta x_j = x_n - x_j$$
$$\Delta y_j = y_n - y_j$$

Hence, Eq. (9.8) can be represented as

$$B_L = A_L X + \Delta_L \tag{9.9}$$

where

$$A_L = \begin{bmatrix} 2(x_n - x_1) & 2(y_n - y_1) \\ 2(x_n - x_2) & 2(y_n - y_2) \\ \vdots & \vdots \\ 2(x_n - x_{n-1}) & 2(y_n - y_{n-1}) \end{bmatrix} \tag{9.10}$$

$$B_L = \begin{bmatrix} l_1^2 - l_n^2 - (x_1^2 + y_1^2) + (x_n^2 + y_n^2) \\ l_2^2 - l_n^2 - (x_2^2 + y_2^2) + (x_n^2 + y_n^2) \\ \vdots \\ l_{n-1}^2 - l_n^2 - (x_{n-1}^2 + y_{n-1}^2) + (x_n^2 + y_n^2) \end{bmatrix} \tag{9.11}$$

$$\Delta_L = \begin{bmatrix} \varepsilon_n^2 - \varepsilon_1^2 + 2l_1\varepsilon_1 - 2l_n\varepsilon_n \\ \varepsilon_n^2 - \varepsilon_2^2 + 2l_2\varepsilon_2 - 2l_n\varepsilon_n \\ \vdots \\ \varepsilon_n^2 - \varepsilon_{n-1}^2 + 2l_{n-1}\varepsilon_{n-1} - 2l_n\varepsilon_n \end{bmatrix} = \begin{bmatrix} \varepsilon_1^2 - \varepsilon_n^2 + 2d_1\varepsilon_1 - 2d_n\varepsilon_n \\ \varepsilon_2^2 - \varepsilon_n^2 + 2d_2\varepsilon_2 - 2d_n\varepsilon_n \\ \vdots \\ \varepsilon_{n-1}^2 - \varepsilon_n^2 + 2d_{n-1}\varepsilon_{n-1} - 2d_n\varepsilon_n \end{bmatrix} \tag{9.12}$$

$$X = [x, y]^T \tag{9.13}$$

Then the position of the target can be obtained using least squares algorithm:

$$\hat{X} = ({A_L}^T A_L)^{-1} {A_L}^T B_L \tag{9.14}$$

Angle model: The following equations can be obtained:

$$\begin{cases} \dfrac{y - y_1}{x - x_1} = \tan\theta_1 \\ \dfrac{y - y_2}{x - x_2} = \tan\theta_2 \\ \quad\vdots \\ \dfrac{y - y_n}{x - x_n} = \tan\theta_n \end{cases} \tag{9.15}$$

$$\begin{cases} x\tan\theta_1 - y = x_1\tan\theta_1 - y_1 \\ x\tan\theta_2 - y = x_2\tan\theta_2 - y_2 \\ \quad\vdots \\ x\tan\theta_n - y = x_n\tan\theta_n - y_n \end{cases} \tag{9.16}$$

It can be represented as

$$A_\theta X = B_\theta \tag{9.17}$$

while

$$A_\theta = \begin{bmatrix} \tan\theta_1 & -1 \\ \tan\theta_2 & -1 \\ \vdots & \vdots \\ \tan\theta_i & -1 \end{bmatrix} \tag{9.18}$$

$$B_\theta = \begin{bmatrix} x_1 \tan\theta_1 - y_1 \\ x_2 \tan\theta_2 - y_2 \\ \vdots \\ x_i \tan\theta_i - y_i \end{bmatrix} \tag{9.19}$$

$$X = \begin{bmatrix} x, y \end{bmatrix}^T \tag{9.20}$$

The position of the target can be obtained using least squares algorithm:

$$\hat{X} = ({A_\theta}^T A_\theta)^{-1} {A_\theta}^T B_\theta \tag{9.21}$$

Distance and angle model: For this model, both the distance and angle measurements can be obtained. We can combine the two methods as described above:

$$A = \begin{bmatrix} A_L \\ A_\theta \end{bmatrix} \tag{9.22}$$

$$B = \begin{bmatrix} B_L \\ B_\theta \end{bmatrix} \tag{9.23}$$

The estimation of the position is

$$\hat{X} = (A^T A)^{-1} A^T B \tag{9.24}$$

9.1.2.2 Kalman Filter Method

Distance model We assume that the distance measurement L can be only obtained. The state model of the system is

$$X_{k+1} = AX_k + w_k \tag{9.25}$$

where $X_k = [x_{rk}, y_{rk}]^T$ represents the position vector of the target which is calculated by node k. Here $A = I_2 = \begin{bmatrix} 1 & 0 \\ 0 & 1 \end{bmatrix}$. And the observation model of the system can be represented by

$$Z_k = h(X_k) + v_k \tag{9.26}$$

where $Z_k = l_k$, the actual measurement of node k, and v_k is measurement noise of node k whose covariance is R_k. The observation function $h(X_k)$ and corresponding Jacobian H_k derived from Eq. (9.2) are given as follows:

$$h(X_k) = \sqrt{(x_{rk} - x_k)^2 + (y_{rk} - y_k)^2} \tag{9.27}$$

$$H_k = \left[\frac{x_{rk}-x_k}{\sqrt{(x_{rk}-x_k)^2+(y_{rk}-y_l)^2}} \; \frac{y_{rk}-y_k}{\sqrt{(y_{rk}-y_k)^2+(y_{rk}-y_l)^2}} \right] \tag{9.28}$$

For this state-space model, it is ready to apply the extended Kalman filter (EKF) which consists of two consequently stages at step k ($k = 1, 2, \ldots, n$).

Before the new measurement Z_k arrives, we can make predictions as follows.

Predicted state vector:

$$\hat{X}_{k|k-1} = A\hat{X}_{k-1|k-1} \tag{9.29}$$

Predicted error covariance:

$$P_{k|k-1} = AP_{k-1|k-1}A^T \tag{9.30}$$

With new measurement Z_k, we can update $\hat{X}_{k|k}$ and $P_{k|k}$ as follows.

Measurement residual:

$$\tilde{y}_k = Z_k - h(\hat{X}_{k|k-1}) \tag{9.31}$$

Residual covariance:

$$S_k = H_k P_{k|k-1} H_k^T + R_k \tag{9.32}$$

Kalman gain:

$$K_k = P_{k|k-1} H_k^T S_k^{-1} \tag{9.33}$$

Updated state estimate:

$$\hat{X}_{k|k} = \hat{X}_{k|k-1} + K_k \tilde{y}_k \tag{9.34}$$

Updated estimate covariance:

$$P_{k|k} = (I - K_k H_k) P_{k|k-1} \tag{9.35}$$

Angle model Suppose that θ is the only measurement that can be obtained. We also need to locate the target through the observation of the n sensor nodes whose positions are a priori known.

The initial value of the target $X_2 = [x_{r2}, y_{r2}]^T$ is determined by the measurements θ_1 and θ_2 from the first two sensors:

$$\tan\theta_1 = \frac{y_{r2} - y_r}{x_{r2} - y_1} \tag{9.36}$$

$$\tan\theta_2 = \frac{y_{r2} - y_r}{x_{r2} - y_2} \tag{9.37}$$

then x_{r2} and y_{r2} can be given by

$$x_{r2} = \frac{y_1 - y_2 - (x_1 \tan\theta_{r1} - x_2 \tan\theta_{r2})}{(\tan\theta_{r2} - \tan\theta_{r1})} \tag{9.38}$$

$$y_{r2} = \frac{\tan\theta_{r1}\tan\theta_{r2}(x_2 - x_1) + \tan\theta_{r2} y_1 - \tan\theta_{r1} y_2}{\tan\theta_{r2} - \tan\theta_{r1}} \tag{9.39}$$

where (x_1, y_1) and (x_2, y_2) are the coordinates of the first two sensor nodes.

The state model of the target and the observation model of sensor k ($k \geq 3$) can be described by

$$X_k = AX_{k-1} \tag{9.40}$$

$$Z_k = h(X_k) + v_k \tag{9.41}$$

where $X_k = [x_{rk}, y_{rk}]^T$ is the position of the target that is calculated by node k, $Z_k = \theta_k$ denotes the measurement of node k, v_k is the measurement noise of node k with covariance Q_k, and $A = I_2$. The observation function $h_k(X_k)$ and corresponding Jacobian H_k are

$$h(X_k) = \arctan(y_{rk} - y_k, x_{rk} - x_k) \tag{9.42}$$

$$H_k = \left[\frac{y_k - y_{rk}}{(x_{rk}-x_k)^2 + (y_{rk}-y_k)^2} \;\; \frac{x_{rk}-x_k}{(x_{rk}-x_k)^2 + (y_{rk}-y_k)^2} \right] \tag{9.43}$$

Then Kalman filter can be used for state estimation through Eqs. (9.29)–(9.35).

Distance and angle model We assume that both of the distance and angle can be measured. The initial value of the target $X_1 = [x_{r1}, y_{r1}]^T$ is

$$\begin{cases} x_{r1} = x_1 + l_1 \cos\theta_1 \\ y_{r1} = y_1 + l_1 \sin\theta_1 \end{cases} \tag{9.44}$$

where (x_1, y_1) is the coordinate of the first sensor node whose measurements are l_1 and θ_1.

The state model of the target and the observation model of sensor k ($k \geq 2$) are

$$X_k = AX_{k-1} \tag{9.45}$$

$$Z_k = h(X_k) + v_k \tag{9.46}$$

where $X_k = [x_{rk}, y_{rk}]^T$ is the position of the target that is calculated by node k, $Z_k = [l_k, \theta_k]^T$ denotes the measurement of node k, v_k is the measurement noise of node k with covariance Q_k, and $A = I_2$. The observation function $h_k(X_k)$ and corresponding Jacobian H_k are

$$h(X_k) = \begin{bmatrix} \sqrt{(x_{rk} - x_k)^2 + (y_{rk} - y_k)^2} \\ \arctan(y_{rk} - y_k, x_{rk} - x_k) \end{bmatrix}$$

$$H_k = \begin{bmatrix} \frac{x_{rk}-x_k}{\sqrt{(x_{rk}-x_k)^2+(y_{rk}-y_l)^2}} & \frac{y_{rk}-y_k}{\sqrt{(y_{rk}-y_k)^2+(y_{rk}-y_l)^2}} \\ \frac{y_k-y_{rk}}{(x_{rk}-x_k)^2+(y_{rk}-y_k)^2} & \frac{x_{rk}-x_k}{(x_{rk}-x_k)^2+(y_{rk}-y_k)^2} \end{bmatrix}$$

Then Eqs. (9.29)–(9.35) can be used for position estimation.

9.1.3 Theoretical Analysis

For convenience, we use $\tilde{X} = \hat{X} - X$ to denote the estimation error of the position of the unknown target.

9.1.3.1 Least Squares

For simplicity, we only present some theoretical results for the *Distance model* with the least squares algorithm. More complete discussions can be found in [2]. These results enable us to investigate how the estimation accuracy depends on the sensor noise and the sensor network; in other words, by properly choosing the available sensor types and designing the locations of sensor nodes, we may improve the localization accuracy of wireless sensor networks.

Theorem 9.1.1 *Consider the distance model, and suppose that the noise ε_i of sensor node i is Gaussian noise with zero mean and covariance σ_i^2, i.e., $\varepsilon_i \sim N(0, \sigma_i^2)$. Then for the least squares algorithm presented in Sect. 9.1.2.1, we have*

$$E\Delta_L = [\sigma_1^2 - \sigma_n^2, \sigma_2^2 - \sigma_n^2, \ldots, \sigma_{n-1}^2 - \sigma_n^2]^T \tag{9.47}$$

$$E\tilde{X} = (A_L^T A_L)^{-1} A_L^T E(\Delta_L) \tag{9.48}$$

$$E[\tilde{X}\tilde{X}^T] = (A_L^T A_L)^{-1} A_L^T R A_L (A_L^T A_L)^{-1} \tag{9.49}$$

where Δ_L is defined in Eq. (9.12), A_L is defined in Eq. (9.10), and $R = (R_{ij})_{(n-1)\times(n-1)}$ with

$$R_{ij} = \begin{cases} 3\sigma_i^4 - 2\sigma_i^2\sigma_n^2 + 4d_i^2\sigma_i^2 + 4d_n^2\sigma_n^2 + 3\sigma_n^4 & i = j \\ \sigma_i^2\sigma_j^2 - (\sigma_i^2 + \sigma_j^2)\sigma_n^2 + 4d_n^2\sigma_n^2 + 3\sigma_n^3 & i \neq j \end{cases} \tag{9.50}$$

Furthermore, if all sensors are identical with the same noise covariance σ^2, then we have $E\Delta_L = 0$, and hence $E\tilde{X} = 0$, which means that the position estimation of the unknown target is unbiased.

Theorem 9.1.2 *Consider the distance model, and suppose that the noise ε_i of sensor node i is uniformly distributed in $[-a_i, a_i]$, i.e., $\varepsilon_i \sim U(-a_i, a_i)$. Then for the least* *squares algorithm presented in Sect. 9.1.2.1, we have*

$$E\Delta_L = \left[\frac{a_1^2}{3} - \frac{a_n^2}{3}, \frac{a_2^2}{3} - \frac{a_n^2}{3}, \dots, \frac{a_{n-1}^2}{3} - \frac{a_n^2}{3}\right]^T \tag{9.51}$$

$$E\tilde{X} = \left(A_L^T A_L\right)^{-1} A_L^T E(\Delta_L) \tag{9.52}$$

$$E[\tilde{X}\tilde{X}^T] = \left(A_L^T A_L\right)^{-1} A_L^T R A_L \left(A_L^T A_L\right)^{-1} \tag{9.53}$$

where Δ_L is defined in Eq. (9.12), A_L is defined in Eq. (9.10), and $R = (R_{ij})_{(n-1)\times(n-1)}$ with

$$R_{ij} = \begin{cases} \frac{1}{5}a_i^4 - \frac{2}{9}a_i^2 a_n^2 + \frac{4}{3}d_i^2 a_i^2 + \frac{4}{3}d_n^2 a_n^2 + \frac{1}{5}a_n^4 & i = j \\ \frac{1}{9}a_i^2 a_j^2 - \frac{1}{9}\left(a_j^2 + a_i^2\right) a_n^2 + \frac{1}{5}a_n^4 + \frac{4}{3}d_n^2 a_n^2 & i \neq j \end{cases} \tag{9.54}$$

Furthermore, if all sensors are identical with the same noise covariance σ^2, then we have $E\Delta_L = 0$, and hence $E\tilde{X} = 0$, which means that the position estimation of the unknown target is unbiased.

Theorem 9.1.3 *Consider the distance model, and suppose that the noise ε_i of sensor node i is bounded such that $|\varepsilon_i| \leq |\delta_i|$. Then for the least squares algorithm presented in Sect. 9.1.2.1, we have*

$$|\Delta_{L,j}| \leq \delta_j^2 + 2d_j\delta_j + 2d_n\delta_n \tag{9.55}$$

$$\tilde{X} = \left(A_L^T A_L\right)^{-1} A_L^T(\Delta_L) \tag{9.56}$$

where $\Delta_L = [\Delta_{L,1}, \Delta_{L,2}, \dots, \Delta_{L,n-1}]^T$ is defined in Eq. (9.12), and A_L is defined in Eq. (9.10).

9.1.3.2 Kalman Filter

Theorem 9.1.4 *In the state model presented in Sect. 9.1.2.2, there is no state noise and $A = I_2$. Then, for the extended Kalman filter given in Eqs. (9.29)–(9.35), the updated state estimation and covariance can be represented as*

$$\hat{X}_{k|k} = \hat{X}_{k-1|k-1} + K_k\tilde{y}_k \tag{9.57}$$

$$P_{k|k} = (I - K_k H_k)P_{k-1|k-1} \tag{9.58}$$

$$= g_k(P_{k-1|k-1}) \tag{9.59}$$

where function $g_k(\cdot)$ is a Riccati iteration defined by

$$g_k(P) = P - PH_k^T\left(H_k P_k H_k^T + R_k\right)^{-1} H_k P \tag{9.60}$$

Consequently, for given initial conditions $\hat{X}_{0|0} = X_0$ *and* $P_{0|0} = P_0 > 0$, *the nth sensor node will yield the following estimation:*

$$\hat{X}_{n|n} = X_0 + \sum_{k=1}^{n} K_k \tilde{y}_k \tag{9.61}$$

$$P_{n|n} = P_0 \prod_{k=1}^{n} (I - K_k H_k) \tag{9.62}$$

$$= G_n(P_0) \tag{9.63}$$

where $G_n = g_n \circ g_{n-1} \circ \cdots \circ g_1$.

9.1.4 Simulation Studies

The algorithms described above are implemented in MATLAB to evaluate their performance and make comparison. The simulated sensor network which is shown in Fig. 9.4 consists of $N = 100$ nodes, which is well-distributed on a square area of 20 m × 20 m. A random unknown target is placed in this area; then we use the measurements of sensor nodes to estimate the target's position and calculate the root mean squared errors (RMSE) via Monte Carlo simulations:

$$J = \frac{1}{M} \sum_{m=1}^{M} \sqrt{[(\hat{x}^{(m)} - x_{true})^2 + (\hat{y}^{(m)} - y_{true})^2]} \tag{9.64}$$

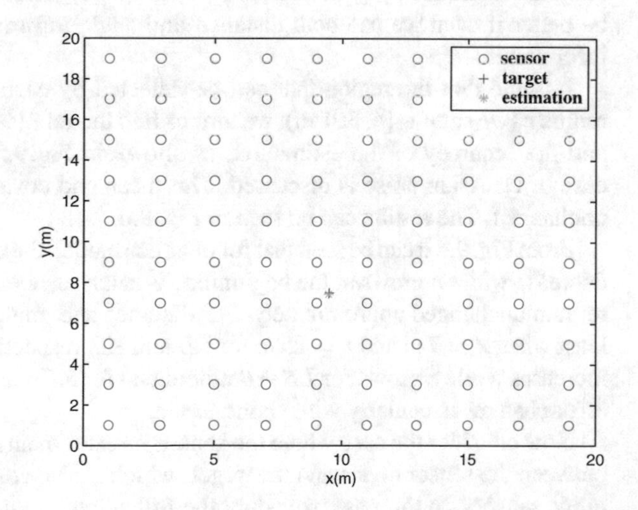

Fig. 9.4 The simulated sensor network

Table 9.1 Gaussian noise

	$LS-l$	$LS-\theta$	$LS-l\&\theta$	$EKF-l$	$EKF-\theta$	$EKF-l\&\theta$
J	0.0787	0.2247	0.1139	0.0196	11.6343	0.0412

Table 9.2 Uniform noise

	$LS-l$	$LS-\theta$	$LS-l\&\theta$	$EKF-l$	$EKF-\theta$	$EKF-l\&\theta$
J	0.0473	0.1397	0.0650	0.0140	4.1048	0.0273

Table 9.3 Bounded noise

	$LS-l$	$LS-\theta$	$LS-l\&\theta$	$EKF-l$	$EKF-\theta$	$EKF-l\&\theta$
J	0.0408	0.1340	0.0533	0.0147	2.7314	0.0182

where (x_{true}, y_{true}) is the true position of the node, $(\hat{x}^{(m)}, \hat{y}^{(m)})$ is the position estimate in the m-th simulation, and M is the total number of random simulations.

In the case of Gaussian sensor noise, we set $\sigma_{Li} = 0.1$, $\sigma_{\theta i} = \pi/180$; for uniform sensor noise, we set $a_{Li} = 0.1$, $a_{\theta i} = \pi/180$; for bounded sensor noise, and we set $\delta_{Li} = 0.1$, $\delta_{\theta i} = \pi/180$.

Assume that the target can be detected by all the sensor nodes, and the RMSEs of the different methods are shown in Tables 9.1, 9.2, and 9.3, respectively.

From the simulation results we can see that for each kind of noise, the estimation error tends to be smaller by using Kalman filter through the distance measurement and both distance and angle measurements. When we only use the angle measurement, the results of Kalman filter are obviously unsatisfying, which may be explained by some discussions in [3]. In addition, for the two algorithms, the results seem not to be better if even we use both distance and angle information than single distance information.

Assume that the region that can be detected by each sensor node is limited to radius r (we set $r \in [4, 30]$ m), we aim to find the relationship between the distance and the accuracy of the estimation, as shown in Fig. 9.5. For simplicity, only the case of Gaussian noise is discussed. The mean and covariance of ε_i and ϵ_i remain unchanged. The results can be seen in Fig. 9.6.

From Fig. 9.6 it can be seen that for distance model, the errors of the two algorithms decrease while r grows at the beginning. When r reaches a certain value, the errors remain unchanged approximately. For distance and angle model, the errors are the least when $r = 7$ m and $r = 25$ m for LS and KF, respectively. In addition, the error increases while r grows for $LS-\theta$ model and for $KF-\theta$ model, and the estimation error changes irregularly while r increases.

Now consider the case where the sensor measurements are related to the distance between the sensor nodes and the target, which is also common in practice for some range sensors. In this case, we adopt the following measurement model:

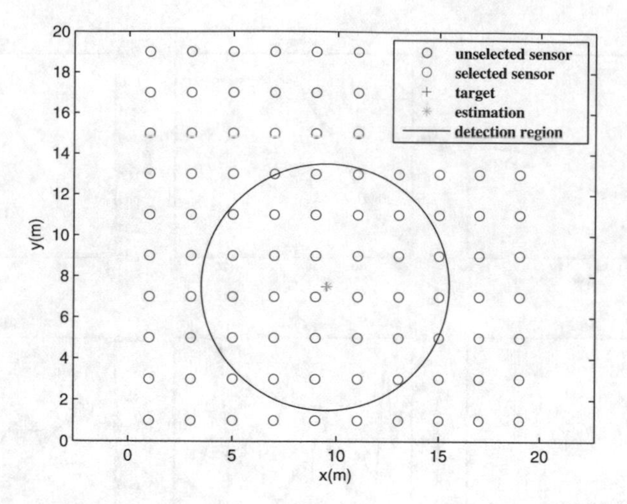

Fig. 9.5 Sensors with detection region

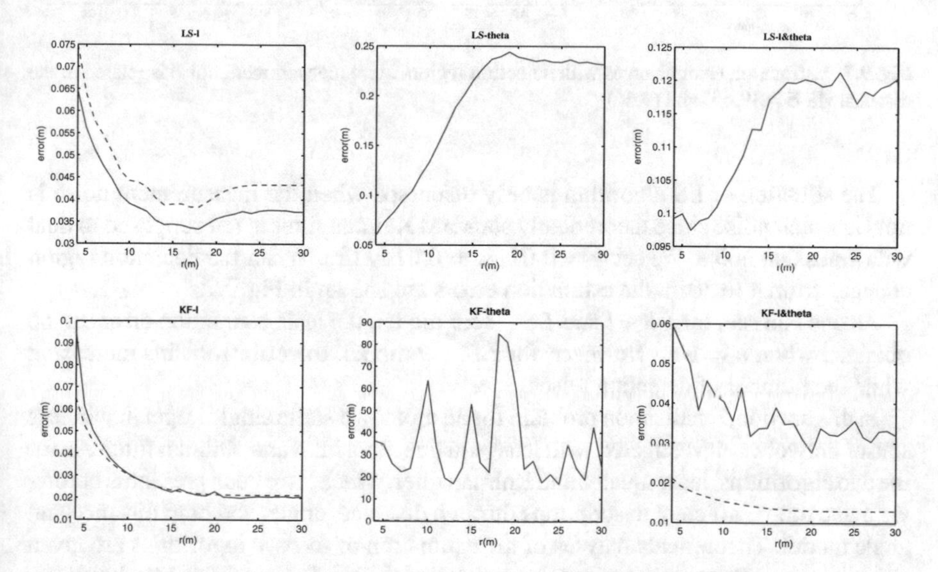

Fig. 9.6 Estimation error changes with detection region. Here measurement noise is unrelated to the distance

$$l_i = (1+\gamma)d_i + \varepsilon_i \tag{9.65}$$

$$\theta_i = \varphi_i + \mu d_i + \epsilon_i \tag{9.66}$$

which means that a new linear part of the true distance will be added to the distance and angle measurements, that is to say, the measurement errors $l_i - d_i$ and $\theta_i - \phi_i$ are roughly proportional to the distance between the sensor nodes and the target.

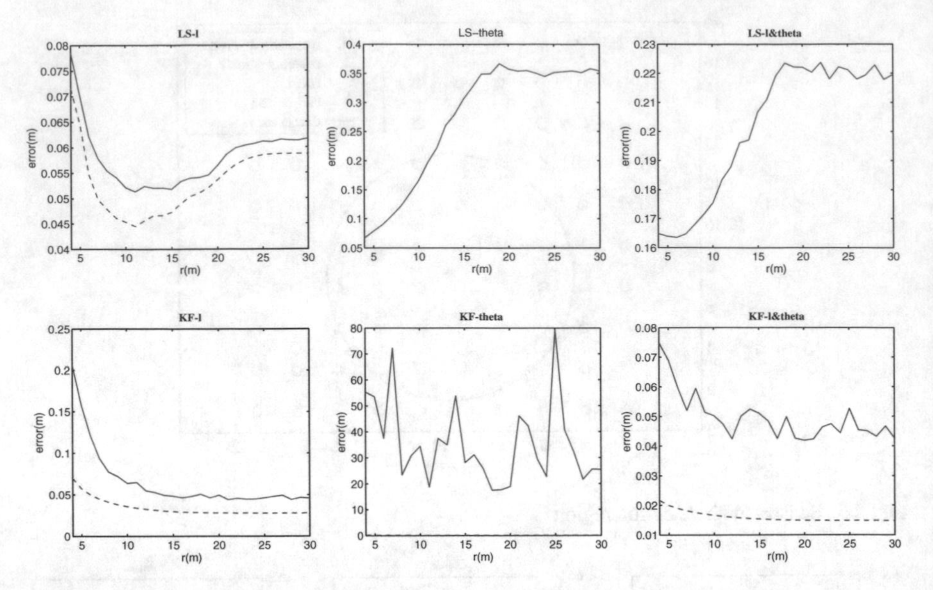

Fig. 9.7 Estimation error changes with detection region. Here measurement noise is related to the distance via Eqs. (9.65) and (9.66)

The situation of LS algorithm is only discussed when the measurement noise is not Gaussian noise since theoretically standard Kalman filter is not supposed to deal with Gaussian noise. We set $\gamma = 0.01$, $\mu = 0.01\pi/180$. When the detection region changes from 4 to 30 m, the estimation errors are shown in Fig. 9.7.

As one can see, for $LS - l$ and $LS - l\&\theta$ model, the least estimation error can be obtained when $r = 8$ m. However, for $LS - \theta$ model, the error remains increasing while the number of detection raises.

In this section, localization problem for an unknown static single target in wireless sensor network is investigated with least squares algorithm and Kalman filter. As for the two algorithms, least squares and Kalman filter, which have been presented before, we make more sufficient descriptions through distance, angle, and both distance and angle model. Theoretical analyses of the estimation of the two algorithms are given and especially three cases of measurement noise are discussed for LS algorithm. Through the simulation, we can find that the $KF - l$ model performs well and the $KF - \theta$ model seems a bad choice because of its large errors. In addition, we also simulate two more realistic cases that the detection region of the sensor is limited and the measurement noise is related to the true distance between the sensor and the target. By changing the detection region, the estimation error of each model changes differently.

9.2 RFID-based Indoor Localization Using Interval Kalman Filter

Recall last section, localization of a single target is achieved using WSNs. But in some place, the high energy consumption and low precision of wireless sensor may limit its application. As an alternative, RFID is of low cost and the ability of high-speed contactless identification and are widely used in establishments such as the airport lobby, supermarkets, libraries, underground parking, and so on.

This section reports a passive RFID indoor positioning algorithm based on interval Kalman filter which can reduce the positioning errors. This localization algorithm aims to seek a breakthrough to guarantee positioning accuracy, while taking into account the costs.

9.2.1 Interval Kalman Filter for RFID Indoor Positioning

9.2.1.1 Description of Positioning Scene and Issue

In the localization, we need to locate the target in an ordinary room in which there are some tables, chairs, and other obstacles. RFID reference tags will be posted on the ground according to certain rules which ensure that there are at least two tags that will respond when the reader is located in tag array as shown in Fig. 9.8. When the target carrying the reader goes into the room, there will be a different sequence of tags responding at each moment and the target is positioned based on it.

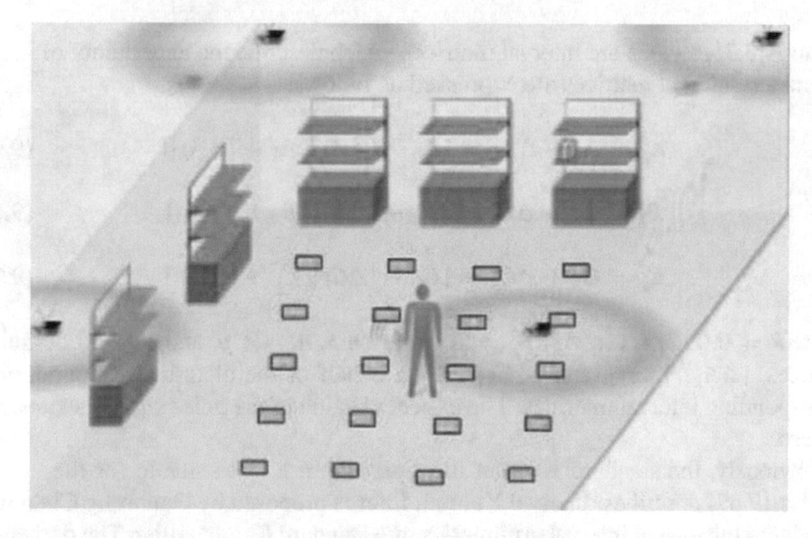

Fig. 9.8 Arrangement of RFID tags in the room

It should be noted that the positioning device is in general ultra-high frequency (UHF) reader with omnidirectional antenna and passive tags in the market. The reader is unable to measure tag's signal intensity, that is, we can only know which tag has responded and cannot calculate the distance between the reader and response tag in the positioning process. Besides, the interference from the environment and deficiencies of hardware itself will affect the positioning result. In theory the antenna is isotropic antenna, but in practice the detection range in each direction is different [4]. Obstructions will affect signal transmission, which may make the reader to receive the wrong response sequence. All these will lead to increasing positioning error. Hardware performance is limited, and environment impact cannot be changed; hence, we can only focus on the positioning algorithm and make use of mathematical optimization to improve the positioning accuracy. According to the description of positioning problem, the estimation result is an interval. It should be considered that how to deal with interval data in positioning algorithm.

9.2.1.2 Interval Kalman Filter

Due to external noise interference, the positioning results will have the fluctuation which can be regarded as interval data. The basic rules of interval arithmetic should be understood and interval Kalman filter algorithm is introduced to deal with interval data [5, 6].

Consider the following class of uncertain linear interval system [7]:

$$\begin{cases} x_{k+1} = A_k^I x_k + B_k^I \zeta_k \\ y_k = C_k^I x_k + \eta_k \end{cases} \tag{9.67}$$

where A_k^I, B_k^I, and C_k^I are interval matrices which describe the uncertainty of model parameter. Interval matrices are expressed as follows:

$$A_k^I = A_k + \Delta A_k = [A_k - |\Delta A_k|, A_k + |\Delta A_k|] \tag{9.68}$$

$$B_k^I = B_k + \Delta B_k = [B_k - |\Delta B_k|, B_k + |\Delta B_k|] \tag{9.69}$$

$$C_k^I = C_k + \Delta C_k = [C_k - |\Delta C_k|, C_k + |\Delta C_k|] \tag{9.70}$$

where $k = 0, 1, 2, 3, \ldots$, A_k^I, B_k^I, and C_k^I are $n \times n$, $n \times p$, and $q \times p$ coefficient matrices, $|\Delta A_k|$, $|\Delta B_k|$, and $|\Delta C_k|$ are on behalf of the disturbance boundary of corresponding interval matrices. The expected value of the noise signal is known to be zero.

Obviously, the standard Kalman filter algorithm is not suitable for the system that Eq. (9.67) describes. Interval Kalman filter is proposed by Guanrong Chen who introduces the idea of interval arithmetics into standard Kalman filter. The derivation process of interval Kalman filter is similar with standard Kalman filter and both have the same part of nature [7, 8].

9.2.2 Mathematical Model and Positioning Algorithm

As described in the previous paragraph, the target can be located in a certain area according to response tags if the tags are laid as the way of rectangular as shown in Fig. 9.9. The coordinate we obtain is in fact a range; however, it is noisy. We need to deal with a series of interval data and filter out noise to optimize positioning results. This section tries to establish mathematical model of positioning process and introduce the interval Kalman filter algorithm into estimating the location coordinates.

Suppose at time k, the reader's abscissa, the speed on abscissa direction, and the acceleration on abscissa direction are, respectively, denoted as $x(k)$, $\dot{x}(k)$, $\ddot{x}(k)$, the vertical axis, and the ordinate, the speed on ordinate direction, and the acceleration on ordinate direction are, respectively, denoted as $y(k)$, $\dot{y}(k)$, $\ddot{y}(k)$.

From kinematic state equation, we can obtain

$$\begin{bmatrix} x(k+1) \\ \dot{x}(k+1) \\ \ddot{x}(k+1) \\ y(k+1) \\ \dot{y}(k+1) \\ \ddot{y}(k+1) \end{bmatrix} = \begin{bmatrix} 1 & t & \frac{t^2}{2} & 0 & 0 & 0 \\ 0 & 1 & t & 0 & 0 & 0 \\ 0 & 0 & 1 & 0 & 0 & 0 \\ 0 & 0 & 0 & 1 & t & \frac{t^2}{2} \\ 0 & 0 & 0 & 0 & 1 & t \\ 0 & 0 & 0 & 0 & 0 & 1 \end{bmatrix} \begin{bmatrix} x(k) \\ \dot{x}(k) \\ \ddot{x}(k) \\ y(k) \\ \dot{y}(k) \\ \ddot{y}(k) \end{bmatrix} + V(k) \tag{9.71}$$

According to Eq. (9.67), define

$$A_k^I = \begin{bmatrix} 1 & t & \frac{t^2}{2} & 0 & 0 & 0 \\ 0 & 1 & t & 0 & 0 & 0 \\ 0 & 0 & 1 & 0 & 0 & 0 \\ 0 & 0 & 0 & 1 & t & \frac{t^2}{2} \\ 0 & 0 & 0 & 0 & 1 & t \\ 0 & 0 & 0 & 0 & 0 & 1 \end{bmatrix} \tag{9.72}$$

$$V(k) = B_k^I \zeta_k \tag{9.73}$$

We take B_k^I as unit matrix and obtain the system equation; then the observation equation is given based on the relationship between reader coordinate and response

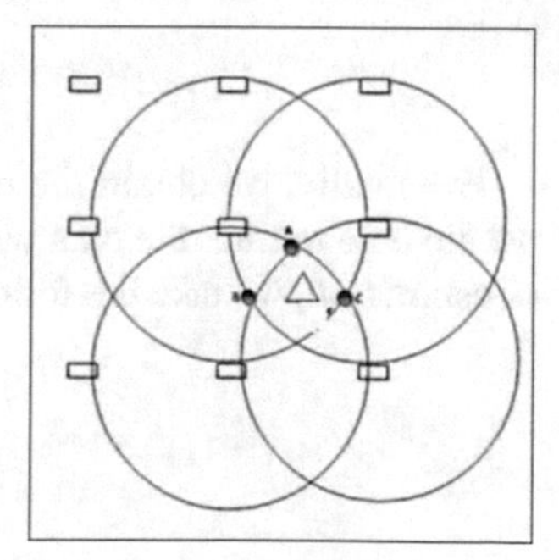

Fig. 9.9 The intersection of response tag sequence

tag coordinate. There is a set of response tag sequence at each time, and the reader can be positioned in an area according to the coordinates of response tags. We can calculate the boundary points through the intersection operation. Suppose the number of all tags is n and reading distance is r. At time k, there are m tags responding to the reader and the ith response tag coordinate is (x_i, y_i); the boundary point coordinate is (x_b, y_b).

$$\begin{cases} (x_b - x_1)^2 + (y_b - y_1)^2 = r^2 \\ (x_b - x_2)^2 + (y_b - y_2)^2 = r^2 \\ \vdots \\ (x_b - x_m)^2 + (y_b - y_m)^2 = r^2 \\ (x_b - x_{m+1})^2 + (y_b - y_{m+1})^2 > r^2 \\ \vdots \\ (x_b - x_n)^2 + (y_b - y_n)^2 > r^2 \end{cases} \tag{9.74}$$

Through the above equation, Fig. 9.9 can be calculated. Here we suppose that the average values of boundary point coordinates are regarded as the center of the region, approximately valid in most situations. Suppose that the center coordinate is (x_c, y_c). Then, the observation equation can be expressed as follows:

$$\begin{bmatrix} x_c(k) \\ y_c(k) \end{bmatrix} = \begin{bmatrix} 1+s & 0 & 0 & 0 & 0 & 0 \\ 0 & 0 & 0 & 1+s & 0 & 0 \end{bmatrix} \begin{bmatrix} x(k) \\ \dot{x}(k) \\ \ddot{x}(k) \\ y(k) \\ \dot{y}(k) \\ \ddot{y}(k) \end{bmatrix} + U(k) \tag{9.75}$$

Among the above equation, $(x_c(k), y_c(k))$ is regional center coordinate, and s is the adjustment range which represents the approximation degree between reader and center point. Define

$$C_k^I = \begin{bmatrix} 1+s & 0 & 0 & 0 & 0 & 0 \\ 0 & 0 & 0 & 1+s & 0 & 0 \end{bmatrix} \tag{9.76}$$

$$U(k) = \eta_k \tag{9.77}$$

Eventually, we obtain the basic equations of interval Kalman filtering and then we are able to filter the data with the interval Kalman filter. To ensure unbiasedness of estimation, we take the following initial values:

$$\hat{x}_0^I = E\left(x_0^I\right) \tag{9.78}$$

$$P_0^I = \mathrm{Cov}\left(x_0^I\right) \tag{9.79}$$

with the following one-step prediction covariance matrix M_k^I and interval Kalman gain matrix G_{k+1}^I:

$$M_k^I = A_k^I P_k^I \left[A_k^I\right]^T + Q_k^I \tag{9.80}$$

$$G_{k+1}^I = M_k^I \left[C_{k+1}^I\right] \left[C_{k+1}^I M_k^I \left[C_{k+1}^I\right]^T + R_{k+1}^I\right]^{-1} \tag{9.81}$$

Then the interval states and covariance matrices are updated by

$$\hat{x}_{k+1}^I = A_k^I \hat{x}_k^I + G_{k+1}^I \left[y_{k+1}^I - C_{k+1}^I A_k^I \hat{x}_k^I\right] \tag{9.82}$$

$$\begin{aligned} P_{k+1}^I = &\left[I - G_{k+1}^I C_{k+1}^I\right] M_k^I \left[I - G_{k+1}^I C_{k+1}^I\right]^T \\ &+ \left[C_{k+1}^I\right] R_{k+1}^I \left[C_{k+1}^I\right]^T \end{aligned} \tag{9.83}$$

This shows that interval Kalman filter is similar with standard Kalman filter, except that the interval matrix operation in filter process. In Eq. (9.81) the calculation of gain matrix G_{k+1}^I contains interval matrix inversion which takes a long time and is not conducive to real-time filtering. In order to improve computational efficiency, we simplify the inverse process and substitute conventional matrix $[C_{k+1}M_k[C_{k+1}]^T + \Delta Rk + 1]^{-1}$ for interval matrix $[C_{k+1}^I M_k^I [C_{k+1}^I]^T + R_{k+1}^I]^{-1}$ [7].

9.2.3 Simulation Studies

In the simulation, we use the 920 MHz UHF reader with an omnidirectional antenna and rectangular passive tags with side length 2 cm. Now we make a simulation experiment of positioning moving object in a 5 m × 5 m indoor room. As shown in Fig. 9.9, the tags are arranged 20 cm apart, and every tag's coordinate is known a priori. The target is equipped with the reader and its reading range distance is set as 25 cm. Besides, noise signal will be added to simulate the influence of environment change and antenna performance.

We conduct extensive simulation experiments according to the above assumptions. The average value of upper and lower bounds represents estimated coordinates in

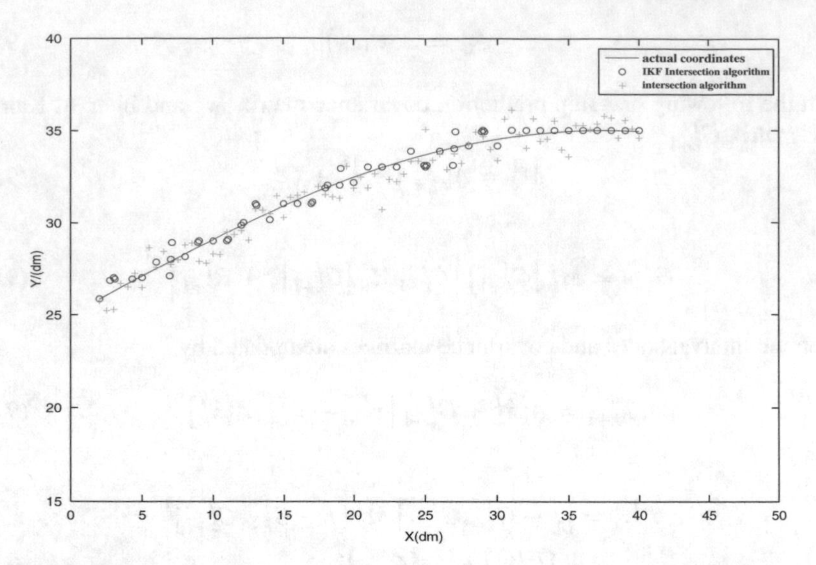

Fig. 9.10 The comparison chart of position tracking

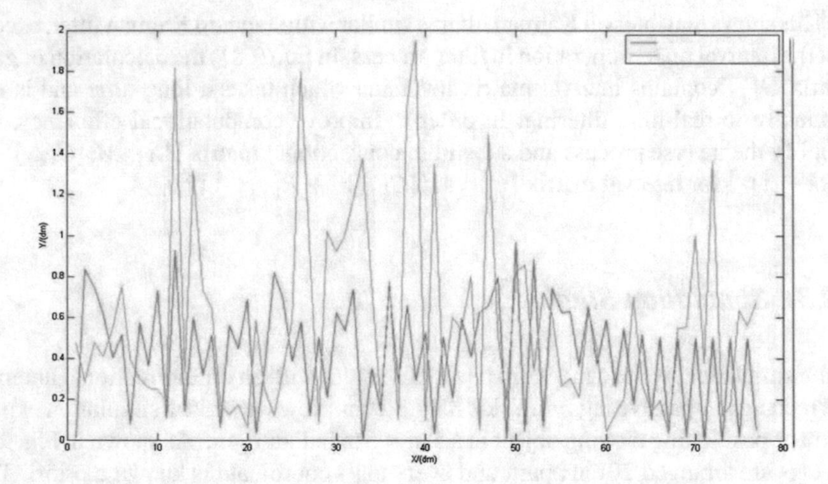

Fig. 9.11 The comparison chart of positioning error

interval Kalman filter algorithm. The position tracking is shown in Fig. 9.10, and the positioning error is plotted in Fig. 9.11.

Blue circle represents the position estimation after the interval Kalman filter algorithm, and the green is generated by intersection. Compared with the green point, the blue track is closer to actual coordinates curve, and the positioning error becomes smaller and stable with time goes on. Then we change object's motion path, and correspondingly, the simulation result is shown in Fig. 9.12. Obviously, comparing Fig. 9.10 with Fig. 9.12, the positioning effect is better when object is in linear motion.

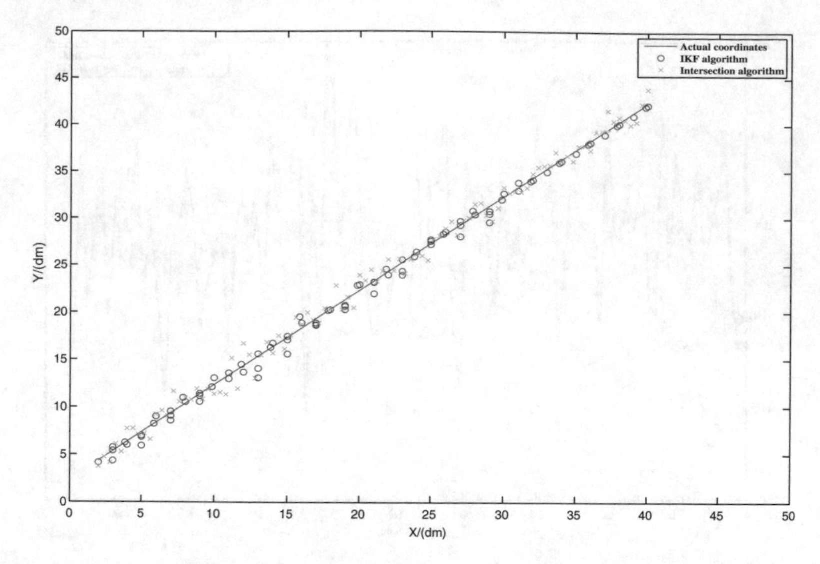

Fig. 9.12 Position tracking at linear motion

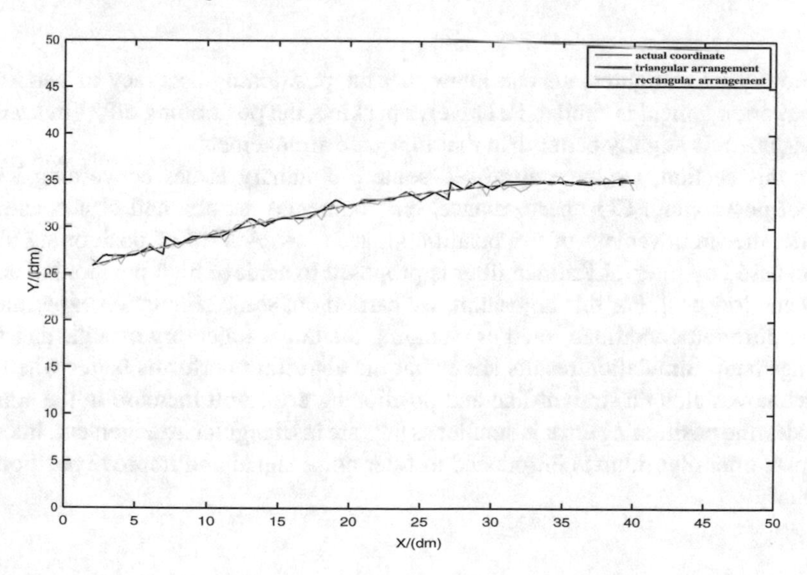

Fig. 9.13 Position tracking in different arrangements

Different tag arrangements will also affect the positioning accuracy. Two arrangements are adopted for doing simulation experiments, among which one is the arrangement of square, and another is the arrangement of equilateral triangle, where the distance between adjacent tags is 20 cm. The simulation result is shown in Figs. 9.13 and 9.14.

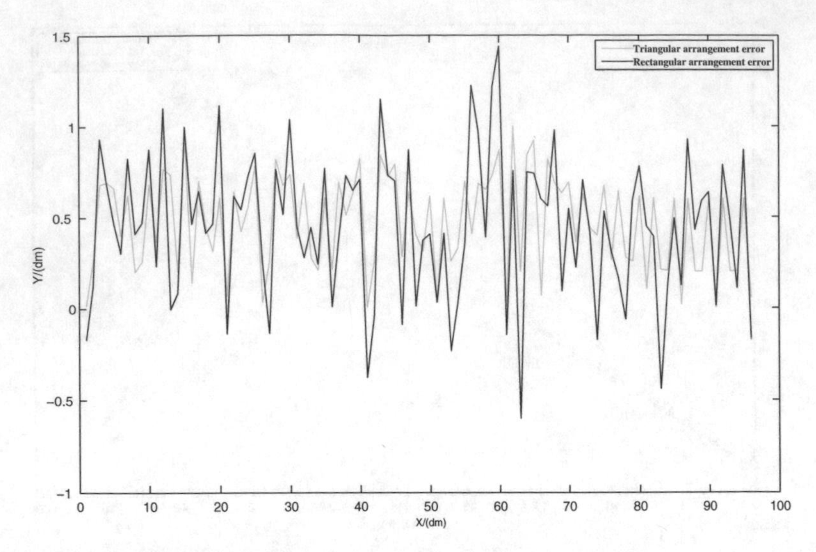

Fig. 9.14 Positioning error in different arrangements

From the two figures, we can know that the positioning accuracy of two kinds of tags arrangement is similar. Relatively speaking, the positioning effect in triangle arrangement is slightly better than that in square arrangement.

In this section, we have discussed some preliminary issues concerning RFID indoor positioning. Chip performance, environmental factors, and cost constraint restrict the improvement of the positioning accuracy. A kind of positioning algorithm based on interval Kalman filter is proposed to achieve high positioning accuracy and low cost. For this algorithm, we carried out some simulation experiments under different conditions, such as changing the target trajectory or adjusting tags arrangement. Simulation results show that the algorithm performs better when the target moves along a straight line and positioning error will increase in the corner. Besides, the positioning error is smaller as tags are in triangular arrangement. Interval Kalman filter algorithm is introduced to filter noise signal and improve positioning accuracy.

9.3 Particle Filter-Based Simultaneous Localization and Mapping (PF-SLAM)

In the last two sections, we investigated the localization with the WSNs and RFID. Information of the environment map is useful to the localization. When robot is situated in unknown environment, it is possible to determine its localization while building a consistent map of the environment by observations of sensors at the same time. This is called the simultaneous localization and mapping (SLAM) techniques.

In this section, we will investigate the indoor SLAM problem based on a particle filter with a low-cost laser LIDAR [9]. The robot could realize simultaneous localization and mapping using the data from a low-cost laser scanner RPLIDAR in an unknown environment.

9.3.1 Model of Particle Filter (PF) SLAM Using Landmarks

9.3.1.1 Coordinate System of SLAM

The coordinate system of the global map is the global coordinate system (global map), and it is spanned by axes x^r and y^r. Different from the local coordinate system (robot-centered coordinate, local map), the global coordinate system is spanned by axes x^i and y^i. We can describe landmark z in both coordinate systems: vector z_m^r describes the m-th landmark in the global coordinate system while vector z_m^i describes the m-th landmark in the local coordinate system.

In the global coordinate system, the position of the robot can be described by the location (x^r, y^r) and the orientation θ^r of the robot, where θ^r is the angle between the global map coordinate axis X^r and the local map coordinate axis X^i.

The global map is unknown at first, and thus the first local map is utilized as the initial global map. The initial position of the robot is (0, 0, 0) in global coordinate system. The x^r axis is the direction angle of the robot. After the first scan, the present local map is generated by the latest scan data. When the robot moves to a new position, the position of the robot is (x^r, y^r), and the x^r axis is the orientation of the robot. The relationship between local coordinate system and global coordinate system is shown in Fig. 9.15.

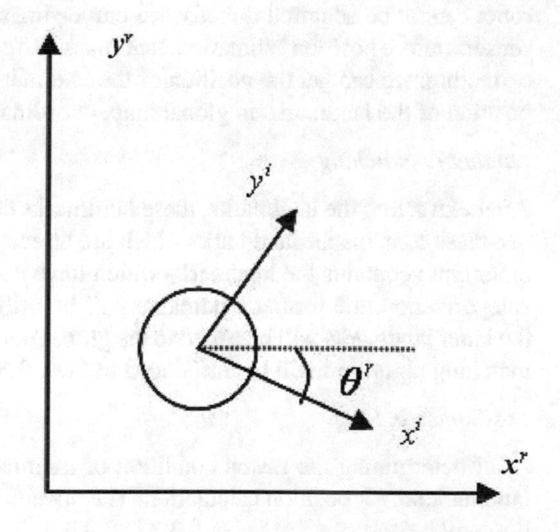

Fig. 9.15 The relationship between local coordinate system and global coordinate system. Reprinted from Ref. [9], with permission of Springer

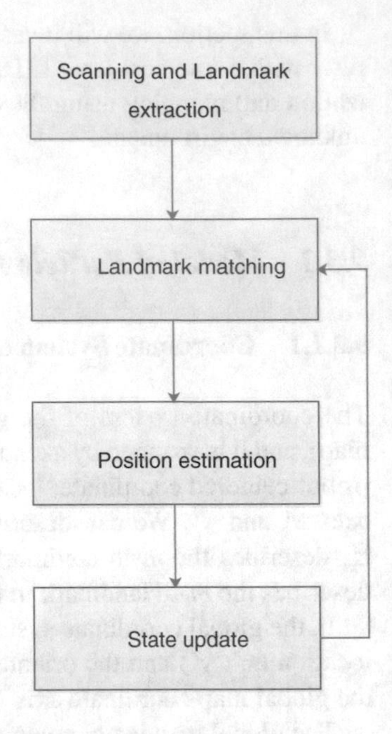

Fig. 9.16 The structure of the PF-SLAM using landmarks. Reprinted from Ref. [9], with permission of Springer

9.3.1.2 Structure of PF-SLAM Using Landmarks

The structure of PF-SLAM using landmarks is shown in Fig. 9.16, which illustrates the following main steps of SLAM:

Scanning and landmark extraction

Scanning and landmark extraction is an important step in SLAM. In this step, the sensor returns a serial of data from the environment. Although the position of the robot cannot be acquired directly, we can estimate it from the data returned by the sensor. Before position estimation, landmark extracting is necessary. After landmark extracting, we can get the position of the landmarks in the local map. To obtain the position of the landmarks in global map, coordinate conversion is inevitable.

Landmark matching

After extracting the landmarks, these landmarks can be divided into two categories: one class contains the landmarks which are already stored in the global map, and the other class contains the landmarks which have not been observed previously. After categorization, the former landmarks will be utilized in position estimation, while the latter landmarks will be added to the global map. A particle filter-based landmark matching algorithm will be introduced in Sect. 9.3.2.

Position estimation

After determining the match condition of landmarks, the next step is to select the landmark set for position calculation. The specific algorithm of the position estimation will be described in Sects. 9.3.3 and 9.3.4.

State update

Update the state of the robot and the global map.

9.3.2 Particle Filter Matching Algorithm

The most common matching algorithm in SLAM is nearest-neighbor (NN) matching [10]. The criterion of NN is based on the position deviation between the landmarks observed and the landmarks stored in the global map. If the position deviation is lower than a certain threshold, for example 0.03 m, the landmark is matched with the landmark stored in the global map. We use the term *landmark set* to refer to the set of matched landmarks. However, the observation position of these landmarks relies on the estimation position of the robot. If the estimation position of the robot is wildly inaccurate, the NN algorithm will not work well.

To solve this problem, some algorithms [11] are introduced to find the landmark set by comparing the relative position of the landmarks observed and the landmarks stored in the map. These algorithms can work under the assumed conditions. However, we would like to highlight that both the deviation and the relative position of landmarks are useful in landmark matching. In order to use both of these criteria mentioned above, the particle filter matching algorithm is introduced.

Particle filter is introduced to find these landmark set in this section. Particle filter is a filtering method based on Monte Carlo Bias and recursive estimation. The basic idea of particle filter algorithm is essentially a set of online posterior density estimation algorithms that estimate the posterior density of the state space by directly implementing the Bayesian recursion equations with a set of particles to represent the posterior density.

In particle filter, the prior conditional probability is $p(\mathbf{x}_0)$, where $\{x^i_{0:k}, \omega^i_k\}^{N_s}_{i=1}$ is an approximation to the posterior probability distribution $\mathbf{x}_k$, $p(\mathbf{x}_{0:k}|\mathbf{z}_{1:k})$. Here $\{x^i_{0:k}, i = 0, 1, \ldots, N_s\}$ is the particle set, and $\{\omega^i_k, i = 0, 1, \ldots, N_s\}$ are the weights of particle sets. And $x_{0:k} = \{x_j, j = 0, 1, \ldots, k\}$ is the sequence of states from time 0 to time k.

For completeness, the updating process of the posterior probability can be summarized as follows [12, 13].

- The importance weight can be recursively updated by

$$\bar{\omega}^i_k = \omega^i_{k-1} p(\mathbf{z}_k|\mathbf{x}^i_k) \tag{9.84}$$

- After normalizing the weight ω^i_k by Eq. (9.85), we can get the weight ω^i_k for posterior probability calculation.

$$\omega^i_k = \bar{\omega}^i_k / \sum_{i=1}^{N_s} \bar{\omega}^i_k \tag{9.85}$$

- The posterior probability $p(\mathbf{x}_k|\mathbf{z}_{1:k})$ can be expressed as Eq. (9.86):

$$p(\mathbf{x}_k|\mathbf{z}_{1:k}) = \sum_{i=1}^{N_s} \omega_k^i \delta(\mathbf{x}_k - \mathbf{x}_k^i) \tag{9.86}$$

If $N_s \rightarrow \infty$, Eq. (9.86) is approximating to the true posterior probability $p(\mathbf{x}_k|\mathbf{z}_{1:k})$.

It is hypothesized that the speed of the robot is slow enough and the position changes between each scan are under the range of the distribution of these particles. Particle filter is adopted in landmark matching, under this hypothesis. The position of the robot calculated in last scan is the prior information of the particle filter, which can generate the distribution $\{x_{0:k}^i, i = 0, 1, \ldots, N_s\}$ and weights $\{\omega_{0:k}^i, i = 0, 1, \ldots, N_s\}$ of these particles. The posterior information $p(\mathbf{x}_k|\mathbf{z}_{1:k}) = \sum_{i=1}^{N_s} \omega_k^i \delta(\mathbf{x}_k - \mathbf{x}_k^i)$ of these particles are calculated by these position deviations of landmark set; the less the value of the position deviation, the more the weight of the particle. After calculating the weights of these particles, we can choose the matching condition of the particle with the highest weight.

9.3.3 Landmark Set Selection Method

Different from the selection algorithm proposed by [14] which takes the angle between observations as selection criterion, the improved method takes the quality of observation into selection criterion too.

In this section, two key elements are determined in landmark set selection. One element is the angle between observations, while the other is the quality of observation. In the associated method introduced in this section, these two elements mentioned above are added to the selection criterion together.

It is assumed that the universal set of all landmarks is denoted by $land(m)$, where m is the number of landmark set. This section mainly focuses on the situation of $m \geq 2$, which is the common case in indoor environment. One subset of $land(m)$ utilized in localization is $sets(n)$, and the number of these landmark set needed in localization is 2. Thus the number of landmark set in $sets(n)$ is 2 and there are totally C_m^2 kinds of $sets(n)$. The most credible $sets(n)$ is chosen by the selection criterion for localization.

The angle between these two observations in $sets(n)$ is one issue in the selection criterion. When the angle between these landmarks is approaching 90°, the influence of the position error is smaller. And if the angle between these landmarks is around 180°, the distance between two possible positions of the robot will become very small, which may cause some difficulties in localization.

The quality of the observation is another issue in the selection criterion. It is estimated by the position deviation of the landmark set. The greater the value of the deviation is, the fewer the quality of the observation is.

Both these two issues are considered in the selection method. The formula used in landmark set selection is introduced as follows.

Variable β is defined as the weight of the quality element, which satisfies Eq. (9.87):

$$0 < \beta < 1 \tag{9.87}$$

The credibility of the landmark set can be calculated by Eq. (9.88), where e stands for the quality of the observation, a stands for the angle coefficient of $sets(n)$, and r represents the credibility of $sets(n)$.

$$\beta e + (1 - \beta)a = r \tag{9.88}$$

After calculating the credibility of each landmark in the landmark set, the $sets(n)$ with the highest r is chosen for position calculation.

9.3.4 Advanced Position Calculation Method

This subsection is mainly about how to apply advanced triangulation technique in position calculation.

From the conclusion of [15], if the visual angle between two landmarks Z_1^r and Z_2^r measured at an unknown position p and the distance between the landmarks are known, then position p lies on an arc of a circle spanned by landmarks Z_1^r and Z_2^r, shown in Fig. 9.17.

As to the laser LIDAR, the distance between p to Z_1^r and p to Z_2^r are known, and the point p should satisfy two range constraints, as shown in Fig. 9.18. Hence, the

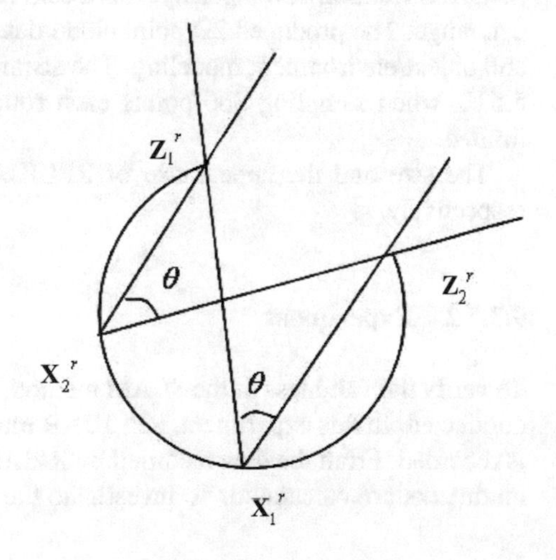

Fig. 9.17 Triangulation approach. Reprinted from Ref. [9], with permission of Springer

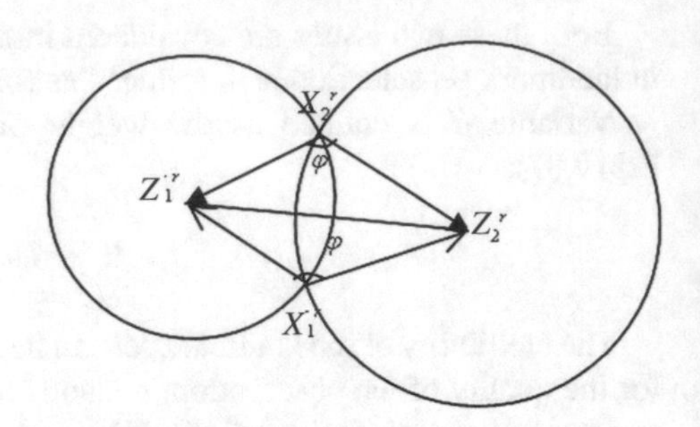

Fig. 9.18 Localization with range sensor. Reprinted from Ref. [9], with permission of Springer

intersections of these two circles determine possible positions of p. Since two circles can usually have two intersection points, a third landmark is needed to determine the position p uniquely. With the idea introduced in [16], by making full use of temporal and spatial data provided by the laser LIDAR, the number of landmarks needed can be decreased to 2.

Based on the hypothesis given in Sect. 11.2.2, the position changing of the robot is under a certain range, and the selection method mentioned in Sect. 9.3.3 assured that the distance between these two intersections is large enough. From the position calculated in the last scan and the position rage of the robot, we can select the right position from these two intersections.

9.3.5 *Experiment Study*

9.3.5.1 RPLIDAR

The sensor used in the experiment is RPLIDAR, a low-cost 360° 2D laser scanner (LIDAR) solution developed by RoboPeak. RPLIDAR can perform 360° scan within 6 m range. The produced 2D point cloud data can be used in mapping, localization, and object/environment modeling. The scanning frequency of RPLIDAR can reach 5.5 Hz when sampling 360 points each round. Hence, the accuracy of its data is limited.

The size and the appearance of RPLIDAR are shown in Figs. 9.19 and 9.20, respectively.

9.3.5.2 Experiment

To verify the validness of the SLAM method, one simple yet nontrivial experiment is conducted. In this experiment, RPLIDAR moves around in a box and a serial of data is recorded. From the data recorded by RPLIDAR, the positions of the robot and the landmarks are calculated. To investigate the performance of the proposed method,

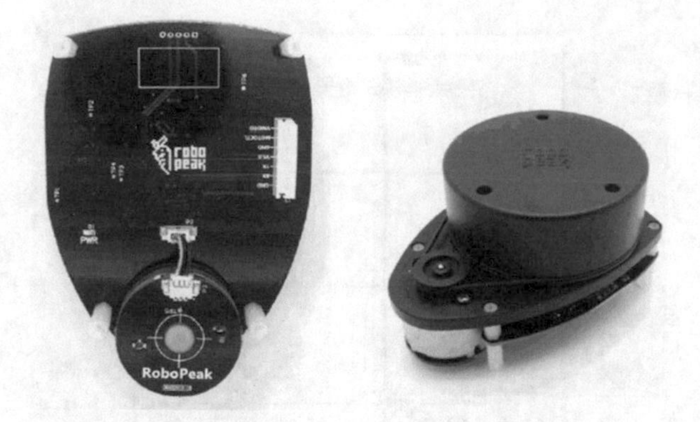

Fig. 9.19 The appearance of RPLIDAR. Reprinted from Ref. [9], with permission of Springer

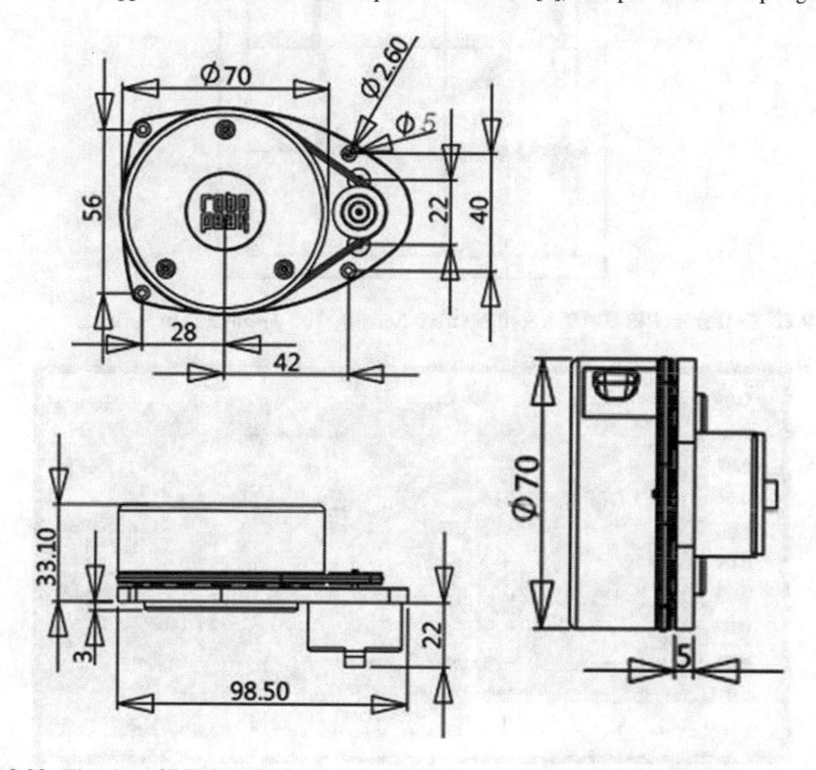

Fig. 9.20 The size of RPLIDAR. Reprinted from Ref. [9], with permission of Springer

in the following experiment we compare our approach with the previous approach, which is detailed presented in the literature [14].

The experiment result is shown in Fig. 9.21. The region in this experiment is $0.4\,\text{m} \times 1.0\,\text{m}$ large, where the stars represent the feature points. The points in the

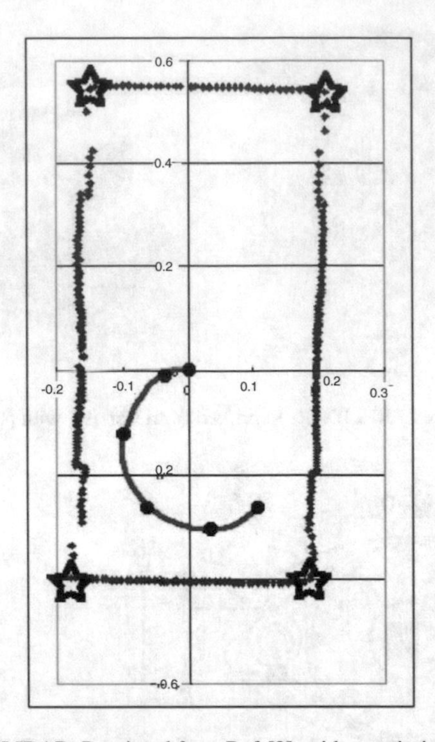

Fig. 9.21 Data from RPLIDAR. Reprinted from Ref. [9], with permission of Springer

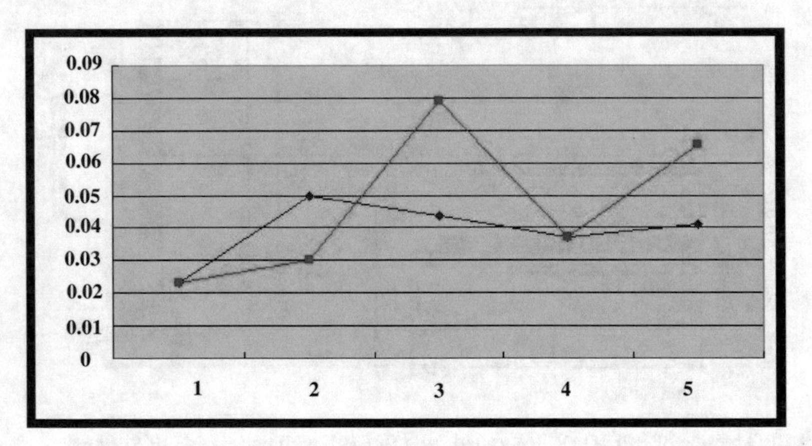

Fig. 9.22 Position errors of the landmarks. Reprinted from Ref. [9], with permission of Springer

picture are the estimation result of the robot and the red line is the approximate path of the robot.

The position error of the improved method and the previous approach are shown in Fig. 9.22, where the blue line represents the result of the improved method and the pink line represents the result of the previous approach. It is easy to find that the estimation result of the improved method has a better tolerance.

The algorithm adopted in the section improves the tolerance of the calculation result, by advanced landmark set selection method and position calculation method. Experiment result shows that the method has a better tolerance than the previous approach.

In this section, a novel approach of PF-SLAM using landmark is proposed. After introducing the structure of the system, a novel selection method has been described and the proposed method takes the quality of observation into consideration too while previous approach only take the angle between the landmarks into consideration. And one experiment has been conducted to test whether the proposed new algorithm can work well with the low-cost radar sensor RPLIDAR, which has been proved to be useful in the robot navigation and the efficiency of the algorithm introduced in this contribution has been verified from the experiment result.

9.4 Integrated INS/VMS Navigation System

Global positioning system (GPS) can provide accurate velocity and position information for land vehicle navigation. But GPS cannot achieve continuous localization in urban canyons, tunnels, and other environments where satellite signals are blocked. Therefore, GPS is usually combined with inertial navigation system (INS) and vehicle motion sensor (VMS) as both systems are complementary.

In this section, we investigate the problems of VMS scale factor calibration and INS-to-VBF alignment, and present two INS/VMS integration architectures [17].

9.4.1 Introduction of INS/VMS Navigation System

In GPS/INS integrated systems, GPS information can be used to correct INS errors and improve the long-term accuracy of INS. The INS providing position information during GPS outages can assist GPS signal reacquisition after an outage and reduce the search domain required for detecting and correcting GPS cycle slips [18, 19]. For brief GPS outage periods, INS can navigate with acceptable accuracy through the outage. For long outages, as inertial navigation errors are rapidly growing, certain mechanisms should be applied to improve the navigation accuracy, many of which have been proposed in the last decade [18–24]. Generally, these methods can be classified into two categories: (1) filtering methods and (2) multisensor integrated navigation methods.

The filtering methods improve the accuracy and reliability of GPS/INS integrated navigation system mainly based on filtering and system identification by utilizing appropriate algorithms. The Kalman filter has been widely adopted as the standard optimal estimation tool for GPS/INS integration scheme [22, 23].

Among the various land vehicle navigation devices, VMS is getting widely used due to its cost-effective and conveniently deployed characteristics [25]. The VMS can detect the rotation of vehicle's drive shaft and generate an electrical pulse stream. Each pulse represents a fixed number of rotations of vehicle's drive shaft. The navigation software multiplies the number of pulses input from the VMS by a scale factor so as to convert the number of drive shaft revolutions into the traveled distance. Due to the advantages of VMS, VMS has been widely integrated with the use of GPS and INS for the purpose of navigation, and their integration can usually be utilized in two navigation modes [26, 27]: GPS/INS mode and INS/VMS mode. When GPS signals are available, the GPS/INS mode works, in which GPS not only provides the navigation results but also calibrates the VMS errors, such as scale factor error and INS-to-VBF misalignment angles. When GPS signals are unavailable, INS/VMS mode is switched on, in which the calibration of VMS scale factor and matching of INS body axis with VBF are critical for the following reasons.

(i) **Calibration of VMS scale factor**. The operation of VMS requires pinpoint scale factor to assure navigation accuracy. In real life, the scale factors of VMS are affected by skidding, temperature, inflation and abrasion of vehicle tires, road surface condition, etc. In other words, the scale factor of VMS is highly environmental-dependent, and thus is hard to determine.
(ii) **INS-to-VBF alignment**. Besides the calibration of VMS scale factor, the alignment of INS body frame with VBF is also difficult. Unlike in the case of a flight vehicle, the motion of a land vehicle on a surface of the Earth is governed by two nonholonomic constraints [28], which as well as VMS outputs can effectively suppress position drift of INS when the INS body frame is well-aligned with the VBF [29, 30]. However, the INS body frame is often misaligned with the vehicle frame because of installation mistakes and vehicle deformation in real systems.

The difficulties of VMS scale factor determination and INS-to-VBF alignment undermine the performance of INS/VMS integration, which is working when long-term GPS signals are unavailable. Therefore, in this contribution, the problems of VMS scale factor calibration and INS-to-VBF alignment are investigated. According to the degree to which the VMS scale factor calibration and INS-to-VBF alignment alter the internal workings of each subsystem (i.e., INS or VMS) in its stand-alone configuration, we classify the INS/VMS integration architectures into two categories: *loosely coupled* and *tightly coupled* modes.

(i) The loosely coupled implementation treats the standard outputs from INS and VMS as separate measurements. To provide high-quality VMS data for such implementation, the orientation between the INS body frame and VBF as well as the VMS scale factor is determined through VMS error analysis in advance.
(ii) The tightly coupled implementation adopts a single Kalman filter to correct the errors of INS and VMS. With this scheme, the orientation between the INS body frame and the vehicle chassis is automatically estimated using a 19-dimensional state system model where both the inertial errors and the VMS errors are coupled together through the measurement equations.

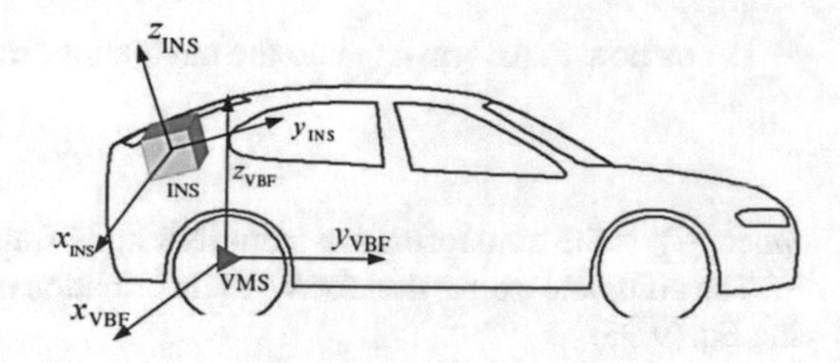

Fig. 9.23 Misaligned INS and vehicle axes

9.4.2 Analysis of VMS Errors

This section analyzes the VMS errors which are caused by VMS scale factor error, and axes mismatch between the INS body frame and the VBF (Fig. 9.23).

As illustrated in Fig. 9.23, the VBF (b' frame) locates at the center of the rear axle of the vehicle with the pitch axis (x_{VBF}) pointing out the right-hand side, the roll axis (y_{VBF}) pointing forward, and the yaw axis (z_{VBF}) pointing upward, all with respect to the vehicle itself. The INS is rigidly attached to the vehicle and the body frame (b frame) is implicitly defined by configuration of the three gyroscopes/accelerometers. The geographic frame (n frame; E, N, U axes) is chosen as the navigation reference frame, and the conventional Earth-centered Earth-fixed coordinate is taken as the Earth frame (e frame).

The transformation from b frame to b' frame can be carried out through three successive rotations about different axes. The first rotation is about axis-z_{INS} by ϕ, the second rotation is about the new x_{INS}-axis by θ, and the final rotation is about the new y_{INS}-axis by γ. The direction cosine matrix for this transformation is

$$C_b^{b'} = \begin{bmatrix} c\gamma c\phi - s\gamma s\phi s\theta & c\gamma s\phi + s\gamma c\phi s\theta & -s\gamma c\theta \\ -s\phi c\theta & c\phi c\theta & s\theta \\ s\gamma c\phi + c\gamma s\phi s\theta & s\gamma s\phi - c\gamma c\phi s\theta & c\gamma c\theta \end{bmatrix} \tag{9.89}$$

herein s and c are short for circular functions sin and cos, respectively.

With nonholonomic constraints, the incremental VMS output vector is defined by

$$S_{od}^{b'} = \left[0, s_{od}, 0\right]^T \tag{9.90}$$

where s_{od} is the VMS-derived position change.

Expressing the VMS output vector in the INS body frame yields

$$S_{od}^{b} = \left(C_b^{b'}\right)^T S_{od}^{b'} = M s_{od} \tag{9.91}$$

with

$$M = \left[-s\phi c\theta, c\phi c\theta, s\theta\right]^T \tag{9.92}$$

Let us now transform S_{od}^{b} into the navigation frame

$$S_{od}^{n} = C_{b}^{n} M s_{od} \tag{9.93}$$

where C_{b}^{n} is the transformation from INS body frame to navigation frame.

The complete expression for VMS information error is found by directly perturbing Eq. (9.93)

$$\begin{aligned} \delta S_{od}^{n} &= \delta C_{b}^{n} \hat{M} \hat{s}_{od} + \hat{C}_{b}^{n} \delta M \hat{s}_{od} + \hat{C}_{b}^{n} \hat{M} \delta s_{od} & (9.94) \\ &= \left(C_{b}^{n} \hat{M} \hat{s}_{od} \right) \times \zeta + \left(\hat{C}_{b}^{n} N_{mis} \hat{s}_{od} \right) \begin{bmatrix} \delta\phi \\ \delta\theta \end{bmatrix} + \left(\hat{C}_{b}^{n} \hat{M} \hat{s}_{od} \right) \delta k & (9.95) \end{aligned}$$

where symbol $\hat{\cdot}$ is the designator for a system calculated quantity, hence containing error. The quantity without the $\hat{}$ designation is by definition error free.

9.4.3 Loosely Coupled INS/VMS

In this subsection, based on Eq. (9.95), the equations of VMS errors (i.e., $\delta\phi$, $\delta\theta$, and δk) are explicitly presented, which inspire a geometrical calibration method which is proposed by Yan [31], and then the loosely coupled INS/VMS architecture will be introduced with some discussions on its implementation.

9.4.3.1 Calibration of VMS Errors

The calibration is required to be implemented on a planar surface; hence, the vehicle pitch and roll can be assumed to be zeros. The INS horizontal alignment errors (i.e., $\zeta(1)$ and $\zeta(2)$), and the initial Euler angles $\hat{\phi}$ and $\hat{\theta}$ are assumed to be small. With the above assumptions, the VMS error Eq. (9.95) can be simplified as

$$\begin{aligned} \delta S_{od}^{n} \approx\ & (C_{b}^{n} \hat{M} \hat{s}_{od}) \times \begin{bmatrix} 0 \\ 0 \\ \zeta(3) \end{bmatrix} + (C_{b}^{n} \hat{M} \hat{s}_{od}) \times \begin{bmatrix} 0 \\ 0 \\ \delta\phi \end{bmatrix} \\ & + \hat{s}_{od} \begin{bmatrix} 0 \\ 0 \\ \delta\theta \end{bmatrix} + (\hat{C}_{b}^{n} \hat{M} \hat{s}_{od}) \delta k \\ =\ & (C_{b}^{n} \hat{M} \hat{s}_{od}) \times \begin{bmatrix} 0 \\ 0 \\ \zeta(3) + \delta\phi \end{bmatrix} + \hat{s}_{od} \begin{bmatrix} 0 \\ 0 \\ \delta\theta \end{bmatrix} \\ & + (\hat{C}_{b}^{n} \hat{M} \hat{s}_{od}) \delta k \end{aligned} \tag{9.96}$$

Substituting Eq. (9.93) into Eq. (9.96) yields

$$\delta S_{od}^n = \hat{S}_{od}^n \times \begin{bmatrix} 0 \\ 0 \\ \zeta(3) + \delta\phi \end{bmatrix} + \hat{s}_{od} \begin{bmatrix} 0 \\ 0 \\ \delta\theta \end{bmatrix} + \hat{S}_{od}^n \delta k \tag{9.97}$$

According to Eq. (9.97), the horizontal and vertical components of δS_{od}^n can be given as follows:

$$\left(\delta S_{od}^n\right)_H = \left(\hat{S}_{od}^n\right)_H \times \begin{bmatrix} 0 \\ 0 \\ \zeta(3) + \delta\phi \end{bmatrix} + \left(\hat{S}_{od}^n\right)_H \delta k \tag{9.98}$$

$$\left(\delta S_{od}^n\right)_\perp = \hat{s}_{od}\delta\theta + \hat{S}_{od}^n(3)\delta k \tag{9.99}$$

where $()_H$ denotes the horizontal component of () (i.e., vertical component is zero), and $()_\perp$ denotes the vertical component of ().

In Eq. (9.98), the error term $(\hat{S}_{od}^n)_H \times \begin{bmatrix} 0 \\ 0 \\ \zeta(3) + \delta\phi \end{bmatrix}$ implies that $\delta\phi$ and $\zeta(3)$ cannot be separated from each other. Hence, it requires improvement of the accuracy of INS fine alignment as far as possible in order to calibrate $\delta\phi$ more precisely. In addition, this term is in the horizontal plane and perpendicular to the other term $(\hat{S}_{od}^n)_H \delta k$, which is also in the horizontal plane with direction along the horizontal trajectory.

Based on Eqs. (9.98) and (9.99), $\zeta(3) + \delta\phi$, $\delta\theta$ and δk can be determined with the aid of two landmarks, and compensated in the next phase of navigation. The procedure of calibration is described as follows [31].

The vehicle starts from the landmark **A** and ends at the landmark **C**. Because of the uncalibrated factor, the dead-reckoning algorithm (see the next subsection for more details) derived position result **C′** at the terminal point will deviate from **C**. The actual horizontal displacement $(\Delta S)_H$ and computed horizontal displacement $(\Delta\hat{S})_H$ are utilized to achieve the calibration through a geometric relationship.

As shown in Fig. 9.24, on one hand, area formula of triangle ACC' can be expressed as $\frac{|(\Delta S)_H||(\Delta\hat{S})_H|\sin(\delta\phi+\zeta(3))}{2}$; on the other hand, the area can be also expressed as the form of vector cross-product $\frac{((\Delta\hat{S})_H \times (\Delta S)_H)_3}{2}$ (the third element after cross-product). Therefore,

$$\frac{|(\Delta S)_H||(\Delta\hat{S})_H|\sin(\delta\phi + \zeta(3))}{2} = \frac{((\Delta\hat{S})_H \times (\Delta S)_H)_3}{2} \tag{9.100}$$

Thus $\delta\phi + \zeta(3)$ can be expressed as

$$\delta\phi + \zeta(3) = \arcsin\frac{((\Delta\hat{S})_H \times (\Delta S)_H)_3}{|(\Delta S)_H||(\Delta\hat{S})_H|} \tag{9.101}$$

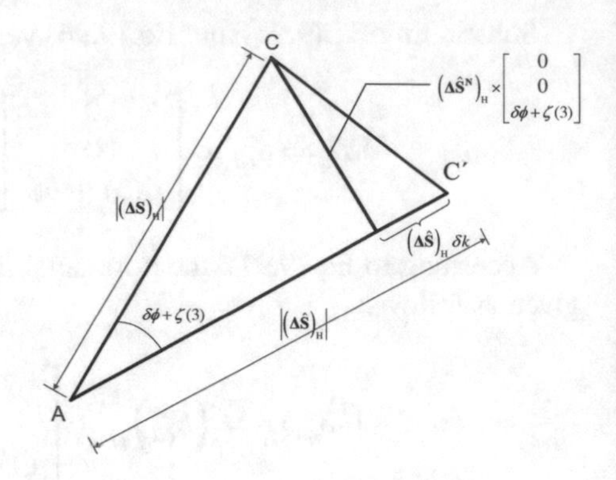

Fig. 9.24 The geometric relationship for calibration

The relationship between the true displacement and dead-reckoning displacement satisfies

$$|(\Delta S)_H| = (1 - \delta k)|(\Delta \hat{S})_H| \tag{9.102}$$

Thus, the calibration equation of VMS scale factor error is as follows:

$$\delta k = 1 - \frac{|(\Delta S)_H|}{|(\Delta \hat{S})_H|} \tag{9.103}$$

Rearranging Eq. (9.99) yields

$$\delta\theta = \frac{(\delta S_{od}^n)_\perp - \hat{S}_{od}^n(3)\delta k}{\hat{s}_{od}} \tag{9.104}$$

Finally, ϕ, θ, and k are compensated as follows:

$$\phi = \hat{\phi} - \delta\phi, \quad \theta = \hat{\theta} - \delta\theta, \quad k = (1 - \delta k)\hat{k} \tag{9.105}$$

9.4.3.2 Loosely Coupled Implementation

With Eq. (9.105) and the form of M from (9.92), the navigation equations for loosely coupled INS/VMS integrated system are given as follows:

$$\dot{C}_b^n = C_b^n(\omega_{ib}^b \times) - (\omega_{in}^n \times)C_b^n \tag{9.106}$$

$$S_{od}^n = \begin{bmatrix} S_{od}^E \\ S_{od}^N \\ S_{od}^U \end{bmatrix} = C_b^n M s_{od} \tag{9.107}$$

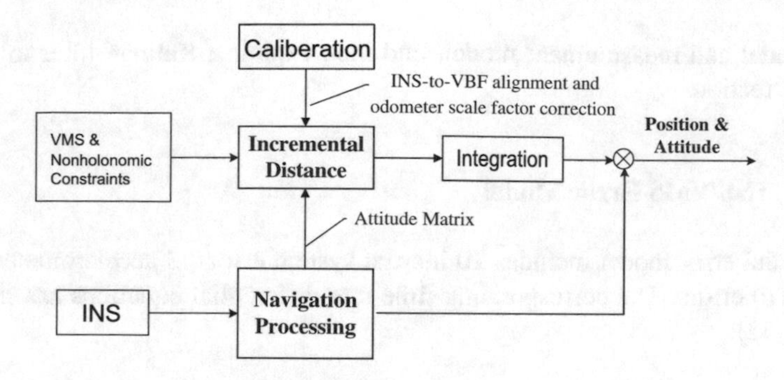

Fig. 9.25 Loosely coupled implementation

$$dL_{inte} = \frac{S_{od}^{N}}{R_m + h} \tag{9.108}$$

$$dl_{inte} = \frac{S_{od}^{E} \sec L}{R_t + h} \tag{9.109}$$

$$dh_{inte} = S_{od}^{U} \tag{9.110}$$

where dL_{inte}, dl_{inte}, and dh_{inte} are changes in latitude, longitude, and altitude, respectively. Here R_m and R_t are, respectively, the meridian radius of curvature and the transverse radius of curvature.

These changes in latitude, longitude, and altitude are accumulated, starting from initialized values, to maintain the navigation solution's position. Figure 9.25 illustrates the configuration of loosely coupled INS/VMS.

Loosely coupled mode has the advantage of an easy-to-build architecture. But its shortcomings are obvious:

(i) The INS errors such as attitude error, velocity error, and position error cannot be corrected using VMS information.
(ii) In practice, the VMS scale factor k and Euler angle θ are dynamically varied; hence, the constant values derived from the calibration method cannot always guarantee the positioning accuracy.
(iii) Two landmarks are required to aid in achieving the calibration, and thus the self-contained characteristics of INS/VMS is destroyed.

9.4.4 Tightly Coupled INS/VMS

Unlike the calibration method which is introduced in the previous section, the orientation between the INS body frame and the vehicle chassis is automatically estimated in the tightly coupled INS/VMS. In this section, we first derive INS/VMS dynamic error model and measurement model, and then exploit a Kalman filter to achieve cross-correction.

9.4.4.1 INS/VMS Error Model

The inertial error model includes 10 inertial system errors, 3 accelerometer errors, and 3 gyro errors. The corresponding time rate differential equations are shown as follows [32]:

$$\dot{\zeta} = -\omega_{in}^{n} \times \zeta + \omega_{ie}^{n} \times \varrho - C_{b}^{n} \delta\omega_{ib}^{b} + \delta\omega_{en}^{n} \tag{9.111}$$

$$\begin{aligned}\delta\dot{v}^{n} &= -(2\omega_{ie}^{n} + \omega_{en}^{n}) \times \delta v^{n} + (C_{b}^{n} f^{b}) \times \zeta \\ &\quad + C_{b}^{n} \delta f^{b} - (2\omega_{ie}^{n} \times \varrho + \delta\omega_{en}^{n}) \times v^{n}\end{aligned} \tag{9.112}$$

$$\dot{\varrho} = -\omega_{en}^{n} \times \zeta + \delta\omega_{en}^{n} \tag{9.113}$$

$$\delta\dot{h} = u_{z} \cdot \delta v^{n} \tag{9.114}$$

with

$$\delta\omega_{en}^{n} = \delta\rho u_{z} + \frac{1}{r_{l}}(u_{z} \times \delta v^{n}) - \frac{1}{r_{l}^{2}}(u_{z} \times v^{n})\delta h \tag{9.115}$$

$$u_{z} = \left[0,\ 0,\ 1\right]^{T} \tag{9.116}$$

$$\delta\rho = -(\omega_{ie}^{n} \times \varrho) \cdot u_{z} \tag{9.117}$$

where ϱ denotes the angle error vector associated with the position direction cosine matrix $\hat{C}_{n}^{e}$, δv^{n} is the velocity error vector, and δh is the altitude error. Definitions for other terms are given in Ref. [32].

The accelerometer bias ∇ and gyro bias ε are modeled as first-order Markov processes as follows:

$$\dot{\nabla}(t) = -\frac{1}{\tau_{a}}\nabla(t) + w_{a}(t), \quad \dot{\varepsilon}(t) = -\frac{1}{\tau_{g}}\varepsilon(t) + w_{g}(t) \tag{9.118}$$

where τ_{a} and τ_{g} are the correlation time constants and $w_{a}(t)$, $w_{g}(t)$ are zero mean Gaussian white noise processes.

The error model for the VMS includes three errors: scale factor error and INS-to-VBF misalignment angles. The relationship between the INS body frame and the VBF is fixed in sense that the INS is rigidly mounted on the vehicle. Therefore, the misalignment angles $\delta\phi$ and $\delta\theta$ can be modeled as unknown constants with a slow random buildup as follows:

$$\delta\dot{\phi}(t) = w_{\phi}, \quad \delta\dot{\theta}(t) = w_{\theta} \tag{9.119}$$

where w_ϕ and w_θ denote white noise sequences representing the rate of change of $\delta\phi$ and $\delta\theta$.

VMS scale factor error is modeled in the same way

$$\delta\dot{k} = w_{od} \tag{9.120}$$

Equations (9.111)–(9.120) can be combined to form a single matrix error equation for the VMS aided INS as follows:

$$\dot{x}(t) = F(t)x(t) + w(t), \quad w(t) \sim N(0, Q(t)) \tag{9.121}$$

where

$$x = [\zeta^T \ (\delta v^n)^T \ \varrho^T \ \delta h \ \nabla^T \ \varepsilon^T \ \delta\phi \ \delta\theta \ \delta k]^T$$

Here, the system model is represented by the dynamic error models of the INS and VMS. The state $x(t)$ is a 19-dimensional vector to be estimated, and $F(t)$ is the continuous transition matrix of the system.

9.4.4.2 Measurement Model

The measurements of integrated velocity provided by the INS and the VMS are compared at each measurement update to generate the filter measurement differences. The measurements provided by the VMS are resolved into computed navigation frame n' before comparison takes place. The nth observation is structured as follows:

$$Z_n = \int_{t_{n-1}}^{t_n} \hat{v}^n(t)dt - \int_{t_{n-1}}^{t_n} \hat{C}_b^n(t)\hat{M}(t)\hat{s}_{od}(t)/T_{od}dt \tag{9.122}$$

where T_{od} denotes the sample time of VMS output, t_n is the current Kalman filter update time, and t_{n-1} is the previous Kalman filter update time.

The corresponding measurement equation is obtained as the differential of observation Eq. (9.122)

$$z_n = \int_{t_{n-1}}^{t_n} \delta v^n(t)dt - \int_{t_{n-1}}^{t_n} \delta\left(\hat{C}_b^n(t)\hat{M}(t)\hat{s}_{od}(t)\right)/T_{od}dt \tag{9.123}$$

With Eq. (9.95) in Eq. (9.123), we obtain the equivalent measurement equation in the form

$$z_n = \int_{t_{n-1}}^{t_n} \delta v^n(t)dt - \int_{t_{n-1}}^{t_n} \left(C_b^n\hat{M}\hat{s}_{od}/T_{od}\right) \times \zeta dt$$

$$
-\int_{t_{n-1}}^{t_n} \left(\hat{C}_b^n N_{mis} \hat{s}_{od} / T_{od} \right) \begin{bmatrix} \delta\phi \\ \delta\theta \end{bmatrix} dt
$$
$$
-\int_{t_{n-1}}^{t_n} \left(\hat{C}_b^n \hat{M} \hat{s}_{od} / T_{od} \right) \delta k dt \tag{9.124}
$$

Based on the system model Eq. (9.121) and measurement equation (9.124), a Kalman filter is constructed to estimate and control the system state errors, and the tightly coupled configuration is depicted in Fig. 9.26.

9.4.5 Experiment Study

In this experiment, ground-based navigation experiments have been carried out to evaluate the performance of these two different kinds of architectures.

In our tests, the VMS is installed in the vehicle chassis while the INS is installed inside the vehicle. A receiver of GPS with high accuracy, which is installed on the top of the vehicle, is employed to provide reference values. The INS is within an accuracy of $1\text{nmi/h}(1\sigma)$, and some major errors of the sensors considered in the tests are shown in Table 9.4. The initial VMS scale factor is $0.012\,\text{m/pulse}$. The sampling rates of INS and that of VMS are both 200 Hz. The Kalman filter update time interval ΔT is set to 1 s.

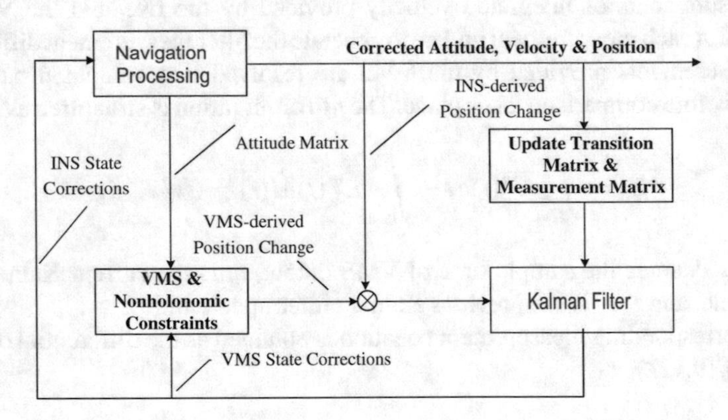

Fig. 9.26 Tightly coupled implementation

Table 9.4 Specification of inertial sensors

	(Ring Laser) Gyro	Accelerometer
Bias	$0.01°/\text{h}\ (1\sigma)$	$100\,\mu\text{g}\ (1\sigma)$
Random walk	$0.005°/\sqrt{\text{h}}$	–
Scale factor	$50\,\text{ppm}\ (1\sigma)$	$80\,\text{ppm}\ (1\sigma)$

9.4.5.1 Calibration

As discussed in Sect. 9.4.3, for loosely coupled INS/VMS, the INS-to-VBF misalignment angles ($\delta\phi$ and $\delta\theta$), as well as the VMS scale factor error (δk), should be preliminarily determined, so as to get an acceptable position accuracy. In the tests, landmarks **A**(39.800342, 116.166874, 40.97) and **C**(39.791720, 116.118074, 70.63) are used to achieve the calibration, and this process lasts approximately 400 s. The initial values of ϕ, θ, and k are set to 0 rad, 0 rad, and 0.0120 m /pulse, respectively. The loosely coupled algorithm (9.106)–(9.110) is then implemented to calculate the position of **C**. The computed result is **C'**(39.791662, 116.121010, 35.06). Figure 9.27 shows the ground track history, where the real trajectory means the trajectory through the GPS-indicated positions and the computed trajectory means the one computed using the loosely coupled algorithm.

It is clear that the computed horizontal trajectory is similar to the actual trajectory, although the misalignment angles $\delta\phi$ and $\delta\theta$ cause the computed trajectory to deviate from the actual trajectory and δk plays a role in shortening or lengthening the trajectory. The final calibration results shown in Table 9.5 are obtained using the calibration method presented in Sect. 9.4.3.

These final calibration results will be used in the next subsection to improve the positioning accuracy when the loosely coupled INS/VMS works.

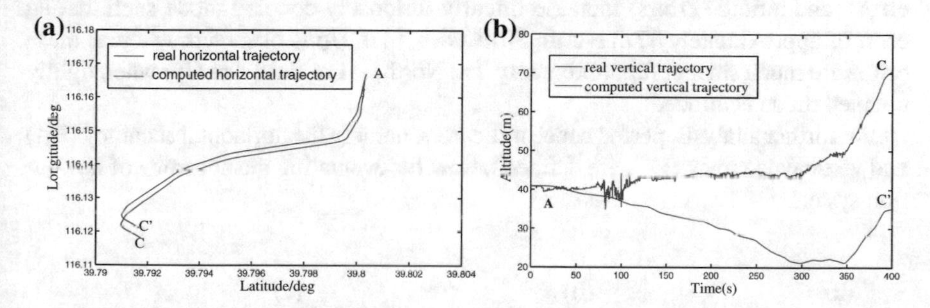

Fig. 9.27 Actual and computed trajectories. **a** Horizontal trajectories. **b** Vertical trajectories

Table 9.5 Final calibration results

Initial values	$\hat{\phi}$	$\hat{\theta}$	$\hat{k}$
	0	0	0.0120
Calibration results	$\delta\phi$	$\delta\theta$	δk
	0.0152	−0.0078	−0.0617
Final results	ϕ	θ	k
	−0.0152	0.0078	0.01274

9.4.5.2 Comparison of Loosely Coupled and Tightly Coupled INS/VMS

This subsection tries to compare the performance of loosely coupled and tightly coupled INS/VMS through experiment results. Both architectures are run with the same datasets.

Figure 9.28a–c presents the Kalman filter estimates for the Euler angles that translate the VBF into the INS body frame in terms of rotation, and VMS scale factor, which are corrected automatically during the INS/VMS system working in tightly coupled mode. From the perspective of control theory, these three states, whose rapidity of convergence depends on the vehicle maneuvers, are unobservable under the stationary situation. In practical circumstance, affected by x-axial vibration and unbalanced weight of the vehicle, the misalignment in pitch is time varying. In the case where the accuracy of altitude is strictly required, θ should be modeled more accurately. In addition, these parameters computed by the geometric method are also plotted for the purpose of comparison.

With the INS-to-VBF misalignment corrections and VMS scale factor being derived by the Kalman filter, the VMS-derived distance traveled over a measurement update interval may be used to aid the INS. Figure 9.29a–c shows the errors in latitude, longitude, and altitude of the vehicle estimated using direct integration of INS/VMS data (loosely coupled mode) and using the Kalman filter (tightly coupled mode). As shown in Fig. 9.29, the position errors (include latitude errors, longitude errors, and altitude errors) increase linearly in loosely coupled mode such that an error of approximately 37 m North, 30 m East, 14 m Up is observed, whereas these errors are much smaller (approximately 7 m North, −1 m East, 4 m Up) when tightly coupled mode is utilized.

For further analysis, performance indicators, namely the horizontal accuracy (*HA*) and vertical accuracy (*VA*), are defined below for evaluating the accuracy of navigation systems:

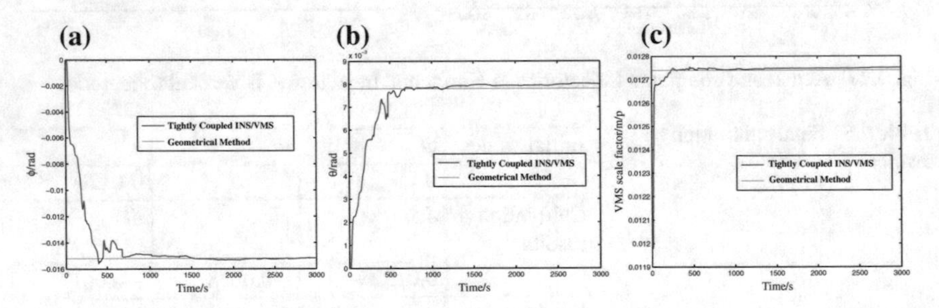

Fig. 9.28 Comparison of VMS error estimates. **a** Misalignment angle ϕ. **b** Misalignment angle θ. **c** VMS scale factor

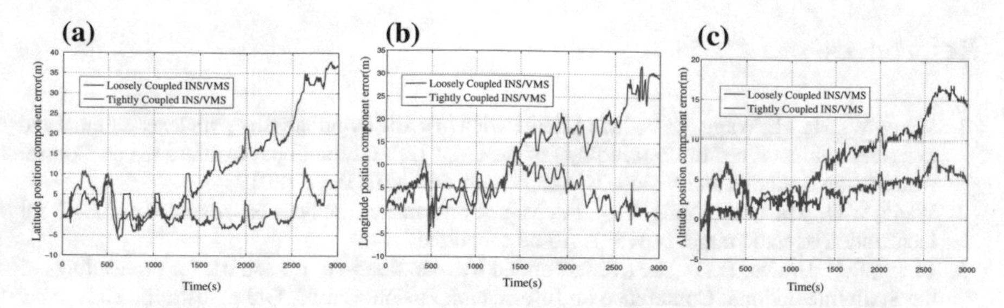

Fig. 9.29 Comparison of position errors. **a** Latitude errors. **b** Longitude errors. **c** Altitude errors

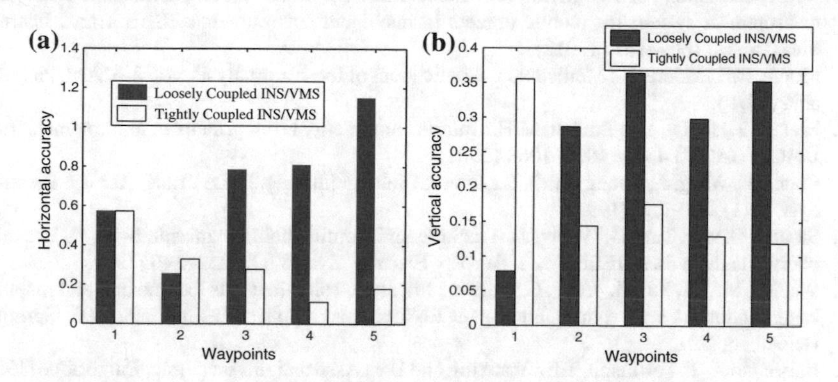

Fig. 9.30 Comparison of positioning accuracy. **a** Horizontal accuracy. **b** Vertical accuracy

$$HA = \frac{\sqrt{\mathrm{Err}_{latitude}^2 + \mathrm{Err}_{longitude}^2}}{S_{OD}} \times 1000 \tag{9.125}$$

$$VA = \frac{\mathrm{Err}_{altitude}}{S_{OD}} \times 1000 \tag{9.126}$$

where $\mathrm{Err}_{latitude}$, $\mathrm{Err}_{longitude}$, and $\mathrm{Err}_{altitude}$ denote the latitude error, longitude error, and altitude error, respectively. The travel distance is denoted by S_{OD}.

We select five way points at time 600, 1200, 1800, 2400, and 3000 s to evaluate the INS/VMS performance. The corresponding results are shown in Fig. 9.30a, b.

By examining Figs. 9.28a–c and 9.30a, b carefully, it can be concluded that the tightly coupled INS/VMS provides lower HA and VA when ϕ, θ, and k reach steady states. The tightly coupled INS/VMS generally outperforms loosely coupled mode, and this fact may intuitively benefit from the optimal estimation mechanism provided by the Kalman filter for tightly coupled mode which can dynamically provide the corrections for both INS and VMS.

References

1. Wang, W., Ma, H., Wang, Y., Fu, M.: Localization of static target in wsns with least-squares and extended kalman filter. In: Proceedings of the 2012 12th International Conference on Control Automation Robotics and Vision (ICARCV), pp. 602–607. IEEE (2012)
2. Wang, W.D., Ma, H.B., Wang, Y.Q., Fu, M.Y.: Performance analysis based on LS and EKF for localization of static target in WSNs (to be submitted)
3. Bizup, D.F., Brown, D.E.: The over-extended Kalman filter—don't use it!. In: Proceedings of the Sixth International Conference on Information Fusion, Cairns, Qld., Australia, Univ. New Mexico, pp. 227–233 (2003)
4. Nazari Shirehjini, A., Yassine, A., Shirmohammadi, S.: An rfid-based position and orientation measurement system for mobile objects in intelligent environments. IEEE Trans. Instrum. Meas. **61**(6), 1664–1675 (2012)
5. Moore, R.E., Moore, R.: Methods and Applications of Interval Analysis, vol. 2. SIAM, Philadelphia (1979)
6. Hickey, T., Ju, Q., Van Emden, M.H.: Interval arithmetic: From principles to implementation. J. ACM (JACM) **48**(5), 1038–1068 (2001)
7. Chen, G., Wang, J., Leang Shieh, S.: Interval kalman filtering. IEEE Trans. Aerosp. Electron. Syst. **33**(1), 250–259 (1997)
8. Siouris, G.M., Chen, G., Wang, J.: Tracking an incoming ballistic missile using an extended interval kalman filter. IEEE Trans. Aerosp. Electron. Syst. **33**(1), 232–240 (1997)
9. Wu, M., Ma, H., Fu, M., Yang, C.: Particle filter based simultaneous localization and mapping using landmarks with rplidar. Intelligent Robotics and Applications, pp. 592–603. Springer, Heidelberg (2015)
10. Bar-Shalom, Y., Fortmann, T.E.: Tracking and Data Association. Academic, Cambridge (1988)
11. Ding, S., Chen, X., Han, J.D.: A new solution to slam problem based on local map matching. Robots **31**(4), 296–303 (2009)
12. Julier, Simon J., and Jeffrey K. Uhlmann. "Unscented filtering and nonlinear estimation." Proceedings of the IEEE 92.3 (2004): 401–422
13. Ma, Y., Ju, H., Cui, P.: Research on localization and mapping for lunar rover based on rbpf-slam. Intell. Hum. Mach. Syst. Cybern. **2**, 2880–2892 (2009)
14. Hao, Y.M., Dong, D.L., Zhu, F., Wei, F.: Landmarks optimal selecting for global location of mobile robot. High Technol. Lett. **11**(8), 82–85 (2001)
15. Gellert, W., Köstner, H., Hellwich, M., Kästner, H.: The VNR Concise Encyclopedia of Mathematics. Van Nostrand Reinhold, New York (1977)
16. Ma, H., Lum, K.: Adaptive estimation and control for systems with parametric and nonparametric uncertainties. Adaptive Control, pp. 15–64. I-Tech Education and Publishing, Vienna, Austria (2009)
17. Qingzhe, W., Mengyin, F., Xuan, X., Zhihong, D.: Automatic calibration and in-motion alignment of an odometer-aided ins. In: Proceedings of the 2012 31st Chinese Control Conference (CCC), pp. 2024–2028. IEEE (2012)
18. Noureldin, A., El-Shafie, A., Bayoumi, M.: GPS/INS integration utilizing dynamic neural networks for vehicular navigation. Inf. Fusion **12**, 48–57 (2011)
19. Zhang, H., Zhao, Y.: The performance comparison and analysis of extended kalman filters for GPS/DR navigation. Optik **122**, 777–781 (2011)
20. Kaygisiz, B., Erkmen, I., Erkmen, A.: GPS/INS enhancement for land navigation using neural network. J. Navig. **57**, 297–310 (2005)
21. Stančić, R., Graovac, S.: The integration of strap-down INS and GPS based on adaptive error damping. Robot. Auton. Syst. doi:10.1016/j.robot.2010.06.004
22. Grewal, M.S., Weill, L.R., Andrews, A.P.: Global Positioning Systems, Inertial Navigaiton, and Integration. Wiley, New York (2001)
23. Rogers, R.M.: Applied Mathematics in Integrated Navigation Systems, 3rd edn. American Institute of Aeronautics and Astronautics Inc, Virginia (2003)

24. Chiang, K., Huang, Y.: An intelligent navigator for seamless INS/GPS integratede land vehicle navigation applications. Appl. Soft Comput. **8**, 722–733 (2008)
25. Toledo-Moreo, R., Btaille, D., Peyret, F., Laneurit, J.: Fusing GNSS, dead-reckoning, and enhanced maps for road vehicle lane-level navigation. IEEE J. STSP **3**(5), 798–809 (2009)
26. Kubo, Y., Kindo, T., Ito, A., Sugimoto, S.: DGPS/INS/wheel sensor integration for high accuracy land-vehicle positioning. In: Proceedings of the ION GPS, Nashville, TN, pp. 555–564, Sept 1999
27. Cunha, S., Bastos, L., Cunha, T., Tomé, P.: On the integration of inertial and GPS data with an odometer for land vehicles navigation. In: Proceedings of the ION GPS/GNSS, Portland, OR, pp. 940–944, Sept 2003
28. Dissanayake, G., Sukkarieh, S., Nebot, E., Durrant-Whyte, H.: The aiding of a low-cost strapdown inertial measurement unit using vehicle model constraints for land vehicle applications. IEEE Trans. Robot. Autom. **17**(5), 731–747 (2001)
29. Syed, Z.F., Aggarwal, P., Niu, X., EI-Sheimy, N.: Civilian vehicle navigation: required alignment of the inertial sensors for acceptable navigation accuracies. IEEE Trans. Veh. Technol. **57**(6), 3402–3412 (2008)
30. Wu, Y.X., Wu, M.P., Hu, X.P., Hu, D.W.: Self-calibration for land navigation using inertial sensors and odometer: Observability analysis. In: Proceedings of the AIAA Conference on Guidance, Navigation, and Control, Chicago, Illinois, pp. 1–10, Aug 2009
31. Yan, G.M., Qin, Y.Y., Yang, B.: On error compensation technology for vehicular dead reckoning (DR) system. J. Northwest. Polytech. Univ. **24**(1), 26–30 (2006)
32. Savage, P.G.: Strapdown Analytics. Strapdown Associates, Minnesota (2000)

Chapter 10
Multiagent Robot Systems

Abstract In this chapter, we will first give an introduction of multiagent systems, which can serve as abstraction or simplified models for vast real-life complex systems, where local interactions among agents lead to complex global behaviors such as coordination, synchronization, formation, and so on. Then, as simple yet nontrivial examples of cooperation among robots, two typical cases of three-robot line formations are investigated and illustrated in a general mathematical framework of optimal multirobot formation. The robots moving with different speeds are expected to row on one straight line with the minimum formation time, so that the formation can be formulated in the most efficient way. Next, we investigate the hunting issue of a multirobot system in a dynamic environment. The proposed geometry-based strategy has the advantages of fast calculation and can be applied to three-dimensional space easily. At the end of the chapter, we investigate a few important problems in multirobot cooperative lifting control, and present an simulation study showing an example of four arms lifting one desk.

10.1 Introduction to Multiagent System

A multiagent system in the area of dynamic systems and control refers to a system consisting of a number of independent dynamic subsystems named agents. Controlling a multiagent system so that all the agents work cooperatively to accomplish certain task is usually called the coordinated/cooperative control problem.

Multiagent systems have received great attention since 1980s in many areas such as physics, biology, bionics, engineering, artificial intelligence, and so on. On one hand, this is driven by our increasing demand for higher efficiency, greater flexibility, and lower cost in applications. On the other hand, multiagent systems could offer many significant advantages.

Various control strategies developed for multiagent systems can be roughly assorted into two architectures: centralized and decentralized. In the decentralized control, local control for each agent is designed only using locally available information so it requires less computational effort and is relatively more scalable with respect to the swarm size. In recent years, especially since the so-called Vicsek

© Science Press and Springer Science+Business Media Singapore 2016

C. Yang et al., *Advanced Technologies in Modern Robotic Applications*,
DOI 10.1007/978-981-10-0830-6_10

model was reported in [1], decentralized control of multiagent system has received much attention in the research community (e.g., [2, 3]). In the (discrete-time) Vicsek model, there are n agents and all the agents move in the plane with the same speed but with different headings, which are updated by averaging the heading angles of neighbor agents. By exploring matrix and graph properties, a theoretical explanation for the consensus behavior of the Vicsek model has been provided in [2]. In [4], a discrete-time multiagent system model has been studied with fixed undirected topology and all the agents are assumed to transmit their state information in turn. In [5], some sufficient conditions for the solvability of consensus problems for discrete-time multiagent systems with switching topology and time-varying delays have been presented by using matrix theories. In [3], a discrete-time network model of agents interacting via time-dependent communication links has been investigated. The result in [3] has been extended to the case with time-varying delays by set-value Lyapunov theory in [6].

Consider a multiagent system consisting of N agents. Each individual agent may have different dynamics, which can be described as:

$$\begin{aligned} \dot{x}_i(t) &= f_i(x_i(t), \mu_i(t)) \\ y_i(t) &= g_i(x_i(t), \mu_i(t)), \quad i = 1, \ldots, N \end{aligned} \tag{10.1}$$

where $x_i \in \mathbb{R}^{n_i}$, $\mu_i \in \mathbb{R}^{m_i}$, and $y_i \in \mathbb{R}^p$ represent the state, the input, and the output of the ith agent, respectively. $\mu_i \in \mathbb{R}^{q_i}$ is an uncertain parameter vector which represents either the system parameter variation or external disturbance or both. N is the number of the agents in the system. The communication among all the agents can be described as a topological graph. For convenience, we introduce the following definitions in graph theory.

A graph is usually defined as $\mathcal{G}(t) = (\mathcal{V}, \mathcal{E}, \mathcal{A}(t))$ with vertex set $\mathcal{V} = \{v_1, \ldots, v_N\}$, edge set $\mathcal{E}(t) \subseteq \mathcal{V} \times \mathcal{V}$ and adjacency matrix $\mathcal{A}(t)$ if the edge set $\mathcal{E}(t)$ contains an edge (v_i, v_j), it means v_j can receive the information from v_i at time t and v_i is called a neighbor of v_j. The graph $\mathcal{G}(t)$ is undirected if $(v_i, v_j) \in \mathcal{E}(t) \Leftrightarrow (v_j, v_i) \in \mathcal{E}(t)$; and is called fixed if the topology does not change over time. The graph $\mathcal{G}(t)$ can be undirected or directed, fixed or time-varying, depending on specific applications and circumstances. The adjacency matrix is thus defined as $\mathcal{A}(t) = [a_{ij}(t)]_{N \times N}$ with $a_{ij}(t) = 1$ if and only $(v_i, v_j) \in \mathcal{E}(t)$ and $a_{ij}(t) = 0$ otherwise. The neighborhood of v_i can be described as $\mathcal{N}_i(t) = \{v_j \in \mathcal{V} | (v_j, v_i) \in \mathcal{E}(t)\}$. As seen from Eq. (10.1), each agent in the multiagent system usually has independent dynamics, which is different from traditional decentralized large-scale systems.

Moreover, instead of using centralized or decentralized control, distributed control is usually employed. This means that the controller of each agent uses the information of itself and its neighbors only. For example, a class of distributed dynamic control laws can be described as follows:

$$\begin{aligned} \dot{\theta}_i(t) &= \Theta_i(y_i(t), y_j(t), \theta_i(t)) \\ \mu_i(t) &= \Gamma_i(y_i(t), y_j(t), \theta_i(t)), \quad i = 1, \ldots, N, \; j = \mathcal{N}_i \end{aligned} \tag{10.2}$$

where $\theta_i \in \mathbb{R}^{n_\theta}$ is the controller state vector to be determined by specific problems, and functions Θ_i and Γ_i are sufficiently smooth.

In practice, it is common for the simple embedded microprocessors and low-cost communication and actuation modules to be used in implementation of multiagent control systems because they offer great advantages such as general programming versatility, signal processing flexibility, and low cost. They are in charge of communicating with its neighbors, processing the information, computing the control algorithm, and driving the agent. Normally such a low-cost hardware would lead to limited communication capacity and limited onboard energy resource. Therefore, it is of great theoretical and practical significance to consider these issues in controller design and implementation for multiagent systems. Next, we will show several applications of multiagents in robot systems.

10.2 Optimal Multirobot Formation

In many applications of multiple robots such as cleaning, mine sweeping, security patrols, and so on, robot formation plays one important and basic role in sense that the robots in a team are usually controlled as a whole to follow required formations to accomplish tasks with satisfied overall performance. In robot fundamental problems, we are to explore the answers of the following two basic problems:

- *How to define a formation?*
- *How to define the 'optimal' formation?*

Motivated by the above questions and many practical demands for efficient robot formation, a novel framework of general optimal formation problem has been introduced to lay a solid foundation for future research on multirobot optimal formation in [7].

In this section, we will investigate three-robot optimal line formation and present mathematical results for the optimal solutions in two typical cases [7], i.e., raw line formation, and fixed-angle line formation, respectively. Line formation can serve as a basis to investigate more general formations of multiple robots.

The theoretical work in this section contains the following issues:

(i) The minimum-time raw line formation of three-robot group is analytically discussed with optimal formation explicitly given, based on geometric and algebraic analysis.
(ii) Optimal solution to the minimum-time fixed-angle line formation problem of three-robot group is also explicitly presented.
(iii) Simulation results have verified our nontrivial theoretical results for three-robot line formation, whose proofs are omitted to save space.

10.2.1 Concepts and Framework of Multirobot Formation

10.2.1.1 Preliminary Concepts

For brevity and convenience, we sketch some preliminary concepts and the framework of optimal multirobot formation in this section.

Definition 10.2.1 Suppose that the robot moves in one subset Ω of the whole euclidean space $\mathcal{R}^m$, where m is usually taken as 2 or 3 in practical life if we consider position only as state of each robot. Here $\Omega \subseteq \mathcal{R}^m$ is called *free space* of the robot. If $\Omega = \mathcal{R}^m$, then we say that the robot is *moving unconstrainedly*. Furthermore, if there are n robot agents which share the same free space, we can define $\mathcal{S} \triangleq \Omega^n = \Omega \times \Omega \times \cdots \Omega$. Any vector $S(t) \in \mathcal{S}$ is called the grout status of robots (at time t).

Definition 10.2.2 Any mapping $f : \mathcal{S} \to \mathcal{R}^1$ is called a scalar function of group status. Any mapping $f : \mathcal{S} \to \mathcal{R}^d$ is called a vector function of group status.

Given a function f of group status, and suppose that robot R_i move to S_i at time t ($i = 1, 2, \ldots, n$), then $f(S(t))$ is in fact a value depending on all robots' positions. For convenience, we use $d(P, Q)$ to denote the distance between points P and Q in euclidean space $\mathcal{R}^m$. That is to say, in case of $m = 2$,

$$d(P, Q) \triangleq \sqrt{(x_P - x_Q)^2 + (y_P - y_Q)^2} \tag{10.3}$$

where (x_P, y_P) and (x_Q, y_Q) are coordinates of points P and Q, respectively. In case of $m = 2$, we also use $\alpha(P, Q)$ to denote the slope angle of vector $\mathbf{PQ}$, which can be defined by

$$\alpha(P, Q) \triangleq \arctan(x_Q - x_P, y_Q - y_P) \tag{10.4}$$

where $\arctan(x, y) \in \mathcal{R}[2\pi]$ denotes the arctangent angle in the correct quadrant determined by the coordination (x, y).

Function $d_{ij}(\cdot) : \mathcal{S} \to \mathcal{R}^1$ is defined by

$$d_{ij}(S(t)) \triangleq d(S_i(t), S_j(t)) \tag{10.5}$$

Besides, we need the follow notations:

$$\begin{aligned}
\alpha_i(S(t); Q) &\triangleq \alpha(Q, \S_i(t)) \\
\alpha_{ij}(S(t)) &\triangleq \alpha(S_i(t), \S_j(t)) \\
\beta_{ij}(S(t); Q) &\triangleq \alpha(Q, \S_j(t)) - \alpha(Q, S_i(t)) \\
\beta_{ikj}(S(t)) &\triangleq \alpha(S_k(t), \S_j(t)) - \alpha(S_k(t), S_i(t))
\end{aligned}$$

Definition 10.2.3 For a group of robots $R_1, R_2, \ldots, R_n$, suppose that the group status $S(t) \in \Omega^n$ is determined by a time-dependent mapping $G_t : \mathcal{S} \times \Theta \to \mathcal{S}$:

$$S(t) = G_t(S(0), \theta), \theta \in \Theta_{\mathcal{G}} \tag{10.6}$$

where $S(0)$ is the initial group status of robots, and $\theta \in \Theta_{\mathcal{G}}$ is a vector holding parameters of the robots motion. Then, we say that G_t is a *group motion model* of the robots and θ is called the *motion parameter vector* of the robots.

Definition 10.2.4 For a group of robots $R_1, R_2, \ldots, R_n$, suppose that all robots share the same motion model, that is to say, each robot's position $\mathcal{S}(t) \in \Omega$ at time t is determined by a time-dependent mapping $F_t : \Omega \times \Theta \to \Omega$:

$$\mathcal{S}(t) = F_t(\mathcal{S}(0), \theta_i), \theta_i \in \Theta \tag{10.7}$$

where Θ is a common set of parameter vectors for all robots and $\mathcal{S}(0)$ is the initial position of Robot R_i. In this case, F_t is a *homogeneous motion model* of the robots and $\theta_i \in \Theta$ is the *motion parameter* of Robot R_i.

Definition 10.2.5 For a group of robots $R_1, R_2, \ldots, R_n$, suppose that all robots share the same motion model and each robot will not move any longer once stop at certain time instant, that is to say, each robot's position $\mathcal{S}(t) \in \Omega$ at time t is determined by a mapping $F : \mathcal{R} \times \Omega \times \Theta \to \Omega$ such that

$$S_i(t) = \begin{cases} F(t; \mathcal{S}(0), \theta_i), & \text{if } \ 0 \le t \le T_i \\ F(T_i; \mathcal{S}(0), \theta_i), & \text{if } \ t \ge T_i \end{cases} \tag{10.8}$$

where Θ is a common set of parameter vectors for all robots and $\mathcal{S}(0)$ is the initial position of Robot R_i. In this case, F is called a *truncatable homogeneous motion model* of the robots and $\theta_i \in \Theta$ is the *motion parameter* of Robot R_i. Note that the truncation time (or termination time) T_i generally may be dependent of θ_i.

10.2.1.2 What Is a Formation?

Definition 10.2.6 For a group of robots $R_1, R_2, \ldots, R_n$, a prescribed *formation set* $\mathcal{F}$ is essentially a set of group status which satisfy certain specific constraints. Mathematically speaking, suppose that

$$g_1, g_2, \ldots, g_p, h_1, h_2, \ldots, h_q : \mathcal{S} \to \mathcal{R} \tag{10.9}$$

are several given functions of group status, then the set

$$\mathcal{F} = \left\{ S = (P_1, \dots, P_n) \in \mathcal{S} \;\middle|\; \begin{array}{l} g_1(S) = 0 \\ g_2(S) = 0 \\ \vdots \\ g_p(S) = 0 \\ h_1(S) \geq 0 \\ h_2(S) \geq 0 \\ \vdots \\ h_q(S) \geq 0 \end{array} \right\} \tag{10.10}$$

is called a *formation set*, which is defined by equality constraints $g_j(S) = 0$ $(j = 1, 2, \dots, p)$ and inequality constraints $h_j(S) = 0$ $(j = 1, 2, \dots, q)$. And any element of $\mathcal{F}$ is called a *desired formation.*

10.2.1.3 What Is Optimal Formation?

Definition 10.2.7 Mathematically speaking, a *formation process* $\mathcal{P}$ is an indexed collection $\{S(t), t \geq 0\}$ of group status $S(t)$, ending with or approaching to a desired formation. Intuitively speaking, a formation process is the whole process of a group of robots starting from their initial group status until they formulate a desired formation.

Definition 10.2.8 Given a formation set $\mathcal{F}$ and a formation process $\mathcal{P} = \{S(t), t \geq 0\}$, let

$$d_{\text{close}}(t) \triangleq \inf_{S_F \in \mathcal{F}} d(S(t), S_F) \tag{10.11}$$

where $d(S(t), S_F)$ denotes the distance between group status $S(t)$ and S_F, which will be discussed later and which intuitively measures the degree of closeness to the desired formation.

Definition 10.2.9 Let $S = (P_1, P_2, \dots, P_n) \in \Omega^n$ and $S' = (P_1', P_2', \dots, P_n') \in \Omega^n$ be two group status. Then the distance between S and S' can be defined as

$$d(S, S') \triangleq \max(d(P_1, P_1'), d(P_2, P_2'), \dots, d(P_n, P_n')) \tag{10.12}$$

where $d(P_i, P_i')$ denotes the Euclidean distance between point P_i and P_i' in $\mathcal{R}^m$. It is easy to verify that

$$d(S, S) = 0; d(S, S') = d(S', S); d(S, S') = 0 \text{ iff } S = S' \tag{10.13}$$

and the triangular inequality

$$d(S, S') + d(S', S'') \geq d(S, S'') \tag{10.14}$$

holds for any group status $S, S', S'' \in \Omega^n$. Hence $d(\cdot, \cdot)$ defines a distance well in Ω^n.

Definition 10.2.10 Any scalar function $I(\mathcal{P})$ of a formation process $\mathcal{P}$ can be regarded as a *performance index* of the formation process, which presents a criterion for quantitatively evaluating the formation process.

Definition 10.2.11 Given a formation set $\mathcal{F}$, a performance index $I(\cdot)$, a group motion model G_t, and a set $\mathbb{P}$ of formation processes (following the specified motion model) starting from initial group status S_0, then the purpose of general optimal formation problem is to find an *optimal formation process* $\mathcal{P}_{opt}$ which minimizes the performance index $I(\mathcal{P})$, i.e.

$$\mathcal{P}_{opt} = \arg\min_{\mathcal{P}\in\mathbb{P}} I(\mathcal{P}) \tag{10.15}$$

And

$$I^* \triangleq I(\mathcal{P}_{opt}) \tag{10.16}$$

is called the *optimal performance index* of this optimal formation problem.

When we consider parametric motion models and formation processes, the optimal formation process is indeed to find the *optimal formation parameters* $\theta^* \in \Theta$ such that

$$\theta^* = \arg\min_{\theta\in\Theta} I(S_0; \Theta) \tag{10.17}$$

where

$$I(S_0; \theta) \triangleq I(\mathcal{P}(S_0; \theta)) \tag{10.18}$$

is essentially determined by the initial group status as well as the motion parameters $\theta \in \Theta$, and consequently the optimal performance index is

$$I^* = I(S_0; \theta). \tag{10.19}$$

Definition 10.2.12 In Definition 10.2.11, if the performance index is taken as the formation time $T(\mathcal{P})$ given in Example 10.2.1, then the optimal formation problem is called *minimum-time optimal formation problem*, or simply *minimum-time formation problem*.

Example 10.2.1 Formation time T is a scalar determined by the formation process, hence the formation time $T(\mathcal{P})$ can be regarded as a performance index of the formation process $\mathcal{P}$. For a finite-time formation process generated by a truncatable homogeneous motion model F, if all the robots finally stay at a group status which is a desired formation, then we have

$$T = \max(T_1, T_2, \ldots, T_n) \tag{10.20}$$

where T_i is the termination time of Robot $R_i, i = 1, 2, \ldots, n$. This performance index is meaningful since the smaller the T is, the more efficient the formation process is.

The formulation of optimal formation problems looks like very simple from Definition 10.2.11, however, from this section, we can see that such a class of formation problems are generally very nontrivial even for the most simple problems of three-robot optimal line formations. Based on the concepts given above, we are ready to present the minimum-time three-robot line formation problem.

10.2.2 Minimum-Time Three-Robot Line Formation

In this part, we focus on the following two typical types of line formations.

Definition 10.2.13 (*(Raw) Line Formation*) A *(raw) line formation* refers to the group status such that all robots are exactly located on the same straight line. Such a formation set can be mathematically defined by

$$\mathcal{F} = \{S|\beta_{ijk}(S) = 0, \forall 1 \le i < j < k \le n\} \tag{10.21}$$

or, explicitly, the constraints in the plane can be reduced to

$$(x_1 - x_2)(y_1 - y_i) - (x_1 - x_i)(y_1 - y_2) = 0, \quad i = 3, 4, \ldots, n \tag{10.22}$$

where (x_i, y_i) denotes the desired position of Robot R_i.

In many situations, the desired line formation needs more additional constraints, some of which have been described in [7], besides raw line formation, here we consider only *fixed-angle line formation* in this note.

Definition 10.2.14 (*Fixed-Angle Line Formation*) Suppose that all robots are expected to follow a line with certain slope angle $\kappa_{\mathcal{F}} = \tan \gamma_{\mathcal{F}}$, where $\gamma_{\mathcal{F}}$ is the desired slope angle. Then, the formation set can be mathematically defined by

$$\mathcal{F} = \left\{S \in \mathcal{S}|\alpha_{ij}(S) = \kappa_{\mathcal{F}}, \forall 1 \le i < j < k \le n\right\} \tag{10.23}$$

In this subsection, following our discussions given in [8], where the mini-max travel distance optimal line formation problem for three robots is investigated, minimum time optimal line formation for three robots will be discussed. Optimal line formation of a three-robot group, which is the basis for understanding optimal formation of more robots, looks easy and simple, however, the nontrivial analysis given in [8] and this section indicates that this problem is far from obvious.

10.2.2.1 Notations and Lemmas for Analysis

As to the performance index, we consider the time $T(\mathcal{P})$, which intuitively reflects maximal cost for robots to reach a desired formation. For the two typical desired line formations given in Definitions 10.2.13 and 10.2.14, we are ready to investigate the corresponding minimum-time three-robot line formation with any initial group status $S(0)$, and we will present its optimal solution by stating explicitly the desired target position of each robot and the optimal performance index, i.e., minimal formation time, will be analytically figured out.

For the line formation problems considered here, the optimal formation process can be generated as follows:

(i) Compute every robot's optimal target position in the optimal formation, then we can get the set $\mathcal{F}^*$ of desired optimal formations and the corresponding optimal formation parameters θ^*.
(ii) Robots go straight to their optimal target position, and stop after reaching their optimal target position.
(iii) Once the last robot reaches its optimal target position, the minimum-time formation has formed.

From the process above, the formation time $T_{\max}$ can be calculated by

$$\begin{aligned} T_{\max} &= \max\{t_1, t_2, \ldots, t_n\} \\ &= \max\{\frac{d_1}{v_1}, \frac{d_2}{v_2}, \ldots, \frac{d_n}{v_n}\} \end{aligned} \tag{10.24}$$

where t_i is the time cost by robot R_i and d_i is the distance moved by robot R_i. Obviously, $T_{\max}$ depends on the desired line formation f explicitly. We hope to find the optimal line formation f^* such that it minimizes $T_{\max}$.

To facilitate our theoretical analysis, we need to introduce some notations first.

(i) Three robots are denoted by R_1, R_2, and R_3, respectively. Initially, three robots usually constitute a triangle, whose vertices are also named as R_i $(i = 1, 2, 3)$ without confusion. Without loss of generality, for convenience of later discussions, we suppose that initial position of R_2 is the origin, initial position of R_3 is on the x-axis with positive x-coordinate, and initial position of R_1 has a positive y-coordinate.
(ii) Let $f \in \mathcal{F}$ denote any straight line, i.e. one possible desired formation, then the so-called optimal line formation problem is in fact to determine an optimal line f^* such that three robots can locate on the line f^* by moving for some time. Sometimes, we also need to consider the slope angle of the line f, and name the slope angle as γ, which is always a positive value ($\gamma \in [0, \pi)$).
(iii) Let $\bar{d}_i(f)$ be the initial *signed* distance between R_i and line f, which is positive if R_i is above the line f, or negative if R_i is below the line f. Here the term "above" ("below") refers to the case where there exists a point Z on line f with the same x-coordinate as the point R_i and the larger (smaller) y-coordinate than

the point R_i. When the line f is parallel to y-axis, i.e. its slope angle is $\frac{\pi}{2}$, then point R_i is "above" ("below") the line f if x-coordinate of R_i is larger (smaller) than x-coordinate of any point on line f.

(iv) Let

$$l_1 = d_{23}(S(0)), l_2 = d_{13}(S(0)), l_3 = d_{12}(S(0)) \tag{10.25}$$

be the initial distances of each two robots.

(v) Denote three points by D_1, D_2, and D_3, where the point D_i is a point on the line segment $R_j R_k$ ($i \neq j \neq k, i, j, k \in 1, 2, 3$) such that

$$\frac{d(R_j, D_i)}{d(R_k, D_i)} = \frac{v_j}{v_k}. \tag{10.26}$$

(vi) Denote the line f which passes the point D_i by f_i, as indicated in Fig. 10.1b. And denote the line f which passes the point D_i and the point D_j by f_{ij}, and the slope angle of f_{ij} is denoted by γ_{ij}, as indicated in Fig. 10.1c.

With the notations above, given a line f, let

$$T_{\max}(f) = \max\{\frac{|\bar{d}_1(f)|}{v_1}, \frac{|\bar{d}_2(f)|}{v_2}, \frac{|\bar{d}_3(f)|}{v_3}\} \tag{10.27}$$

Therefore, to obtain the optimal formation time T^*, we should find a line f^* such that

$$T^* = \min_{f \in \mathcal{F}} T_{\max}(f) = T_{\max}(f^*) \tag{10.28}$$

which consequently determines the optimal formation parameters θ^* for three robots, in the way that the robots should go to the desired line with the shortest path, as indicated by $d_i(f)$, $i = 1, 2, 3$.

For convenience, we introduce the following symbols:

Definition 10.2.15

$$t_1' = \frac{l_2}{v_1}|\sin(\angle R_3 + \gamma)| - \frac{l_1 v_3}{(v_2 + v_3)v_1} \sin\gamma$$

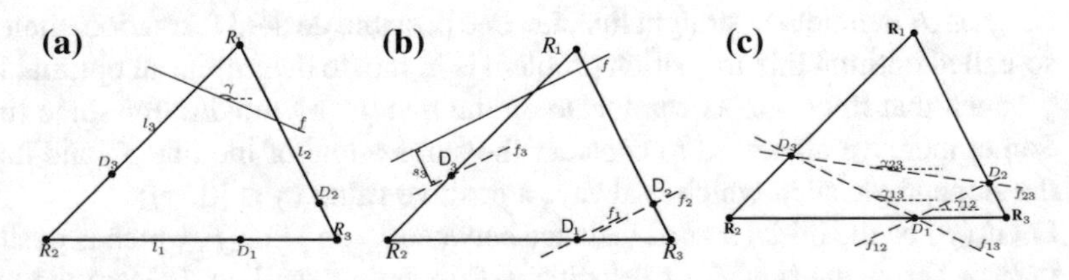

Fig. 10.1 The case of line formation problem

$$t_{23} = \frac{l_1}{v_2 + v_3} \sin\gamma$$
$$t_2^{'} = \frac{l_3}{v_2}|\sin(\angle R_2 - \gamma)| \quad - \frac{v_2 l_3}{(v_1 + v_3)v_2}|\sin(\angle R_3 + \gamma)|$$
$$t_{13} = \frac{l_2}{v_1 + v_3}|\sin(\angle R_3 + \gamma)|$$
$$t_3^{'} = \frac{l_2}{v_3}|\sin(\angle R_3 + \gamma)| \quad - \frac{v_1 l_3}{(v_1 + v_2)v_3}|\sin(\angle R_2 - \gamma)|$$
$$t_{12} = \frac{l_3}{v_1 + v_2}|\sin(\angle R_2 - \gamma)|$$
$$\bar{t}_{12} = \frac{l_1}{v_2 + v_3} \sin\gamma_{12}$$
$$\bar{t}_{13} = \frac{l_2}{v_2 + v_3}|\sin(\gamma_{23} + \angle R_3)|$$
$$\bar{t}_{23} = \frac{l_1}{v_2 + v_3} \sin\gamma_{23}$$

Lemma 10.2.1 (Special Line Passing D_1, D_2 or D_3)

(i) Robot R_j and robot R_k cost the same time to move to the line f_i, and the time that they cost is t_{jk}, which is defined in Definition 10.2.15, and the time that robot R_i cost is $t_i^{'}$, which is defined in Definition 10.2.15.

(ii) All the three robots R_1, R_2 and R_3 cost the same time to move to the line f_{ij}, and the time that they cost is $\bar{t}_{ij}$, which is defined in Definition 10.2.15.

Definition 10.2.16 For a given line f, we can translate it to pass any specified point, and we denote the *signed* distance of translation by s, which is negative (positive) when line f is translated by positive (negative) direction along the y-axis.

Lemma 10.2.2 (General Line Not Passing D_1, D_2 and D_3)

(i) For any given line f with a certain slope angle γ, f can be translated to pass the point $D_j (j = 1, 2, 3)$ with the same slope angle γ, and the signed *distance of translation is s_j as indicated in Fig. 10.1b, then the time that robot R_i costs to move to the line f can be calculated by*

$$t_i = \frac{|\bar{d}_i(f)|}{v_i} = \frac{|\bar{d}_i(f_j) + s_j|}{v_i}, \quad i \in 1, 2, 3. \tag{10.29}$$

(ii) For any given line f with a certain slope angle γ, which does not pass the points D_1, D_2 and D_3, we have

$$\min\{t_1, t_2, t_3\} > \min\{t_{12}, t_{13}, t_{23}\}. \tag{10.30}$$

where t_i's and t_{ij}'s are defined in Eq. (10.29) and Definition 10.2.15, respectively.

Next, with Lemmas 10.2.1 and 10.2.2, we may present our theoretical results for the three-robot optimal line formation problems, whose proofs are all omitted to save space.

10.2.2.2 Fixed-Angle Line Formation

Theorem 10.2.1 *For a three-robot group with any initial group status $S(0)$, if their desired formation is* fixed-angle line formation *given in Definition 10.2.14, suppose that the desired slope angle is γ, the points D_1, D_2, and D_3 are defined in Definition 10.2.15, then*

(i) The minimal formation time T^* is

$$T^* = \min\{t_{12}, t_{13}, t_{23}\}. \tag{10.31}$$

(ii) The corresponding optimal formation f^* is the line f_1 if

$$T^* = t_{23} \tag{10.32}$$

or the line f_2 if

$$T^* = t_{13} \tag{10.33}$$

or the line f_3 if

$$T^* = t_{12} \tag{10.34}$$

10.2.2.3 Line (Raw) Formation

Theorem 10.2.2 *For a three-robot group with any initial group status $S(0)$, if their desired formation is* raw line formation *given in Definition 10.2.13, then*

(i) The minimal formation time T^* is

$$T^* = \min\{\bar{t}_{12}, \bar{t}_{23}, \bar{t}_{13}\}. \tag{10.35}$$

(ii) The corresponding optimal formation f^* is the line f_{12} if

$$T^* = \bar{t}_{12} \tag{10.36}$$

or the line f_{23} if

$$T^* = \bar{t}_{23} \tag{10.37}$$

or the line f_{13} if

$$T^* = \bar{t}_{13} \tag{10.38}$$

Remark 4 *Raw line formation* is the most basic line formation, which contains least constraints in the formation set $\mathcal{F}$, hence the corresponding formation set $\mathcal{F}$ is much larger than other line formations.

10.2.3 Simulation Results

In last subsection, we have presented two theoretical results about minimum-time three-robot line formation problems. We will conduct some simulation experiments to verify these conclusions in this part.

In the plane, the initial positions of three robots can be arbitrarily set to make them form different types of triangles. After setting the initial groups status, we may use the theoretical results given in last subsection to calculate optimal T^* and the corresponding optimal desired line formation f^*. To verify the validity of the theoretical results, for other arbitrarily chosen line f, whose equation can be expressed in the two-dimensional Cartesian coordinate system with two parameters γ and c as follows

$$x \sin \gamma - y \cos \gamma + c = 0, \tag{10.39}$$

we can calculate the maximal $T_{\max}(f) = \max\{t_1, t_2, t_3\}$, and then check whether $T_{\max}(f) \geq T^*$. By changing the parameters γ and c in a large range, we need only verify that $T_{\max}(f) \geq T^*$ always hold and the value T^* can be achieved for some parameters c^* and γ^* (obviously $\gamma^* = \gamma$ in case of fixed-angle line formation).

To illustrate Theorems 10.2.1 and 10.2.2, we suppose that

$$v_1 = 4\,\text{m/s}, \quad v_2 = 2\,\text{m/s}, \quad v_3 = 1\,\text{m/s} \tag{10.40}$$

where v_i is robot R_i's speed. In this section, we let the three robots' initial positions be $R_1(50, 86.60)$, $R_2(0, 0)$ and $R_3(136.60, 0)$. From the initial status, we can calculate the length of three sides as follows

$$l_1 = 136.60\,\text{m}, \quad l_2 = 122.47\,\text{m}, \quad l_3 = 100\,\text{m} \tag{10.41}$$

Then we obtain the three points' coordinates: $D_1(91.07, 0)$, $D_2(119.28, 17.32)$, and $D_3(16.67, 28.87)$ according to Eq. (10.26). Next, we will do some simulations for two cases of optimal line formation problems.

Fixed-Angle Line Formation

In this experiment, we take the slope angle of the desired optimal formation as $\gamma = 71.57°$. To find the optimal line formation f^* and corresponding minimum time T^*, since the slope angle of the line f^* is fixed as γ, we need only to find

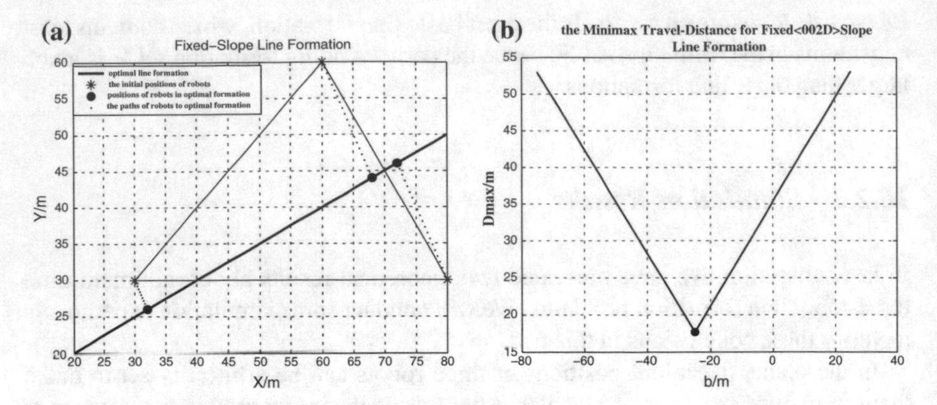

Fig. 10.2 The case of fixed-angle line formation problem. **a** Optimal line formation. **b** Curve of formation time

the corresponding parameter c^* and then calculate T^*. By using Theorem 10.2.1, we obtain that the line f^* is the line f_1, whose parameter c is -87.40. So the line f^* is

$$0.95x - 0.32y - 87.40 = 0 \tag{10.42}$$

and the minimal formation time T^* is 43.20 s.

In Fig. 10.2a, the solid thick line illustrates the optimal line f^*, and the dot lines illustrate the paths on which three robots move to their optimal positions. For comparison, Fig. 10.2b depicts the curve of formation time $T_{\max}$ with respect to arbitrary desired line f with the certain slope angle $\gamma = 71.57°$, which is characterized by a single parameter c since parameter γ is fixed. In Fig. 10.2b, the big point with coordinate $(-87.40, 43.20)$ is just the value of (c^*, T^*) obtained from Theorem 10.2.1. As is seen, T^* has the smallest value of $T_{\max}$ in the figure, which coincides with the theoretical results given in Theorem 10.2.1.

Raw Line Formation

In this experiment, we also use two ways to find the line f^* and get the minimal formation time T^*. By using Theorem 10.2.2, we know that the optimal line f^* is line f_{13}, whose slope angle γ^* is 158.69°, and parameter c^* is -32.91. Then we conclude that the optimal line formation f^* is given by

$$0.36x + 0.93y - 32.91 = 0, \tag{10.43}$$

and the corresponding formation time T^* is 16.47 s.

Figure 10.3a depicts the optimal line f^* with the solid thick line and the robots' optimal paths with the dot lines. For comparison, Fig. 10.3b depicts the surface of formation time $T_{\max}$ (z-axis) w.r.t. arbitrary desired line f, which is characterized by two parameters c (x-axis) and γ (y-axis). In Fig. 10.3b, the big point with coordinate $(-32.91, 158.69, 16.47)$ is just the value of (c^*, γ^*, T^*) obtained from The-

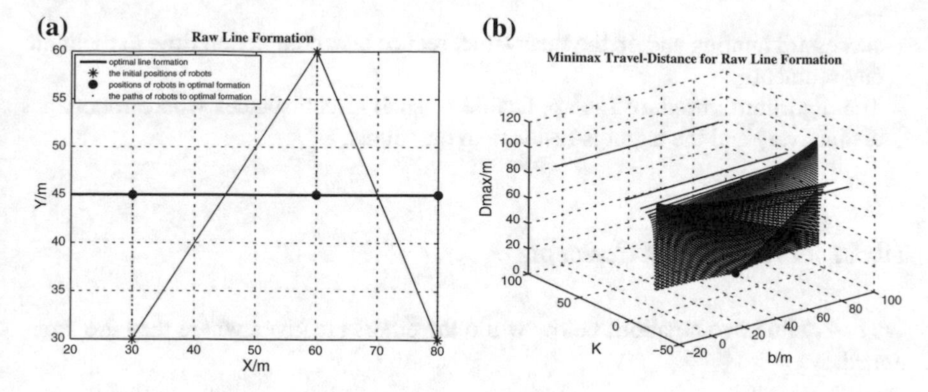

Fig. 10.3 The case of raw line formation problem. **a** Optimal line formation. **b** Surface of formation time

orem 10.2.2. Obviously, T^* is the smallest value of $T_{\max}$, which coincides with the theoretical results given in Theorem 10.2.2.

10.3 Multirobot Cooperative Pursuit

In last section, the optimal formation problems of multirobot systems have been investigated. The formation and hunting of the multirobot system have been extensively researched as formations applications received increasing attention [9–12]. Researches on formation are also biologically-inspired [13, 14]. As is often the case, researchers decompose the process of hunting behaviors into several steps [13, 14], and plenty of approaches have been proposed to cope with the cases of more than one evader [15–18]. In [15], the authors threw light on the alliance conditions and designed one allied hunting strategy on account of the circular-elliptic besieging circles. In [16], task allocation was analyzed in terms of an improved auction algorithm. Problems regarding pursuit also exert a dramatic impact on multirobot systems. In [19], the authors established a couple of essential concepts including detectability, the vision zone, the range and field of the view, etc. And later they developed several associated fundamental concepts including attackability [20]. In-depth discussions of these methods, to a certain extent, give answer to the task allocation of the multitarget systems and their path planning, as well as other multirobot problems.

This section investigates the problems on hunting activities of a multirobot system in dynamic environment [21]. The theoretical work of this section has the following motivations:

- Necessary conditions of launching one successful hunting with a multirobot system and the external conditions of an optimal hunting path are established a mathematical model according to certain geometrical principles.
- An optimized pursuit strategy with two controllable parameters (k and m) is proposed to make a relatively more stable system, which can improve the rate of

successful hunting and, in the meantime, reduce task completion time in dynamic environment.

- The algorithm introduced is applicable to such circumstances where robots act distinctively and the target is faster than the robots.

10.3.1 Preliminary Concepts

We first define two notations below while the others are given where they are mentioned:

- V_i: Maximum speed of robots ($i = 1, 2 \ldots, n$).
- V_t: Maximum speed of the target T.

In order to escape as quickly as possible, the target should always escape at the maximum speed in the algorithm simulation. Therefore, without loss of generality, in the later discussions, we only consider a worst case when $V_t \geq \max\{V_1, V_2, \ldots, V_n\}$.

10.3.1.1 Occupation

Assume that the robots and the target are in an infinite plane Γ. To make it easy to understand, we establish the Cartesian coordinate system with the origin on the target, robot R_i on $(x_i(t), y_i(t))$ at a certain moment. Choose a point (x, y) in Γ that satisfies:

$$\frac{\sqrt{(x - x_i(t))^2 + (y - y_i(t))^2}}{\sqrt{x^2 + y^2}} = \frac{V_0}{V_t}(V_0 \leq V_t) \tag{10.44}$$

Equation (10.44) describes an ellipse surrounding R_i. Points on the ellipse can be reached by R_i and T at the same time, while points inside the ellipse are less time-consuming for R_i, and the situation is opposite to those outside of the ellipse.

Get a random point F on the ellipse. Let $\angle\frac{\theta}{2} = \angle FTR_i$, $\angle\beta = \angle FR_iT$, according to the sine theorem $V_i(t_0)t/\sin(\frac{\theta}{2}) = V_t t/\sin\beta$, which means that $\sin\frac{\theta}{2} = \frac{V_i(t_0)}{V_t}\sin\beta$, hence we have

$$\theta_{\max} = 2\sin^{-1}\frac{V_i(t_0)}{V_t}|_{\beta=\frac{\pi}{2}} \tag{10.45}$$

In this case, line *TF* is tangent to the ellipse. $\theta_{\max}$ is the angle (known as θ) contained by the 2 tangent lines crossing T. It can be proved from Eq. (10.45) that θ is only relevant to the ratio of the two velocities V_i and V_t, while independent from the distance between the robots and the target d_i. Assume $k_i = V_i(t_0)/V_t$. In order to achieve the goal as quickly as possible, the robots must travel at their maximum speed. So we have $V_i(t) = V_i = constant$. Therefore, there is a parameter $k_i = V_i / V_t$ for each robot describing its properties.

Define the robot R_i as $R_i(\alpha_i, \rho_i, k_i, t)$, and (α_i, ρ_i) is the coordinate in the polar coordinate system at time t with the target on the origin.

The whole plane Γ is separated into four zones known as A, B, C, and D shown in Fig. 10.4 by the 2 tangent lines mentioned above. Suppose that R_i is able to make the optimal choice (the procedure of how the choice is made will be discussed in further detail in the following paragraphs). Meanwhile, assume that TR_i is the axis of the polar coordinate system. At the time of t_0, provided that $\angle \overrightarrow{V_t} \in [-\frac{\theta}{2}, \frac{\theta}{2}]$, and the direction of the vector does not change, the target will be captured in zone A, and will never reach zone D in this case, for which it will be ultimately captured by the robot. As for the condition $\angle \overrightarrow{V_t} \notin [-\frac{\theta}{2}, \frac{\theta}{2}]$, the target will be free from being caught.

It is worthy of remark that in the process of the decision-making, modeling is on the basis of the target's being at the origin all the time, so that although the speed of the target varies from time to time, it remains unchanged for $\overrightarrow{V_t}$ at a certain moment.

There are some conclusions as followed:

Occupied criterion: The *hunting angle* of a robot to the target is known as $\theta = 2\sin^{-1} k_i$, where $k_i = V_i(t_0)/V_t(t_0)$ $(V_i > V_t)$, indicates that the robot occupies the area contained by the angle θ at time of t_0.

A necessary but insufficient condition of $R_i (i \in Z^+)$'s absolute capturing of the target can be described as below:

$$\sum_{i=1}^{n} \theta_i \geq 2\pi \tag{10.46}$$

So we can easily obtain:

$$\theta_{\max} = \pi|_{\frac{V_i}{V_t} \to 1^-}$$

$$n_{\min} = 3(V_i < V_t)$$

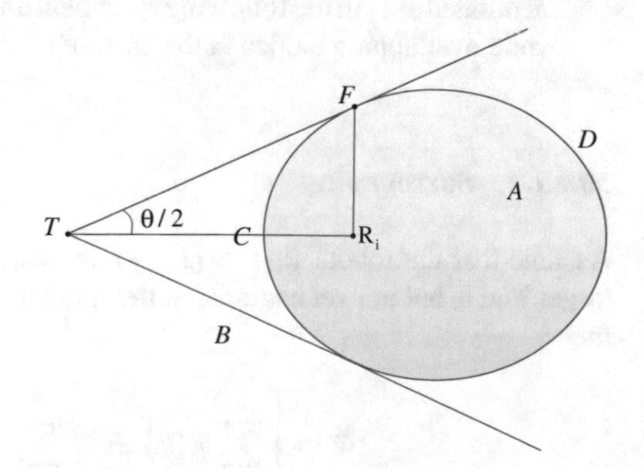

Fig. 10.4 The occupied area of a robot in pursuing the target

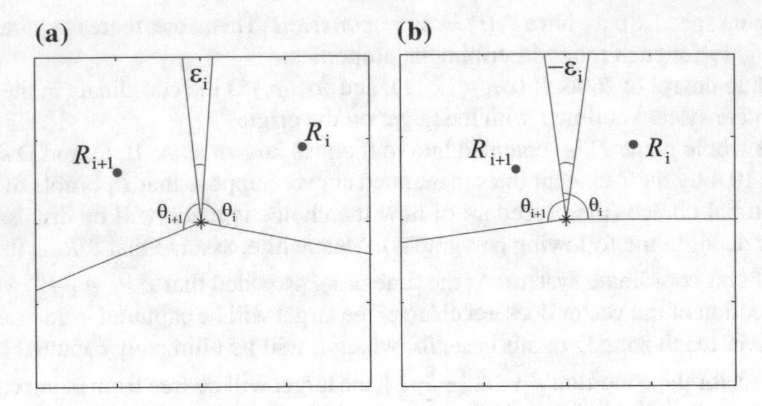

Fig. 10.5 The gap angle and overlapping angle of two robots in pursuing the target

10.3.1.2 Overlapping Angle

Suppose that at one point, two robots adjacent to each other (in terms of angle α, counter-clockwise, $\alpha \neq \alpha_{i+1}$), $R_i(\alpha_i, \rho_i, k_i, t_0)$ and $R_{i+1}(\alpha_{i+1}, \rho_{i+1}, k_{i+1}, t_0)$ are to trap the target within the down-half plane, conditions of their hunting angles θ_i and θ_{i+1} are as followed:

(i) $|\alpha_i - \alpha_{i+1}| > |\frac{\theta_i}{2} + \frac{\theta_{i+1}}{2}|$
Two gaps are formed between θ_i and θ_{i+1} as shown in Fig. 10.5a, with the gap angle $\varepsilon_i = |\alpha_i - \alpha_{i+1}| - |\frac{\theta_i}{2} + \frac{\theta_{i+1}}{2}| > 0$, In order that the target gets restricted within the down-half plane, measures should be taken to diminish ε_i, which will be discussed in full detail in the following chapter.

(ii) $|\alpha_i - \alpha_{i+1}| < |\frac{\theta_i}{2} + \frac{\theta_{i+1}}{2}|$
In this case (shown in Fig. 10.5b), there is an overlap between θ_i and θ_{i+1}, and $\upsilon_i = |\alpha_i - \alpha_{i+1}| - |\frac{\theta_i}{2} + \frac{\theta_{i+1}}{2}| < 0$ which should be lessened to ensure maximum area R_i and R_{i+1} occupy. ε_i and υ_i are defined by same way, and they are both mentioned as ε_i in the following statement and are called gap angles when $\varepsilon_i > 0$, while overlapping angles in the case of $\varepsilon_i < 0$.

10.3.1.3 Surrounding

Assume that the robots $R_i(i \in [1, n]\ n \geq n_{\min})$ have successfully surrounded the target T at t_0 but not yet captured at the moment. The following condition should be met:

$$\Phi = \left[\sum_{i=1}^{n} \theta_i(t_0) + \sum_{i \in \{\varepsilon_i(t_0) < 0\}} \varepsilon_i(t_0) \right] \geq 2\pi \tag{10.47}$$

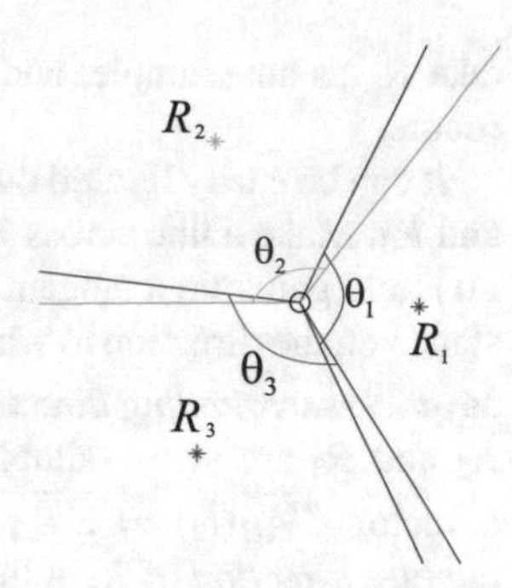

Fig. 10.6 The target is surrounded by three robots.

If the condition above, Eq. (10.47), is not satisfied, all the positive ε_i ought to be reduced, even to below zero. R_i's hunting angle θ_i is indicated in the Fig. 10.6 which also shows that the three robots have already surrounded the target. Φ is known as the angle given in Eq. (10.47).

10.3.2 Hunting Strategy

10.3.2.1 Quick-Surrounding and Quick-Capture Direction

Suppose that the position of R_1, R_2, R_3, and T at the time of t_0 are as shown in Fig. 10.7. After a time period Δt, which is infinitely short, the target $T(t_0 + \Delta t)$ will be situated where the square stands. At t_0 point, R_1, R_2, R_3 possess the speed of V_1, V_2, V_3 for which at $t_0 + \Delta t$, R_i will definitely be located on the circle with the center of (α_i, ρ_i) (the center will be referred to as $R_i(t_0)$ in the statement below) and a radius of $V_i \cdot \Delta t$ as well. To develop our analysis in a brief and effective way, we

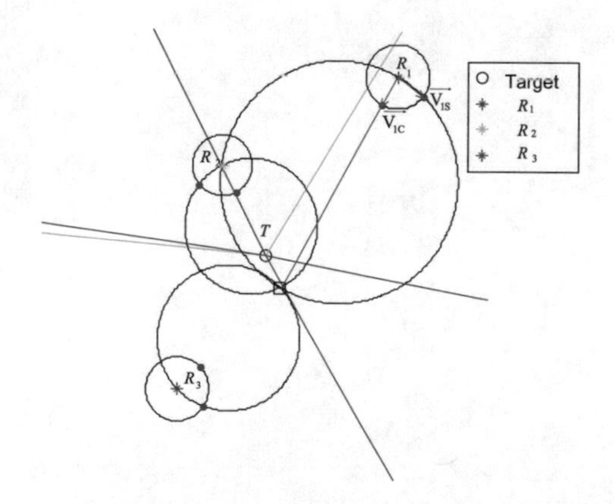

Fig. 10.7 Quick-surrounding and quick-capture direction of robot R_1

take R_1 as an example, and the same conclusions can be shared with all the other robots.

It can be easily figured out that at the time of t_0, gap angle $\varepsilon(t_0)$ exists for robot R_1 and R_3. Make a line across $T(t_0 + \Delta t)$ which is tangent to the circle $\bigcirc_1(R_1(t_0), V_1 \cdot \Delta t)$, and generate a tangent point $R_{1s}(t_0 + \Delta t)$ on the side of the gap angle, which stands on the direction to which $|\varepsilon_1(t_0)|$ shrink most rapidly. This direction is defined as *quick-surrounding direction* $\overrightarrow{V_S}$. Besides R_1, the surrounding directions of robots R_2 and R_3 are also available in Fig. 10.7. Link the points $R_i(t_0)$, $T(t_0 + \Delta t)$ with a vector $\angle\overrightarrow{V_{iC}}(t_0) = \angle\overrightarrow{R_i, T(t_0 + \Delta t)}$ $(i = 1, 2 \ldots n)$ that introduces the *quick-capture direction* of R_i at the time of t_0.

10.3.2.2 Optimization Decisions

As the directions of $\overrightarrow{V_C}$ and $\overrightarrow{V_S}$ have already been obtained, it is rather simple to demonstrate that $\lambda = \angle(\overrightarrow{V_C}, \overrightarrow{V_S}) < \pi/2$, and R_i's optimal decision at t_0 point satisfied:

$$\angle\overrightarrow{V_i}(t_0) \in \begin{cases} (\angle\overrightarrow{V_C}, \angle\overrightarrow{V_S}) & \angle\overrightarrow{V_S} > \angle\overrightarrow{V_C} \\ (\angle\overrightarrow{V_S}, \angle\overrightarrow{V_C}) & \angle\overrightarrow{V_S} < \angle\overrightarrow{V_C} \end{cases}$$

Simplify the model by positioning $\bigcirc_i(R_i(t_0), V_i \cdot \Delta t)$ together with $\overrightarrow{V_{iC}}(t_0)$, $\overrightarrow{V_{iS}}(t_0)$ and $\overrightarrow{V_i}(t_0)$ in a single Fig. 10.8, and define $\beta = \angle(\overrightarrow{V_i}(t_0), \overrightarrow{V_{iC}}(t_0))$, the intersection angle of $\overrightarrow{V_i}(t_0)$ and $\overrightarrow{V_{iC}}(t_0)$, as the *deviation angle* of decision.

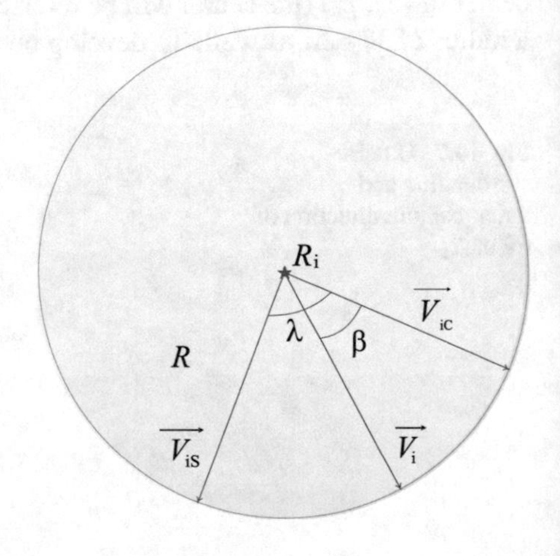

Fig. 10.8 Decision V_i based on V_{iC} and V_{iS}

The optimized decision is given as follows:

$$\begin{array}{ll} \beta = \lambda(1 - e^{-k\delta_i\gamma_i}) & k \geq 0 \\ \delta_i = |\frac{\varepsilon_i - \varepsilon_{i-1}}{2\pi}|^{\frac{1}{m}} & m > 1 \\ \gamma_i = \sin[\pi(\frac{d_i}{nEd})^{\log_n^2}] & (\exists j)d_j \neq 0 \end{array} \quad (10.48)$$

where the parameters are listed below

k: Decision Influencing Factor
δ: The surrounded factor under the current situation of the robot
γ: Capture factor under the current situation of the robot
m: Factor that has an impact on the surrounded factor
$\varepsilon_i - \varepsilon_{i-1}$: The difference of the gap angles between R_i and its two adjacent robots
Ed: Expectations of distance between all robots and the target in current situation
n: Number of robots.
This decision meets with the following qualities:

- The value of $\overrightarrow{V_C}$ should peak by the time the target becomes surrounded by the robots. Otherwise, $\overrightarrow{V_S}$ will obtain a high value if Φ is relatively too small.
- It is suitable on condition that every robot's final state position is requested critically, meaning it is necessary for the robots to spread evenly around the target with their mission terminated. If it is not necessary for robots to achieve this, δ_i can be taken as:

$$\delta_i = \begin{cases} |\frac{\varepsilon_i - \varepsilon_{i-1}}{2\pi}|^{\frac{1}{m}} & \Phi < 2\pi, m > 1 \\ 0 & \Phi = 2\pi \end{cases}$$

- The value of $\overrightarrow{V_C}$ and $\overrightarrow{V_S}$ is concerned with the properties of the task. If the rate of success counts more than the task completion time, $\overrightarrow{V_S}$ will account for a relatively bigger proportion with a larger k.
- If the distances to the targets vary drastically from one robot to another, which means a big variance of ρ_i, the $\overrightarrow{V_C}$ of the robots much further from the target should be raised higher than those closer to it.
- When every robot performs distinctively to accomplish separate tasks, for example, one of the robots is responsible for outflanking the target from outside the periphery, and we are able to achieve this distinction by varying k. However, the k of each robot will be set the same if such requirement does not exist and that all robots share the same quality.
- We can change the value of m when only the surrounded factor's impact on the decision is to be adjusted. A bigger m represents a smaller influence by the surrounded factor.
- When a robot is located too far or too close from the target, $|\overrightarrow{V_C}|$ will be larger automatically.
- To the robot R_i, if $d_i / \sum_{j=1}^{n} d_j \rightarrow \frac{1}{n}$ is satisfied, then $\overrightarrow{V_S}$ will attain an approximately peak value.

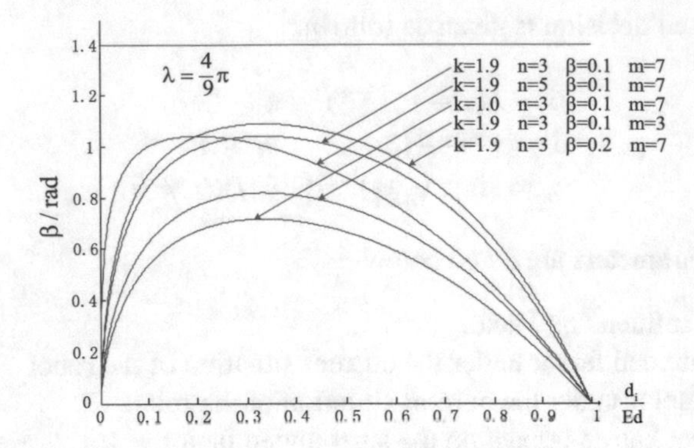

Fig. 10.9 Parameters' influence on decision

The decision (Eq. (10.48)) shows that the variations of the factors have an influence on decision. Figure 10.9 illustrates how the deviation angle changes with the variation of the surrounded factor, the capture factor, as well as factor k:

The theoretical range of k is $[0, \infty]$, and simulation indicates a better result with $k \in [1, 3]$. However, this range may not remain the same due to the diversity of the missions. A bigger k is needed when the rate of success is more essential than the completion time.

The influence on the determination by a single factor m is shown in Fig. 10.10. The curves give the trend how the *deviation angle* varies with the change of γ. Two situations are taken as an example when $m = 1$ and $m = 7$, and the curves provides a changing value of $(\varepsilon_i - \varepsilon_{i-1})/2\pi$ from 0.3 to 0.9. Factor m rising, as is seen in Fig. 10.10, the surrounded factor affect the determination less strongly.

10.3.3 Simulation Study

During the simulation, several simplification and assumptions are carried out:

- To ensure that the task accomplishment is excellent and efficient, all robots travel at its maximum speed V_i.

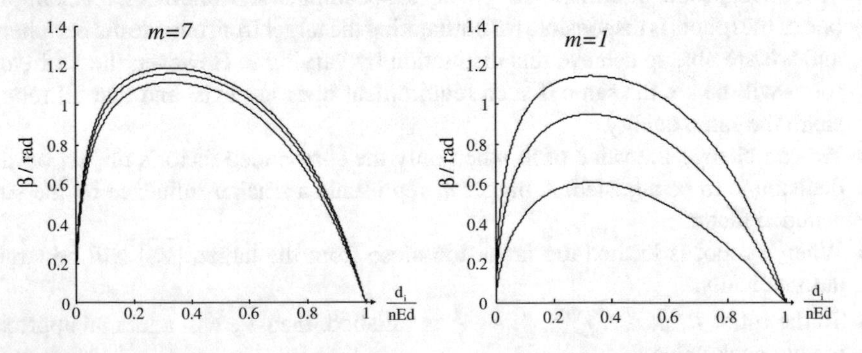

Fig. 10.10 The influence of m on the decision

- Changes to all directions are available and are not time consuming.
- The target and the robots are in an infinite plane. So failure is inevitable when the target's performance and intelligence are absolutely superior to the robots'.
- The robots and the target are both regarded as particles. The target is respected as captured when its distance to the robots decrease to less than the value of d_0. During the simulation, Δt is set as 0.2 s, and $d_0 = V_i \cdot \Delta t$.
- Mission is accomplished when one of the robots approaches the target within the range of d_0.

The robots make their decisions dynamically that are still discrete no matter how fast they calculate and move, for which a gap of Δt will appear between decisions, when the robot cover a distance of $d_0 = V_i \cdot \Delta t$. d_0 is named as step length, and it has an effect on the real-time performance of the system. Usually the target will not abruptly change its velocity, making it possible to forecast its position to some extent.

Also, the step length is allowed to vary with the distinctive motivations of different targets, for example, a man with a strong motive, and a car that picks its path at random. In the case of $d_0^{car} < d_0^{man}$, quick decision will not make any sense. The paths of the target and robots during the simulation are shown in Fig. 10.11a, and the robots' initial positions are as below:

$$R_1(-0.32, 3.16, 0.90, 0)$$
$$R_2(+2.03, 2.20, 0.85, 0)$$
$$R_3(+3.14, 1.00, 0.90, 0)$$

Parameters of decision: $k = 1.9, m = 7$. Although R_1, R_2 and R_3 adopt the same strategy provided in formula Eq. (10.48), the three robots are capable of taking different actions according to their circumstances. At the initial state, R_3 is closest to the target. Therefore, its duty is to drive the target within a certain range of area where R_1 and R_2 outflank the target. And the target is eventually caught by R_1.

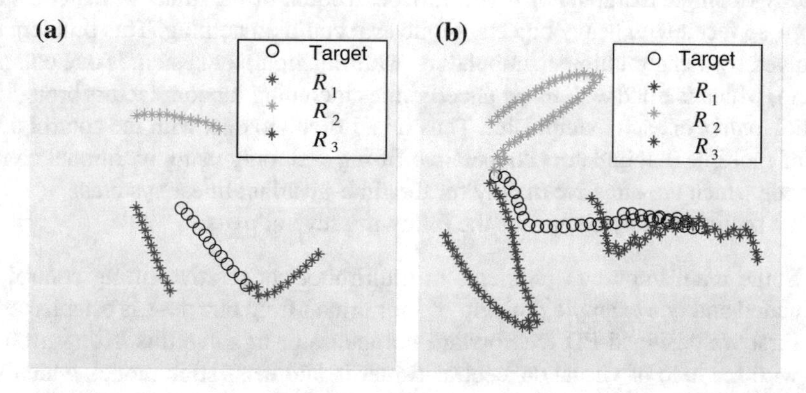

Fig. 10.11 Algorithm simulation in dynamical environment

If the target's intelligence increases, its decision will be based on the robots' dynamic positions, as is stated in the formula Eq. (10.49). The routes presented in the Fig. 10.11b indicate that the robots will ultimately hunt down the target even if the environment is dynamically changing.

$$\angle \overrightarrow{V_t}(t) = \angle\left(\frac{\overrightarrow{V_1}(t)}{d_1(t)} + \frac{\overrightarrow{V_2}(t)}{d_2(t)} + \frac{\overrightarrow{V_3}(t)}{d_3(t)}\right) \tag{10.49}$$

This section elucidates the hunting issue of a multirobot system in dynamic environment. First, we define the concepts of *occupy*, *overlapping angle*, etc., then we provides the necessary condition of developing a successful surround-and-capture process with the robots' velocities different from each other.

It also substantiates the range of an optimal path based on the quick-capture direction and quick-surrounding direction obtained, as well as gives an optimized solution effectiveness is proved by simulation.

This geometry-based strategy has the advantages of fast calculation and can be applied to three-dimensional space easily. For example, we can establish a cone with its apex locating on the target and its center line crossing R_i according to the conception of occupied criterion. We are able to get the occupied space in the cone whose leaning angle of generating line is relevant to the velocity ratio while has nothing to do with the distance between the target and the air vehicle. In this way, problems in plane can be transferred to three-dimensional space easily.

10.4 Multirobot Cooperative Lifting

Nowadays, manipulators have been widely used in manufacturing industry and the manipulator can do many repetitive jobs such as carrying and lifting. The loading capacity of single manipulator is still low on account of the limits of materials and energy, so faced with heavy objects, single arm could do nothing. This problem can be solved by using multiple manipulators. Multimanipulator system is one complex system, which is filled with many uncertainties including random factors brought by inertial matrix of each manipulator. Thus it is difficult to deal with the control problem of multiple manipulators cooperative lifting well only using traditional control methods which are effective mainly for the time-invariant linear systems.

This section mainly comprises the following several parts:

(i) Some main important problems in multirobot cooperative lifting control are stated and one example problem of four arms lifting one desk is established.
(ii) First we designed PD feedforward compensator to solve this lifting problem with the help of virtual concentrated inertia and mass stick model, which was in fact single-arm control and then make simulations on this control method in MATLAB.

(iii) Applying adaptive thoughts into multiarm cooperatively lifting, we design the adaptive cooperative controller to finish this lifting task which has been proved asymptotic stable in the sense of Lyapunov stability. Simulations for the adaptive controller were also carried out in MATLAB, which turned out to be more efficient than the above PD controller in terms of average error and variance.

10.4.1 Problem Formation

Multirobot cooperative lifting has attracted great attention in research field. Many scholars have made deep research on the control methods of this problem. Generally speaking, the object that need to be lifted is one rigid body.

There are two main important problems concerning this lifting issue. One is how to distribute the size of the acting force applied on the rigid body among all concerning manipulators before the implementation of control. One direct method is to directly measure the forces of each manipulator by force sensors. The scheme of conducting hybrid force-position control for two cooperative manipulators by force sensors is proposed in the literature [22]. Force sensor is convenient to use but expensive and easily broken, so one compromising method is to establish one model which is used to reasonably distribute force acting on the rigid body among the arms. When the manipulators have been holding the rigid body, all of them form one closed-loop kinematic chain, where the stress and motion are intercoupling, which results in the fact that the solution of inverse kinematics is not unique. In the literature [23], one model called virtual concentrated inertia and mass stick model was proposed, and this article derived the explicit solution of thc forcc distribution for each manipulator by using the principle that the sum of inner forces at the mass center is zero for rigid body. The other problem is the estimation of the position of manipulators and one good method is to adopt vision. In the literature [24], the position information is obtained by vision, and then the task of people and HPR-2 robot lifting one desk cooperatively will be finished as shown in Fig. 10.12.

As an example, the circumstances of four manipulators lifting one rigid desk cooperatively is considered, as shown is Fig. 10.13 and the classic manipulator PUMA560 that has six degrees is chosen. The arm1 connected with Novint Falcon can receive position information from Falcon, and so to speak, it is the host arm. With reasonable control, the other arms can respond to the host arm and lift the desk finally, as we expected. Generally it is necessary to adopt vision process for position information of manipulators and rigid object, however, vision is not the main issue of this section. The whole lifting process can be divided into two stages. In the first stage, arm1 slowly lifts the desk upward with arm3 being one fulcrum like one lever. When arm1 is moving upward, some forces act on the end effectors of arm2 and arm4, because of the motion of the rigid body pulled by arm1. Then the virtual lumped mass model can measure the force and feed back to controller for controlling. In the second stage, arm3 in turn lift up its corresponding corner with arm1 being the

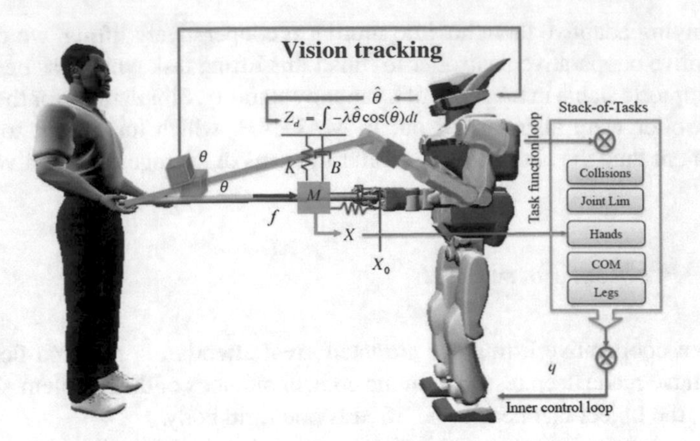

Fig. 10.12 Human subject and HPR-2 robot lifting one desk cooperatively

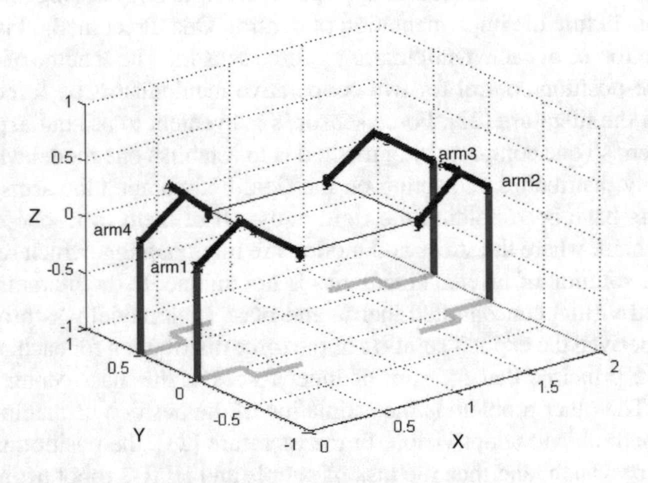

Fig. 10.13 Four manipulators lifting one rigid desk cooperatively

fulcrum. Then similarly, due to the force imposed on arm2 and arm4 as a result of rigid desk moving, according control for these two arms are implemented.

10.4.2 PD Feedforward Compensation Control

In order to resolving this lifting task, we first use one conventional kind of PD controller with feedforward complementation.

10.4.2.1 Dynamics Model of Rigid Body

Considering the mass center, the dynamics equations of the lifted object can be obtained by Newton–Euler equation as follow:

$$M\ddot{X}_c = \sum_{i=1}^{4} F_i + Mg \tag{10.50}$$

$$I_c\dot{\omega} = -\omega \times (I_c\omega) + \sum_{i=1}^{4} N_i + \sum_{i=1}^{4} L_i \times F_i \tag{10.51}$$

The main notations are listed in the Table 10.1. The last two equations are used to obtain F_i and N_i. If we directly solve the equations on the basis of given position information (X_c, ω) of mass center, the unique and explicit solution could not be obtained. Hence the model called virtual concentrated inertia and mass stick model is proposed, and by it the load forces are distributed among all the manipulators.

The so-called virtual concentrated inertia and mass stick model, depicted in Fig. 10.14 is to assume that one zero-mass rigid rod has one mass-lumped and tensor-lumped end and all the external forces and torques act on the other no-mass end. Let

Table 10.1 Symbol meaning

Symbol	Description
M	The rigid object mass
I_c	The inertial tensor
$\ddot{X}_c$	The acceleration of rigid body
ω	The angle speed of rigid body
L_i	The vector from mass center of object to the grasp point of the ith arm
F_i	The applied force on object by the i arm
N_i	The applied torque on object by the i arm

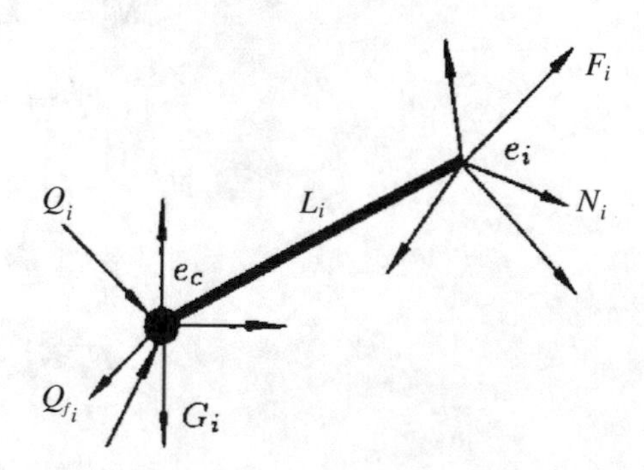

Fig. 10.14 The virtual concentrated inertia and mass stick model under external action

inertia force $Q_{f_i} = -m_i \ddot{X}_c$ and inertial torque $Q_{N_i} = -I_{ci}\dot{\omega} - \omega \times (I_{ci}\omega)$ and then according to d'alembert's principle, the dynamic equilibrium equation of the virtual concentrated inertia and mass stick model under external action in Fig. 10.14 is as follows:

$$Q_{fi} + G_i + F_i = 0 \tag{10.52}$$

$$Q_{N_i} + N_i + L_i \times F_i = 0 \tag{10.53}$$

where m_i is the rigid body mass that is assigned to the ith mass rod. Obviously the two equations $\sum_{i=1}^{4} m_i = M$ and $G_i = m_i g$ hold. The distributed mass and inertial tensor of rigid body are, respectively, concentrated on mass center, which result that the rigid body is equivalent to one point with mass and inertial tensor. Then by the above virtual concentrated inertia and mass stick model, we can put multiarm lifting a rigid body cooperatively into one model of multiarm holding a no-mass rigid rod connected to the centroid.

After simplification, we get

$$m_i = \xi_i M \tag{10.54}$$

$$I_{c_i} = \xi_i I_c \tag{10.55}$$

where ξ_i is the load capacity factor of the ith virtual lumped mass rod and obviously $\sum_{i=1}^{4} \xi_i = 1,\ i = 1, 2, 3, 4$. In this section, we choose four same arms, or four PUMA560, which have the same load capacity, hence $\xi_i = 0.25$.

According to the given motion information of the rigid body, the acting force F_i and acting torque N_i can be computed for each arm. This computation can be divided into two parts, one is the static force F_i^s and torque N_i^s for only balancing the gravity load G_i; the other part is F_i^d and N_i^d for not only balancing inertial force and inertial torque but also moving the object. Then the following equation set, or the force and torque distribution model for the problem of multimanipulator lifting object cooperatively can be obtained as follow:

$$F_i^s = -G_i \tag{10.56}$$

$$N_i^s = -L_i \times F_i^s \tag{10.57}$$

$$F_i^d = -Q_{f_i} \tag{10.58}$$

$$N_i^d = -L_i \times F_i^d - Q_{N_i} \tag{10.59}$$

$$F_i = F_i^s + F_i^d \tag{10.60}$$

$$N_i = N_i^s + N_i^d \tag{10.61}$$

See Fig. 10.15 for illustration of some quantities.

10.4.2.2 Controller Design

This subsection will mainly discuss the design of PD feedforward complementation controller, which is composed of linear PD terms and feedforward terms for desired trajectory of manipulators. By means of dynamics equation of rigid manipulator, the same controller for arm2 and arm4 can be designed as:

$$\tau_i = D_i(q_i)\ddot{q}_i + C_i(q_i, \dot{q}_i)\dot{q}_i + G_{r_i}(q_i) + J_{v_i}(q_i)^T F_{e_i} + \tau_{pd_i} \tag{10.62}$$

where $q_i \in R^6$ is the joint vector $D_i \in R^{6\times6}$ is the joint-space inertial matrix, $C_i \in R^{6\times6}$ is the Coriolis and centripetal coupling matrix, $G_{r_i} \in R^6$ is the gravity loading, and $J_{v_i}^T(q_i)F_{e_i} \in R^6$ is the joint forces due to external wrench and force applied at the end effector from the object. All of them are for the ith arm. Given the desired joint q_i^d, τ_{pd_i} is defined as:

$$\tau_{pd_i} = K_{p_i} e + K_{d_i} \dot{e},$$

$$e = q_i^d - q_i,$$

$$\dot{e} = \dot{q}_i^d - \dot{q}_i.$$

According to Newton's Third Law, we get $F_{e_i} = -[F_i^T \quad N_i^T]^T$.

As the previous subsection, in the first stage, when receiving position information from Novint Falcon, arm1 is lifting the desk with the static arm3 being the fulcrum, and due to geometrical relationship we have $X_c = \frac{1}{2}X_1$, $\dot{X}_2^d = \dot{X}_4^d = \frac{1}{2}\dot{X}_1$, where $\dot{X}_2^d$, $\dot{X}_4^d$ are the desired Cartesian-space velocity of arm2 and arm4. By the given initial joint $q_i(0)$, the q_i^d at the arbitrary time t during the stage can be obtained:

$$q_i^d(t) = \int_{t_0}^{t} \left(J_{v_i}(q_i)^+ \dot{X}_i^d\right) dt + q_i(0) \tag{10.63}$$

As a result, each term of the control torque in Eq. (10.62) can be determined. The control torque of the second stage can be obtained using the same method.

10.4.2.3 Simulation Studies

In what follows, we conduct the simulation for PD feedforward complementation controller in the environment MATLAB using RVC toolbox, or Robotics, Vision, and Control Toolbox, which is developed by Peter Corke, one professor of Queensland University of Technology. RVC toolbox is practical for kinematics model, robotic vision, and dynamics simulation.

The process was that first, through Novint Falcon, we gave arm1 one position information $X_1 = [0\ 0\ 0.5]$, which was only in the positive z-axis, that is to say, straight up. The proportional coefficient K_p and derivative coefficient K_d was determined to be $K_p = K_d = 5$ through a set of trial and error The overall control block diagram of arm2 in SIMULINK is shown in Fig. 10.16 as one example.

The trajectory errors were defined as:

$$E = X_1 - 2X_2 \tag{10.64}$$

and simulation results were given in Figs. 10.17, 10.18, 10.19, 10.20 and 10.21.

From the above simulation figures, we conclude that the trajectories of arm2 and arm4 were similar to the ones of arm1, while the difference was the magnitude and the former was half the latter, which was expected in the previous theory. The trajectory deviation between arm2 and arm1 did not exceed 6.5 % if neglecting the coefficient 2, which reflected multirobot cooperation, the subject of this section. The end scenes of two lifting stages are, respectively, depicted in Figs. 10.22 and 10.23.

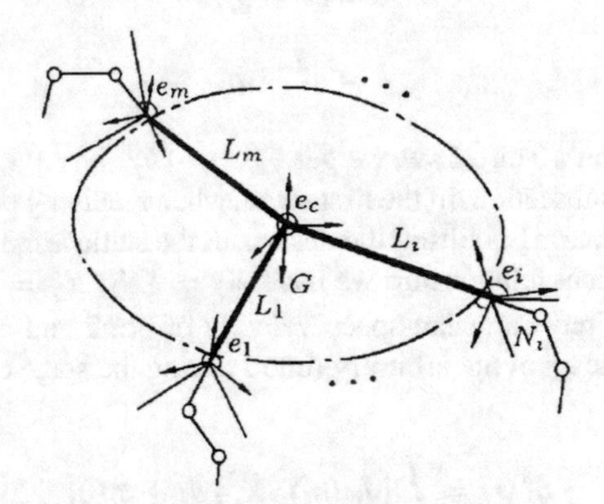

Fig. 10.15 The system applied force figure after simplification

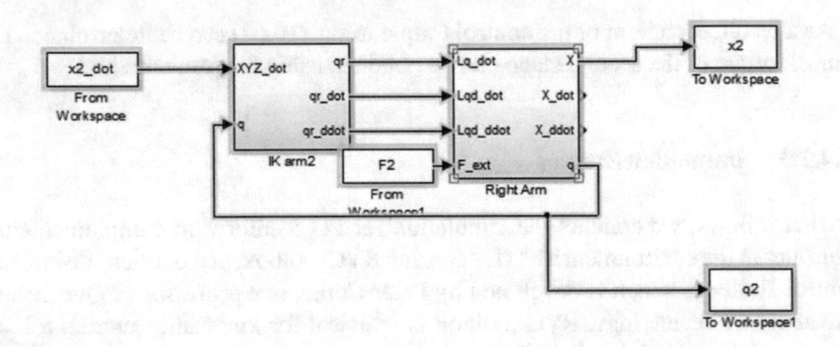

Fig. 10.16 The overall control block diagram of arm2 in SIMULINK

Fig. 10.17 The position trajectory of arm1

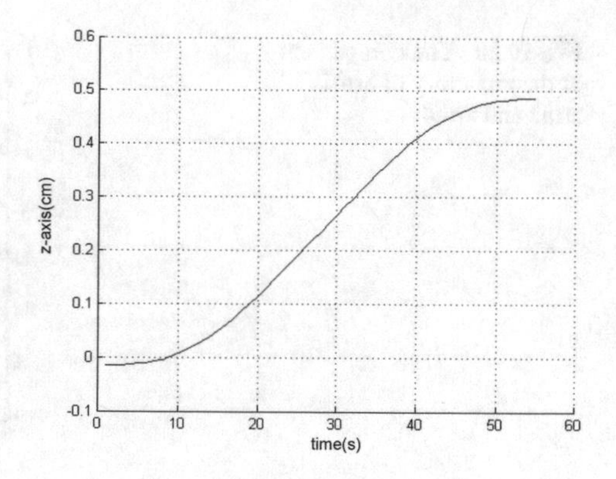

Fig. 10.18 The position trajectory of arm2

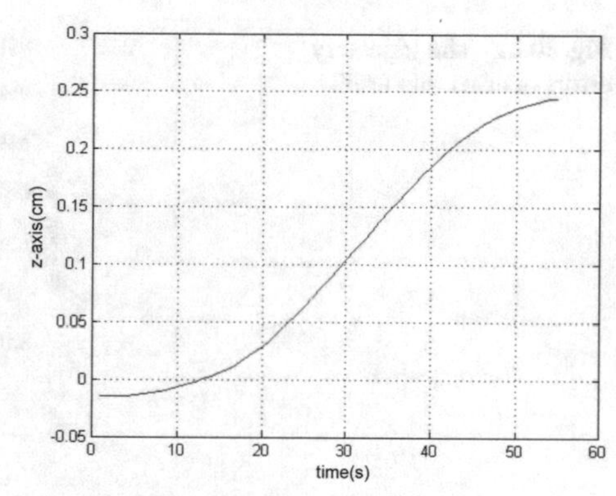

Fig. 10.19 The position trajectory of arm4

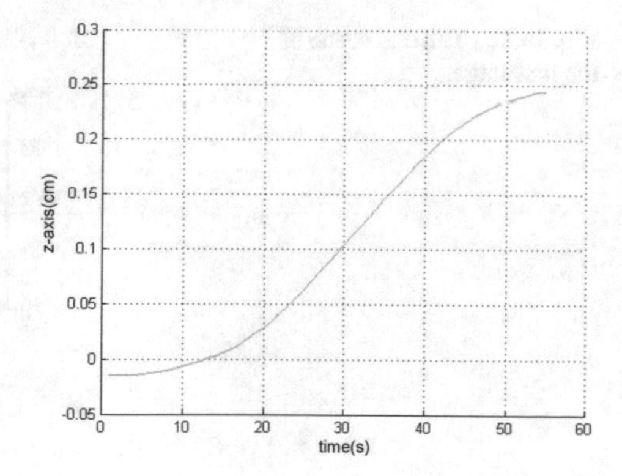

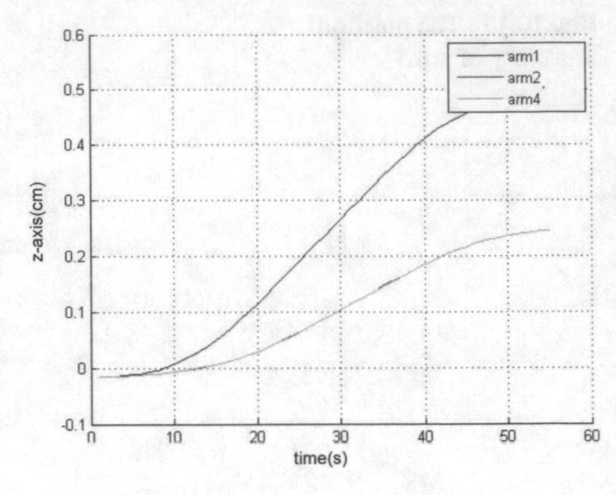

Fig. 10.20 The comparison of the trajectory of arm1, arm2 and arm4

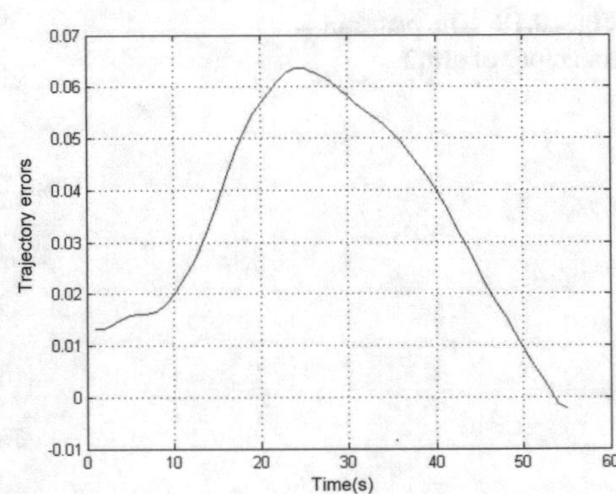

Fig. 10.21 The trajectory errors of arm1 and arm2

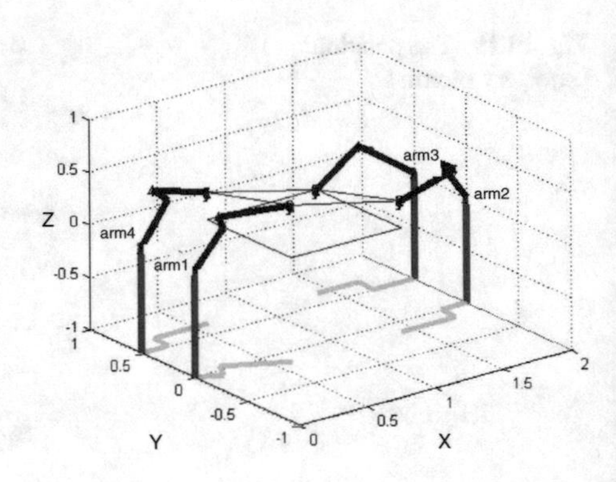

Fig. 10.22 The end scene of the first stage

10.4.3 Adaptive Control

In the previous subsection, we have adopted one classical controller—PD feedforward complementation to solve the multimanipulator lifting problem. This kind of method is essentially open-loop and assumes that the parameters of arms are of constant values available to the designer. Facing one manipulator system with some

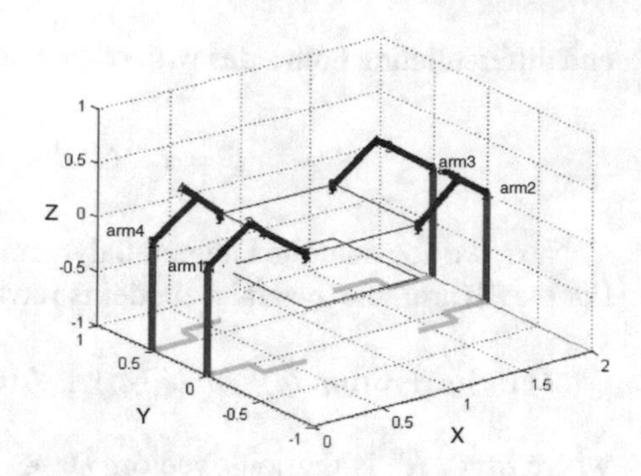

Fig. 10.23 The end scene of the second stage

time-variant parameters, the traditional controller is supposed to be replaced by one effective controller, where we use adaptive controller.

10.4.3.1 Integrated Model of Multiarm Lifting

In fact during the design of PD controller, each arm was separately handled with, that is to say, the dynamics model of each arm was independent. Generally speaking, there are coupling between each arm and hence it is more reasonable to treat all arms as well as the rigid-body object—the desk, as one integrity, which is one complexed and nonlinear system and need one novel mathematical model.

At first, we consider the kinematics model of multiarm. For one single manipulator, by Jacobian matrix the following velocity relationship can be obtained

$$\dot{x}_{e_i} = J_i(q_i)\dot{q}_i, \quad i = 1, 2, 3, 4 \tag{10.65}$$

where $x_{e_i} \in R^{6\times 6}$ is the pose of the ith arm and the pose of all the manipulators can be rewritten in vector form as

$$v_c = J\dot{q} \tag{10.66}$$

where $v_c = [\dot{x}_{e_1}^T \ \dot{x}_{e_2}^T \ \dot{x}_{e_3}^T \ \dot{x}_{e_4}^T]^T$ and J is one block-diagonal matrix, whose block element is $J_i(q_i)$. By the formula Eq. (10.75), we get

$$F_o = GF_e \tag{10.67}$$

and the duality of force and velocity [25], we get

$$G^T\dot{x} = v_c \tag{10.68}$$

Combining Eq. (10.66), hence $G^T\dot{x} = J\dot{q}$. Then the $\dot{q}$ is

$$\dot{q} = J^{-1}G^T\dot{x} \tag{10.69}$$

and differentiating both sides with respect to time for the above equation, we obtain

$$\ddot{q} = J^{-1}G^T\ddot{x} + \frac{d}{dt}(J^{-1}G^T)\dot{x} \tag{10.70}$$

Next, we consider the integrated dynamics model for multirobot and rigid body. For the ith arm, the dynamics model is rewritten as follows

$$D_i(q_i)\ddot{q}_i + C_i(q_i, \dot{q}_i)\dot{q}_i + G_i(q_i) + J_i(q_i)^T F_{e_i} = \tau_i, \quad i = 1, 2, 3, 4 \tag{10.71}$$

where $q_i \in R^6$ is the joint vector, $D_i \in R^{6\times6}$ is the joint-space inertial matrix, $C_i \in R^{6\times6}$ is the Coriolis and centripetal coupling matrix, $G_{r_i} \in R^6$ is the gravity loading, $J_{v_i}^T(q_i)F_{e_i} \in R^6$ is the joint forces due to external wrench and force applied at the end effector from the rigid object, and $\tau_i \in R^6$ is the control input torque. The term $F_{e_i} \in R^6$ is composed of $f_{e_i} \in R^3$ and $\eta_{e_i} \in R^3$. All of them are for the ith arm. All the arms can be combined into one equation in diagonal matrix form as follows:

$$D(q)\ddot{q} + C(q, \dot{q})\dot{q} + G_r(q) + J(q)^T F_e = \tau \tag{10.72}$$

where $D \in R^{24\times24}$, $C \in R^{24\times24}$, and $J \in R^{24\times24}$ are all diagonal square matrices, and their diagonal element are, respectively, D_i, C_i, and J_i.

$$q = \begin{bmatrix} q_1 \\ \cdot \\ \cdot \\ \cdot q_4 \end{bmatrix}, \quad G_r = \begin{bmatrix} G_1 \\ \cdot \\ \cdot \\ \cdot G_4 \end{bmatrix}, \quad \tau = \begin{bmatrix} \tau_1 \\ \cdot \\ \cdot \\ \cdot \tau_4 \end{bmatrix}, \quad F_e = \begin{bmatrix} F_{e_1} \\ \cdot \\ \cdot \\ \cdot F_{e_4} \end{bmatrix}$$

Assuming that the manipulators hold the rigid body without slipping, and by Newton–Euler equation the following equation can be obtained easily

$$M_1\ddot{z} + M_1 g = \sum_{i=1}^{i=4} f_{e_i} \tag{10.73}$$

$$I\dot{\omega} + \omega \times (I\omega) = \sum_{i=1}^{i=4} (\eta_{e_i} + r_i \times f_{e_i}) \tag{10.74}$$

where $z \in R^3$ is the mass center of the object and $\omega \in R^3$ is the angle speed of the object, the diagonal matrix $M_1 \in R^{3\times3}$ is mass matrix, whose diagonal element is the mass of object, $I \in R^{3\times3}$ is the inertial matrix, and $r_i = [r_{ix}, r_{iy}, r_{iz}]^T$ is the vector from the mass center of object to the touch point between each arm and the object.

Define $\dot{x} = [\dot{z}^T\ \omega^T]^T$, then the dynamics given in Eq. (10.74) can be rewritten as

$$M\ddot{x} + N_2\dot{x} + G_l = GF_e = F_o \tag{10.75}$$

where $G \in R^{6\times 24}$ is grasp matrix, which is defined as

$$G = [T_1\ T_2\ T_3\ T_4] \tag{10.76}$$

The matrix $T_i \in R^{6\times 6}$ is defined as

$$T_i = \begin{bmatrix} I_{3\times 3} & 0 \\ \Omega(r_i) & I_{3\times 3} \end{bmatrix}, \quad \Omega(r_i) = \begin{bmatrix} 0 & -r_{iz} & r_{iy} \\ r_{iz} & 0 & -r_{ix} \\ -r_{iy} & r_{ix} & 0 \end{bmatrix}$$

where $I_{3\times 3} \in R^{3\times 3}$ is the identity matrix and

$$M = \begin{bmatrix} M_1 & 0 \\ 0 & I \end{bmatrix}, \quad N_2\dot{x} = \begin{bmatrix} 0 \\ \omega \times (I\omega) \end{bmatrix}, \quad G_l = \begin{bmatrix} M_1 g \\ 0 \end{bmatrix}$$

From Eq. (10.72), we get F_e

$$F_e = J^{-T}[\tau - D\ddot{q} - C\dot{q} - G_r] \tag{10.77}$$

Substituting it into Eq. (10.75), then

$$M\ddot{x} + N_2\dot{x} + G_l = GJ^{-T}[\tau - D\ddot{q} - C\dot{q} - G_r] \tag{10.78}$$

Replacing $\dot{q}$ and $\ddot{q}$ by Eqs. (10.69) and (10.70), respectively, we have

$$\begin{aligned} &M\ddot{x} + N_2\dot{x} + G_l \\ &= GJ^{-T}\tau - GJ^{-T}D(J^{-1}G^T\ddot{x} + \frac{d}{dt}(J^{-1}G^T)\dot{x}) - GJ^{-T}CJ^{-1}G^T\dot{x} - GJ^{-T}G_r \end{aligned} \tag{10.79}$$

or

$$\begin{aligned} (GJ^{-T}DJ^{-1}G^T + M)\ddot{x} + (GJ^{-T}D\frac{d}{dt}(J^{-1}G^T) + GJ^{-T}CJ^{-1}G^T + N_2)\dot{x} \\ + (GJ^{-T}G_r + G_l) = GJ^{-T}\tau \end{aligned} \tag{10.80}$$

By defining the following equations

$$D^* = GJ^{-T}DJ^{-1}G^T + M \tag{10.81}$$

$$C^* = GJ^{-T}D\frac{d}{dt}(J^{-1}G^T) + GJ^{-T}CJ^{-1}G^T + N_2 \tag{10.82}$$

$$G^* = GJ^{-T}G_r + G_l \quad (10.83)$$

$$\tau^* = GJ^{-T}\tau \quad (10.84)$$

the formula Eq. (10.80) can be rewritten as

$$D^*\ddot{x} + C^*\dot{x} + G^* = \tau^* \quad (10.85)$$

which is the integrated dynamics model of multimanipulator and rigid body.

10.4.3.2 Design of Adaptive Controller

For manipulator torque control, traditional impedance controller can be expressed as

$$\tau^* = (F_{ext} + C^*\dot{x} + G^*) + D^*\{\ddot{x}_d - M^{-1}[B(\dot{x} - \dot{x}_d) + K(x - x_d) + F_{ext}]\} \quad (10.86)$$

where $x_d \in R^n$ is the desired Cartesian-space motion trajectory, $M \in R^{n\times n}$ the inertial matrix, $B \in R^{n\times n}$ the damped coefficient matrix, $K \in R^{n\times n}$ the stiffness coefficient matrix, and F_{ext} the external impedance force of manipulators. Substituting the controller Eq. (10.86) into the integrated model, we can obtain closed-loop dynamics equation:

$$M(\ddot{x} - \ddot{x_d}) + B(\dot{x} - \dot{x_d}) + K(x - x_d) = -F_{ext} \quad (10.87)$$

The above controller could only handle with time-invariant and known dynamics parameters, hence in order to deal with time-varying or unknown parameters, we design the following adaptive controller strategy:

$$\tau^* = (F_{ext} + \hat{C}^*\dot{x} + \hat{G}^*) + \hat{D}^*\{\ddot{x}_d - M^{-1}[B(\dot{x} - \dot{x}_d) + K(x - x_d) + F_{ext}]\} + \tau_1^* \quad (10.88)$$

which is the control input torque, where $\hat{C}^*$, $\hat{G}^*$ and $\hat{D}^*$ are respectively, the estimation of C^*, G^* and D^*, and τ_1^* is the control input torque to be designed.

In what follows, we will prove that the above adaptive controller can ensure the trajectory of multimanipulator lifting shall be asymptotically stable.

First, some preparatory work will be done. Substituting Eq. (10.85) further, we get

$$\begin{aligned} &M(\ddot{x} - \ddot{x_d}) + B(\dot{x} - \dot{x_d}) + K(x - x_d) \\ &\quad = M\hat{D}^{*-1}[(\hat{D}^* - D^*)\ddot{x} + (\hat{C}^* - C^*)\dot{x} + (\hat{G}^* - G^*)] - F_{ext} + M\hat{D}^{*-1}\tau_1^* \end{aligned} \quad (10.89)$$

where we define $e = x - x_d$ and assume $\tau_1^* = \hat{D}^* M^{-1} F_{ext}$. Then Eq. (10.89) can be rewritten into

$$\ddot{e} + M^{-1}B\dot{e} + M^{-1}Ker = -\hat{D}^{*-1}[(\hat{D}^* - D^*)\ddot{x} + (\hat{C}^* - C^*)\dot{x} + (\hat{G}^* - G^*)] \quad (10.90)$$

Linearizing the left side of Eq. (10.85), we have

$$D^*\ddot{x} + C^*\dot{x} + G^* = Y^* p_x = \tau^* \quad (10.91)$$

where Y^* is the linear regression matrix, function of x, $\dot{x}$ and $\ddot{x}$, and p_x is unknown parameter vector.

Note that Eq. (10.90) can be expressed as

$$\ddot{e} + M^{-1}B\dot{e} + M^{-1}Ke = -\hat{D}^{*-1}Y(x, \dot{x}, \ddot{x})(p_x - \hat{p}_x) \quad (10.92)$$

where $\hat{p}_x$is estimation of p_x. Assuming that $x = [e^T \; \dot{e}^T]^T \in R^{2\times n}$, Eq. (10.85) can be rewritten as

$$\dot{x} = A_x x - B_x \hat{D}^{*-1} Y(p_x - \hat{p}_x) \quad (10.93)$$

where

$$A_x = \begin{bmatrix} 0_{n\times n} & I_n \\ -M^{-1}K & -M^{-1}B \end{bmatrix} \in R^{2n\times 2n}$$

$$B_x = \begin{bmatrix} 0_{n\times n} \\ I_n \end{bmatrix} \in R^{2n\times n}$$

Then, we adopt the following Lyapunov function

$$V(x, \widetilde{p}_x) = \frac{1}{2}x^T P x + \frac{1}{2}\widetilde{p}_x^T \Gamma \widetilde{p}_x \quad (10.94)$$

where $\widetilde{p}_x = p_x - \hat{p}_x$, and matrices $\Gamma \in R^{n\times n}$ and $P = P^T \in R^{2n\times 2n}$ are both positive defined.

$$\dot{V} = -\frac{1}{2}x^T Q x - \widetilde{p}_x^T[(\hat{D}^{*-1}Y)^T B_x^T P x + \Gamma \dot{\widetilde{p}}_x] + x^T P B_x M^{-1} F_{ext} \quad (10.95)$$

where $Q = Q^T$ is known and positive defined matrix, and $A_x^T P + P A_x = -Q$ holds.

The updated estimation of unknown parameter p_x is

$$\dot{\hat{p}}_x = -\Gamma^{-1}(\hat{D}^{*-1}Y)^T B_x^T P x \quad (10.96)$$

As a result, Eq. (10.95) can be rewritten into

$$\dot{V} = -\frac{1}{2}x^T Qx \leq 0 \tag{10.97}$$

while the final motion of the whole system would converge to the following second-order differential equation

$$M(\ddot{x} - \ddot{x_d}) + B(\dot{x} - \dot{x_d}) + K(x - x_d) = 0 \tag{10.98}$$

which is not, however, what we expected in Eq. (10.87). Hence some amendments are supposed to be done.

Define $s = \dot{e} + \Lambda e$ and $v = \dot{x}_d - \Lambda e$ as augmented trajectory error and reference trajectory error, respectively, where $\Lambda = \text{diag}(\lambda_1, \lambda_2, \ldots, \lambda_n), \forall\lambda_i > 0$ and then Eq. (10.85) could be rewritten as this form

$$D^*\dot{s} + C^*\dot{s} + G^* + D^*\dot{v} + C^*v = \tau^* \tag{10.99}$$

Adaptive controller can be changed into the following torque

$$\tau^* = \hat{G}^* + \hat{D}^*\dot{v} + \hat{C}^*v - Ks \tag{10.100}$$

Similarly, substitute the above adaptive controller into the integrated dynamics model and the closed-loop dynamics equation will be

$$D^*\dot{s} + C^*\dot{s} + Ks = -(D^* - \hat{D}^*)\dot{v} - (C^* - \hat{C}^*v) - (G^* - \hat{G}^*) \tag{10.101}$$

where $K \in R^{n\times n}$ is one positive defined matrix and the right side of which can be linearized into

$$D^*\dot{s} + C^*\dot{s} + Ks = -Y(x, \dot{x}, v, \dot{v})\widetilde{p}_x \tag{10.102}$$

Define another Lyapunov function

$$V(s, \widetilde{p}_x) = \frac{1}{2}s^T Ps + \frac{1}{2}\widetilde{p}_x^T \Lambda\widetilde{p}_x \tag{10.103}$$

$$\dot{V} = -s^T Ks - \widetilde{p}_x^T(\Lambda\widetilde{p}_x + Y^T s) \tag{10.104}$$

Likewise, the updated estimation p_x is redesigned as

$$\dot{\hat{p}}_x = -\Lambda^{-1}Y^T s \tag{10.105}$$

which result in

$$\dot{V} = -s^T Ks \leq 0 \tag{10.106}$$

In this way, the closed-loop is asymptotically stable using Lyapunov stability theory.

10.4.3.3 Improvement of Adaptive Controller

In the above adaptive controller, the computation of linear regression matrix is time-consuming, whereas this process can be simplified by using (Function Approximation Techniques—FAT) [26]. That is to say, we need to make some improvement in the original control method.

In the integrated model Eq. (10.101), let $\hat{D}^* \to D^*$, $\hat{C}^* \to C^*$, and $\hat{G}^* \to G^*$, then Eq. (10.101) can be rewritten as

$$D^*\dot{s} + C^*s + Ks = 0 \tag{10.107}$$

Hence, we need to prove the above system given in Eq. (10.107) is stable such that we can avoid the complexed computation of linear regression matrix.

Adopt the following Lyapunov function:

$$V = \frac{1}{2}s^T D^* s \tag{10.108}$$

then

$$\dot{V} = -s^T Ks + \frac{1}{2}s^T(\dot{D}^* - 2C^*)s \tag{10.109}$$

As a result of the antisymmetry of $\dot{D}^* - 2C^*$, we have

$$\dot{V} = -s^T Ks \le 0 \tag{10.110}$$

Design the following Function Approximation Techniques:

$$D^* = W_{D^*}^T Z_{D^*} \tag{10.111}$$

$$C^* = W_{C^*}^T Z_{C^*} \tag{10.112}$$

$$G^* = W_{G^*}^T Z_{G^*} \tag{10.113}$$

where W_{D^*}, W_{C^*} and W_{G^*} are all weight matrices, while Z_{D^*}, Z_{C^*} and Z_{G^*} are primary functions.

We choose the following equations as parameter estimations of the integrated model

$$\hat{D}^* = \hat{W}_{D^*}^T Z_{D^*} \tag{10.114}$$

$$\hat{C}^* = \hat{W}_{C^*}^T Z_{C^*} \tag{10.115}$$

$$\hat{G}^* = \hat{W}_{G^*}^T Z_{G^*} \tag{10.116}$$

Substituting them into Eq. (10.101), we obtain

$$D^*\dot{s} + C^*s + Ks = -\widetilde{W}_{D^*}Z_{D^*}\dot{v} - \widetilde{W}_{C^*}Z_{C^*}v - \widetilde{W}_{G^*}^T Z_{G^*} \quad (10.117)$$

where $\widetilde{W}_{D^*} = W_{D^*} - \hat{W}_{D^*}$, $\widetilde{W}_{C^*} = W_{C^*} - \hat{W}_{C^*}$, $\widetilde{W}_{G^*} = W_{G^*} - \hat{W}_{G^*}$.

Adopt the following expression as Lyapunov function:

$$\begin{aligned} &V(s, \widetilde{W}_{D^*}, \widetilde{W}_{C^*}, \widetilde{W}_{G^*}) \\ &= \frac{1}{2}s^T D^* s + \frac{1}{2}trace(\widetilde{W}_{D^*}^T Q_{D^*}\widetilde{W}_{D^*} + \widetilde{W}_{C^*}^T Q_{C^*}\widetilde{W}_{C^*} + \widetilde{W}_{G^*}^T Q_{G^*}\widetilde{W}_{G^*}) \end{aligned} \quad (10.118)$$

where Q_{D^*}, Q_{C^*} and Q_{G^*} are all positive defined. and differentiating it with respect to time, we obtain

$$\begin{aligned} \dot{V} = &-s^T Ks - [trace\widetilde{W}_{D^*}^T(Z_{D^*}\dot{v}s^T + Q_{D^*}\dot{\hat{W}}_{D^*}) \\ &+ trace\widetilde{W}_{C^*}^T(Z_{C^*}\dot{v}s^T + Q_{C^*}\dot{\hat{W}}_{C^*}) + trace\widetilde{W}_{G^*}^T(Z_{G^*}\dot{v}s^T + Q_{G^*}\dot{\hat{W}}_{G^*})] \end{aligned} \quad (10.119)$$

The updating law for unknown parameters are the following expressions

$$\dot{\hat{W}}_{D^*} = -Q_{D^*}^{-1}Z_{D^*}\dot{v}s^T \quad (10.120)$$

$$\dot{\hat{W}}_{C^*} = -Q_{C^*}^{-1}Z_{C^*}vs^T \quad (10.121)$$

$$\dot{\hat{W}}_{G^*} = -Q_{G^*}^{-1}Z_{G^*}s^T \quad (10.122)$$

By substituting them into Eq. (10.119), we can verify

$$\dot{V} \leq 0 \quad (10.123)$$

10.4.3.4 Simulation Studies

Likewise, we carry out the simulation for the just-adaptive controller in MATLAB using RVC toolbox. The first stage is depicted in Figs. 10.24, 10.25, 10.26, 10.27, and 10.28 and the end scene was shown in Fig. 10.29.

Since the considered task, lifting one desk upward, was relatively easy, PD feedforward complementation controller and adaptive controller both could finish it, however, the average error of adaptive controller, 2.75 % was much lower than the one of PD controller, 6.5 %, which could be concluded by comparing Figs. 10.21 and 10.28. In addition, the variance of adaptive controller is also far lower than the one

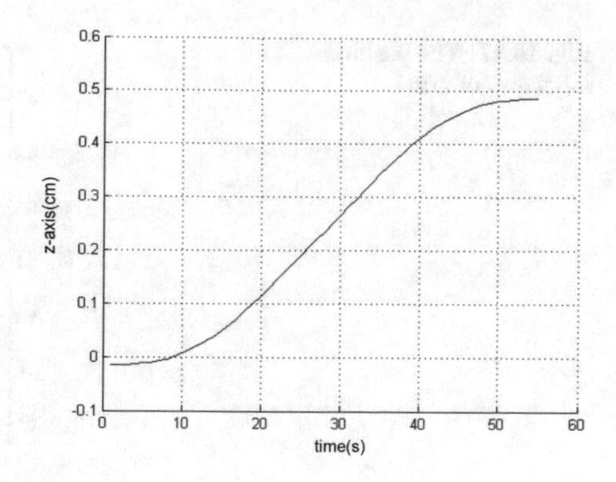

Fig. 10.24 The position trajectory of arm1

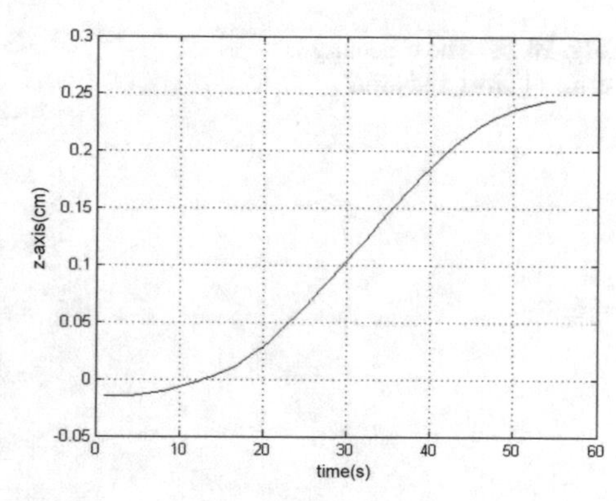

Fig. 10.25 The position trajectory of arm2

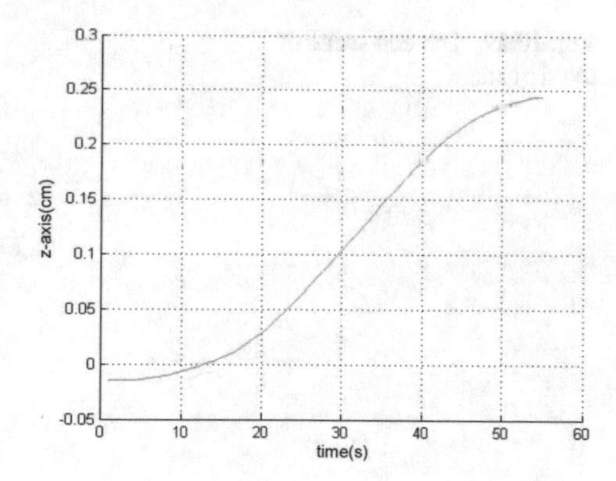

Fig. 10.26 The position trajectory of arm3

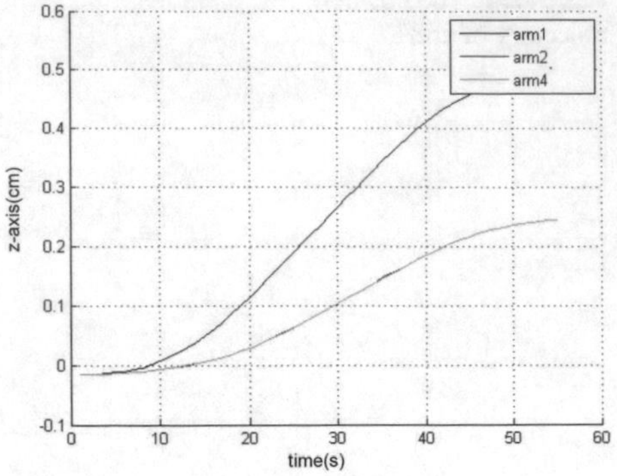

Fig. 10.27 The position trajectory of arm4

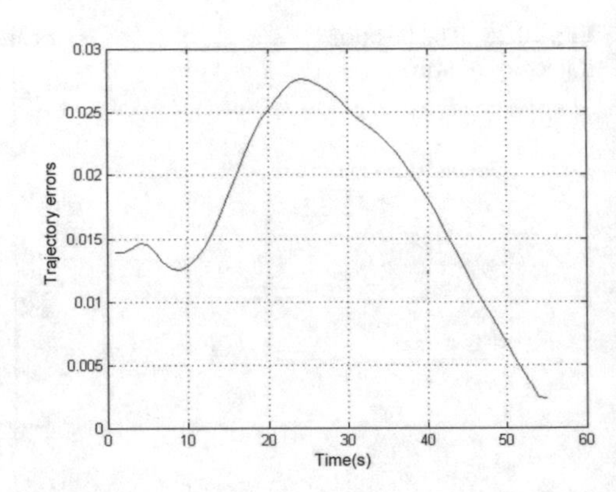

Fig. 10.28 The trajectory error of arm1 and arm2

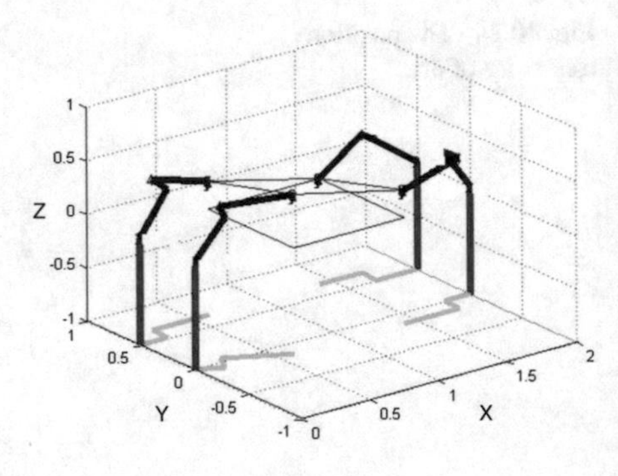

Fig. 10.29 The end scene of the first stage

of PD controller. In view of these results, we can come to a conclusion that adaptive controller just designed is much more efficient for multimanipulator cooperative lifting.

References

1. Vicsek, T., Czirók, A., Ben-Jacob, E., Cohen, I., Shochet, O.: Novel type of phase transition in a system of self-driven particles. Phys. Rev. Lett. **75**(6), 1226 (1995)
2. Jadbabaie, A., Lin, J., et al.: Coordination of groups of mobile autonomous agents using nearest neighbor rules. IEEE Trans. Autom. Control **48**(6), 988–1001 (2003)
3. Moreau, L.: Stability of multiagent systems with time-dependent communication links. IEEE Trans. Autom. Control **50**(2), 169–182 (2005)
4. Tanner, H.G., Christodoulakis, D.K.: State synchronization in local-interaction networks is robust with respect to time delays. In: 44th IEEE Conference on Decision and Control, 2005 and 2005 European Control Conference. CDC-ECC'05, pp. 4945–4950. IEEE (2005)
5. Xiao, F., Wang, L.: State consensus for multi-agent systems with switching topologies and time-varying delays. Int. J. Control **79**(10), 1277–1284 (2006)
6. Angeli, D., Bliman, P.-A.: Stability of leaderless discrete-time multi-agent systems. Math. Control Signals Syst. **18**(4), 293–322 (2006)
7. Hongbin, M., Meiling, W., Zhenchao, J., Chenguang, Y.: A new framework of optimal multi-robot formation problem. In: Control Conference (CCC), 2011 30th Chinese, pp. 4139–4144. IEEE (2011)
8. Jia, Z., Ma, H., Yang, C., Wang, M.: Three-robot minimax travel-distance optimal formation. In: 50th IEEE Conference on Decision and Control and European Control Conference (CDC-ECC), pp. 7641–7646. IEEE (2011)
9. Jia, Z.C., Ma, H.B., Yang, C.G., Wang, M.L.: Three-robot minimax travel-distance optimal formation. In: 2011 50th Ieee Conference on Decision and Control and European Control Conference (Cdc-Ecc), pp. 7641–7646 (2011)
10. Wang, M.L., Jia, Z.C., Ma, H.B., Fu, M.: Three-robot minimum-time optimal line formation. In: 2011 9th IEEE International Conference on Control and Automation (ICCA 2011), pp. 1326–31. IEEE (2011)
11. Jose, G., Gabriel, O.: Multi-robot coalition formation in real-time scenarios. Robot. Auton. Syst. **60**(10), 1295–1307 (2012)
12. Wang, H., Guo, Y., IEEE.: Minimal persistence control on dynamic directed graphs for multi-robot formation. In: 2012 IEEE International Conference on Robotics and Automation (2012)
13. Madden, J., Arkin, R.C., MacNulty, D.R.: Multi-robot system based on model of wolf hunting behavior to emulate wolf and elk interactions. In: 2010 IEEE International Conference on Robotics and Biomimetics (ROBIO), pp. 1043–1050 (2010)
14. Weitzenfeld, A., Vallesa, A., Flores, H.: A biologically-inspired wolf pack multiple robot hunting model, pp. 90–97 (2006)
15. Ma, Y., Cao, Z.Q., Dong, X., Zhou, C., Tan, M.: A multi-robot coordinated hunting strategy with dynamic alliance. In: 21st Chinese Control and Decision Conference, pp. 2338–2342 (2009)
16. Sun, W., Dou, L.H., Fang, H., Zhang, H.Q.: Task allocation for multi-robot cooperative hunting behavior based on improved auction algorithm, pp. 435–440 (2008)
17. Gong, J.W., Qi, J.Y., Xiong, G.M., Chen, H.Y., Huang, W.N.: A GA based combinatorial auction algorithm for multi-robot cooperative hunting. In: International Conference on Computational Intelligence and Security (2007)
18. Li, J., Pan, Q.S., Hong, B.R.: A new approach of multi-robot cooperative pursuit based on association rule data mining. Int. J. Adv. Robot. Syst. **6**(4), 329–336 (2009)
19. Ge, S.S., Ma, H.B., Lum, K.Y.: Detectability in games of pursuit evasion with antagonizing players. In: Proceedings of the 46th IEEE Conference on Decision and Control, 12–14 Dec. 2007, pp. 1404–1409. IEEE (2007)
20. Ma, H.B., Ge, S.S., Lum, K.Y.: Attackability in games of pursuit and evasion with antagonizing players. In: Proceedings of the 17th World Congress, International Federation of Automatic Control. IFAC Proceedings Volumes (IFAC-PapersOnline), vol. 17. Elsevier (2008)
21. Wang, C., Zhang, T., Wang, K., Lv, S., Ma, H.B.: A new approach of multi-robot cooperative pursuit. In: Control Conference (CCC), 2013 32nd Chinese, pp. 7252–7256. IEEE (2013)

22. Choi, H.S., Ro, P.I.: A force/position control for two-arm motion coordination and its stability robustness analysis. KSME J. **8**(3), 293–302 (1994)
23. Wang, X., Qin, J., Han, S., Shao, C., et al.: Coordinated dynamic load distribution for multiple robot manipulators carrying a common object system. Acta Mechanica Sinica **15**(1), 119–125 (1999)
24. Agravante, D.J., Cherubini, A., Bussy, A., Kheddar, A.: Human-humanoid joint haptic table carrying task with height stabilization using vision. In: 2013 IEEE/RSJ International Conference on Intelligent Robots and Systems (IROS), (Tokyo,Japan), pp. 4609–4614. IEEE (2013)
25. Pouli, R.: Robot manipulators mathematics. In: Programming and Control (1981)
26. Chien, M.C., Huang, A.C.: Adaptive impedance control of robot manipulators based on function approximation technique. Robotica **22**(04), 395–403 (2004)

Chapter 11
Technologies for Other Robot Applications

Abstract This chapter presents some robot applications and technologies which are not covered in the previous chapters. At first, we introduce the robot kicking and describe the inverse kinematics and software used in order to reach the best kick that a small humanoid can give to a ball. Next, a computational model of human motor reference trajectory adaptation has been developed. This adaptation model aims to satisfy a desired impedance model to minimize interaction force and performance error. It can also be used as a robotic motion planner when robot interacts with objects of unknown shapes.

11.1 Investigation of Robot Kicking

The robots investigated in this section were made by Plymouth University based on the Bioloid humanoid robot from Robotis. They were modified to match the requirements to play football. The legs have been improved in such a way that robots can kick a ball staying stood. The fact is that the ball slowly follows a path closely to the ground and reaches no more than 60 cm of distance. One interesting and challenging problem is how to design *the best robot kick* so that the robot can kick the ball in an efficient and effective way. To this end, based on the current humanoid robot platform developed in UoP, we have made some attempts to optimize the path motion and to use compliance coefficient added to joint motion available on the leg's RX-28 servomotors in order to improve the kick. As a result of our preliminary research and development, this section will report what we have done toward the goal of implementing the best robot kick [1].

11.1.1 Kinematics

Recall the content in Chap. 2, as one of the most important disciplines in robots, kinematics describes the motion or gait of the actuated robots. In order to create a new behavior, forward kinematics has to be computed first and then inverse kinematics.

© Science Press and Springer Science+Business Media Singapore 2016

C. Yang et al., *Advanced Technologies in Modern Robotic Applications*,
DOI 10.1007/978-981-10-0830-6_11

The forward kinematics gives the endpoint's position and orientation according to joint angles while the inverse kinematics (IK) gives the joint angles needed to create the endpoint's position and orientation. The forward kinematics is associated to specific matrices according to the Denavit–Hartenberg convention or to homogeneous 2D transformation matrices. The IK can be computed from several existing methods: Single Value Decomposition (SVD), pure geometric/analytic equations, quaternions, dual-quaternions, or exponential mapping [2]. This part will describe the kinematics applied on the robot leg for a football kick.

11.1.1.1 Mechanism Description

The mechanism studied here is a 3 joints–3 links robot leg (hip, knee, ankle and femur, tibia, foot) from the University of Plymouth robot football team. The first joint on top of Fig. 11.1a corresponds to the hip. The angle between the ground (horizontal) and the femur is called θ_1, the angle between the femur and the tibia is called θ_2 and the one between the tibia and the foot is called θ_3. The femur has a length $L_1 = 75\,\text{mm}$, the tibia $L_2 = 75\,\text{mm}$ and the foot has a height $L_3 = 32\,\text{mm}$ and a length $L_4 = 54\,\text{mm}$.

According to Lees and Nolan [3], the initial and final leg configurations for a kick can be generalized. Thus, for the initial configuration: $\theta_1 = 3\pi/4$, $\theta_2 = \pi/2$ and $\theta_3 = \pi/4$ and the final configuration: $\theta_1 = \pi/4$, $\theta_2 = 0$ and $\theta_3 = \pi/8$. Then, the path chosen is based on a B-spline trajectory [4]:

$$P^T_{traj}[\lambda] = \begin{bmatrix} 1 & \lambda & \lambda^2 \end{bmatrix} \begin{bmatrix} 1 & 1 & 0 \\ -2 & 2 & 0 \\ 1 & -2 & 1 \end{bmatrix} \begin{bmatrix} P^T_{initial} \\ P^T_{inter} \\ P^T_{final} \end{bmatrix} - P^T_{inter}$$

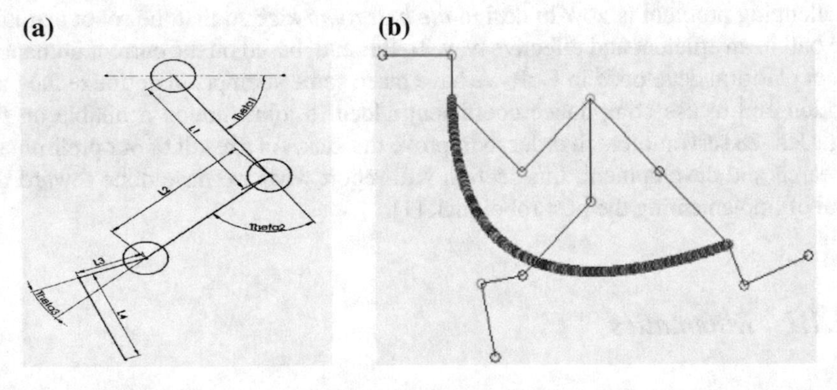

Fig. 11.1 Robot leg configurations and path. **a** Random configuration. **b** Robot leg configurations and path

where λ is the vector step of the path, $P_{initial}$, P_{inter} and P_{final} are the initial, reference and final points, respectively, and P_{traj} is an array corresponding to the path of the ankle (Fig. 11.1b).

11.1.1.2 Forward Kinematics

The forward kinematics helps to create the configurations of robots. The reference coordinate of the robot leg is at the hip joint and the motion is assumed to be planar. So x_1, y_1 are the coordinates of the knee, x_2, y_2 are the coordinates of the ankle, x_3, y_3 are the coordinates of the heel and x_4, y_4 are the coordinates of the end of the foot. With the mechanism description, the following equations are obtained:

$$
\begin{aligned}
x_1 &= L_1 \cos(\theta_1), \quad y_1 = L_1 \sin(\theta_1) \\
x_2 &= x_1 + L_2 \cos(\theta_1 + \theta_2), \quad y_2 = y_1 + L_2 \sin(\theta_1 + \theta_2) \\
x_3 &= x_2 + L_3 \cos(\theta_1 + \theta_2 + \theta_3) \\
y_3 &= y_2 + L_3 \sin(\theta_1 + \theta_2 + \theta_3) \\
x_4 &= x_3 + L_4 \cos(\theta_1 + \theta_2 + \theta_3 + \frac{\pi}{2}) \\
y_4 &= y_3 + L_4 \sin(\theta_1 + \theta_2 + \theta_3 + \frac{\pi}{2})
\end{aligned}
$$

11.1.1.3 Inverse Kinematics

As this robot leg is 3 joints–3 links and is assumed to follow a planar motion, the inverse kinematics can be calculated with pure trigonometrical calculus according to Fig. 11.2:

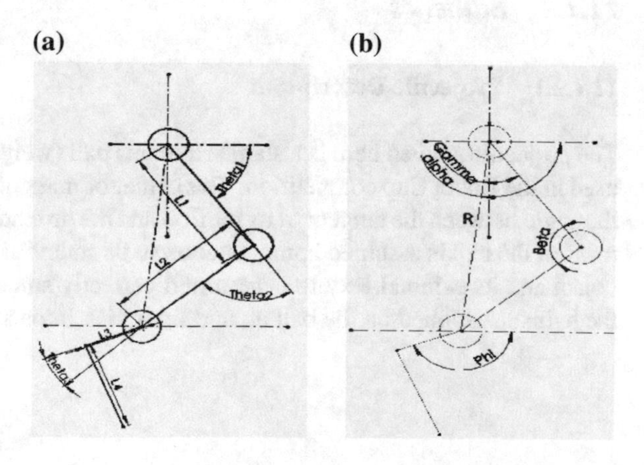

Fig. 11.2 Trigonometry for the leg. **a** Lengths and angles. **b** Additional features

$$\left.\begin{aligned} x_2 &= x_3 - L_3\cos(\phi) \\ y_2 &= y_3 - L_3\sin(\phi) \end{aligned}\right\} \quad \alpha = \arctan\left(\frac{y_2}{x_2}\right)$$

Let

$$R = \sqrt{x_2^2 + y_2^2}$$

Now θ_2 is calculated with:

$$\begin{aligned} L_1^2 + L_2^2 - 2L_1L_2\cos(\beta) &= R^2 \\ \Rightarrow \cos(\beta) &= -\frac{R^2 - L_1^2 - L_2^2}{2L_1L_2} \\ \theta_2 = \pi - \beta &= \pi - \arccos\left(-\frac{R^2 - L_1^2 - L_2^2}{2L_1L_2}\right) \end{aligned}$$

And then θ_1 with:

$$\begin{aligned} R^2 + L_1^2 - 2RL_1\cos(\gamma) &= L_2^2 \\ \Rightarrow \cos(\gamma) &= -\frac{L_2^2 - R^2 - L_1^2}{2RL_2} \\ \theta_1 = \alpha - \gamma &= \alpha - \arccos\left(-\frac{L_2^2 - R^2 - L_1^2}{2RL_2}\right) \end{aligned}$$

And eventually θ_3 with:

$$\theta_3 = \phi - \theta_1 - \theta_2$$

where ϕ is $\pi/2+$ the foot orientation.

11.1.2 Ballistics

11.1.2.1 Projectile Description

The projectile studied here is a standard tennis ball (weight: $58g$ $\varnothing 67$ mm) officially used in the Robot Cup competition. The center of mass of the tennis ball is called m, the angle between the tangent of its motion and the ground is called α (Fig. 11.3). The mass of the ball is assumed homogeneous so its center of mass is at the center of the object and its external texture is assumed perfectly smooth. This part will describe the ballistics applied on the ball at, and after kick impact.

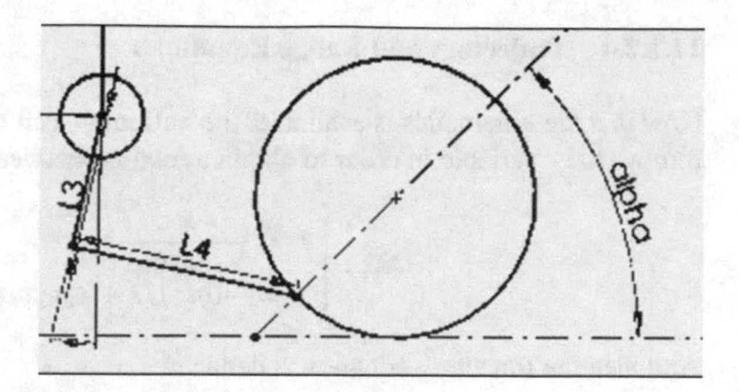

Fig. 11.3 Initial ball's state

11.1.2.2 Dynamic Analysis

Assuming a motion without friction (effect of the air on the ball) and without Archimedes' principle effect, the only force that can be applied is the gravity force: $\mathbf{P} = m\mathbf{g}$ where m is the ball's mass and $\mathbf{g}$ is the gravitational acceleration. According to the second law of Newton, $\sum \mathbf{F}_{ext} = m\mathbf{a}_\mathbf{M}$, $\mathbf{P} = m\mathbf{g} = m\mathbf{a}_\mathbf{M}$ and finally $\mathbf{g} = \mathbf{a}_\mathbf{M}$. As the motion is assumed planar, X axis is horizontal and Y axis is vertical, $\mathbf{g} = \mathbf{a}_\mathbf{M} = g_y = a_{My}$.

11.1.2.3 Kinematics

In order to compute the speed vector, the following relation is used: $\mathbf{a}_\mathbf{M} = d(\mathbf{v}_\mathbf{M})/dt$. From the previous equations:

$$\mathbf{v}_\mathbf{M} : \begin{cases} v_x = v_{0x} \\ v_y = -gt + v_{0y} \end{cases}$$

And according to the initial state:

$$\mathbf{v}_\mathbf{M} : \begin{cases} v_x = v_0 \cos(\alpha) \\ v_y = -gt + v_0 \sin(\alpha) \end{cases}$$

In order to compute the position vector coordinates, the following relation is used: $\mathbf{v}_\mathbf{M} = d(\mathbf{OM})/dt$. From the previous equations:

$$\mathbf{OM} : \begin{cases} x = v_0 t \cos(\alpha) + x_0 \\ y = -(gt^2)/2 + v_0 t \sin(\alpha) + y_0 \end{cases}$$

And according to the initial state, we have

$$\mathbf{v}_\mathbf{M} : \begin{cases} x = v_0 t \cos(\alpha) \\ y = -(gt^2)/2 + v_0 t \sin(\alpha) \end{cases}$$

11.1.2.4 Trajectory and Range Equations

Now that the kinematics is available, the ball range can be raised. The means are to remove the t variable in order to obtain a relation between x and $y : y = f(x)$:

$$\mathbf{OM} : \begin{cases} t = \dfrac{x}{v_0 \cos(\alpha)} \\ y = -(gt^2)/2 + v_0 t \sin(\alpha) \end{cases}$$

And then the trajectory equation is deduced:

$$\begin{aligned} y &= -\frac{1}{2} g \left(\frac{x}{v_0 \cos(\alpha)} \right)^2 + v_0 \left(\frac{x}{v_0 \cos(\alpha)} \right) \sin(\alpha) \\ &= -\frac{1}{2} g \frac{x^2}{v_0^2 \cos^2(\alpha)} + x \tan(\alpha) \end{aligned}$$

Let $k = -\dfrac{2v_0^2 \cos^2(\alpha)}{g}$ and $b = \tan(\alpha)$, then we have

$$y = \frac{x^2}{k} + bx$$

In order to make the ball reach a target T on the horizontal axis, it is enough to solve the previous equation for $y = 0$:

$$\frac{x_T^2}{k} + bx_T = 0$$

The solution $x_T = 0$ is rejected so then $x_T = -kb$ is the only one solution:

$$x_T = \frac{2v_0^2 \cos^2(\alpha) \tan(\alpha)}{g} = \frac{v_0^2}{g} \sin(2\alpha)$$

From the last equation, it can be raised that the maximum range is reached when $\sin 2\alpha = 1$ or $\alpha = \pi/4$.

11.1.2.5 Application

Assuming a maximum speed $v_{\max} = 59.9\,\text{RPM}$ and that all the 3 joints have already reached their maximum speed at the impact, the end-foot speed will be:

$$v_0 = 3v_{\max} = 179.7\text{RPM} = 18.81\ \text{rad/s}$$

For a gravitational acceleration $\mathbf{g} = 9.81\,\text{m/s}$, an angle $\alpha = \pi/4$, $x_{T_{\max}} = 25.5\,\text{m}$. This theoretical result is obtained for a perfect kick in a perfect ball, moving in a perfect environment, without external constraints.

11.1.3 Structure of the Robot

11.1.3.1 CM700 Controller and Dynamixel RX-28

The robot is built with an embedded CM700 controller from Robotis based on an ATMEGA2561 microcontroller from ATMEL. This board is assembled with a sub-board which is composed of the power stage and the connectors needed to control the servomotors. The CPU allows the communication with several servos using the standard buses: TTL (3-pin) or RS485 (4-pin). The compiled code is sent to the embedded boot loader through a terminal by the RS232 bus.

The RX-28 servomotor is designed for high-end robotics applications. It is a smart actuator composed of metal gear reducer, a DC motor and its control circuitry, and a Daisy Chain serial networking functionality fitted in a small-sized package. Several parameters can be read and written in its memory. This last is divided into two parts, one which is an EEPROM area and the other one a RAM area. In the EEPROM one, constant and security related parameters will be set as the ID number, the baud rate, the temperature limit, etc. In the RAM, the LED and motion control parameters are adjustable and will be in this memory as long as the servo is connected to a power supply, otherwise, they are reset to initial values. Among all its features, it can be raised that the servo can be controlled with a resolution of 1024 steps in position, speed (even if limited to 59.9 RPM) and torque. An unusual feature here, the compliance module, is interesting. Five variables, i.e., clockwise (CW) and counterclockwise (CCW) slope and margin, and punch torque, help to set the compliance pattern (Fig. 11.4).

11.1.3.2 Robot Calibration

The robot is based on a small humanoid robot from Robotis. Its structure allows 20 degrees of freedom, 12 for the lower limbs, 6 for the upper limbs, and two for

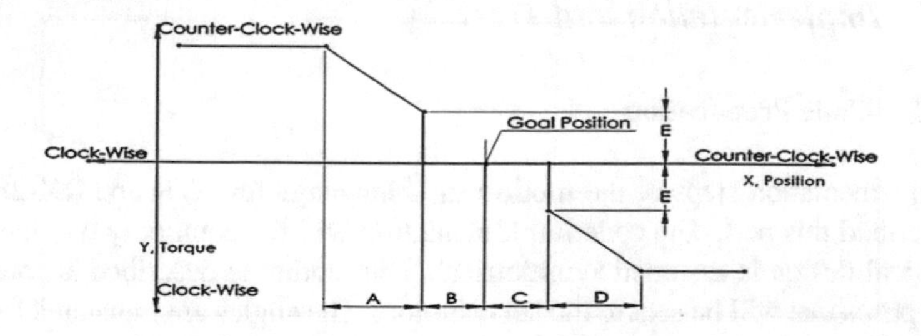

Fig. 11.4 RX-28's compliance pattern

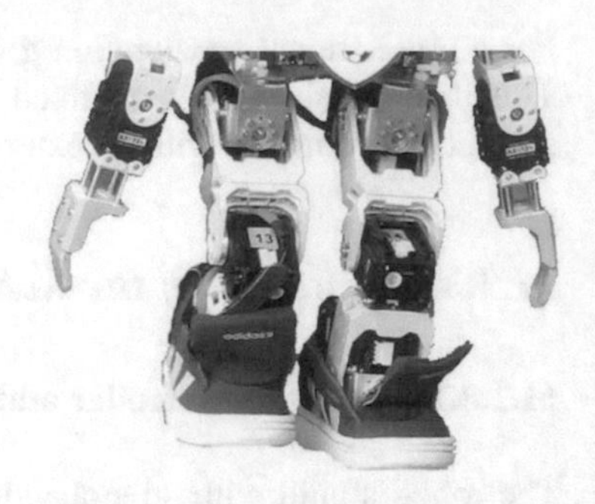

Fig. 11.5 Robot's legs in steady position [Photo taken at Plymouth University]

the head. Each leg is composed of 6 servomotors (RX-28), 4 frames, and mounted thanks to bearings and screws as shown in Fig. 11.5. Every frame, gear, motor, and plastic packages from one leg to the other leg have different dimension tolerances plus different assembling clearances. That is why a calibration, in order to make the control commands be the same on each leg, has to be set on every joints. This long step permits to correct mechanical plays by adding or subtracting offsets in the motion parameters in the code. The calibration is helped with control panel software (from the UoP robot team) on computer linked to the robot. Each parameter (servo position) can be tweaked and then tested without code update. A good calibration is done by checking the parallelism of all frames, starting from the hips (servo #7-#8) to the feet (servo #17-#18). It is then validated when the robot follows a routine of swing and stays on its spot (stride height = 13, stride length = 15). After setting the new parameters, the proper coding step can start.

11.1.4 MATLAB Simulation

Consider the problem introduced in Sect. 11.1.1.1. The MATLAB simulation results are given in Fig. 11.6d, where the inverse kinematics represent the angles for the 60 poses (chronologically from left to right) in degrees, with blue rounds for θ_1, green crosses for θ_2, red lines for θ_3, and turquoise stars for the foot orientation.

11.1.5 Implementation and Tests

11.1.5.1 Code Preparation

The implementation steps of the motion in C language for AVR and RX-28 servo are described this part. The code implementation will focus on only one leg as the mechanical design is assumed symmetrical. The motion is described according to several poses that will be sent to the servomotors. The angles are translated from the

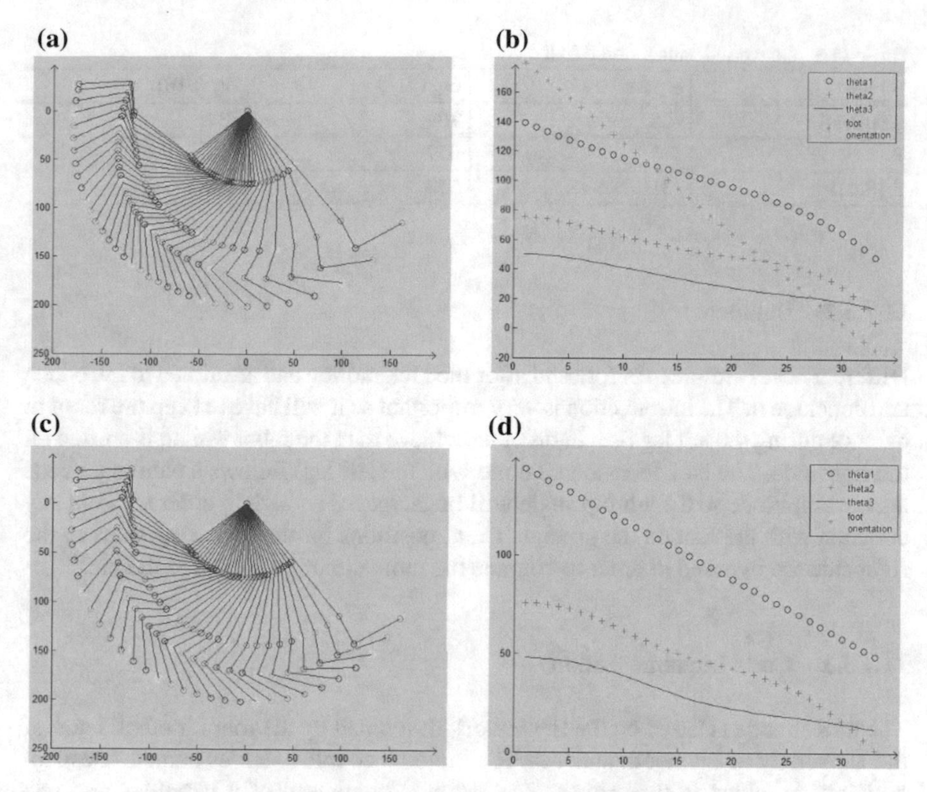

Fig. 11.6 MATLAB results for 60 pose. **a** First series leg steps plot. **b** First joint angle (deg). **c** Final series leg steps plot. **d** Final joint angles (deg)

MATLAB results to servomotor instructions. In fact, the RX-28 needs a position value from 0 to 1023 which correspond to 0–300 degrees. Here again another translation is needed according to the direction of the frame on the servomotor as well as to the orientation of the servo. The angles Θ_1, Θ_2 and Θ_3 are eventually defined as:

$$\Theta_1 = (240 - \theta_1) * 1023/300$$
$$\Theta_2 = (\theta_2 + 150) * 1023/300$$
$$\Theta_3 = (\theta_3 + 150) * 1023/300$$

Only three poses from the MATLAB results are kept as there is no need to send moves of small degrees. Furthermore, according to the baud rate settings [5] and the "empty buffer before send" step that are implemented, a too high number of commands sent can create a bottleneck effect and therefore, it will slow down the robot motion. The selected poses (Table 11.1) are obviously the initial and final poses plus an interim pose (selected at the half of the path).

Table 11.1 Converted values stored in the code

# pose	Θ_1 (servo #12)	Θ_2 (#14)	Θ_3 (#16)
1 (initial)	438	704	631
2	518	657	587
3 (final)	631	552	558

11.1.5.2 Balance

Different poses arc added before and after the kick motion and are aimed to introduce and conclude it. The introduction is very important as it will have to keep the robot in balance during the left leg rise. In order to achieve that, the robot weight is shifted on the right side. The kick (considered done with the left leg) follows a path that needs a large amplitude so the left hip angle will be increased enough in order to avoid any collision with the foot on the ground. Then, positions for the following steps of the right side are tweaked in order to counter the momentum effect due to the kick.

11.1.5.3 Code Implementation

The proper code is based on the framework developed by the robot football team of the University of Plymouth under AVR Studio. A couple c–h files named "BestRobotKick" is added to the project. The c-file is composed of 4 variables and one function. The variables are matrices representing 10 series (10 poses) of 9 values (9 positions or 9 speeds) for each side (leg and arm of right and left). The kick function is composed of two overlapped "for" loops that explore the variables and send the picked-up position value plus the calibration offset and the speed value to the current servomotor. In addition, a specified "Sleep" time is included after the end of each pose in order to wait that the servomotors could reach their final position. This sleep time is adjusted according to the length of the motion in space and in time.

11.1.5.4 Compliance and Distance to Reach

In order to increase the stability criterion (decrease the momentum effect), the left leg speed can be computed to its necessary value, i.e., enough speed to reach a desired value. The final step of this project is to use and test the compliance feature of the servomotors. The margin parameter is skipped as it creates a constant low torque and the module focuses on the slope and punch parameters. The implementation in the code can follow the Algorithm 4 where the variable $distance$ stands for the desired distance (in m), $compliance$ for the selected compliance (in %), LSM & RSM for the speed matrix of the left & right side, LPM & RPM for the position matrix of the left & right side, and LO & RO for the offset matrix of the left & right side. The compliance feature is set for the left hip and knee only. The procedure ResetCompliance() aims to reset the compliance feature at initial values; slope and punch at 32. The ApplyDelay($\#Pose$, $ServoSpeed$) procedure is used for time fitting between the poses.

Algorithm 4 Complete Algorithm

```
procedure BESTROBOTKICKER(distance, compliance)
  RESETCOMPLIANCE()
  speed ← (distance * g) / sin(2α)
  speed ← √speed / 3
  slope ← 255 * compliance + 1
  punch ← 32 / (compliance + 1)
  LSM[Kick][Hip] ← speed
  LSM[Kick][Knee] ← speed
  LSM[Kick][Ankle] ← speed
  for i ← 1, NbPoses do
    for j ← 1, NbServos do
      if i = BeforeKick || i = Kick then
        if j = LeftHip || j = LeftKnee then
          AX12SETCOMPLIANCE(j,slope,punch)
        end if
      else if i = Kick + 1 then
        RESETCOMPLIANCE()
      end if
      AX12SETPOSITION(LPM[i][j]+LO[i][j],LSM[i][j])
      AX12SETPOSITION(RPM[i][j]+RO[i][j],RSM[i][j])
    end for
    APPLYDELAY(i,speed)
  end for
end procedure
```

11.1.5.5 Protocol of Tests

The tests are performed on a carpeted surface divided in 50 cm squares. The tennis ball is manually placed at the robot's left foot, closed to be in contact with the shoe. Because of the modifications introduced by the balance problem, the robot needs to be rotated of about 45° from the desired kick direction. After this preparation step (which can assumed to be done by the robot itself before in a real situation), the kick phase is manually launched by pressing the reset button on the CM700 sub-board of the robot. The distance is measured straight from the initial ball's position to its final position, ignoring the case if the ball delineates a curved or backward motion in its

(a)

(b)

Fig. 11.7 Two tests with high stiffness. **a** A shoot of 2.2 m. **b** A shoot of 3 m [Photo taken at Plymouth University]

path. In fact, even if the ball can be in the goal in the true application (football game) once during its whole path, here the relevant information is the distance that the ball can reach at the end of its trajectory. The slope and punch parameters are modified such as the slope varies according to 5 steps and the punch to 3 steps. These relevant values are chosen checking the robot behavior (e.g., below a slope of 8, the robot shakes).

11.1.5.6 Results

A complete test is done with 20 valid shoots. Figure 11.7 shows two different shoots with the same parameters written in the software but with a different ball's initial position, i.e., the robot's foot is slightly shifted in the second shoot. The red dot is the initial position of the ball on the carpet. Figure 11.8 shows 9 series' distributions of the ball's final position, represented by small rounds and summarized in ellipses. The origin of all the shoots in the series is the *XY* mark and the robot, rotated of 45°. The 9 series differ according to various slope and punch parameters. The arcs in the ellipses epitomizes the average distance reached by the ball (minimum: 1.38 m, maximum: 2.66 m). The ellipses are built according to the following configuration: horizontal (*X* axis) and vertical (*Y* axis) standard deviations for ellipse's half axes.

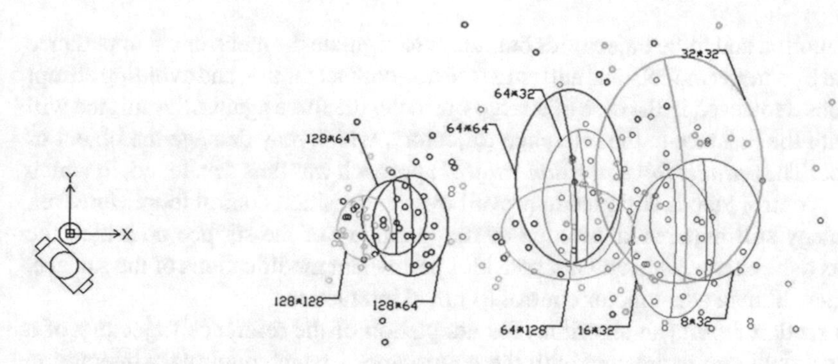

Fig. 11.8 Distribution of the different series

Table 11.2 Results summary (Fig. 11.8)

Slope	Punch	Aver. (in m)	Std. dev. (in m)	Color
128	32	1.56	0.15	Blue
64	32	2.16	0.21	Red
32	32	2.58	0.35	Green
16	32	2.54	0.16	Yellow
8	32	2.66	0.20	Turquoise
128	64	1.50	0.13	Indigo
64	64	2.20	0.30	Light Blue
128	128	1.38	0.10	Light Green
64	128	2.14	0.26	Orange

Table 11.2 shows the statistical results obtained from the trials. The large ellipse's size embodies the fact that the lack of accuracy in the preparation step can be very influential on the measured distance. Furthermore, the tennis ball surface, as well as the carpet surface, does not allow a perfect straight motion. The best kick is attained with a punch value of 32 and a slope value of 8 which corresponds to a high stiffness configuration while a low stiffness set (slope at 128) involves a smoother kick and low reached distance.

11.2 Reference Trajectory Adaptation

The control of robot interacting with obstacle has been addressed mainly using three approaches. In *hybrid force/position control* [6], the control is divided into contact and noncontact, and the transition from one case to the other. This method require accurate model of the contact surface and inaccurate detection of position or timing will lead to jerky transitions. The *impedance control* developed in [7] does not attempt to track motion and force trajectories but rather to regulate the mechanical impedance specified by a target model, thus unifying free and contact motion and avoiding abrupt transitions. However, in the case of a large obstacle, the force against the surface will grow with the distance to the reference trajectory, which may damage the object or the robot . The *parallel force/position control* approach was thus developed, in which the force control loop is designed to prevail over the position control loop. However, this strategy still requires knowledge of the geometry of the surface on which the force has to be exerted, and will not consider permanent modifications of the surface. In contrast, human adapt motor control to novel interactions.

This section reports an algorithm for adaptation of the reference trajectory of a human or robot arm interacting with the environment, which minimizes interaction force and performance error by satisfying a desired impedance [8]. The algorithm establishes a mathematically rigorous model of the underlying mechanism of motion planning adaptation in humans. It also enables robots to adapt reference trajectory so excessive force can be avoided when interacting with an obstacle and to recover back to the originally intended movement in free space in the same manner as human.

11.2.1 Interaction Dynamics

In the following we assume that a robot or a human arm is interacting with a dynamics environment characterized by a force field described by a continuously differentiable function $F(x, \dot{x})$ depending on the end effector position x and velocity $\dot{x}$, where $x \in \mathbb{R}^m$ denotes the position and orientation of the hand's center of mass in the Cartesian space with $\mathbb{R}$ the set of real numbers and $m \leq 6$ is the degree of freedom (DOF). Further $q \in \mathbb{R}^n$ denotes the coordinates of the arm in the joint space.

Let us first compute the dynamics of the arm interacting with an external force field. Differentiating the direct kinematics relation $x = \phi(q)$ with respect to time yields

$$\dot{x} = J(q)\dot{q}\,, \ddot{x} = J(q)\ddot{q} + \dot{J}(q)\dot{q}\,, \tag{11.1}$$

where $J(q) = \left(\frac{\partial \phi(q)}{\partial q_j}\right) \in \mathbb{R}^{m \times n}$ is the Jacobian defined in Eq. (11.1). The rigid body dynamics of an arm is

$$M(q)\ddot{q} + N(q, \dot{q}) = \tau + J_q^T F(x, \dot{x}) \tag{11.2}$$

where $M(q)$ is the (symmetric, positive definite) mass matrix, $N(q, \dot{q})$ is the nonlinear term representing the joint torque vector due to the centrifugal, Coriolis, gravity and friction forces, and τ is the vector of joint torques produced by muscles.

As many studies show, see [9], the joint torque as control input can be simply modeled as a PD control:

$$\tau = -K\left((q - q_r) + \Gamma^{-1}(\dot{q} - \dot{q}_r)\right) \tag{11.3}$$

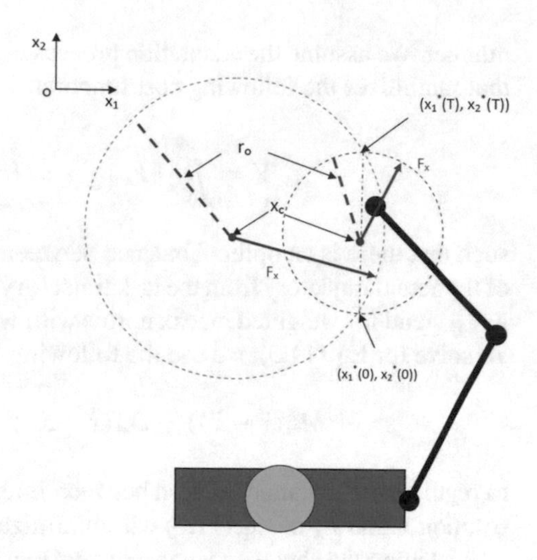

Fig. 11.9 Experiment setup and results

where K is a stiffness matrix and $K\Gamma^{-1}$ a damping matrix, which is usually taken as proportional to the stiffness matrix. As our focus is on the adaptation of the *sensor reference trajectory* q_r, we assume that K and Γ, which are either constant or time varying signals, do not change from trial to trial. Equations (11.2) and (11.3) yield the interaction dynamics (in which we set $J_q^T \equiv J^T(q)$ and $F_x \equiv F(x, \dot{x})$).

$$M\ddot{q} + N(q, \dot{q}) + K\left((q - q_r) + \Gamma^{-1}(\dot{q} - \dot{q}_r)\right) = J_q^T F_x \tag{11.4}$$

11.2.2 Adaptation Model

The experiments in [10], as illustrated in Fig. 11.9, show that when interacting with a compliant environment, humans use a compensatory response tending to recover the free space trajectory, while when it is interacting with a stiff environment they will modify the trajectory to follow the obstacle's shape.

We model this situation by assuming that a balance will be maintained between performance in terms of tracking error, with respect a *task trajectory* $x^*(t)$, and interactive force. Humans thus adapt the reference trajectory from trial to trial to decrease the interaction force, so that we normally bypass a high stiffness obstacle during a reaching movement.

We consider trials of finite time T. For a given task, we assume there is a task trajectory x^* which is normally the trajectory in free space. It then defines task trajectory q^* in joint space (via a pseudoinverse given redundant case), to which the reference trajectory q_r is aligned with initially, i.e., $q_r^0 \equiv q^*$. The reference trajectory q_r^i will thereafter gradually adapt, where superscript $i \geq 0$ is used to specify the trial number. We assume the adaptation process is to seek an optimal reference trajectory that minimizes the following cost function:

$$V = \int_0^T \|F_x\|_Q + \|x(\sigma) - x^*(\sigma)\|_R \, d\sigma \tag{11.5}$$

such that there is an optimal balance between the interaction force F_x and deviation of the actual trajectory from the task trajectory would be achieved. Note that $\|\cdot\|_R$ and $\|\cdot\|_Q$ stand for weighted matrix norms with weighting matrix Q and R, respectively. To solve for Eq. (11.5), we use the following impedance model:

$$M_m(\ddot{x} - \ddot{x}^*) + D_m(\dot{x} - \dot{x}^*) + K_m(x - x^*) = F_x \tag{11.6}$$

to regulate the dynamic relation between interaction force and task error, so that the solution x and F_x of Eq. (11.6) will minimize V of Eq. (11.5) [7].

Matching the above impedance model can be equivalently represented as making

$$w = (\ddot{x} - \ddot{x}^*) + K_D(\dot{x} - \dot{x}^*) + K_P(x - x^*) - K_F F_x \tag{11.7}$$

equal to 0, where $K_D = M_m^{-1} D_m$, $K_P = M_m^{-1} K_m$, $K_F = M_m^{-1}$. In this purpose we propose the following *adaptation algorithm for the reference trajectory*:

$$q_r^0 = q^*, q_r^{i+1} = q_r^i - Lz^i, \quad i = 0, 1, 2, \ldots \tag{11.8}$$

where $z = (\dot{x} - \dot{x}^*) + \Lambda(x - x^*) - f_x$ with matrix Λ being the weight of position error against velocity error and filtered force f_x defined by

$$\dot{f}_x + \Gamma f_x = K_F F_x \tag{11.9}$$

and L a constant matrix satisfying $\|I - K\Gamma^{-1} LJ(q)M^{-1}(q)\| < 1$, which will provide a contraction (see Eq. (11.41)), thus showing convergence of the trajectory adaptation.

In next section, we will show that the trajectory adaptation Eq. (11.8), together with closed-loop human arm dynamics described by model Eq. (11.4) will leads to eventually achieve the desired impedance Eq. (11.6) by making $w \to 0$ in Eq. (11.7), where $K_D = \Gamma + \Lambda$ and $K_P = \dot{\Gamma} + \Gamma\Lambda$.

11.2.3 Convergence Analysis

$\Gamma + \Lambda = K_D$ and $\dot{\Gamma} + \Gamma\Lambda = K_P$ are defined so that

$$\dot{z} + \Gamma z = w\,. \tag{11.10}$$

Further, by defining

$$u_r = K\left(q_r + \Gamma^{-1}\dot{q}_r\right) \tag{11.11}$$

we can combine the adaptation of reference position Eq. (11.8) and its first order derivative into

$$u_r^{i+1} = u_r^i - K\Gamma^{-1}Lw^i. \tag{11.12}$$

Define the *equilibrium trajectory* x_e as the one that satisfies

$$(\ddot{x}_e - \ddot{x}^*) + K_D(\dot{x}_e - \dot{x}^*) + K_P(x_e - x^*) = K_F F_e \tag{11.13}$$

with $F_e \equiv F(x_e, \dot{x}_e)$, such that $w = 0$ when $x = x_e$ according to Eq. (11.7). We assume that $F_x = \mathbf{0}_m$ at $x(0)$ (or $q(0)$) such that $z(0) = \mathbf{0}_n$, $q(0) = q_e(0)$ and $x(0) = x^*(0)$.

Given x_e in Cartesian space, we have the corresponding equilibrium trajectory q_e in joint space. Then, we define the *equilibrium reference trajectory* q_{re} as the trajectory that satisfies

$$\begin{aligned} J^T(q_e)F_e = {} & M(q_e)\ddot{q}_e + N(q_e, \dot{q}_e) \\ & + K\left((q_e - q_{re}) + \Gamma^{-1}(\dot{q}_e - \dot{q}_{re})\right) \end{aligned} \tag{11.14}$$

such that $w = 0$ when $q_r = q_{re}$. Using u_r from Eq. (11.11), we define

$$u_{re} = K\left(q_{re} + \Gamma^{-1}\dot{q}_{re}\right) \tag{11.15}$$

and $h(q, \dot{q}) \equiv N(q, \dot{q}) + K\left(q + \Gamma^{-1}\dot{q}\right)$, such that Eq. (11.14) can be rewritten as

$$M(q_e)\ddot{q}_e + h(q_e, \dot{q}_e) = u_{re} + J^T(q_e)F_e\,. \tag{11.16}$$

Similarly, by using u_r defined in Eq. (11.11) and Eq. (11.14) can be rewritten as

$$M(q)\ddot{q} + h(q, \dot{q}) = u_r + J_q^T F_x\,. \tag{11.17}$$

Defining $\Delta M \equiv M(q) - M(q_e)$, $\Delta h \equiv h(q, \dot{q}) - h(q_e, \dot{q}_e)$, $\Delta q \equiv q - q_e$, $\Delta F \equiv F_x - F_e$, $\Delta u_r = u_r - u_{re}$ and combining Eqs. (11.16) and (11.17), we have

$$\begin{aligned} \Delta\ddot{q} = M^{-1}(q)(&\Delta u_r + J_q^T\Delta F + \Delta J_q^T F_e \\ & - \Delta M(q)\ddot{q}_e - \Delta h). \end{aligned} \tag{11.18}$$

Considering the boundedness of L, $M^{-1}(q)$ and $J(q)$ as well as the continuity of K, Γ^{-1}, K_D, K_P F_e, $\dot{q}_e$, $\ddot{q}_e$, and $\dot{J}$ on the compact time interval $[0, T]$, we introduce $b_L, b_{M_I}, b_J, b_K, b_{\Gamma_I}, b_{K_D}, b_{K_P}, b_{Fe}, b_{qe1}, b_{qe2}$ as their $\|\cdot\|_\infty$-norm. As any continuously derivable function is Lipschitz on a compact set [11], $h(q,\dot{q})$ is Lipschitz in terms of q and $\dot{q}$ on the finite time interval $[0,T]$ as well as $J(q)$, $\dot{J}(q)$, F_x, $M(q)$, and $\phi(q)$. The Lipschitz coefficient of $J(q)$, $\dot{J}(q)$, F_x, $M(q)$, ϕ and $h(q)$ are denoted as l_J, l_{J_1}, l_{F1}, l_{F2}, l_M, l_ϕ and l_h. To facilitate further analysis, we use the α-norm [12], defined for any matrix $A(t)$ $\|A(t)\|_\alpha \equiv \sup_{0\leq t\leq T} e^{-\alpha t}\|A(t)\|$, such that

$$\|a(t)\|_\alpha \leq \|a(t)\|_\infty \leq e^{\alpha T}\|a(t)\|_\alpha . \tag{11.19}$$

We now set $\Delta x \equiv x - x_e$ such that from Eqs. (11.13) and (11.7), w can be rewritten as

$$w = \Delta\ddot{x} + K_D\Delta\dot{x} + K_P\Delta x - K_F\Delta F . \tag{11.20}$$

According to $x = \phi(q)$, Eq. (11.1) and the Lipschitz condition of F_x, we have

$$\begin{aligned}
\Delta x &= \Delta\phi = \phi(q) - \phi(q_e) \\
\Delta\dot{x} &= J_q\Delta\dot{q} + \Delta J\dot{q}_e \\
\Delta\ddot{x} &= J_q\Delta\ddot{q} + \dot{J}_q\Delta\dot{q} + \Delta J\ddot{q}_e + \Delta\dot{J}\dot{q}_e \\
\|\Delta F_x\| &\leq l_F\|[q,\dot{q}]\|
\end{aligned} \tag{11.21}$$

where $\Delta J = J(q) - J(q_e)$ and $l_F = \max\{l_{F1}l_\phi, l_{F2}b_J + l_J b_{qe1}\}$.

We consider the Bellman–Gronwall lemma [13], stating that if signals $f(t)$, $\alpha(t)$ and $b(t)$ satisfy the condition

$$f(t) \leq \alpha(t) + \int_0^t b(\sigma)f(\sigma)\,d\sigma, \tag{11.22}$$

then the following inequality holds:

$$f(t) \leq \alpha(t) + \int_0^t \alpha(\sigma)b(\sigma)e^{\int_\sigma^t b(s)ds}\,d\sigma. \tag{11.23}$$

Consider the case that $b(\sigma) = c_b$ and $\alpha(\sigma) = c_a\int_0^\sigma a(s)ds$, then Eq. (11.22) becomes

$$f(t) \leq \int_0^t c_a a(\sigma) + c_b f(\sigma)\,d\sigma. \tag{11.24}$$

Let $\beta(\sigma) = e^{c_b(t-\sigma)}$. Noting that $\beta(t) = 1$ and $\alpha(0) = 0$, partial integration of Eq. (11.23) yields

$$
\begin{aligned}
f(t) &\leq \alpha(t) + c_b \int_0^t \alpha(\sigma) e^{c_b(t-\sigma)} d\sigma \\
&= \int_0^t \frac{d(\alpha(\sigma)\beta(\sigma))}{d\sigma} d\sigma - \int_0^t \alpha(\sigma) \frac{d\beta(\sigma)}{d\sigma} d\sigma \\
&= \int_0^t \frac{d\alpha(\sigma)}{d\sigma} \beta(\sigma) d\sigma = c_\alpha \int_0^t a(\sigma) e^{c_{b(t-\sigma)}} d\sigma.
\end{aligned} \tag{11.25}
$$

From Eq. (11.17), we have

$$
\ddot{q} = M^{-1}(q) \left(J_q^T F_x - h(q, \dot{q}) \right) + M^{-1}(q) u_r \tag{11.26}
$$

and in particular when $q = q_e$

$$
\ddot{q}_e = M^{-1}(q_e) \left(J^T(q_e) F_e - h(q_e, \dot{q}_e) \right) + M^{-1}(q_e) u_{re}. \tag{11.27}
$$

It follows that

$$
\begin{aligned}
\Delta\ddot{q} = &-M^{-1}(q)\Delta h - \Delta M^{-1}(q) h(q_e, \dot{q}_e) \\
&+ M^{-1}(q) J_q^T \Delta F_x \\
&+ (M^{-1}(q)\Delta J^T + \Delta M^{-1} J^T) F_e \\
&+ M^{-1}(q)\Delta u_r + \Delta M^{-1}(q_e) u_{re},
\end{aligned} \tag{11.28}
$$

$$
\begin{aligned}
\Delta h &\equiv h(q, \dot{q}) - h(q_e, \dot{q}_e), \\
\Delta M^{-1} &\equiv M^{-1}(q) - M^{-1}(q_e) \\
\Delta J &\equiv J(q) - J(q_e), \quad \Delta u_r \equiv u_r - u_{re}.
\end{aligned} \tag{11.29}
$$

Due to the boundedness of $M^{-1}(q)$, $J(q)$, $h(q_e, \dot{q}_e)$, F_e, u_{re} and the Lipschitz condition of $M^{-1}(q)$, $h(q, \dot{q})$, $J(q)$, F_x, it can be shown from Eq. 11.28 that there exists a constant c_0 such that

$$
\|\Delta\ddot{q}\| \leq c_0 \|[\Delta q, \Delta\dot{q}]\| + b_{M_l} \|\Delta u_r\|. \tag{11.30}
$$

Then combining Eqs. (11.28) and (11.30) by integration and using $\Delta q = \mathbf{0}_n$ yields

$$
\|[\Delta q(t), \Delta\dot{q}(t)]\| \leq \int_{\sigma=0}^t c_1 \|[\Delta q(\sigma), \Delta\dot{q}(\sigma)]\| + b_{M_l} \|\Delta u_r(\sigma)\| \, d\sigma, \tag{11.31}
$$

where b_{M_l} is the ∞-norm of M^{-1} and $c_1 = 1 + c_0$. Further, according to Eqs. (11.24) and (11.25) we have

$$
\|[\Delta q, \Delta\dot{q}]\| \leq b_{M_l} \int_0^t e^{c_1(t-\sigma)} \|\Delta u_r(\sigma)\| \, d\sigma. \tag{11.32}
$$

Multiplying by $e^{-\alpha t}$ on both sides yields

$$e^{-\alpha t}\|[\Delta q, \Delta \dot{q}]\| \leq b_{M_I} \int_{\sigma=0}^{t} e^{-\alpha\sigma}\|\Delta u_r\| e^{-(\alpha-c_1)(t-\sigma)}\, d\sigma$$

which leads to

$$\begin{aligned}\|[\Delta q, \Delta \dot{q}]\|_\alpha &\leq b_{M_I}\|\Delta u_r\|_\alpha \int_0^t e^{-(\alpha-c_1)(t-\sigma)} d\sigma \\ &\leq p_1\|\Delta u_r\|_\alpha \end{aligned} \quad (11.33)$$

where $p_1 = \left|\left|\frac{b_{M_I}(1-e^{-(\alpha-c_1)t})}{\alpha-c_1}\right|\right|$. Then, considering Eq. (11.30), we have

$$\|\Delta \ddot{q}\|_\alpha \leq p_2\|\Delta u_r\|_\alpha, \quad (11.34)$$

where $p_2 = b_{M_I} + c_0 p_1$. Combining Eqs. (11.21) and (11.20) and using Eqs. (11.33) and (11.34), it can finally be shown that

$$\|w\|_\alpha \leq c_2\|\Delta \ddot{q}\|_\alpha + c_3\|[\Delta q, \Delta \dot{q}]\|_\alpha \leq p_3\|\Delta u_r\|_\alpha\,, \quad (11.35)$$

where $c_2 = b_J$, $c_3 = b_{qe2}l_j + b_{J1} + b_{qe1}l_{J1} + b_{K_D}b_J + b_{qe1}l_J + b_{K_P} + b_{K_F}(l_{F1} + l_{F2})$ and $p_3 = p_1c_2 + p_2c_3$.

It is shown that there exists constants c_1 and p_3 such that

$$\|[\Delta q, \Delta \dot{q}]\| \leq b_{M_I} \int_0^t e^{c_1(t-\sigma)}\|\Delta u_r(\sigma)\|\, d\sigma \quad (11.36)$$

and

$$\|w\|_\alpha \leq p_3\|\Delta u_d\|_\alpha. \quad (11.37)$$

From Eqs. (11.20), (11.21), (11.18) and (11.12), we have

$$\begin{aligned}\Delta u_r^{i+1} &= \Delta u_r^i - K\Gamma^{-1}L[J_q\Delta\ddot{q} + \dot{J}_q\Delta\dot{q} + \Delta J\ddot{q}_e \\ &\quad + \Delta\dot{J}\dot{q}_e + K_D(J_q\Delta\dot{q} + \Delta J\dot{q}_e) + K_P\Delta\phi - K_F\Delta F] \\ &= (I - K\Gamma^{-1}LJ_qM^{-1})\Delta u_r^i - K\Gamma^{-1}L \\ &\quad \times [J_qM^{-1}\left(J_q^T\Delta F + \Delta J_q^T F_e - \Delta M\ddot{q}_e - \Delta h\right) \\ &\quad + J_q\Delta\ddot{q} + \Delta J\ddot{q}_e + \Delta\dot{J}\dot{q}_e\Delta\dot{q}_e + K_DJ_q\Delta\dot{q} + K_D\Delta J\dot{q}_e \\ &\quad + K_P\Delta\phi - K_F\Delta F]. \end{aligned} \quad (11.38)$$

Taking norm on both sides yields

$$\begin{aligned}
\|\Delta u_r^{i+1}\| \leq\ & \|(I - K\Gamma^{-1}LJ_qM^{-1})\|\|\Delta u_r^i\| \\
& + \|K\Gamma^{-1}L\|[(\|J_qM^{-1}J_q^T\| + \|K_F\|)\|\Delta F\| \\
& + (\|F_e\| + \|\ddot{q}_e\| + \|\dot{q}_e\|)\|\Delta J\| \\
& + \|\dot{q}_e\|\|\Delta \dot{J}\| + \|J_qM^{-1}\ddot{q}_e\|\|\Delta M\| \\
& + \|\|J_qM^{-1}J_q^T\|\Delta N\| + \|K_P\|\|\Delta\phi\| \\
& + (\|\dot{J}\| + \|K_D\|\|J_q\|)\|\Delta\dot{q}\|\|] \\
\leq\ & \rho\|\Delta u_r^i\| + c_2\|[\Delta q, \Delta\dot{q}]\|\,,
\end{aligned} \tag{11.39}$$

where $\rho = \|(I - K\Gamma^{-1}LJ_qM^{-1})\| < 1$ and

$$\begin{aligned}
c_2 = b_Lb_Kb_{\Gamma_l}[&(b_Jb_{M_l}b_J + b_{K_F})(l_{F1} + l_{F2}) \\
&+ (b_{Fe} + b_{qe2} + b_{qe1})l_J + b_{qe1}l_J + b_Jb_{M_l}b_{qe2}l_M \\
&+ b_Jb_{M_l}b_Jl_N + b_{K_P}l_\phi + (b_{J1} + b_{K_D}b_J)]\,.
\end{aligned} \tag{11.40}$$

Combining with Eq. (11.36) and multiplying by $e^{-\alpha t}$ on both sides of Eq. (11.39) yields

$$\begin{aligned}
\|\Delta u_r^{i+1}\|_\alpha \leq\ & \rho\|\Delta u_r^i\|_\alpha \\
& + c\|\Delta u_r\|_\alpha \int_{\sigma=0}^{t} e^{-(\alpha-c)(t-\sigma)}d\sigma \leq \rho_1\|\Delta u_r^i\|_\alpha
\end{aligned} \tag{11.41}$$

where $c = b_{M_l}c_2$ and $\rho_1 = \rho + \frac{c(1-e^{-(\alpha-c)T})}{\alpha-c}$. By choosing large enough α such that $\rho_1 < 1$, $\|\Delta u_r\|_\alpha \to 0$ thus $\|\Delta u_r\| \to 0$ according to Eq. (11.19). Then, from inequalities Eqs. (11.37) and (11.19) follows $w \to 0$.

11.2.4 Simulation Studies

Simulations of the experiment from [10] have been performed to test the trajectory adaptation algorithm, and compare the prediction of this model with experimental results. The setup of the human subject experiments in [10] is illustrated in Fig. 11.9. This simulation is carried out based on a two joints model of human arm used in [14].

The task consists in performing a point to point movement, with minimal jerk nominal task trajectory $x^*(t)$ from $x(0) = [0, 0.31]$ m to $x(T) = [0, 0.56]$ m:

$$\begin{aligned}
x^*(t) &= x(0) + (x(T) - x(0))\,p(t), \\
p(t) &= 10\left(\frac{t}{T}\right)^3 - 15\left(\frac{t}{T}\right)^4 + 6\left(\frac{t}{T}\right)^5,
\end{aligned} \tag{11.42}$$

assuming a movement duration $T = 1.2\,\text{s}$. The interaction dynamics of the task is described in Eq. (11.4). In the controller of Eq. (11.3), the matrix K, Γ and Λ reference adaptation Eq. (11.8) are

$$K = \begin{bmatrix} 126 & 84 \\ 84 & 198 \end{bmatrix},$$
$$\Gamma(t) = \begin{bmatrix} 10 & 0 \\ 0 & 10 \end{bmatrix} + \begin{bmatrix} 30 & 0 \\ 0 & 30 \end{bmatrix} p(t),$$
$$\Lambda(t) = \begin{bmatrix} 5 & 0 \\ 0 & 5 \end{bmatrix} + \begin{bmatrix} 20 & 0 \\ 0 & 20 \end{bmatrix} p(t). \tag{11.43}$$

As in the experiment of [10], two radial force fields are used which correspond to a large circular object centered at $x_c = [-0.2165, 0.4350]\,\text{m}$ with radius $r_0 = 0.25$, and a smaller circular object centered at $x_c = [0, 0.4350]\,\text{m}$ with radius $r_0 = 0.125$, respectively (Fig. 11.9). Let r_0 be the radius of the circle and r the distance from the force field center to the end effector, then the radial force field is computed in real time as:

$$F_x = \begin{cases} k_E(r_0 - r)\,\mathbf{n} & r \le r_0 \\ 0 & r > r_0 \end{cases} \tag{11.44}$$

where $\mathbf{n}$ is the unit vector pointing from circular center to the end effector and k_E is a spring constant. We see that force F_x is always pointing from circle center to outside along the normal direction.

Trajectory adaptation is carried out in different simulations with the two circular obstacles and three stiffness K : 200, 400 and 1000 N/m. 24 trials are computed in each conditions, with force field on during trials 1–12, and off during trials 13–

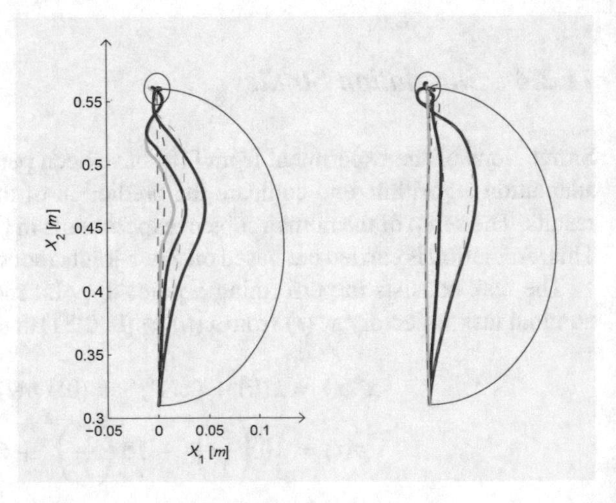

Fig. 11.10 Adaptation of trajectories in the high curvature radial force field with stiffness 200 N/m. *Left* the actual and reference trajectories in the 1st (*cyan* and *magenta*), 4th (*red* and *black*) and 12th (*blue* and *green*) trials, with interaction force. *Right* the actual and reference trajectories in the 13th (*blue* and *green*), 16th (*red* and *black*) and 24th (*cyan* and *magenta*) trials, without interaction force

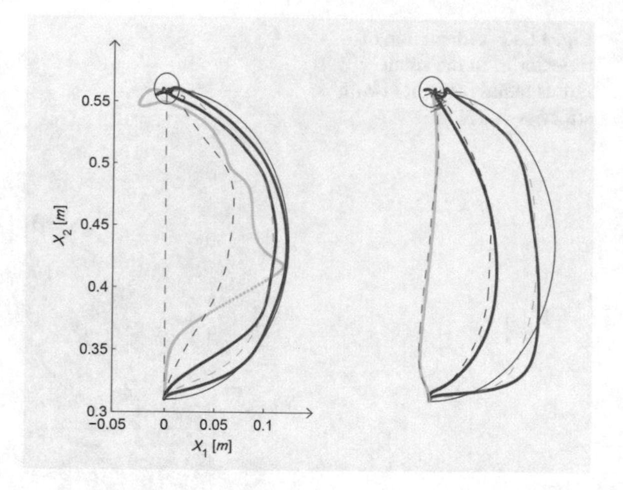

Fig. 11.11 Adaptation of trajectories in the small radius radial force field with stiffness 1000 N/m

24. The simulation results are shown in Figs. 11.10, 11.11 and 11.12 for the high curvature (small circle) and Figs. 11.13, 11.14 and 11.15 for the low curvature (large circle), in which the semicircle boundaries of the force fields are ploted in solid thin blue. In each figure, the left panels show the adaptation (with interaction force) of actual trajectories and reference trajectories in the 1st (cyan and magenta), 4th (red and black) and 12th (blue and green) trials. The right panels show the adaptation (without interaction force) of the actual trajectories and reference trajectories in the 13th (blue and green), 16th (red and black), and 24th (cyan and magenta) trials.

We see that in the presence of the force field (trials 1 to 12) the reference trajectory (dashed lines) drifts to the right direction along which the interaction force decreases as was observed in [10]. Conversely, after removing of the force (trials 13 to 24) the reference trajectories drifts gradually back close to the straight line task trajectory which is identical to the initial reference trajectory. This property of the adaptation algorithm Eq. (11.8) can be explained by studying the equilibrium trajectory and equilibrium reference trajectory in Eqs. (11.13) and (11.14). We see that when there is zero interaction force, the equilibrium trajectory x_e will be identical to x^* and the equilibrium reference trajectory x_{re} will be very close to x^*.

We see in Figs. 11.10 and 11.13 that similar to the experimental results in [10], when the interaction force is of low, the end effector trajectories as well as the reference trajectories remain close to the straight line task trajectory even after adaptation. This is because the low stiffness interaction force will not make the *equilibrium trajectory* deviate much from the *task trajectory*. From the physiological point of view, humans tend to compensate small interaction forces (which will not cost too much effort).

In contrast, as shown in Figs. 11.11 and 11.15, with the high stiffness force field, the reference and actual trajectories after adaptation tend to the circular object realized by the force field (of either small or large curvature). The circular shape of the force field decides the equilibrium trajectory and equilibrium reference trajectory so that

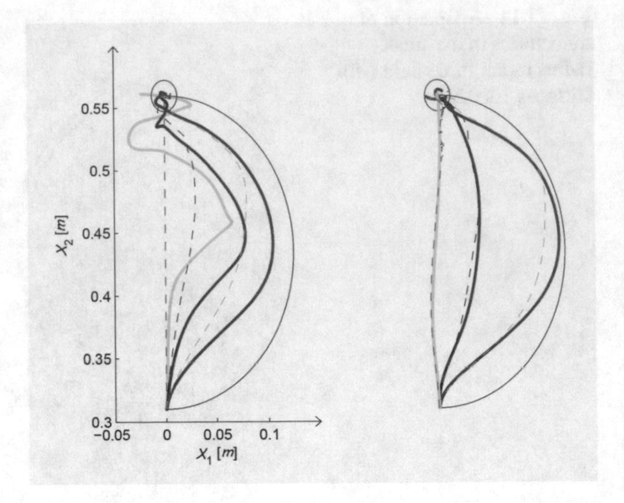

Fig. 11.12 Adaptation of trajectories in the small radius radial force field with stiffness 400 N/m

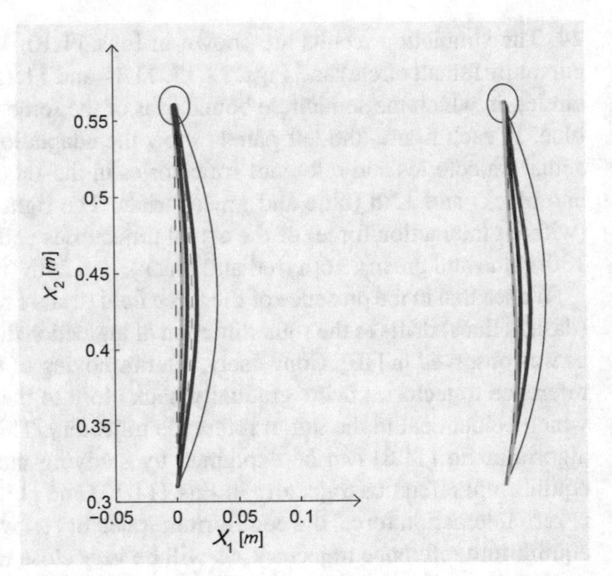

Fig. 11.13 Adaptation of trajectories in the large radius radial force field with stiffness 200 N/m. Explanations are given in the text

the large interaction force (compared to performance error) decreases to a small value. Again, this is similar to the adaptation that was observed in human subjects [10]. The medium stiffness force fields elicits a shift of trajectory between the above two extremes (Figs. 11.12 and 11.14 [10]), as was also observed in the human experiment.

Finally, the *average interaction force* [10]

$$F_I = \frac{1}{T}\int_0^T \|F_x(\sigma)\|d\sigma \tag{11.45}$$

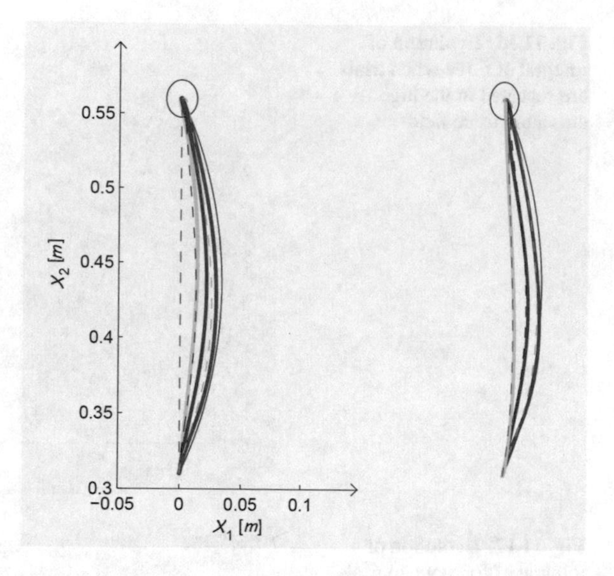

Fig. 11.14 Adaptation of trajectories in the low curvature radial force field with stiffness 400 N/m

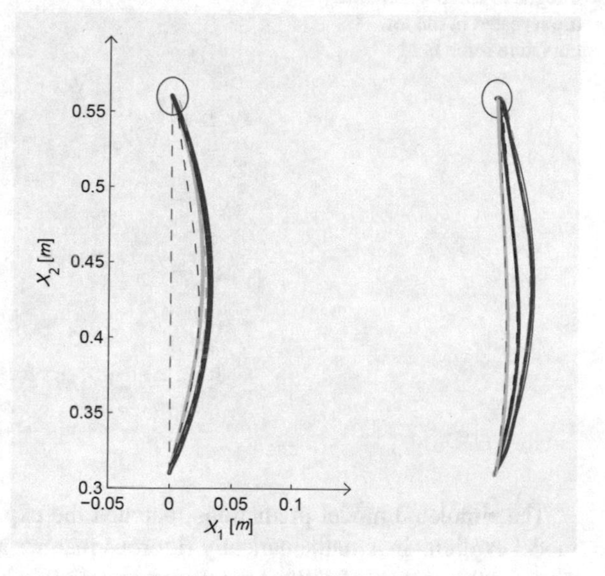

Fig. 11.15 Adaptation of trajectories in the large radius radial force field with stiffness 1000 N/m

was computed, which is the integral of interaction force during one trial. As in the human experiment [10], the force converges to roughly the same level in the three environments with different level of stiffness (Figs. 11.16 and 11.17), both in the large and low curvatures cases. However, the asymptotic force slightly decreases with the environment stiffness, which could be tested in further experiments with human subjects.

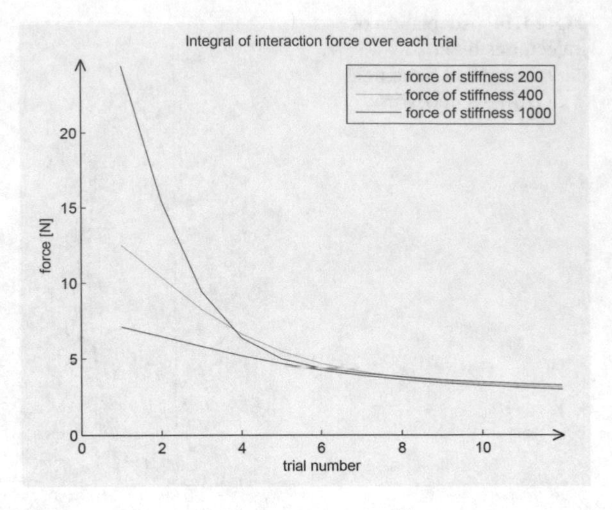

Fig. 11.16 Evolution of integral of force when trials are repeated in the high curvature force field

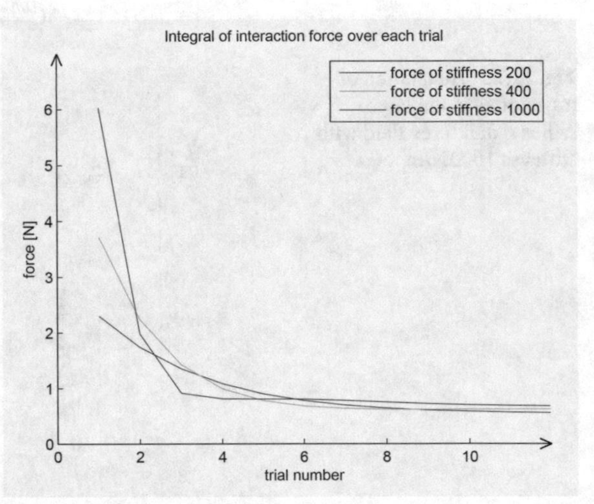

Fig. 11.17 Evolution of integral of force when trials are repeated in the low curvature force field

The simulated model predictions matched the experimental data [10], thus the model explicits in a mathematically rigorous manner a possible underlying mechanism for the sensor reference trajectory adaptation in humans.

References

1. Boue, T., Yang, C., Ma, H., Culverhouse, P.: Investigation of best robot kicking. In: 32nd Chinese Control Conference (CCC) (2013)
2. Sariyildiz, E., Cakiray, E., Temeltas, H.: A comparative study of three inverse kinematic methods of serial industrial robot manipulators in the screw theory framework. Int. J. Adv. Robot. Syst. **8**(5), 9–24 (2011)

3. Lees, A., Nolan, L.: The biomechanics of soccer: a review. J. Sport. Sci. **16**(3), 211–234 (1998)
4. Joy, K.I.: Cubic uniform b-spline curve refinement. In: On-Line Geometric Modeling Notes (2000)
5. Robotis e-manual.: http://support.robotis.com/en/product/dynamixel/rxseries/rx-28.htm
6. Hu, J., Queiroz, M., Burg, T., Dawson, D.: Adaptive position/force control of robot manipulators without velocity measurements. In: Proceedings of the IEEE International Conference on Robotics and Automation, pp. 887–892 (1995)
7. Hogan, N.: Impedance control: an approach to manipulation: part implementation. J. Dyn. Syst. Meas. Control **107**(1), 8–16 (1985)
8. Yang, C., Burdet, E.: A model of reference trajectory adaptation for interaction with objects of arbitrary shape and impedance. In: IEEE International Conference on Intelligent Robots and Systems (IROS), pp. 4121–4126 (2011)
9. Tee, K.P., Franklin, D.W., Kawato, M., Milner, T.E., Burdet, E.: Concurrent adaptation of force and impedance in the redundant muscle system. Biol. Cybern. **102**(1), 31–44 (2010)
10. Chib, V.S., Patton, J.L., Lynch, K.M., Mussa-Ivaldi, F.A.: Haptic identification of surfaces as fields of force. J. Neurophysiol. **95**(2), 2006 (2006)
11. Hirsch, M.W., Smale, S.: Differential equations, dynamical systems and linear algebra. In: A subsidiary of Harcourt Brace Jovanovich, Academic Press Publishers, New York (1974)
12. Wang, D., Cheah, C.C.: An iterative learning control scheme for impedance control of robotic manipulators. Int. J. Robot. Res. **17**(10), 1091–1104 (1998)
13. Tao, G.: Adaptive Control Design and Analysis. Wiley, Hoboken (2003)
14. Burdet, E., Tee, K.P., Mareels, I., Milner, T.E., Chew, C.M., Franklin, D.W., Osu, R., Kawato, M.: Stability and motor adaptation in human arm movements. Biol. Cybern. **94**(1), 20–32 (2006)

Printed in the United States
By Bookmasters